Advances in Disordered Semiconductors — Vol. 2

HOPPING AND RELATED PHENOMENA

ADVANCES IN DISORDERED SEMICONDUCTORS

Editor-in-charge: Hellmut Fritzsche

Volume 1   Amorphous Silicon and Related Materials
                edited by Hellmut Fritzsche

Volume 3   Transport, Correlation and Structural Defects
                edited by Hellmut Fritzsche

Advances in Disordered Semiconductors — Vol. 2

HOPPING AND RELATED PHENOMENA

Edited by

Hellmut Fritzsche
The James Franck Institute
The University of Chicago
Chicago, USA

Michael Pollak
University of California
Riverside, USA

World Scientific
Singapore • New Jersey • London • Hong Kong

Published by

World Scientific Publishing Co. Pte. Ltd.
P O Box 128, Farrer Road, Singapore 9128
USA office: 687 Hartwell Street, Teaneck, NJ 07666
UK office: 73 Lynton Mead, Totteridge, London N20 8DH

Library of Congress Cataloging-in-Publication data is available.

HOPPING AND RELATED PHENOMENA

Copyright © 1990 by World Scientific Publishing Co. Pte. Ltd.

All rights reserved. This book, or parts thereof, may not be reproduced in any form or by any means, electronic or mechanical, including photocopying, recording or any information storage and retrieval system now known or to be invented, without written permission from the Publisher.

ISBN 9971-50-940-7
     9971-50-941-5 (pbk)

Printed in Singapore by JBW Printers and Binders Pte. Ltd.

PREFACE

This volume presents recent developments, new ideas, as well as controversial issues dealing with the general phenomena of hopping transport in disordered systems. Hopping problems occur in a variety of materials such as doped semiconductors, the new high temperature superconducting materials, one- and two-dimensional systems, amorphous semiconductors, oxide glasses, polymers, and biological materials. It felt important therefore that the authors succeed in communicating across barriers of specialization, to help maintain some contact between specialists in different areas, and, more importantly perhaps, to address an audience of generalists, such as graduate students. The degree of success of this book may be judged by how well the authors succeeded to address this last vestige of generalists.

To produce this book, we took advantage of a gathering of a remarkable group of experts at the International Conference on Hopping and Related Phenomena in Chapel Hill, N. C., during August 28−31, 1989. The conference was the third in a series of biennial conferences; the previous two were held in Trieste (1985) and Bratislava (1987), and the next is planned for 1991 in the Federal Republic of Germany.

The program of the Conference was arranged by a program committee chaired by H. Fritzsche (*Chicago*), and an international advisory committee chaired by Sir Nevill Mott (*Cambridge*). M. Pollak (*Riverside*) was chairman of the conference, R. Klein (*Rockwell International*) was treasurer. We owe special thanks to M. Silver for local arrangements in Chapel Hill. As in the past, the conference was a satellite of the International Conference on Amorphous Semiconductors, and we would like to thank their organizing committee for the help extended to us. We also gratefully acknowledge financial support from the University of North Carolina, IBM Thomas J. Watson Research Center, Rockwell International Science Center, Glasstech Solar Inc., and Hughes Aircraft Research Laboratory.

The conference was made successful by the valuable contributions and lively discussions of fifty six participants from twelve countries.

Hellmut Fritzsche
Michael Pollak
October 15, 1989

INTRODUCTION

The term hopping has been applied to an inelastic transfer of an electron between two spacially localized states. Since localized eigenstates are not compatible with the translational symmetry of crystals, hopping is confined to systems lacking translational symmetry, whether due to symmetry breaking (as in the case of a small polaron), or because the system is in a non-equilibrium state (as in the case of amorphous semiconductors). To avoid confusion, we point out that the term hopping has also been used for the classical motion of an atom over a barrier, and sometimes for the elastic tunneling of an electron between two sites. The latter definition is compatible with the traditional motion of electrons in crystals, and is not used in the context of these proceedings.

Hopping is the common mode of transport among localized states in disordered systems, so interest in hopping developed in tandem with the rapid growth of interest in disordered systems. This rapid growth can be attributed to the technological interest in new non-crystalline materials, as well as to fundamental scientific problems inherent in the lack of translational symmetry in noncrystalline materials. Indeed, traditional solid state physics predominantly addressed the properties of crystals, which are governed by their translational symmetry. The properties can be fundamentally different when the symmetry is absent, as is now well documented; the best known example being the localization of electronic eigenstates.

It is of historical interest to note that it was apparently Schrödinger who was the first to have a deep appreciation of the fundamental differences between crystalline and noncrystalline solids (or aperiodic solids, in his nomenclature). In particular, he emphasized the importance of the enormous number of *distinct* states in even small "aperiodic" systems, and the enormously wide spectrum of hopping rates between such states. Schrödinger used these properties as a basis for models for various biological functions, having considered the huge biological molecules as miniature aperiodic solids (now they could perhaps be considered as mesoscopic systems). Today, the same properties are at the heart of models for neural networks. While Schrödinger's models used atomic systems, the discovery of Anderson localization led to the realization that electronic systems can also exhibit similar properties. In particular, the spectra of hopping rates are extremely wide, and can become so slow that glassy behavior of the electrons can be expected.

Early work on hopping focused on ac and dc conductivity and Hall effect, and was followed by work on magnetoresistance, piezo-resistance, and thermo-electric power. It was applied primarily to impurity conduction, and later to amorphous semiconductors. The last decade saw a rapid expansion, in application to new systems, and also in more detailed fundamental studies. Examples of hopping systems of current interest are polymers and biological materials, mesoscopic systems, two- and one-dimensional systems such as MOSFETs (metal-oxide-semiconductor field effect transistors), a large variety of oxide glasses, and the new high temperature superconducting materials (in their normal state). Some of the fundamental problems investigated were effects of static and dynamic interactions with phonons, Coulomb interactions between electrons, new magnetic effects due to coherent scattering, effects of high electric fields, and relaxation phenomena, usually from photoexcited states. A brief overview of the most active areas of research is given below.

Of great current interest are magnetic effects due to virtual transitions through sites between the initial and final sites of a hopping process. The subject, though quite novel, attracted the attention of experimentalists and of theorists alike. While there is general agreement about the importance of such effects on magneto transport, some controversies exist.

The Coulomb interaction problem is another subject of controversy, and given wide attention by experimentalists and theorists. The topic, almost two decades old, is still not adequately understood. One of the unresolved questions usually associated with Coulomb interactions, which arose wide interest, is the possibility of an electronic glass.

Hopping phenomena in mesoscopic effects is an area of growing interest. Modern lithographic technology permits the production of devices small enough to make mesoscopic effects clearly observable. Such experiments, mainly on quasi two- and one-dimensional systems, also stimulated considerable theoretical work on the subject.

Nonlinear hopping transport is another subject of considerable activity. This problem has been developing in two directions, hopping transport in high electric field, and relaxation from photoexcited states. The high field effects which proved potentially important are Frenkel-Poole effects, interesting modifications of the percolation path, and an influence on the wave function. The time resolved photo-excitation experiments proved very useful because they are uniquely suitable to study the extremely wide range of relaxation rates which are characteristic of localized disordered systems.

The editors of this volume believe that it provides a good and lucid representation of recent developments in the various aspects of hopping phenomena.

CONTENTS

Preface   v

Introduction   vii

Chapter 1: COULOMB GAP, VARIABLE RANGE HOPPING

Capacitance Measurements of the Dynamics of Screening in the Electron Glass   3
  Don Monroe

Variable Range Hopping in Just Insulating Si:Sb   25
  Yuichi Ochiai

The Influence of Spin-Spin Interaction on the Low-Temperature Limit of Variable Range Hopping Conductivity   49
  I. S. Shlimak

Hopping Conductivity and Magnetic Transitions of the Cu^{2+} Spins in Single-Crystal La_2CuO_{4+y} 61
  Tineke Thio, R. J. Birgeneau, C. Y. Chen, B. S. Freer, D. R. Gabbe, H. P. Jenssen, M. A. Kastner, P. J. Picone, and N. W. Preyer

Variable Range Hopping in the a-Si:H:Au System   77
  A. R. Long and L. Hansmann

Observation of a Crossover from Mott to Efros-Shklovskii Variable Range Hopping in n-CdSe   85
  Youzhu Zhang, Peihua Dai, Miguel Levy, and M. P. Sarachik

Conduction in Granular Metals, Failure of the Coulomb Gap Model   93
  C. J. Adkins

Simulation Studies of Electronic States in Compensated Si:P System   111
  Mikio Eto and Hiroshi Kamimura

Long-Range Interactions in Systems with Localised States — 121
  M. Ortuño and R. Chicón

Studies of Many-Body Effects in the Coulomb Gap — 129
  M. Mochena and M. Pollak

Chapter 2: QUANTUM INTERFERENCE EFFECTS

Interference Phenomena in Variable Range Hopping Conductivity — 139
  B. I. Shklovskii and B. Z. Spivak

Orbital Magnetoconductance in the Variable Range Hopping Regime — Percolation Approach — 151
  U. Sivan, O. Entin-Wohlman, and Y. Imry

Quantum Interference Effects in the Hopping Conductivity — 169
  Y. Shapir, X. R. Wang, E. Medina, and M. Kardar

Interactions and Quantum Interference in the Variable-Range-Hopping Regime in n-Type GaAs — 181
  F. Tremblay, M. Pepper, R. Newbury, D. A. Ritchie, D. C. Peacock, J. E. F. Frost, G. A. C. Jones, and G. Hill

Orbital Magnetoresistance in the Variable Range Hopping of Indium Oxide Samples — 193
  Z. Ovadyahu

Hopping Processes in Indium Oxide Films — 207
  Meir Nissim and Ralph Rosenbaum

Chapter 3: MESOSCOPIC SYSTEMS

Distribution Function of Conductance of Finite Size Inhomogeneous Barrier Structures — 217
  M. E. Raikh and I. M. Ruzin

1D Variable-Range Hopping in Wide Mesoscopic MOSFETS 243
   Dragana Popović

Computer Simulations of Elastic Tunneling Through Mesoscopic
MOSFETS 263
   M. Green and M. Pollak

Probability Distribution Functions and Wavelength Correlations for
Transmission of Waves Through Random Media: A New Numerical
Method 273
   Itzhak Edrei

Chapter 4: HIGH FIELDS AND FREQUENCY EFFECTS

High Field Hopping and Negative Differential Conductance in Weakly
Compensated Silicon 283
   *D. I. Aladashvili, Z. A. Adamiya, K. G. Lavdovskii, E. I. Levin,
   and B. I. Shklovskii*

Onset of Nonlinear Hopping Conduction in Semiconductors with
Small Compensations 299
   J. Talamantes and M. Pollak

Frequency Dependent Conductivity of Disordered Insulators 309
   A. Hunt and M. Pollak

Hopping Conduction and Localization in High Electric Fields 317
   Harald Böttger and Dieter Wegener

Chapter 5: ELECTRON-PHONON INTERACTIONS AND POLARONS

Large Bipolarons: Formation, Motion, and Superconductivity 349
   David Emin

Small-Polaron Hopping in UO_{2+x} and U_4O_{9-y} Single Crystals 377
   P. Nagels

Polaronic Conduction in Oxide Glasses Containing V_2O_5 ... 385
P. Nagels

Dynamics of the Formation and Migration of the Self-Trapped Hole in Silver Chloride ... 393
L. Rowan and L. Slifkin

Chapter 6: RELAXATION, DRIFT, AND DIFFUSION

Energy and Phase Relaxation in Disordered Semiconductors ... 403
R. Fischer, E. O. Göbel, G. Noll, P. Thomas, and A. Weller

Low Temperature Transport in a-Si:H ... 431
M. Kemp and M. Silver

Hydrogen Glass Behavior in Amorphous Semiconductors ... 441
James Kakalios

Chapter 7: POLYMERS AND BIOLOGICAL SYSTEMS

Ionic Diffusion in Polymer Electrolytes: Dynamic Disorder Hopping Models ... 459
Stephen D. Druger and Mark A. Ratner

Transport and Relaxation of Excitations in Random Organic Solids: Monte Carlo Simulation and Experiment ... 491
Heinz Bässler

Percolation, Hopping, and Dissipative Quantum Tunneling of Protons in Hydrated Protein Powders ... 521
G. Careri

Hopping of Charge Carriers on Quasi-One-Dimensional Chains of Randomly Oriented Proton Spins: Self-Averaging, Cluster, and Finite Size Effects in Paramagnetic Resonance ... 527
J. Köhler, P. Reineker, W. Forst, and M. Schreiber

Subject Index 535

Author Index 539

Chapter 1

COULOMB GAP, VARIABLE RANGE HOPPING

CAPACITANCE MEASUREMENTS OF THE DYNAMICS OF SCREENING IN THE ELECTRON GLASS

Don Monroe

AT&T Bell Laboratories
Murray Hill, New Jersey 07974

ABSTRACT

Electron-electron repulsion in disordered insulators causes a suppression of the effective density of states (DOS) known as the Coulomb gap. Capacitance measurements on specially designed structures provide a direct monitor of the evolution of the DOS with time. Our measurements in a compensated p-type GaAs structure have shown that the relaxation becomes slow at low temperatures, and the failure to return to equilibrium is similar to that in a spin glass. The dominant features of the relaxation are consistent with activation energies of order 0.5meV, but the details of the dynamics cannot be explained even by a distribution of activation energies. The nonlinearity of the DOS is qualitatively in accord with expectations based on simple theories of interactions.

I. INTRODUCTION

Electron-electron interactions should be much more important in insulators than in metals, because localized states are not as effective as extended states in screening the interactions. Indeed, interaction corrections in metals[1] become dominant when the disorder becomes large, and these perturbation theories cannot be extended into the "strongly" localized regime. As described below, there have been important advances in the description of the ground state of interacting disordered systems, which reveal fascinating possibilities.

Transport experiments generally probe these possibilities only indirectly. Even in insulators with noninteracting electrons the dynamics are complicated by the wide range of individual hopping rates, which cover many orders of magnitude in time. The broad distributions both make calculation difficult, and also require experiments covering a large range of parameters. Effective approximations for predicting, for example, the steady-state transport properties of noninteracting disordered insulators have been developed, but because the theories typically involve approximation of averages over large ranges of poorly-known parameters, it is often difficult to work backwards to determine how unique a given explanation is.

We have developed a simple but powerful experiment for probing the interacting electron systems more directly. Capacitance measurements of the screening length give straightforward information about the effective density of states (DOS), which is a result of the unusual screening by the localized states. Since this is also a dynamic measurement, we then watch the screening develop with time. This development becomes so slow at temperatures below a few K that the system is effectively frozen into what has been termed an electron glass state. Other than some early hints of an infrared divergence in the capacitance[2], to date only the capacitance measurements have probed this unusual state. Whether there is a true phase transition to the electron glass remains an open question.

The paper is organized as follows. Section II briefly reviews the theoretical motivation for the Coulomb gap and the experiments demonstrating its existence. The speculations about the collective behavior of the system and the electron glass are also discussed, while Section III shows how the capacitance gives direct information about the screening length and the DOS. Results for the DOS in the linear regime are presented in Section IV and compared with expectations for a Coulomb gap filled in by a processes with a distribution of activation energies. The nonlinear behavior, which provides primary evidence that the effects are a result of interactions, is described in Section V. The primary experimental difficult, that of charging the capacitor through a highly resistive material, is examined in Section VI and shown to be incapable of explaining most of our results. Section VII shows how the capacitance probes an aspect of the dielectric response totally distinct from that probed by resistance, and Section VII summarizes a few of the many open questions.

II. INTERACTIONS AND THE DENSITY OF STATES

The importance of interactions for transport properties was recognized early on by Pollak[3] and also by Srinivasan[4]. Efros and Shklovskii[5] described these effects in terms of a suppression of the one-electron DOS $g_1(E)$, which is the number of states per unit volume per unit energy to bring an electron from outside the system into an empty state with energy $E > 0$, keeping all other carriers *fixed*. Throughout this paper the zero of energy is taken at the Fermi level μ, so for $E < 0$, $-E$ is the energy to remove a electron from a filled state.

If the interacting electron system is in its ground state, no rearrangement of the system can lower the total energy. In particular, the energy must be raised by transfer of an electron from a filled to an empty site. The energy of such an electron-hole excitation from site i to site j is simply the energy to remove the electron (which is minus the one-hole energy $-E_i$) plus the energy to bring it back in (which is the one-electron energy E_j) with one correction: the one-electron energy E_j was calculated with an electron in state i that is no longer there. Because of this overcounting the energy to make the electron-hole excitation is therefore lower than the difference of the one-particle energies by the Coulomb repulsion $e^2/4\pi\kappa\kappa_o r_{ij}$. If i and j are close in energy (and thus just below and above the Fermi level, respectively) they must be far apart, since if they had been close they would already have lowered the system energy by an electron-hole excitation. By assuming that the states of a given energy are distributed randomly in space, Efros and Shklovskii showed that the average density is constrained to be no more than

$$g_1(E) = \alpha_D \left[\frac{4\pi\kappa\kappa_o}{e^2}\right]^D E^{D-1} = \frac{3}{\pi}\left[\frac{4\pi\kappa\kappa_o}{e^2}\right]^3 E^2 \quad (D=3) \quad . \tag{II.1}$$

The last equality uses the result $\alpha_3 = 3/\pi$[6]. The formulas in this paper are written in MKSA units for easier comparison with the experiments. They can be converted into Gaussian units by replacing the permittivity of free space $\kappa_o = 0.0885$ pF/cm with $1/4\pi$. The Efros-Shklovskii DOS is independent of all material parameters except the trivial dependence on the dielectric constant κ. However, systems with a small electron density in the absence of interactions will be constrained by this bound only for a narrow range of energies near the Fermi level.

Later simulations[7] [8] have confirmed these predictions for the one-particle DOS, although Davies, Lee and Rice[8] noticed a marked failure of the underlying assumption of statistical uniformity.

II.A Tunneling experiments

The one-electron DOS g_1 is susceptible to this simple analysis and is easy to calculate in a simulation. In real experiments, however, it is rarely reasonable to describe the other carriers as fixed when a new carrier is injected. The exception to this generalization is tunneling experiments, because the tunneling process is so fast (of order 10^{-14} s) that other carriers do not have a chance to move. Tunneling into granular aluminum[9] and on co-evaporated metal-semiconductor mixtures[10] shows a clear reduction in the apparent DOS dI/dV when biased near zero voltage (energies at the Fermi level). Larger but similar effects had earlier been observed in amorphous semiconductors as well[11]. Another case where the injection process is virtually instantaneous and may show the Coulomb gap is photoemission[12].

II.B Longer times

Most experiments do not directly probe g_1, because other carriers can move to partially screen added carriers during the measurement. There have been many analyses of the longer time behavior, as carriers hop in response to the changed electrostatic environment. These papers for the most part include some subset of possible excitations, based on, for example, the number of particles, or the activation energy alone, rather than a more realistic description of the dynamics. I will not attempt to summarize the calculations; the interested reader is referred to the review articles by Pollak and Ortuño[13] and by Efros and Shklovskii[14].

It is far from clear that a series of simple excitations will bring the interacting electron system to a new ground state after the carriers are added. One can imagine a situation in which the new ground state is so different that large numbers of other electrons must move. Moreover, if the electrons are moved only one or two at a time there may well be large free energy barriers separating the different arrangements; the situation would then be similar to that in a spin glass. Davies, Lee, and Rice[15] coined the phrase "electron glass" to describe such a possible low-temperature phase of the interacting electron system. A mean-field analysis by Grünewald et al.[16] also hinted at such a phase. Indeed, the Hamiltonian for the electron system can be written in a form that is outwardly similar to an Ising spin glass, with the occupancy of a localized state $n_i = 0,1$ analogous to the Ising spin $s_i = \pm\frac{1}{2}$. The differences from spin glasses are probably as great as the similarities. In particular, the large random electrostatic potential at each site (analogous to a random field in the spin problem), has profound effects both experimentally and theoretically[17]. In the absence of a reliable theory, experiments must answer the question of whether a "glassy" regime exists. The present capacitance measurement of the DOS (which as the generalization of $\partial n/\partial \mu$ is analogous to the susceptibility $\chi \equiv \partial M/\partial H$ in the spin system) provides a clear answer in the affirmative. The temperature at which the response becomes slow is substantially lower than the 10K or so that would be predicted for our system by Davies, Lee, and Rice, however.

III. CAPACITANCE MEASUREMENTS

III.A Experimental details

As we shall discuss in detail below, transport measurements do not probe the part of the dielectric response related to the DOS. In contrast, the capacitance is directly related to the DOS, since it measures the response of the system to motion of the Fermi level, that is, its ability to accommodate extra carriers.

Our samples are a parallel-plate capacitors, one plate of which is a metal, while the other is a layer of the disordered semiconductor under study. To insure that the capacitor can be charged quickly even as the semiconductor resistivity grows at low temperatures, the semiconductor layer must be thin, and the entire structure backed with a second metal layer. The charging time of this structure will be discussed in more detail in Section VI.

The entire sample is a single crystal, grown using molecular-beam epitaxy. The structure, listed in Table 1, is similar to a metal-insulator-semiconductor transistor, but when a voltage is applied to the gate the motion of the carriers from the degenerately doped substrate into the semiconductor is perpendicular to the layers. The "metal" is highly doped GaAs, which should remain metallic to low temperature, while the "insulator" is $Ga_{.6}Al_{.4}As$, which presents a potential barrier of ~0.2eV to the holes. The active layer, into which carriers are injected, is p-doped GaAs, with a Be acceptor density, $2\times10^{17}cm^{-3}$, roughly one tenth of the estimated metal-insulator transition density. It is also partially compensated with roughly $1\times10^{17}cm^{-3}$ Si atoms. Since Si can substitute for either Ga or As, as many as 1/3 of these atoms may act as acceptors, so we the compensation $K \equiv N_A/N_D$ is between 1/2 and $(1-1/3)/(2+1/3)=2/7$. The concentrations were based on Hall effect calibrations of the electrical activity in other samples as a function of growth parameters; no additional checks of the nominal composition were performed.

For the data reported here, a mesa of area $0.01cm^2$ was etched lithographically, and In/Zn contacts were alloyed using a roughly five second anneal at 450°C in hydrogen atmosphere. Longer alloying steps produced clearly shorted capacitors, as did some of the mesas for this alloying, but the present mesa had a specific resistivity of at least $10^6 \Omega$–cm^2 at low temperatures. At higher biases Fowler-Nordheim tunneling through the $Ga_{.6}Al_{.4}As$ was observed, but this has not been analyzed in detail.

The sample is mounted on a plastic dual-inline package directly on a copper heat-sinking strap, which is bolted to the cold stage of either a flowing helium dewar (Janis Research, 1.5-300K), a charcoal sorption-pumped ^{3}He cryostat (RMC Cryosystems,

Table 1. Nominal composition of sample.

Layer	Material	Thickness	Doping
Gate	p^+-GaAs	5600 Å	$N_{Be}=2\times10^{18}\,\mathrm{cm}^{-3}$
Insulator	i-Ga$_{.6}$Al$_{.4}$As	560 Å	–
Active	p/n-GaAs	1100 Å	$\begin{cases}N_{Be}=2\times10^{17}\,\mathrm{cm}^{-3}\\ N_{Si}=1\times10^{17}\,\mathrm{cm}^{-3}\end{cases}$
Buffer	p^+-GaAs	5600 Å	$N_{Be}=2\times10^{18}\,\mathrm{cm}^{-3}$
Substrate	p^+-GaAs	~1mm	$N_{Zn}=10^{19}\,\mathrm{cm}^{-3}$

0.29-100K), or for a few experiments, a dilution refrigerator (6mK-2K). Since the sample is primarily a capacitor, the experiment should not involve substantial heating, and no check for electron heating was performed. The signal continues to depend on cold stage temperature down to at least 100mK, so heating can not be too significant, at least for $T > 100$mK.

The capacitance was measured by applying a voltage to the gate of the capacitor through a 50 ohm attenuator. For time domain-measurements, a square wave was applied, whose frequency was made low enough to have no effect on the results. This generally limits the period to at least ten times the longest delay, and therefore makes severe demands on the signal-to-noise ratio. Better signal-to-noise can be obtained using a sine-wave drive, for which the effective delay and the repetition period are comparable. For large drives, however, the signal is nonlinear, and the time-domain data would seem to be the more fundamental.

The charge induced on the opposite plate of the capacitor structure was determined by integrating the current using an op-amp circuit. For the fast response an Analog Devices Model 50J with a integrating capacitor $C_F = 279$pF in parallel with a resistor $R_F = 10$MΩ in the feedback loop was adequate. For the longest time delays an Analog Devices Model 515 electrometer op amp was used. The low input currents of this FET-input device allow the use of an integrating capacitor of 1, 10, or 100nF with a Victoreen 10GΩ shunting resistor yielding ac-coupling time constants (integrating times $R_F C_F$) of hundreds of seconds with the largest capacitor. Since the voltage for a given amount of charge is inversely proportional to the integrating capacitor, however, the signal-to-noise ratio becomes intolerable for the longest integrating times, and the integrator was often operated with a time constant as small as ten times the delay time. The resulting 10% or so distortion was corrected by using the computer to compute the extra charge produced at any time by the deviation of the voltage v on the capacitor from its average $\bar{v}$ and the carefully measured time constant:

$$\Delta Q(t) = \int_0^t \frac{v(t')-\bar{v}}{R_F} e^{-(t-t')/R_F C_F}\, dt' \quad . \tag{III.1}$$

The time-domain data were averaged using either the clumsy but high-repetition-rate

LeCroy model 3500SA, or a more convenient Analogic Data6000 signal averager when high repetition rate was not required. Frequency-domain data were acquired using a Stanford 530 dual-channel lockin. The phase of the signal is always more capacitive than resistive.

III.B Significance of screening length

The capacitance of this structure includes the screening length in the semiconductor layer, which is in turn sensitive to the DOS. The spatial variation of the electrostatic potential is determined by Poisson's equation,

$$\nabla^2 \phi = \frac{\rho(\mathbf{r})}{\kappa \kappa_o} = \frac{\rho[\phi(\mathbf{r})]}{\kappa \kappa_o} \quad . \tag{III.1}$$

We will assume that the DOS has the same functional dependence everywhere in the semiconductor layer, and that we only need to consider variations perpendicular to the layers, in the x direction. Taking $x=0$ as the insulator/semiconductor interface, the semiconductor covers the range $0 < x < L_{sample}$, and the insulator covers $-L_{ins} < x < 0$. If the DOS g is independent of energy, then

$$\rho = (-e) g (-e \phi) = e^2 g \phi \quad , \tag{III.2}$$

so that

$$\frac{d^2 \phi}{dx^2} = \frac{e^2 g}{\kappa \kappa_o} \phi \quad . \tag{III.3}$$

This has solutions

$$\phi = \phi_o \, e^{\pm x/L_D} \quad , \tag{III.4}$$

in which the Debye screening length L_D is defined as

$$L_D \equiv \sqrt{\frac{\kappa \kappa_o}{e^2 g}} \quad . \tag{III.5}$$

The boundary condition that the electric field F be zero deep in the semiconductor requires that we choose the decaying solution (e^{-x/L_D}). Eq. (III.4) applies only within the semiconductor layer: in the insulator the field is constant and equal to

$$F_{ins} = \frac{\kappa}{\kappa_{ins}} \left. \frac{d\phi}{dx} \right|_{x=0} = \frac{\kappa}{\kappa_{ins}} \frac{\phi_o}{L_D} \quad . \tag{III.6}$$

This field induces a charge per unit area on the metal gate of the capacitor of

$$\frac{Q}{A} = \kappa_{ins} \kappa_o F_{ins} = \frac{\kappa \kappa_o \phi_o}{L_D} \quad , \tag{III.7}$$

which is equal to the charge in the semiconductor, if $L_D \ll L_{sample}$.

The total voltage across the sample includes the voltage drop in the insulator $F_{ins} L_{ins}$ and the drop in the semiconductor ϕ_o, that is

$$V = \frac{\kappa}{\kappa_{ins}} \frac{\phi_o}{L_D} L_{ins} + \phi_o = \phi_o \left[1 + \frac{\kappa}{\kappa_{ins}} \frac{L_{ins}}{L_D} \right] . \quad \text{(III.8)}$$

This expression relates the applied voltage to the actual potential at the semiconductor interface, which is important in evaluating the nonlinearity.

It is convenient to express the measured capacitance in terms of the charge separation L which is defined in terms of the measured capacitance Q/V:

$$L \equiv \frac{\kappa \kappa_o A V}{Q} . \quad \text{(III.9)}$$

For the present case,

$$L = \frac{\kappa}{\kappa_{ins}} L_{ins} + L_D . \quad \text{(III.10)}$$

As can be verified by integrating Poisson's equation by parts,

$$L = \frac{\int \rho(x) x \, dx}{\int \rho(x) \, dx} , \quad \text{(III.11)}$$

that is, L is the average spatial separation between the positive and negative charge, provided the integral is extended into regions of zero field on both sides. For example, if all of the positive charge is on the gate, L is the mean distance of the negative charge from the gate.

The measured charge separation L gives a direct indication of the screening length, to the extent that the geometric capacitance of the insulator can be subtracted. Even if this subtraction cannot be performed with high precision, any *changes* in L must reflect changes in L_D (assuming κ_{ins} is not changing). Since L_D is related by Eq. (III.5) to the DOS g, we can therefore determine the DOS directly from the capacitance:

$$g = \frac{\kappa \kappa_o}{e^2 [L - (\kappa/\kappa_{ins}) L_{ins}]^2} . \quad \text{(III.12)}$$

This inversion is sensitive to the exact value of $\kappa L_{ins}/\kappa_{ins}$. In contrast, the transformation to charge displacement L, uses parameters that are easily determined to within 10% or so. Since the dielectric constant of $Ga_{.6}Al_{.4}As$ changes little below 4K, we can compare this length at different temperatures. In particular the "excess length"

$$\Delta L \equiv L - L_{eq} , \quad \text{(III.13)}$$

where L_{eq} is the measured charge separation at high temperatures (e.g. 4K), gives the degree to which the sample has failed to achieve equilibrium. By careful calibration, or by keeping all aspects of the measurement fixed as T is varied, this difference can be determined to within about 0.1%, or a fraction of an Å for this sample. Similarly, the fractional change in the DOS can be determined by inverting Eq. (III.12) and using $L = L_{eq} + \Delta L$:

$$\frac{\Delta g}{g_{eq}} = [1 + \Delta L/L_{eq}]^{-2} - 1 \quad , \tag{III.14}$$

or for small fractional changes

$$\frac{\Delta g}{g_{eq}} \approx -2\frac{\Delta L}{L_{eq}} \quad (\Delta L \ll L_{eq}) \quad . \tag{III.15}$$

$\Delta g/g_{eq}$ can be determined with reasonable accuracy, if L_{eq} is not much smaller than L_{ins}.

In contrast, the absolute screening length $L_D = L - L_{ins}$ and thus the absolute DOS can be determined only by a precise knowledge of the insulator capacitance. While we will show plots of g, they are subject to serious systematic distortions; the plots of ΔL are much less sensitive.

III.C g as a Response Function

The quantity g measured in the capacitance experiment is simply a generalized linear response function that gives the change in charge density for a given modulation of the Fermi level, and requires no qualifiers to determine "which" DOS it is. In general it should be written $g(\mathbf{q},\omega)$ to reflect its dependence on the wavevector and frequency of the modulation. The relevant wavelengths in the capacitance measurement are of order the screening length L_D. If this can be considered long enough to be in the $q=0$ limit, the results can be given in terms of $g(\omega)$ (or $g(t)$) and its nonlinear relatives.

It is the task of theories to explain the observed behavior of $g(\omega,T)$. The high-frequency limit has already been discussed: if carriers are injected rapidly enough that others cannot move, g will be the one-electron DOS g_1. In the opposite limit the DOS takes its equilibrium value $g_{eq} = \partial n/\partial \mu$. To the extent that the system is "glassy," however, this equilibrium DOS will require a long, perhaps infinite, time to be achieved. For kT much less than the random potential responsible for the width of the impurity band, as is the case here, the equilibrium $\partial n/\partial \mu$ must be independent of temperature even in the presence of interactions. Thus any variation of g with temperature must reflect slow relaxation to equilibrium.

IV. RESULTS FOR LINEAR DOS

IV.A Frequency Domain

Our initial observations[18] showed the capacitance anomalies in time domain; this is useful in describing the nonlinear behavior, as discussed in Section V. For the linear regime, however, time and frequency domain results contain the same information (related by a Fourier transform) so we will show results in frequency domain, where the signal-to-noise ratio is generally higher if the relaxation is fast. Fig. 1 shows the evolution of the capacitance with temperature at several frequencies.

On the high temperature side, the capacitance varies little with temperature. In this regime the sample is in equilibrium on the time scale of the measurement, and the capacitance reflects the temperature independence of the equilibrium DOS $\partial n/\partial \mu$. For a given frequency, as the temperature is lowered, the capacitance begins to be reduced from its high temperature value. This observation shows the reduction of the effective

Figure 1. Frequency-domain capacitance versus temperature for several frequencies. Except for the 20kHz data, which may reflect some circuit imperfections, all the curves approach same asymptotic value $C_{eq} \approx 1600$pF, The deviations from C_{eq} at low T reflect the persistence to progressively longer times of the suppressed density of states caused by interactions. The departure from equilibrium gets sharper and moves to lower T with lower frequency.

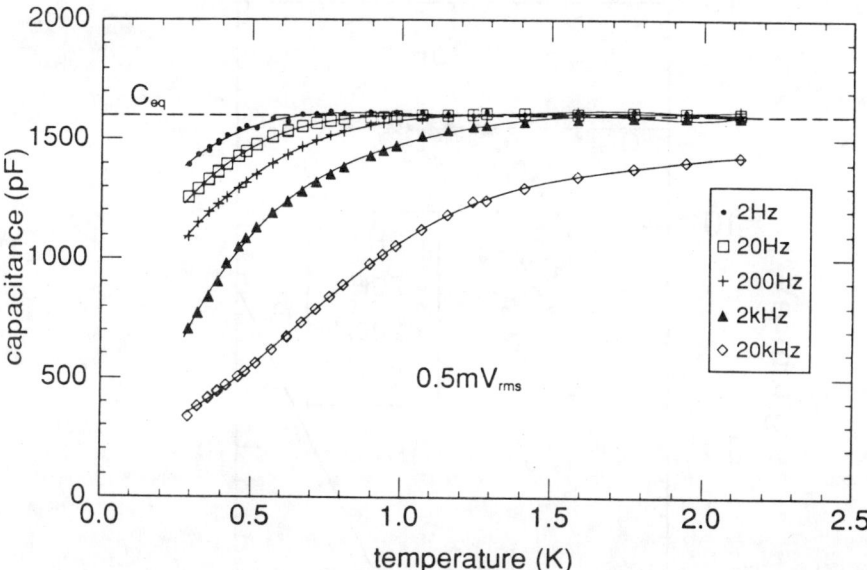

DOS $g(\omega)$. It is clear that the suppression associated with the Coulomb gap is relevant not only on the 10^{-14}s time scale of tunneling measurements, but also for much longer times. For lower frequencies, the deviation of g from its equilibrium value does not occur until lower temperatures. In this intermediate temperature range, the suppressed DOS is returning to equilibrium with a time comparable to the period. The suppression is a dynamic phenomenon, with a relaxation rate that becomes slower as the temperature is lowered.

We present these data directly in terms of g, using Eq. (III.12), in Fig. 2. Since the equilibrium screening length at high temperatures is only of order 100Å and the insulator thickness is $L_{ins} = 550$Å, 10% uncertainties in the absolute magnitude of the capacitance, the dielectric constant of the insulator (taken as 10.8), or the effective area of the sample (0.011 cm^2) cause changes in the magnitude by a factor of as much as two in g, especially when g is large. The parameters used in Fig. 2 represent the best, independent estimate for each of these quantities. The absolute magnitude of the DOS is smaller by a factor of three or so: one would expect the 2×10^{17}cm^{-3} donors to be spread by the Coulomb potentials of the ionized donors and acceptors over an energy

Figure 2. Capacitance data of Fig. 1 replotted as density of states g using Eq. (III.12). The calculated equilibrium DOS at high T is systematically affected by the imprecise knowledge of the geometrical capacitance of the insulator, as is the exact value of the exponent. The dashed line shows the high-T, equilibrium value of the DOS, $g_{eq} = \partial n/\partial \mu$. The solid curves shows the Efros-Shklovskii DOS g_1 from Eq. (II.1), which is expected to apply at very high frequencies.

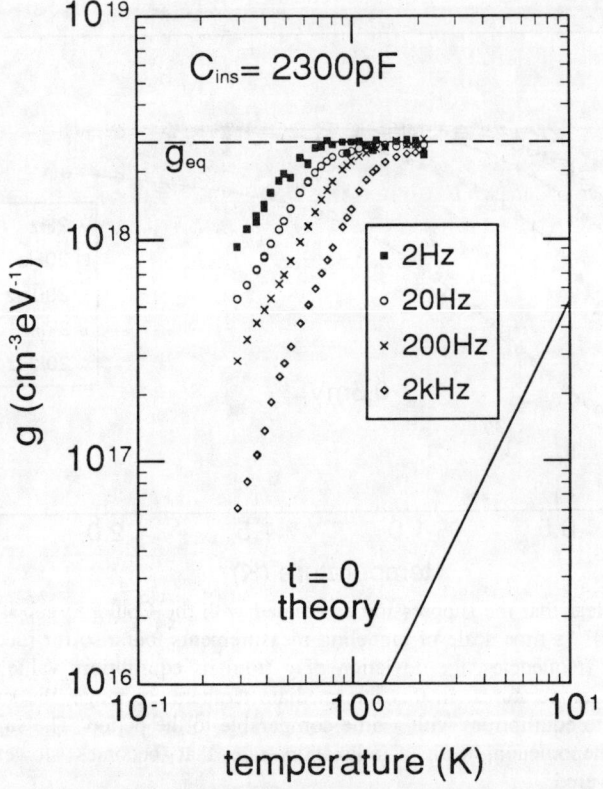

of order 10meV, which would give a DOS in the center of the impurity band of order $10^{19} \mathrm{cm}^{-3} \mathrm{eV}^{-1}$. Nonetheless, the agreement is not bad given no adjustable parameters. In principle the geometric capacitance could be estimated by biasing the capacitor into accumulation, since with the Fermi level in the valence band the screening length will be short. Unfortunately, for the present sample, since the $Ga_{.6}Al_{.4}As$ presents only a 0.2eV barrier to the holes, Fowler-Nordheim tunneling through the barrier makes the capacitance difficult to measure under strong bias. Samples with larger barriers (e.g. AlAs) should reduce this problem. Although the systematic errors in the DOS of Fig. 2 can be substantial, the trends are clear, and in the interesting low-temperature region the data are only slightly affected by the subtraction of L_{ins}. The solid curve showing the Efros and Shklovskii g_1 evaluated at $E = kT$ has a similar parabolic dependence on temperature, but is substantially lower. Indeed, that DOS becomes comparable to the equilibrium DOS g_{eq}, at many tens of Kelvin, which is width of the Coulomb gap, and where we had originally expected to see suppressed screening.

IV.B Time domain

In the linear range, the frequency- and time-domain data contain the same information, but it can be easier to think about relaxation mechanisms for time-domain data. Fig. 3 shows the relaxation of the excess length (the deviation of the screening length from its equilibrium value) as a function of time delay following a voltage step, for various temperatures. Although the decays cover a narrow range of time, only two decades, the time dependence in this region is qualitatively a straight line, that is a power law $\Delta L \propto t^{-\alpha}$. This is only a rough description, and even in this small range the slopes of the curve are decreasing at long delays. More dramatic is the general decrease in slope and to a greater degree in the magnitude, as the temperature is lowered.

Figure 3. Time-domain relaxation of the excess length $\Delta L \equiv L - L_{eq}$, where $L_{eq} = 774$Å, the measured total length at "high" temperature (1.25K and above) is subtracted from all the data.

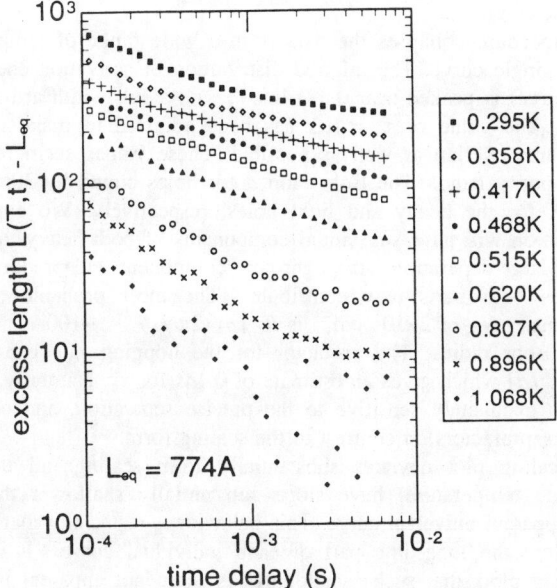

IV.C Scaling description

The decays of Fig. 3 are not simple exponentials $e^{-t/\tau}$, so we must have a distribution of time constants τ. The simplest generalization postulates a distribution of τ governed by a distribution of activation energies $P(E)$, with $\tau_i = \tau_o e^{E_i/kT}$. Then the overall response function is

$$\Delta g(t) = \int_0^\infty dE\, P(E) \exp\left[-t/\tau_o\, e^{-E/kT}\right] \quad . \tag{IV.1}$$

Here $P(E)$ is the density of excitations with an energy barrier E, weighted by their average coupling constant. For $t \gg \tau_o$ the exponential of an exponential is approximately a step function of energy: it is unity for $E > kT \ln(t/\tau_o)$ and zero for smaller energies, so

$$\Delta g(t) \approx \int_{kT\ln(t/\tau_o)}^\infty dE\, P(E) \qquad (t \gg \tau_o) \quad . \tag{IV.2}$$

If we had the signal-to-noise ratio, then the derivative $-\partial g/\partial \ln t$ would show the distribution of activation energies (weighted by the coupling constant). Even with a noiser signal we can check the assumptions on which this equation is based: A plot of $\Delta g(t)$ versus $kT\ln(t/\tau_o)$ should yield a universal curve, independent of the individual values of T and t. Fig 4. shows such a plot. All the data in Fig. 3, including intermediate temperatures (a total of nine) were included in the determination of the best scaling parameter τ_o.

The scaling procedure collapses the data from a wide range of temperatures into what looks like a single curve. The inferred distribution of activation energies (minus the slope of the curve) is peaked near 0.3-0.4meV, with a full width at half maximum of about 0.4meV. The value of $\tau_o = 1\mu s$ has been adjusted to make the curves at different temperatures overlap as well as possible. These values seem reasonable for hopping in the impurity band. The Bohr radii a for holes bound by 30meV in GaAs are 17Å and 40Å for the heavy and light holes, respectively. We expect that the impurity wavefunction will have substantial components of both heavy- and light-hole nature, so that for large separations the light-hole component will predominate, but at moderate distances both parts may contribute. The most probable separation of impurities with density N_A of $2\times10^{17}\text{cm}^{-3}$ is $R_{typ} = (2\pi N_A)^{-1/3} \sim 100$Å, or roughly 2.5 times the longer Bohr radius. The prefactor for the hopping rate can be taken as $10^{12}\text{s}^{-1}\times\exp[-2R_{typ}/a]$ which gives an estimate of 0.1ns for τ_o. Naturally, the value of τ_o for any pair is exquisitely sensitive to the precise separation, and one expects a wide distribution of prefactors, in contrast to the scaling form.

Indeed, the scaling plot deviates substantially from scaling: all the individual curves (at a single temperature) have slope substantially shallower than the local curvature of the apparent universal curve. This behavior is especially marked at higher temperatures and in the long-time part of each individual curve. It is a common weakness of scaling plots that such "scalloping" is often not apparent unless a large range of data is taken, since it must overcome a natural tendency to see a single curve. Since the data presented for this plot clearly deviate even though they cover a narrow

Figure 4. Scaling plot of relaxation. $\Delta g/g_{eq}$ is computed from the full equation (III.14). τ_o has been adjusted to maximize the apparent overlap between successive temperatures. The individual curves do not decay at the same rate as their common trend, suggesting a failure of this simple scaling. If we ignore this fact, we can estimate the derivative of a polynomial fit to the data to give the apparent distribution of activation barriers shown in the inset.

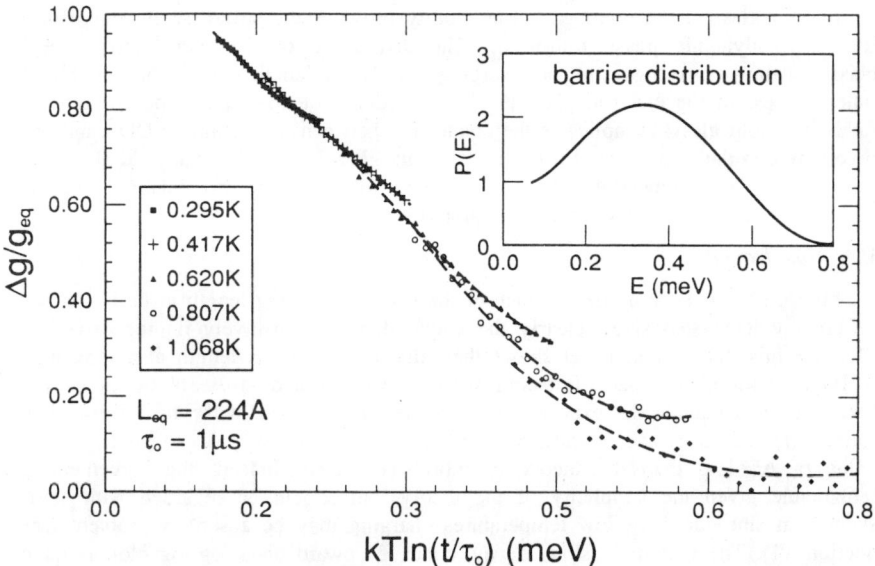

range of time, the failure is quite large.

One would like to know whether the data are consistent with a fixed distribution of both activation energies and prefactors. As for variable-range hopping, this would give a lower apparent activation energy as T goes down. Another possibility might be an increasingly collective behavior as T is lowered, which would give an increasing free energy of activation and perhaps a Vogel-Fulcher ($\exp[-E/(T-T_o)]$) dependence. The relaxations here are broad, however, and it is not clear how to define a characteristic relaxation time.

V. NONLINEAR RESPONSE

So far we have discussed only the linear response g. The observation of slow relaxation in an insulator at low temperatures is perhaps not too surprising, however. We now discuss the nonlinearity, which provides strong evidence that the observed relaxation is indeed a result of correlation effects.

The basic motivation is that the Coulomb gap is a soft gap, with g_1 zero only precisely at the Fermi level, even at zero temperature. For energies away from the Fermi level the density rises until it reaches the equilibrium DOS g_{eq}. The screening experiment probes the states available between μ and $\mu - eV$, where V is the applied voltage. Thus the screening length for high drive voltages should be smaller than at low biases, eventually approaching its equilibrium value. This is indeed observed, and we regard it as a primary confirmation of the collective nature of the effects, since the zero of voltage is defined solely by the occupation of other states.

In spin glasses, the nonlinear susceptibility provides the strongest evidence for a true thermodynamic phase transition. The divergence of the nonlinearity obeys classic scaling laws suggesting a diverging correlation length that is inaccessible to other probes. In the electron glass the large random potential (analogous to random fields in a spin glass) complicates the connection between the nonlinear DOS and any diverging correlation length– for example, a spin glass in even a small field shows no divergence. The connection, if any, between our dynamic nonlinearity and an underlying transition deserves further exploration.

V.A Time Domain

Fig. 5. shows the time dependence of the excess screening length at 6mK for two different voltage step sizes; clearly the length depends sensitively on the drive. To illustrate this, the second panel shows the value of the excess length at a long time (0.1s) for various step sizes. For comparison, the unadorned Efros-Shklovskii g_1 has been used to calculate the expected dependence, as described in the next subsection. Since $g(t)$ should evolve toward the equilibrium value, causing ΔL to go to zero, a value of ΔL less than the theory is entirely expected. Indeed, the agreement is remarkable given the simplicity of the theory, but should be regarded with some skepticism since at these low temperatures charging may be a serious problem (see Section VI). The extremely slow, decay, concave upward on a log-log plot, is quite unusual; it certainly appears that equilibrium will never be reached.

When the size of the voltage step becomes large compared to kT, it is not correct to regard the applied bias as simply "filling up" the available one-electron DOS up to some new Fermi level. Rather, since the gap is actually a reflection of the ordered positions of the other carriers, the DOS itself will be modified by the addition of carriers. Certainly voltage changes of order the size of the gap will totally destroy any remnant of the previous voltages. Exactly how this destruction proceeds as the voltage step is varied is unclear. This problem should be amenable to static calculations, although like the gap itself the relevance of such calculations for the dynamic experiments is unclear.

V.B Frequency domain

Frequency-domain data for the nonlinearity are shown in Fig. 6. This data is taken at a higher temperature (0.29K), where there is a clear transition to a linear behavior at low drive levels. The transition occurs near $1mV_{rms}$ drive, this may be a simple saturation of the screening length at the thickness of the sample.

Figure 5. Time domain decay of excess length at $T=6$mK (from Monroe et al., Ref. 18). The decay here has deviated from a power law to the extent that at long times it appears not to be decaying at all. There is a huge effect of the size of the voltage step on the degree of relaxation.

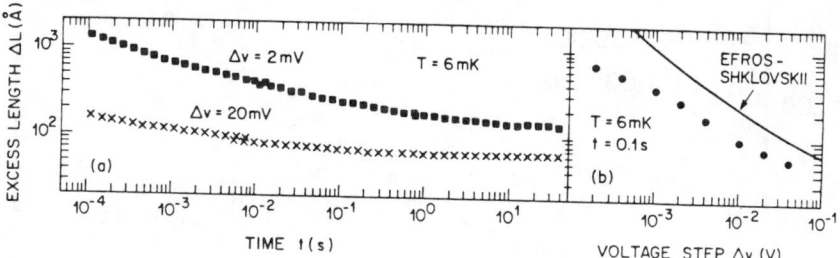

V.C Nonlinearity in the Simple Coulomb Gap Model

The total charge density that would be induced by simply filling all the one-electron states in the Efros-Shklovskii g_1 between the Fermi level and an energy $e\phi$ above it

$$\rho(\phi) = -\frac{e\alpha_D}{D}\left[\frac{4\pi\kappa\kappa_o \phi}{e}\right]^D = -\frac{\alpha_D}{D}\frac{e}{r_\phi{}^D} \quad , \tag{V.1}$$

in which we have defined

$$r_\phi \equiv \frac{e}{4\pi\kappa\kappa_o\phi} \quad , \tag{V.2}$$

which is the length at which the potential induced by a charge e is ϕ. Naturally, this formula should be applied only when $kT \ll e\phi$, if ever.

Once we have an expression for $\rho(\phi)$, we can determine the spatial dependence of the potential $\phi(\mathbf{r})$ and thus of the screening charge by solving Poisson's Equation (III.1):

$$\frac{d^2\phi}{dx^2} = \frac{4\pi\alpha_D}{D}\left[\frac{4\pi\kappa\kappa_o}{e}\right]^{D-1}\phi^D = 4\left[\frac{4\pi\kappa\kappa_o}{e}\right]^2 \phi^3 \quad , \tag{V.3}$$

where in the last equality we again specialize to $D=3$. The solution is

$$\phi(x) = \frac{1}{\sqrt{2}}\frac{e}{4\pi\kappa\kappa_o}\frac{1}{(x+x_o)} \quad . \tag{V.4}$$

x_o is an adjustable constant whose value is determined by the boundary (bias) conditions. The total charge on the gate of the capacitor can be determined from the field in the insulator, which is also the field in the active region at $x=0$. The total

Figure 6. Real and imaginary parts of the capacitance at 2kHz as the drive level is swept from a linear range at low voltages. The capacitance appears to approach its equilibrium value of about 1600pF when the drive is high.

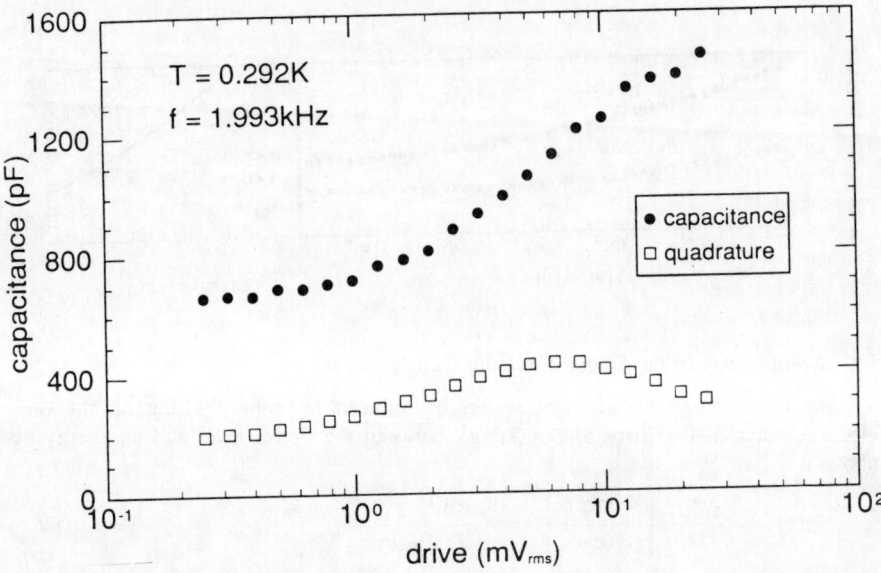

charge per unit area Q/A is just

$$Q/A = \kappa_{ins}\kappa_o \left.\frac{d\phi}{dx}\right|_{x=0} = \frac{\kappa_{ins}}{\kappa} \frac{1}{4\sqrt{2}\pi} \frac{e}{x_o^2} \quad . \tag{V.5}$$

In words, this means that x_o must be roughly the lateral distance between induced charges on the capacitor. This means that the discreteness of the screening charge causes lateral fluctuations comparable to the perpendicular variations[19], which may be important.

The total voltage drop across the capacitor is the drop across the insulator, $(Q/\kappa_{ins}\kappa_o A) L_{ins}$, plus the drop in the sample, $\phi(x=0) - \phi(x=L_{sample})$. With some rearrangement we find for the total charge displacement L,

$$L \equiv \kappa_{ins}\kappa_o A \frac{V}{Q} = L_{ins} + \frac{\kappa_{ins}}{\kappa} \frac{x_o}{1+x_o/L_{sample}} \quad . \tag{V.6}$$

For $x_o \ll L_{sample}$ the measured charge separation L is approximately $L_{ins} + x_o$, or

$$L - L_{ins} \approx \frac{\kappa_{ins}}{\kappa} \left[\frac{eA}{4\sqrt{2}\pi Q} \right]^{\frac{1}{2}} \quad . \tag{V.7}$$

Thus for large sample thicknesses L_{sample} the quantity $L - L_{ins}$ would have a universal inverse-square-root dependence on the added charge per unit area if the Efros-Shklovskii density of states governed the screening.

At nonzero temperature, the gap in g_1 is washed out. For $E < kT$ the effective density of states can be taken to be the zero temperature value at $E = f_T kT$ (with f_T a factor of order unity), that is

$$g_1(\mu) = \frac{3}{\pi} \left[\frac{4\pi\kappa\kappa_o}{e^2} \right]^3 (f_T kT)^2 \quad , \tag{V.8}$$

Thus if the potential throughout the active region satisfies $|e\phi| < kT$, the charge density will be

$$\rho(\phi) = \frac{3}{\pi} \frac{4\pi\kappa\kappa_o \phi}{r_T^2} = \frac{3}{\pi} \frac{e}{r_\phi r_T^2} \quad , \tag{V.9}$$

in which

$$r_T \equiv \frac{e^2}{f_T 4\pi\kappa\kappa_o kT} \quad . \tag{V.10}$$

This is simple Debye-like screening, and the solution to the one-dimensional Poisson's equation is

$$\phi(x) = (constant) \times e^{\pm x/L_D} \quad , \tag{V.11}$$

with

$$L_D = \frac{1}{2\sqrt{3}} r_T = \sqrt{\frac{e^2}{\kappa\kappa_o} g_1(\mu)} \quad . \tag{V.12}$$

This temperature dependent length r_T provides an upper limit to the screening length as the amount of added charge is reduced. In this linear range, the screening length gives a direct indication of the density of states at the Fermi level, as a function of temperature.

VI. CHARGING TIME

VI.A Linear resistivity

The capacitance measurement is designed to explore the relaxation of the density of states, a local quantity, following a change in the chemical potential. Unlike applying a magnetic field to a spin system, however, it is not possible to apply a *uniform* change in chemical potential throughout a bulk sample. What we have available are chemical potential *gradients*. Together with the conserved nature of electronic charge, this implies that any experiments probing the density of states must rely on transport for the supply of extra carriers. Since transport in insulators becomes slow at low temperatures, this question must be addressed carefully.

Our geometry is chosen to make charging the capacitor as easy as possible, since the charge must flow through only a thin semiconductor layer to charge the capacitor. The time required is $R_{sample}C$, $R_{sample} = L_{sample}/A\sigma$, where σ is the conductivity of the semiconductor, while C can be no larger than the insulator capacitance $\kappa_{ins}\kappa_o A/L_{ins}$. Thus the charging time is no longer than

$$\tau_{RC} = \frac{\kappa_{ins}\kappa_o L_{sample}}{\sigma L_{ins}} = \left(\frac{L_{sample}}{L_{ins}}\right)\left(\frac{\kappa_{ins}}{\kappa}\right)\tau_M \quad , \tag{VI.1}$$

where

$$\tau_M \equiv \frac{\kappa\kappa_o}{\sigma} \tag{VI.2}$$

is the Maxwell dielectric relaxation time, the time any excess charge in the bulk of a conductor will be neutralized. Recalling that $\kappa_o = 8.85 \times 10^{-14}$ F/cm $\approx .1$ ps/Ω–cm and since the two factors in parentheses are close to unity, and $\kappa_{GaAs} \approx 13$, the resistivity must approach $10^6 \Omega$–cm before the charging time reaches 1μs.

Fig. 7 shows two-terminal resistance measurements on a thin film of GaAs that is nominally identical in composition to the active layer of our sample. The resistivity rises rapidly at low temperatures, reaching a value of $10^{11}\Omega$ at 1K, so using the length of 0.6mm, width 0.3mm, and thickness 4000Å, a resistivity of $2\times10^6\Omega$–cm, corresponding to a charging time of order 2μs. There is clearly a suppression of the capacitance at times much longer than this, although the suppression is still a small effect above 1K. At lower temperatures where the effect becomes most dramatic, the charging time is not so clearly shorter than the measurement time.

Of course, the geometry of this sample is exactly wrong for making resistance measurements. A much superior procedure would be to selectively etch through the Ga$_{.6}$Al$_{.4}$As to the semiconductor and then form shallow contacts to the active layer and make measurements through the thickness of the film. In addition to allowing measurements further into the interesting temperature range, this measurement can be done on the same layer, unlike the in-plane measurement, which requires an insulating substrate. Shallow contacts should be achievable, and such measurements are planned for the near future.

Even with an ideal measurement of the resistance, there remains one fundamental uncertainty, however. If there are some acceptors that are anomalously far from all others, either because of statistical fluctuations or more serious inhomogeneities, the current in the resistance measurement will simply flow around them. Although there will be a slight geometrical correction for the excluded volume, the resistance is not sensitive to anomalously slow hops. In contrast, the capacitance is more democratic: any state that should be filled in equilibrium will contribute to the excess screening length until it is filled. The capacitance is more sensitive to anomalously slow hops.

Such concerns would seem more significant if the observed effects were small, since they might then be attributed to a few slow hops. At the lowest temperatures the effect is so large that carriers can hardly enter any of the states. This seems unlikely to be caused by anomalously slow centers.

Figure 7. Resistance versus temperature of 3737Å-thick layer of compensated *p*-type GaAs nominally identical in composition to the active layer of our capacitor. The log of resistance versus $1/T$, $1/T^{1/2}$, and $1/T^{1/4}$ should give a straight line for activated hopping, and for variable-range hopping with and without the Coulomb gap, respectively. The latter two are much better straight lines.

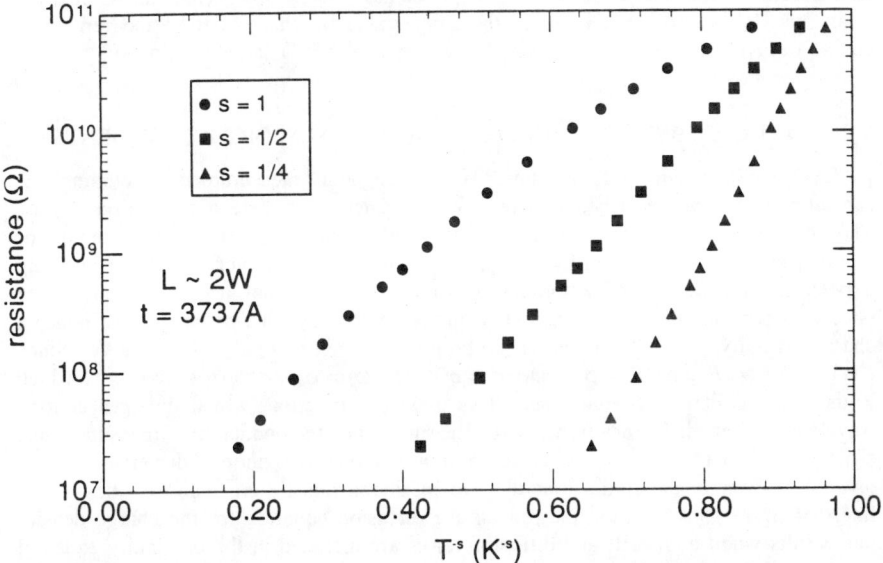

VI.B Nonlinear Charging

Actually, the sample will discharge significantly in a time much shorter than that derived above, because, the fields present in the insulator are large enough to cause nonohmic hopping conductivity. The perpendicular conductivity measurements will be ideal for exploring this phenomenon too, but the attentive reader will note that the effect of such nonlinearity might indeed mimic many of the nonlinearities seen in the capacitance.

Let us consider the extreme case in which there are *no* nonlinearities in the DOS, and the only nonlinearity is the dependence of the conductivity of the sample on field. The equivalent circuit then consists of the insulator capacitance in series with a parallel combination of the sample capacitance and the nonlinear resistance of the sample. In a step-response (time-domain) experiment, the voltage across the entire sample is held constant after the step, so a single parameter, the voltage difference Δv across the semiconductor, characterizes all the internal degrees of freedom. Δv is what determines the nonlinearity, so the current flowing in this model should depend only on Δv, no matter what the voltage V across the sample as whole. Moreover, because Δv must be zero when the sample is fully charged, and under the assumption that the capacitance C (which is the parallel combination of sample capacitance and that of the insulator including the screening length) is constant, Δv must be proportional to the charge deficit from the fully-charged value CV

$$\Delta v(t) = CV - Q(t) \qquad (VI.3)$$

This means that if we plot the charge deficit $CV-Q$ for various voltage steps, all the data should follow the same curve, except for the zero of time. This can be checked easily and is far from the case: at the same value of charge deficit the remaining charge decays much faster for a small drive, where it is a large fraction of the total charge.

VII. CONTRAST BETWEEN SCREENING AND TRANSPORT EXPERIMENTS

It is well known that any linear electromagnetic measurement, including the capacitance, can be described in terms of the generalized dielectric function $\varepsilon(\mathbf{q},\omega)$. The screening measurement probes the dielectric function in a different region of (q,ω)-space from that probed by transport measurements, however. Consider applying a potential modulation, of amplitude V with a q and ω not necessarily along the photon dispersion curve (experimentally this is nontrivial). This potential will induce a charge density ρ, which in equilibrium will be simply $\rho = e^2 g V$. Since $\varepsilon \equiv \varepsilon_\infty + P/\kappa\kappa_o F$, and $P = \rho/q$ and $F = qV$, $\varepsilon(\mathbf{q},\omega) = \varepsilon_\infty + e^2 g(\mathbf{q},\omega)/\kappa\kappa_o q^2$, which is the Thomas-Fermi screening law. This dielectric function, which diverges at long wavelengths, is different from that measured in ac-conductivity measurements, $\varepsilon(\mathbf{q},\omega) = \varepsilon_\infty + \sigma/i\omega$, which are close to $q = 0$, along the photon dispersion curve, $\omega = c/\sqrt{\kappa}$. Even in the static limit the two measurements lie on opposite sides of the line $\omega = (\sigma/e^2 g) q^2$. This line reflects the diffusion equation for the charge density that results when both drift and diffusion terms are included in the continuity equation $\partial\rho/\partial t = -\nabla\cdot\mathbf{J}$ and the charge density is set to $\rho = -e g \phi$, and defines the frequency below which the change in the Fermi level screens comparable to the applied potential. For higher frequencies, the field is reversed before substantial charge can be transferred over the wavelength of the modulation, and so transport can proceed by a "bucket brigade" process, without charge accumulating in any region of the sample.

In contrast, if the potential is applied for longer times, substantial extra charge can be injected and the density of states for added charge becomes relevant. In our screening experiments it is the boundary condition at the insulator/semiconductor interface that forces charge to accumulate, and the dielectric function is probed on a length scale of order the screening length. This is short enough that there may be some q dependence.

VIII. CONCLUSIONS AND SUGGESTIONS FOR THE FUTURE

The role of correlations in the *transport* properties of disordered semiconductors remains a subject of debate, both theoretical and experimental. Our experiments show that the capacitance, which probes the DOS much more directly, is qualitatively changed by the development of correlations. The dynamics of the return of the DOS to equilibrium is subject to complications similar to those in transport, but $(g-g_{eq})/g_{eq}$ can be regarded as an order parameter for this system, and its relaxation is of fundamental interest. Indeed, the analogous quantity in a spin glass, $1 - (\chi T)/(\chi T)_{eq}$, is equal to the Edwards-Anderson order parameter that reflects the freezing of the disordered state. Although the random potential destroys the simple connection between g and the freezing, there seems to be a close relationship between the two, which is much more obvious than in transport measurements.

The demonstration of the sensitivity of the screening length measurement to correlation effects is only the beginning. A detailed analysis of the dynamics is now in order, as the fundamental issue of the role of collective excitations, and thus of a true thermodynamic transition to an electron glass *phase* remains unresolved. At present the theoretical framework for characterizing the dynamics is sparse. Since the localized electronic system lends itself to concrete modeling in terms of well-defined excitations of simple degrees of freedom, perhaps this system will provide some clarification in the realm of glassy relaxation.

There also remains a host of other electronic materials systems in which such effects may be observable. The systematic dependence of the relaxation on materials properties such as composition should certainly be explored. The Coulomb gap is almost certainly universal; is the electron glass as well? In particular, one might argue that compensation is necessary so that the number of possible configurations of the electrons is large; is this true? One may be able to take advantage of the *electrical* tuning of the effective compensation by DC bias conditions to probe these issues. This extra degree of freedom remains relatively untouched.

Finally, it has been known for many years that the understanding of metal-insulator transitions will not be complete without including electron-electron correlations in a fundamental way. Perhaps knowing more about the effect of correlations on the insulating side of the transition will clarify the nature of the transition itself.

Acknowledgements

I would like to express thanks A.C. Gossard for providing samples, to B. Golding for ongoing encouragement and the use of his dilution refrigerator, and to M.A. Kastner, D. Huse, P.A. Lee, and many others for useful discussions.

REFERENCES

1. See, for example, the review by B.L. Altshuler and A.G. Aronov, *Electron-Electron Interactions in Disordered Systems* A.L. Efros and M. Pollak, eds. (North Holland, New York, 1985) p. 1.

2. M.A. Paalanen, T.F. Rosenbaum, G.A. Thomas, and R.N. Bhatt, *Phys. Rev. Lett.* **51**, 1896 (1983); R.N. Bhatt and T.V. Ramakrishnan, *J. Phys. C* **17**, L639 (1984).

3. M. Pollak, *Discuss. Faraday Soc.* **50**, 13 (1970).

4. G. Srinivasan, *Phys. Rev.* **B4**, 2581 (1971).

5. A.L. Efros, and B.I. Shklovskii, *J. Phys. C* **8**, L49 (1975).

6. A.L. Efros, *J. Phys. C* **9**, 2021 (1976).

7. S.D. Baranovskii, A.L. Efros, B.L. Gelmont, and B.I. Shklovskii, *J. Phys. C* **12**, 1023 (1979).

8. J.H. Davies, P.A. Lee, and T.M. Rice, *Phys. Rev.* **29**, 4260 (1984).

9. R.C. Dynes and J.P. Garno, *Phys. Rev. Lett.*, **46**, 137 (1981).

10. W.L. McMillan and J. Mochel, *Phys. Rev. Lett.*, **46**, 556 (1981), and many more recent experiments.

11. H. Fritzsche, in *Electronic and Structural Properties of Amorphous Semiconductors*, (Academic, New York, 1973) p. 55.

12. John H. Davies and Judy R. Franz, *Phys. Rev. Lett.* **57**, 475 (1986).

13. M. Pollak and M. Ortuño, in *Electron-Electron Interactions in Disordered Systems* A.L. Efros and M. Pollak, eds. (North Holland, New York, 1985), p. 287.

14. A.L. Efros and B.I. Shklovskii, in *Electron-Electron Interactions in Disordered Systems* A.L. Efros and M. Pollak, eds. (North Holland, New York, 1985), p. 409.

15. J.H. Davies, P.A. Lee, and T.M. Rice, *Phys. Rev. Lett.* **49**, 758 (1982).

16. M. Grünewald, B. Pohlmann, L. Schweitzer, and D. Würtz, *J. Phys. C* **15**, L1153 (1982).

17. Don Monroe, to be published

18. Don Monroe, A.C. Gossard, J.H. English, B. Golding, W.H. Haemmerle, and M.A. Kastner, *Phys. Rev. Lett.* **59**, 1148 (1987).

19. S.D. Baranovskii, B.I. Shklovskii, and A.L. Efros, *Sov. Phys. JETP* **60**, 1031 (1985).

VRH IN JUST INSULATING Si:Sb

Yuichi Ochiai
Institute of Materials Science
University of Tsukuba
Tsukuba, Ibaraki 305, Japan

ABSTRACT

The temperature dependence of hopping conduction in just insulating Si:Sb has been examined. Mott's variable range hopping (VRH) has been observed in the low temperature conduction very near the metal insulator transition. Magnetic field dependence of the VRH and the Hall coefficient in Si:Sb have been compared with those results of Si:As. In the sample of 2.27×10^{18} Sbcm^{-3}, the resistivity strongly increases below 0.5 K and deviates from the relation of three-dimensional VRH and a certain gap or dip is observed near 0.5 K in the temperature dependence of the Hall concentration. A similar resistivity increase or an anomaly of the low temperature conductions have been reported in Si:P and Si:Sb, so that a new conduction process should be expected. The magnetoresistance has been analyzed using a theoretical calculation based on percolation model.

1. INTRODUCTION

In the non-crystalline materials, Mott's variable range hopping (VRH) is an important electrical conduction to study the disordered electronic states.[1,2] For the Anderson localization states[3] where the density of states $D(E_F)$ at the Fermi level E_F is finite and the states around E_F are localized with a localization length a_L, Mott[4] pointed that a new low-temperature hopping takes place between those states and the conductivity σ is given by

$$\sigma \propto \exp[-(T_0/T)^{1/4}]. \qquad (1)$$

with $k_B T_0 = 18/D(E_F)(a_L^3)$ in the case of three-dimension. Equation (1) is called Mott's VRH and a recent detailed description is reviewed by Kamimura and Aoki[5]. Mott's VRH is considered in one-electron system of a finite $D(E_F)$. However, in the case that Coulombic correlation cannot be neglected, the density of states at the E_F changes because of the long range nature of Coulomb potential. Pollak and Knotek[6] showed that the density of states must have a minimum near the E_F and a gap exists between the filled and empty states. Also Efros and Shklovskii showed that owing to the long range nature of the potential, the single-particle density states vanishes at the E_F.[7] Minimum or vanishing nature of the density of states is called the Coulomb gap. Efros, Nguyen and Shklovskii[8] made a computer simulation and derived the following expression for the conductivity

$$\sigma \propto \exp[-(T'/T)^{1/2}] \quad (2)$$

where T' is a similar constant as well as T_0 as shown in Eq.(1). On the other hand, Pollak pointed out that excitation with less than the gap energy gives rise to multi-electron hopping process.[9] And he suggested that instead of the usual $T^{-1/4}$ behavior a simple power law should be observed in the multi-electron hopping.[10] The power law of conductivity is given by

$$\sigma \propto T^{-Rv/a\ln 2} \quad (3)$$

where Rv is the variance of the distance r_{ij} between site i and j, and a is the localization radius.

Mott[11] suggested that VRH is expected also in the insulator side of the metal-insulator transition (MIT) in doped semiconductors. And for zero compensation, as the concentration increases the activation energy ε_2 of impurity conduction[12] decreases and the $T^{-1/4}$ behavior appears just before the MIT. As for doped Si, such as Si:P and Si:As, VRH conduction has been observed in the just insulating samples near the MIT. However the temperature and the donor concentration regions where VRH is clearly apparent are limited in a narrow range. With a small temperature increase or a slight decrease of the concentration, the just insulating conduction shows a thermal excitation to a high mobility band (including ε_2 activation) or a phonon-assisted hopping between the nearest neighbor impurity sites. Near the MIT the electronic state is not strictly localized, so that correlation effect associated with the overlap of neighboring states cannot be ignored. Electrical resistivity and Hall effect on Si:P have been measured with the compensation effect by Sasaki.[13] At low temperatures, the resistivity deviates from VRH and becomes higher than such an usual $T^{-1/4}$ behavior. The results have been interpreted in terms of VRH assuming a certain structure around the E_F in hopping band. The origin of the Hall effect related to VRH has been shown to be consistent with a percolation approach to VRH. Recently Koon and Castner[14] made conductivity and Hall measurements on Si:As samples on the barely-insulating side of the MIT. They have demonstrated that the Hall coefficient of those insulating samples shows VRH behavior in excellent agreement with the theoretical prediction of Gruenewald et al.[15] The theoretical treatment of Hall mobility in the hopping conduction is based on an averaging procedure on the calculation of the Hall mobility for spatial and energetical disordered hopping systems in the framework of percolation theory. Conductivity and Hall effect in Si:Sb near the MIT were observed by Ochiai, Mizuno and Matsuura.[16] They found that the Hall concentration in just insulating has a peak below 0.5 K. A similar anomaly in the low temperature conductivity has been reported in just insulating Si:Sb.[17] Such an anomaly in the insulating sample near the MIT was also reported in Si:As. Shafarman, Koon and Castner reported that the low-temperature conductivity strongly deviates from VRH and indicates the crossover between the VRH and the much steeper unexplained temperature dependence.[18]

Measurement of magnetoresistance is a very important tool to determine the scattering mechanisms of conducting carriers in doped semiconductors. In metallic samples, the conductivity has a power law dependence on the magnetic field often accompanied with a negative-magnetoresistance. On the other hand, a large positive-magnetoresistance can be seen in the case of nearest neighbor hopping in insulating samples. Shafarman and Castner have observed a positive-magnetoresistance in Si:As in barely insulating side of the MIT.[19] The field dependence of the resistance for the just metallic samples is shown to be proportional to $H^{1/2}$ or H^2 for low or high magnetic fields, respectively. In the intermediate concentration range of impurity conduction on doped semiconductors, the magnetoresistance related to VRH has been investigated by Kurobe and Kamimura.[20] Effects of intra-state correlation on the VRH are studied in the absence and presence of the magnetic field. The magnetoresistance shows a linear dependence on a magnetic field in the lower fields and saturates above a certain field. The positive-magnetoresistance has been explained to be due to the intra-state correlation.

In this paper, the VRH and the related transport properties on just insulating Si:Sb of 2.27×10^{18} cm^{-3} will be mainly studied. The VRH, the Hall coefficient and their magnetic field dependence in Si:Sb will be compared with other doped Si, such as Si:As and Si:P. And finally magnetoresistance near the MIT will be discussed by comparing with theoretical calculations.

2. EXPERIMENTAL DETAILS

Si samples used were prepared from uncompensated commercial quality Czochøralski-grown crystals doped with Sb. A wafer of Si crystal was cut into a Hall-bar shape with eight side arms of the width 0.15cm. Two of unpaired Hall contacts were used as voltage electrodes for four-probe resistance measurement. The length, the width and the depth for a typical resistance measurement were about 0.5, 0.15 and 0.03cm, respectively. After mechanically hand lapping of the surface of the sample by carborandom powder, the sample was rinsed with pure water and then chemically etched using HF, HNO_3 and CH_3COOH solution. Thermal annealing was performed in a dry N_2 gas atmosphere during 30min at 800 C. The lead contact positions of the arms were coated by Sb vacuum evapolation and Au lead wires were welded to the contact position.

The samples were cooled by a ^{3}He cryostat system down to 0.3K or by a ^{3}He-^{4}He dilution refrigerator down to 10mK. In both two systems magnetic field applied to the sample was up to 8T. A calibrated Ge thermometer was used to monitor the temperature of the sample. However, in the case of non-zero magnetic field the temperature was monitored with a calibrated carbon resistor thermometer. The lower temperatures than 50mK in dilution refrigerator were determined by nuclear orientation thermometer consisting of ^{60}Co single crystal. Si sample and the thermometers were mounted in a chamber which was made of oxygen-free Cu and was thermally anchored to a mixing chamber. In ^{3}He cryostat system excellent thermal contact between the sample and ^{3}He liquid was obtained because of direct contact with the liquid. We made low power ac resistance measurements by using an automatic resistance bridge at 25Hz. Input powers were lowered to the range of 10^{-12} to 10^{-15}W and sample self-heating due to ac excitation was not observed. In order to obtain high resistance data, dc resistance measurements were made parallel together with ac method by using a digital volt-meter and a precise current source. In the dc resistance and Hall voltage measurements, the current was flowed with both its polarity so as to eliminate thermal e.m.f. and the current to voltage characteristics was also often measured to monitor a non-ohmic effect. Measurements of specific heat were made in a conventional ^{3}He cryostat system consisting of mainly a thermally isolated sample holder, a electrical heater and an exchange gas line. For a piece of Si crystal which is cut from the same wafer carring out the electrical measurements, the specific heat was measured using a thermal relaxation method from 0.5K to 4.2K at H=0.

3. EXPERIMENTAL RESULTS

The sample studied are listed in Table I along with the values of impurity concentration n, room temperature resistivity ρ_{RT}, $k_F l$, T_0 obtained from the slope of the VRH relation in Eq.(1) and $(T_{H0}/T_0)^{1/4}$. For convenience we have used impurity concentration n which is determined from room temperature Hall voltage by supposing that the Hall coefficient factor is equal to unity. Such n, we believe, would be underestimate because of a slight large Hall coefficient factor even in Si:Sb.[21] The critical concentration of the MIT in Si:Sb[22] has been determined to be 3.0×10^{18} cm^{-3} and is slightly lower than that of 3.74×10^{18} cm^{-3} in Si:P.[23] For the samples near the MIT in Table I, it is difficult to determine whether those belong to metallic or insulating side on the MIT. Here we define the boundary of the MIT in Si:Sb by criterion of $k_F l = 1$. Fermi wave vector k_F in Table I is deduced from the impurity concentration n and mean free path l is estimated from the zero-temperature conductivity from the $T^{1/2}$ dependence of the conductivity at H=0. While the most concentrated sample of 3.33×10^{18} cm^{-3} is clearly metallic, the following two more concentrated samples (2.71 and 2.63×10^{18} cm^{-3}) are difficult to determine whether those are metallic or not. Because of a gradual increase of the resistivity with decreasing temperature even down to 10 mK, it is too difficult to conclude on such a determination with only above criterion for the two samples. A similar situation has been reported in Si:As very near the MIT.[18] In the above mentioned two samples, the low temperature conduction near 1 K almost agrees with the $T^{1/2}$ dependence discussed in quantum correction of conductivity in weak-localized regime.[23] However, below around 0.5 K, the conductivity deviates from the weak-localized conduction.

Table I Sample Characteristics

n (10^{18} cm^{-3})	ρ_{RT} (10^{-2} ohm cm)	$k_F l$	T_0 (K)	$(T_{H0}/T_0)^{1/4}$
3.33	1.04	3.02	-	-
2.71	1.12	1.02	0.000063[##]	0.46[#]
2.63	1.16	0.66	0.02	-
2.51	1.21	-	0.2	0.40[#]
2.27	1.25	-	0.33	0.60[#]
2.1	1.34	-	4700	-
1.95	1.38	-	33000	-
1.9	1.41	-	74000	-

[#]T_{H0} is obtained from the slope of log R_H versus $T^{-1/4}$ at H = 1.5 T.
[##]this small value is obtained from the slope at H = 0 T in Fig. 2.

Below 0.5 K the resistivity steeply increases and the deviation from $T^{1/2}$ law depends on preparation processes of the sample surface.[24] The steep increase of the resistivity below about 1 K has been reported to have T^{-2} dependence[25] or log-T dependence.[24]

Figure 1 shows log ρ versus $T^{-1/4}$ for four Si:Sb samples listed in Table I. The low temperature conduction of the samples can apparently fit Mott's VRH law of Eq.(1) except for the most concentrated sample (3.33). The most dilute sample (1.9) may exhibit the $T^{-1/4}$ behavior in a narrow temperature range. The conduction of other three samples in Fig. 1 fits Eq.(1) for $T^{-1/4}$ >0.6. Even near $k_F l=1$ in the sample of 2.71×10^{18} cm^{-3}, Mott's VRH can be fit below about 0.5 K or less as shown in Fig. 2. This low temperature conduction seems to be fit with rather Eq.(1) than Eq.(2) and than log-T dependence.[24] On the

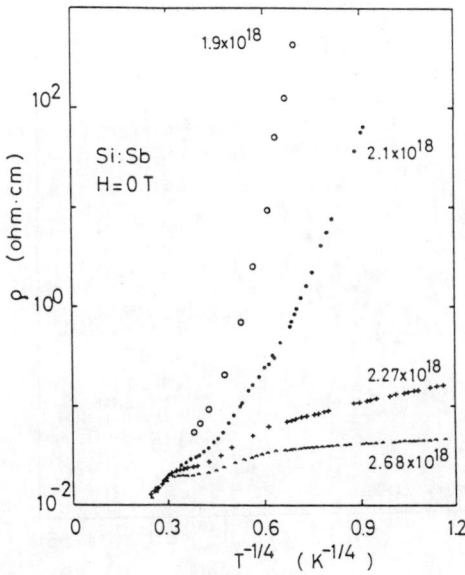

Fig.1 Temperature dependence of the low temperature resistivity for insulating Si:Sb. Each impurity concentration in units of cm^{-3} is indicated beside the curve.

other hand, in the samples of 2.51 and 2.27×10^{18} cm^{-3}, the resistivity shows a steep increase below about 0.5 K or less. The deviation from the VRH behavior in the sample of 2.27×10^{18} cm^{-3} is seen at 0.2 K as shown in Fig. 3a and the steep increase of the resistance in Fig. 3b is almost close to T^{-2} law.

Hall coefficient R_H and Hall mobility for Si:Sb samples are plotted against $T^{-1/4}$ and T in Figs. 4a and 4b, respectively. The logarism of R_H seems to give a linear fit against $T^{-1/4}$ except for the most concentrated sample (3.33). In the sample of 2.27×10^{18} cm^{-3}, R_H deviates from the linear fit and increases below 0.5 K. Hall mobility has been obtained from the resistivity in Fig. 1 and the R_H at 1.5 T in Fig. 4a. The low temperature Hall mobility decreases with lowering

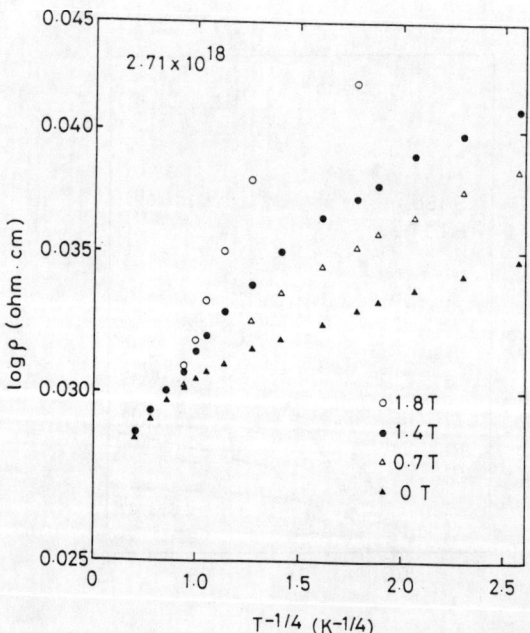

Fig.2 Resistivity versus $T^{-1/4}$ for Si:Sb of 2.71×10^{18} cm^{-3} at H = 0, 0.7, 1/4 and 1.8 T.

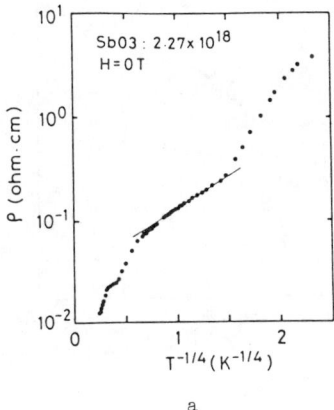

a

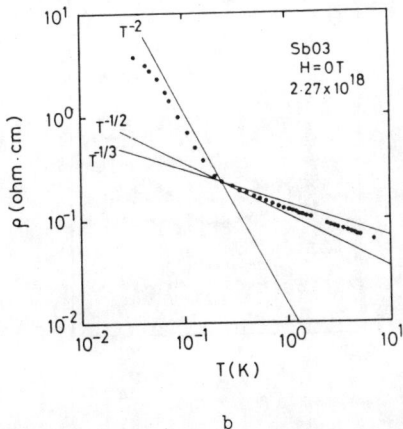

b

Fig.3 Resistivity versus $T^{-1/4}$ (a) and T (b) for Si:Sb of 2.27×10^{18} cm^{-3} at H = 0 T. The straight line in Fig. 3a indicates a fit with Eq.(1). In Fig. 3b three straight lines indicate T^{-2}, $T^{-1/2}$ and $T^{-1/3}$ fits and the resistivity above 0.2 K shows a small curvature.

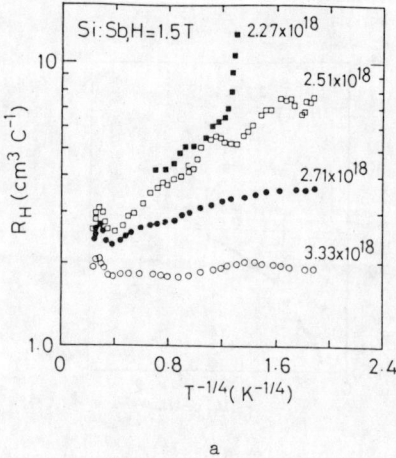

a

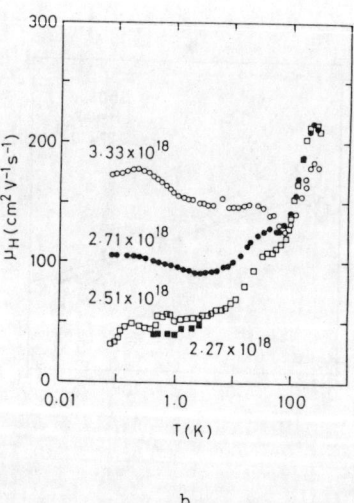

b

Fig.4 Temperature dependences of R_H (a) and the Hall mobility for Si:Sb plotted versus $T^{-1/4}$ and T, respectively. Numbers beside the curves indicate the impurity concentration in units of cm^{-3}.

impurity concentration and seems to stop to decrease and finally to reach a certain boundary value around 2.51×10^{18} cm^{-3}. The boundary value of the Hall mobility is close to the theoretical estimation, 10 cm^2/Vsec, which is calculated with RPA method[26] near l=a when a is considered to be equal to the effective Bohr radius in Si:Sb.

Magnetic field depndence of the resistance and R_H in the sample of 2.27×10^{18} cm^{-3} are shown in Figs 5a and 5b, respectively. The logarism of the resistance and R_H give an almost linear fit against $T^{-1/4}$ for the same temperature region (4.2 to 0.6 K) as studied in Si:As by Koon and Castner.[14] However, we find a few anomaly in our results below about 0.5 K; a steep increase of the resistivity at high magnetic fields, a similar increase in R_H at H=0 and a certain hump in R_H at high magnetic fields. We have also analyzed the ratio $(T_{H0}/T_0)^{1/4}$ after the method of Koon and Castner.[14] They suppose that R_H has the same temperature dependence as in Eq.(1). Then T_{H0} is

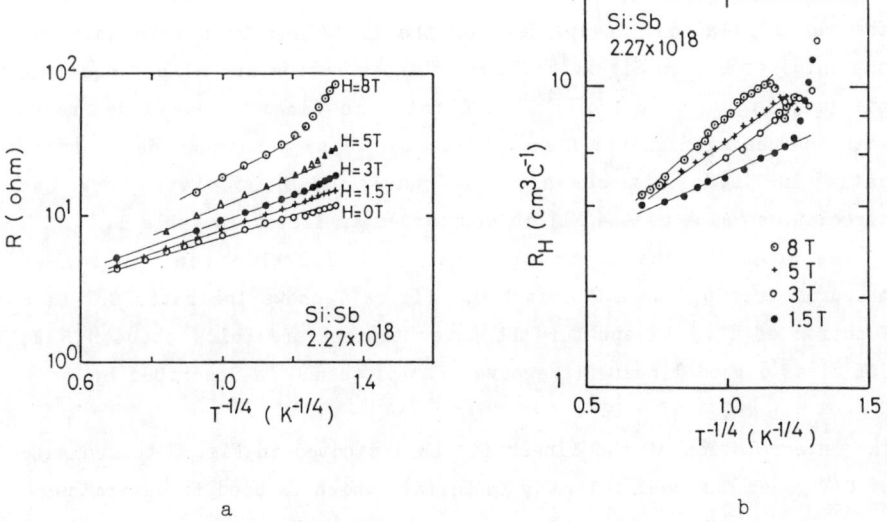

Fig.5 Resistance (a) and R_H (b) versus $T^{-1/4}$ for Si:Sb of 2.27×10^{18} cm^{-3} at H = 0, 1.5, 3, 5 and 8 T.

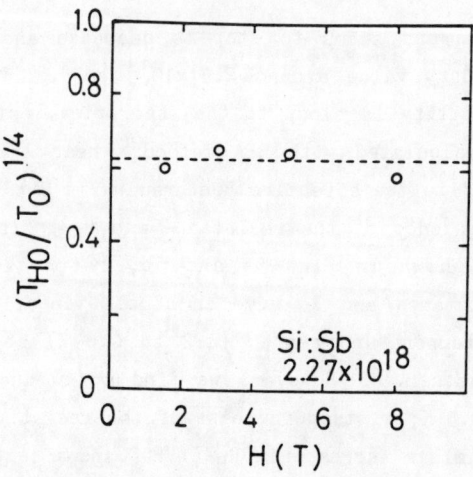

Fig.6 The ratio $(T_{HO}/T_0)^{1/4}$ as a function of magnetic field for Si:Sb of 2.27×10^{18} cm^{-3}. The dotted line shows the value 5/8.[15]

obtained from the slope of R_H in Fig. 5b. Magnetic field dependence of $(T_{HO}/T_0)^{1/4}$ in the sample of 2.27×10^{18} cm^{-3} is shown in Fig. 6. The ratio is almost independent of the field and is nearly equal to the value at H=0 in Si:As.[14] However those ratios are almost equal to the percolation value 5/8.[15] The ratio in Si:As[14] weakly decreases with increasing magnetic field. Although we have no many data on the ratio in Si:Sb, it seems to decrease and to deviate from the percolation value as the MIT is approached as listed in Table I.

The specific heat C in the sample of 2.27×10^{18} cm^{-3} has been measured from 0.5 to 4.2 K at H=0. Figure 7 shows the ratio C/T as a function of T^2. Except for the lower temperature below about 1.5 K, C/T gives a good linear fit against T^2 and then C is described by

$$C = \gamma T + bT^3. \qquad (4)$$

The interpolation of the linear fit as indicated in Fig. 7 to ordinate of C/T gives the coefficient γ in Eq.(4), which is used to determine

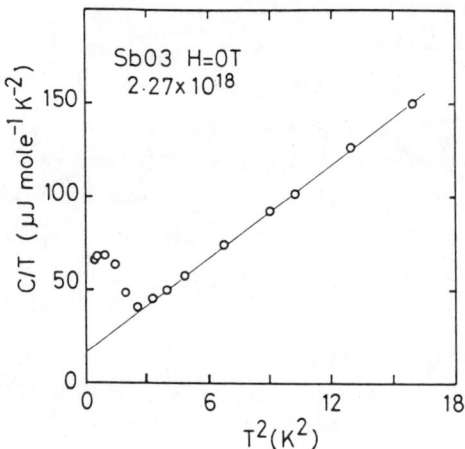

Fig.7   Specific heat for Si:Sb of 2.27×10^{18} cm^{-3}. The straight line corresponds to fit of Eq.(4).

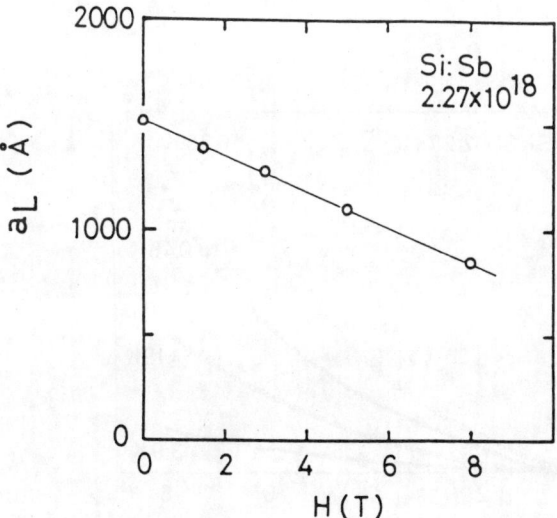

Fig.8   Localization length a_L for Si:Sb of 2.27×10^{18} cm^{-3} as a function of magnetic field.

the density of states at E_F with
$$D(E_F) = 3\gamma/2\pi^2 k_B^2. \qquad (5)$$
For the sample in Fig. 7, we obtain $D(E_F)=1.8 \times 10^{20}$ eV^{-1}cm^{-3}. If $D(E_F)$ depend strongly neither on magnetic field nor on temperature, we can deduce the field dependence of a_L in Eq.(1) from the result of Fig. 5a. The field dependence of a_L is shown in Fig. 8. And a_L decreases linearly against the field. Because magnetic length l_H (cyclotron radius;(he/cH)$^{1/2}$) is much larger than the effective Bohr radius even at 8 T, electronic state belongs to a weak feild limit. From this limit in the framework of isolated donors, we cannot explain why electronic state is gradualy localized as the field increases.

Magnetoresistance of the sample of 2.27×10^{18} cm^{-3} is shown in Fig. 9. The positive-magnetoresistances at various temperatures almost agree with H^2 dependence. $H^{1/2}$ dependence which is reported in weak-localization regime of Si:P[27] at low fields does not appear even at 0.3 K. Then we can observe no inflection point between both above two dependences as studied in Si:As.[18] We have compared the result of

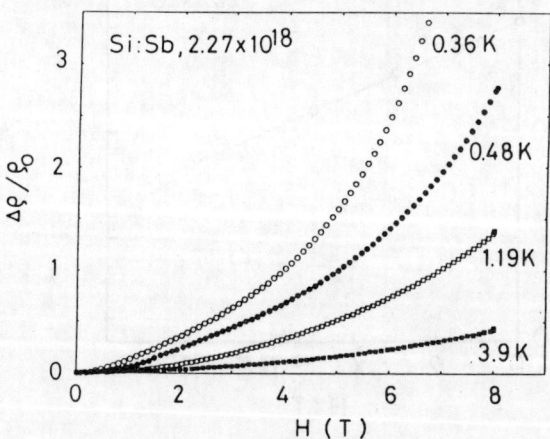

Fig. 9   The magnetoresistance versus magnetic field for Si:Sb of 2.27×10^{18} cm^{-3} at T = 0.36, 0.48, 1.19 and 3.9 K.

Fig. 9 with a theoretical calculation[20] based on percalation model. The theoretical calculation gives $T^{-1/4}$ dependence as in Eq.(1) and the coefficient of the temperature dependence in $T^{-1/4}$ depends on two localization lengths a_{L1} and a_{L2} which are related to singly and doubly ocupied states, respectively. The coefficient is determined by the following relation,[20]

$$T_0 = \frac{1}{k_B} \frac{2268}{115\pi} \frac{s}{D(E_F)} \frac{a_{L1}^3 + 3a_{L2}^3}{(a_{L1}^3 + a_{L2}^3)^2 + 4a_{L2}^6} \qquad (6)$$

where s is the value giving percolation threshold. For simplicity

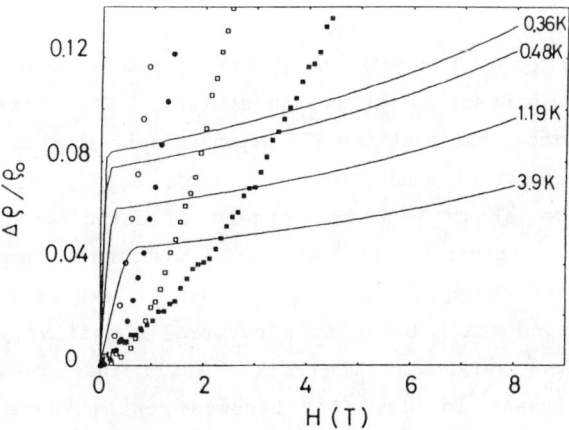

Fig. 10   Calculated results (solid lines) of magnetoresistance as a function of magnetic field for various temperatures and the observed magnetoresistances are plotted together with the same symbol as indicated in Fig. 9.

flat density of states has been assumed and energy dependence of envelope functions in Anderson localized states has been neglected. Here we use magnetic field dependence of a_L in Fig. 8 instead of energy dependence. We put $a_{L1}=a_L$, $2a_{L1}=a_{L2}$.[27] $D(E_F)$ is obtained from the result of Fig. 7 and the intra-state correlation energy U is estimated from ESR spin susceptivility of Si:Sb.[28] Although we could not obtain a good quantitative agreement with the calcutated result, we have succeeded in a qualitative explanation of the positive-magnetoresistance taking into account the magnetic field dependence of a_L. The result of the comparison is plotted in Fig. 10 together with the observed values.

4. DISCUSSION

In the sample of 2.27×10^{18} cm^{-3}, we observe a steep increase of the resistivity below 0.2 K. A similar behavior in the low temperature conduction has been reported and discussed in Si:P[30,31] and Si:As.[18] Kobayashi, Monden and Sasaki[30] have observed T^{-2} dependence in the low temperature conduction in insulating Si:P. And they have explained the power law with random matrix theory taking into account the level correlation of Anderson localization states. After this report, Ootuka et al. pointed out that the T^{-2} dependence is attributed to the surface layer electrical conduction. In fact, such just insulating conduction below about 1 K is largely affected with thermal annealing[24] or with particle irradiation.[32] However, as discussed in the following, the above effect is considered to come from not only surface layer conduction but also structural sensitivity of bulk electronic states very near the MIT. While the slope of the resistivity increase in Si:P[25,30] becomes gentle where the T^{-2} dependence is observable, the slope at the low temperature side in Si:Sb of 2.27×10^{18} cm^{-3} is rather steeper than that of VRH and the steep slope is almost fit with T^{-2} behavior as shown in Fig. 3b. The difference is probably due to the degree of nearness of the MIT and is important to determine whether parallel path currents, such as a surface layer conduction, is effective or not. Electronic state in

Si:Sb of 2.27×10^{18} cm^{-3} is relatively near the MIT when comparing it with that in Si:P.[25,30] In the case very close to the MIT, it is difficult to assume parallel path currents at lower temperatures where such high resistive conduction appears. Similar high resistivity has been observed and discussed in Si:P very near the MIT.[13] Therefore, for Si:Sb, we must consider certain bulk conduction for the steep increase of the low temperature resistivity.

R_H can be observed and fits $T^{-1/4}$ dependence in the temperature region where Mott's VRH is observable as shown in Figs. 5a and 5b. Similar temperature dependence of the resistivity or the R_H has been reported in Si:P[13] and Si:As.[14] After the nearest neighbor hopping theory,[33] Hall effect is expected not to appear. However Hall mobility due to hopping conduction has been studied with three or four mutually nearest neighbor configurations and has been obtained for the case of ac applied fields in the impurity conduction regime of doped semiconductors.[34] Friedman and Pollak[35] calculated the Hall mobility due to hopping-type transport in spatially random systems by means of percolation arguments. For the samples near the MIT in Si:Sb of Fig. 4b, the low temperature mobility is found to be close to the evaluated value by Friedman.[26] This indicates that the evaluation of the critical value of the Hall mobility can be applicable to the case of Si:Sb. Reflecting a percolative nature of hopping conduction, Sasaki[13] discussed on the VRH of Si:P with resistance network which consists of the transport electricity among the random array of localized states. We consider that the VRH can be qualitatively explained in such a framework of the random network model. Moreover we must carry out a lot of quantitative approaches so as to compare with theoretical estimations related to the interacting electron system near the MIT.

Koon and Castner[14] discussed on the ratio $(T_{HO}/T_0)^{1/4}$ where the VRH holds for resistivity in Si:As. The ratios in Si:As at H=0 and Si:Sb in Fig. 6 almost agree with the theoretical prediction.[15] As the concentration approaches to the critical concentration of MIT, the ratio seems to deviate from the prediction value of 5/8.[15] Except for a small difference between the observed and predicted values, this percolative picture is found to be applicable to the just insulating transport near the MIT. However we cannot conclude whether the difference in the ratio comes from an anomaly in Hall coefficient factor near the MIT[14,21] or other effects.

The temperature dependence of R_H at various fields has a steep increase or a hump below around 0.5 K as shown in Fig. 5b. The hump of R_H around 120 K in Fig. 4a is expected to indicate a transition between two different transports, such as excited drift or hopping-type conduction. Then we can consider that the hump near 0.5 K shows a certain boundary between the VRH at high temperature region and another new conduction process at low temperature regions. The new conduction process has a power law dependence on temperature as shown in Fig. 3b. And the temperature dependence of the new conduction process fits neither Eq.(1) nor Eq.(2). One possible explanation of the new process is multi-electron hopping[10] in Eq.(3). The new process is considered to be observable when thermal smearing of Coulomb gap disappears with decreasing temperature. However, its experimental verification may become difficult, since the validity of Eq.(3) is restricted to very low temperature when the resistivity becomes very hard to measure.[10] And usually the power of Eq.(3) is estimated to be about 20.[10] Another possible explanation is T^{-2} dependence deduced by random matrix method.[30] Both two possible explanations have been based on a strong correlation effect in Anderson localization states near the MIT;level correlation picture is appropriate or Coulomb gap plays a important role in the just insulating doped Si. Although we cannot determine which explanation is applicable to the new conduction process, it is clear that the T^{-2} dependent process in the just insulating Si is observed following after VRH with lowering temperature and is related to a strongly correlated electron system.

From the result of R_H in Fig. 5b, we can obtain the temperature dependence of the Hall concentration for various magnetic fields as shown in Fig. 11.[16] At higher temperatures than 0.5 K the concentration decreases as the field increases. However, below 0.5 K, the concentrations have a peak near 0.4 K at higher fields than 3 T.

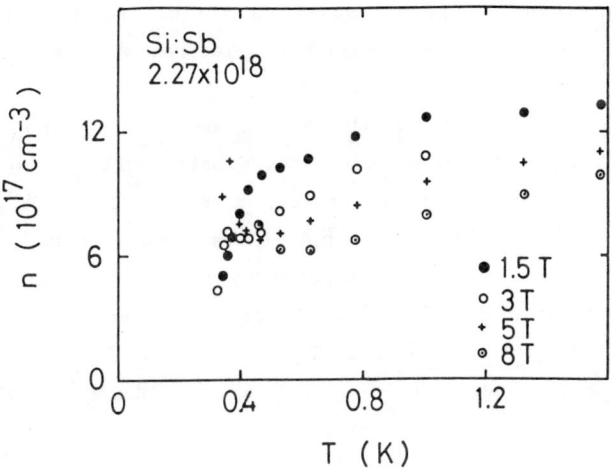

Fig.11   Temperature dependence of Hall concentrations for Si:Sb of 2.27x10^{18} cm^{-3} at H = 1.5, 3, 5 and 8 T.

This means that the decrease of a_L in Fig. 8 can be explained with the decrease of Hall concentration. Therefore this also indicates suppression of hopping conduction due to localization effect of magnetic field. Moreover, it seems to be field-independent as the temperature decreases and the concentration drops near 0.3 K. Although, it is not clear whether such a dip in the Hall concentration is originated from the pseudogap structure of the density of states or not, at nearly the same temperature range, a similar hump behavior has been reported in the low temperature conductivity for various magnetic fields in the just insulating Si:Sb.[17] The anomaly is explained by an electronic phase transition, such as a redistribution of electron due to a change in the valley degeneracy. On the other hand, another explanation is suggested, it is due to an enhancement of conductivity by destructive interference of the scattered waves in conjunction with spin-orbit interaction. Mott[36] suggested that an effect of spin-orbit interaction must be important in Si:Sb and the diffusion length for spin-orbit scattering may be less than the inelastic scattering length at low temperatures in the just insulating Si:Sb. With lowering temperature the screening due to conduction electron will gradually become weaker because of the decrease of carrier concentration as shown in Fig. 11. Then the spin-orbit coupling becomes strong at low temperatures. This means an enhancement of the spin-orbit interaction near the MIT. Therefore the low temperature transport in Si:Sb must be substaintially influenced by spin-orbit interaction. Actually the spin-orbit interaction in Si:Sb is expected to become stronger than other doped Si, such as Si:P and Si:As. If we can compare the result of Fig. 11 with that in other doped Si, an effect of spin-orbit interaction will be discussed in terms of an electron correlation associated with the pseudogap structure of the Hall concentration.

As for a magnetic field effect on VRH, following two possible cases can be considered. One is a change of the occupation number of localized electrons because of Zeeman shift among interacting Anderson localization states. Another is due to shrink of wavefunction in a magnetic field. For the non-magnetic semiconductors the shallow wavefunction shrinks with the field. For doped Si under the usual field strength around 10 T, it is not possible to make the magnetic length l_H sufficiently smaller than the Bohr radius as reported in InSb.[37] Then we consider that the effect in localized donor at a high field limit is not so important for the just insulating Si:Sb below 10 T. Near the MIT, it is easy to make the magnetic length much less than a_L[19], so that a_L will depend on the field strength as shown in Fig. 8. While the reason for the linear dependence on magnetic field in Fig. 8 has not been clarified yet, a large a_L and its field dependence gives evidences of Anderson localization states near the MIT. After theoretical calculation by Kurobe and Kamimura[20], the magnetoresistance saturates when the suppression of hopping process between a doubly-occupied and an empty states is complete in strong magnetic fields. Experimental results in Fig. 9 do not show such a saturation but a strong positive-magnetoresistance. Taking into account the field dependence of a_L, we can suppress the saturation and obtain a positive contribution of the magnetoresistance as shown in Fig. 9. Kamimura and Aoki[5] pointed out an energy dependence of a_L, which is proposed by Fukuyama and Yosida.[38] If the a_L is an increasing function of energy as in the case for E_F below the mobility edge E_c, a_L increases when H is applied. And this effect gives rise to a negative-magnetoresistance. As shown in Fig. 8 a_L observed is a decreasing function of magnetic field and the magnetoresistance is positive, contrary to the above considerations. They assumed that electrons with the spin parallel to the field are majority carriers and the density of states is almost flat near E_F. Since there is no exact information on electron correlation in the Anderson localization states near the MIT in doped Si at the present time, we cannot conclude whether those assumptions are applicable to Si:Sb or not. As for flat density of states near E_F, it seems to be too simple to explain the elctrical transport in correlated electron system. If the density of states has a gap near E_F, a negative contribution of the magnetoresistance will suppress or disappear. In Eq.(6), although we put $2a_{L1}=a_{L2}$ for simplicity, we cannot neglect the magnetic field dependence of the ratio of a_{L2}/a_{L1}. In fact, the positive contribution of the magnetoresistance is enhanced taking into account the field dependence of the ratio. Nevertheless, including magnetic field dependence, we must carry on more precise experiments and analysises in order to investigate correlation effects under the Anderson localization near the MIT.

5. CONCLUSION

The data on insulaing Si:Sb of 2.27×10^{18} cm^{-3} show Mott's VRH at low temperatures below about 4 K and the Hall coefficients show also $T^{-1/4}$ dependence at the same temperature range. We have obtained magnetic field dependence of localization length a_L which is almost propotional to magnetic field. The field dependence of a_L and a large a_L show evidences of the existence of Anderson localization states near the MIT. Especially in the just insulating sample very near the MIT, Mott's VRH can be observed down to 10 mK. It seems to be difficult to fit with Efros's $T^{-1/2}$ behavior. This gives an evidence of almost flat $D(E_F)$ very near the MIT. However, with a slight decrease of impurity concentration, a certain gap or dip is observed in the Hall concentration and growth of pseudogap can be considered in the density of states near E_F. As for a positive-magnetoresistance in the just insulating Si:Sb, it is found that energy dependence of localization length is an important effect in order to clarify the field dependence.

ACKNOWLEDGEMENTS

The authors are very much indebted to Prof. M. Pollak for variable discussions on the VRH and the MIT of Si:Sb. We are grateful to Prof. H. Kamimura, Prof. H. Aoki and Dr. A. Kurobe for variable discussions and useful advices on the theory of magnetoresistance in doped Si and we would like to acknowledge Prof. T. G. Castner for sending us many papers related to transport and resonance experiments on Si:As. We are grateful to Prof. E. Matsuura and Mr. M. Mizuno for discussions and low temperature measurements of this work. Almost all of this work was performed at the Cryogenics Center of University of Tsukuba.

REFERENCES

1. M.Pollak, M.L.Knotek, H.Kurtzman and H.Glick, Phys.Rev.Lett. **30**, 856 (1973).
2. Y.Ochiai, Phys.Stat.Sol.(a) **99**, K87 (1987).
3. P.W.Anderson, Phys.Rev. 109, 1492 (1958).
4. N.F.Mott, J.Non-Cryst.Solids **1**, 1 (1969).
5. H.Kamimura and H.Aoki, The Physics of Interacting Electrons in Disordered Systems, (Clarendon Press, Oxford),121 (1989). to be published.
6. M.Pollak and M.L.Knotek, J.Non-Cryst.Solids **32**, 141 (1979).
7. A.L.Efros and B.I.Shklovskii, J.Phys.C **8**, 149 (1975).
8. A.L.Efros, L.Nguyen and B.I.Shklovskii, Solid State Commun. **32**, 851 (1979).
9. M.Pollak, Philos.Mag. **B42**, 782 (1980).
10. M.Pollak and M.Ortuno, Electron-Electron Interactions in

Disordered Systems, ed. by A.L.Efros and M.Pollak (North Holland Pub.,Amsterdam), 381 (1985).
11. N.F.Mott and E.A.Davis, Electronic Processes in Non-Crystalline Materials, (Clarendon Press, Oxford), 119 (1979).
12. H.Fritzsche, J.Phys.Chem.Solids **6**, 69 (1958).
13. W.Sasaki, Philos.Mag. **B52**, 427 (1985).
14. D.W.Koon and T.G.Castner, Solid State Commun. **64**, 11 (1987).
15. M.Gruenwald, H.Mueller, P.Thomas and D.Wuertz, Solid State Commun. **38**, 1011 (1981).
16. see, for example, Y.Ochiai, M.Mizuno and E.Matsuura in Proceedings of International Conference on the Application of High Magnetic Fields in Semicondoctor Physics 1988, edited by G.Landwehr.
17. A.P.Long and M.Pepper, J.Phys.C **17**, L425 (1984).
18. W.N.Shafarman, D.W.Koon and T.G.Castner, Phys.Rev. **B40**, 1216 (1989).
19. W.N.Shafarman and T.G.Castner, Phys.Rev.Lett. **56**, 980 (1986).
20. A.Kurobe and H.Kamimura, J.Phys.Soc.Jpn. **51**, 1904 (1982).
21. F.Mousty, P.Ostoja and L.Passari, J.Appl.Phys. **45**, 4576 (1974).
22. T.G.Castner, N.K.Lee, G.S.Cieloszyk and G.L.Salinger, Phys. Rev.Lett. **34**, 1627 (1975).
23. T.F.Rosenbaum, K.Andres, G.A.Thomas and R.N.Bhatt, Phys.Rev. Lett. **45**, 1723 (1980).
24. Y.Ochiai and E.Matsuura, Solid State Commun. **49**, 441 (1984).
25. Y.Ootuka, F.Komori, Y.Monden, S.Kobayashi and W.Sasaki, Solid State Commun. **36**, 827 (1980).
26. L.Friedman, J.Non-Cryst.Solids **6**, 329 (1971).
27. A.Natori and H.Kamimura, J.Phys.Soc.Jpn. **43**, 1270 (1977).
28. measurement of the spin susceptibility were done by ESR method, see, for example, Y.Ochiai and E.Matsuura, Phys. Stat. Sol.(a) **38**, 243 (1976).
29. T.F.Rosenbaum, R.F.Milligan, G.A.Thomas, P.A.Lee, T.V. Ramakrishnan and R.N.Bhatt, Phys.Rev.Lett. **47**, 4009

(1982).
30. S.Kobayashi, Y.Monden and W.Sasaki, Solid State Commun. **30**, 661 (1979).
31. T.F.Rosenbaum, R.F.Milligan, M.A.Paalanen, G.A.Thomas and R.N. Bhatt, Phys.Rev. **B27**, 7509 (1982).
32. Y.Nishio, Y.Ootuka, K.Kajita, T.Iwata and W.Sasaki, Phys. Stat.Sol.(a) **91**, 725 (1985).
33. A.Miller and E.Abrahams, Phys.Rev. **120**, 745 (1960).
34. T.Holstein, Phys.Rev. **124**, 1329 (1961).
35. L.Friedman and M.Pollak, Philos.Mag. B **38**, 178 (1978).
36. N.F.Mott, J.Phys.C **20**, 3975 (1987).
37. R.Mansfield, M.Abdul-Gader and P.Fozooni, Solid State Electron. **28**, 109 (1985).
38. H.Fukuyama and K.Yosida, J.Phys.Soc.Jpn. **46**, 102 (1979).

THE INFLUENCE OF SPIN-SPIN INTERACTION ON THE LOW-TEMPERATURE LIMIT OF VHR-CONDUCTIVITY

I. S. Shlimak

A.F. Ioffe Physico-Technical Institute,
Academy of Sciences of the USSR,
Leningrad, 194021

ABSTRACT

We predict that there exists a low temperature limit of the range in which the Coulomb gap governs variable range hopping of charge carriers among localized states at low temperatures. We present experimental evidence for this limit in amorphous $Ge_{1-x}Cr_x$ films, irradiated polyimide, and transmutation-doped n-type Ge.

1. EFFECT OF COULOMB GAP ON VARIABLE RANGE HOPPING AND ITS LOW-TEMPERATURE LIMIT.

In doped semiconductors with impurity concentrations on the dielectric side near the metal-insulator transition (MIT) the Fermi level is situated in the localized part of the density of states (DOS). According to present theoretical understanding, electronic conduction at $T \to 0$ in such systems is by hopping of charge carriers among localized states having energies close to the Fermi level. The characteristic feature of such variable-range hopping (VRH) conductivity σ is a continuous decrease of the conduction activation energy with decreasing temperature yielding

$$\sigma(T) = \sigma_o \exp(-\frac{T_o}{T})^m \quad \text{with } m<1 \qquad (1)$$

These ideas have been confirmed by several experimental observations in different semiconductors at low and super-low temperatures. The exponent m can have different values in 3-dimensional systems, depending on the density of states $g(E_F)$ at the Fermi level E_F. One expects $m=1/4$ when $g(E_F)$ is constant or slowly varying and $m=1/2$ when there is a parabolic gap at the Fermi level and $g(E_F)=0$.

The theoretical explanation of this gap is based on taking into account the Coulomb interaction between the localized carriers. As a result of this interaction a strong correlation appears between energies and distances of any two states above and below the Fermi level[1]

$$\Delta_{ij} = E_i - E_j \; \frac{e^2}{r_{ij}} > 0. \qquad (2)$$

Here Δ_{ij} is the energy difference of the transition between two states, one of them, i, is initially occupied below the Fermi level and the other, j, is initially empty above the Fermi level; r_{ij} is the distance between these localized states.

One can see from Eq. (2) that occupied and empty states must be separated energetically by at least the Coulomb energy corresponding to their distance. This is the reason for the appearance of a parabolic gap in the DOS at E_F. In contrast, the case of $m = 1/4$ is obtained by assuming that there are no correlations and that all localized states have a random distribution in space and energy.

A number of simplifications were made in deriving Eq. (2). For example, the effects of screening and exchange interaction between localized states were neglected. This is acceptable for small concentrations of impurities because the exchange interaction drops very rapidly with increasing distance in contrast to the Coulomb interaction which decreases as a power-law. However, the VRH-conductivity at low and superlow temperatures has been studied on samples with high impurity concentrations near the metal-insulator transition, because the resistivity of more lightly doped semiconductors is enormously high at low temperatures and practically immeasurable.

It is known that near the MIT the exchange interaction represents a considerable part of the binding energy of electrons on impurity atoms. Moreover at $T \to 0$ the hopping distance increases continuously and it is possible that eventually this distance will become larger than the screening radius. When that happens the Coulomb interaction will decrease very sharply. As a consequence, the temperature in which VRH-hopping with a Coulomb gap and $m = 1/2$ is observed should have a low-temperature limit.

We believe that there are two mechanisms which change the $T^{-1/2}$ law at very low temperatures:

1) A decrease of Coulomb interaction due to screening will lead to a disappearance of the parabolic gap (Fig. 1a), and therefore the $T^{-1/2}$ law will change to $T^{-1/4}$ in Eq. (1).

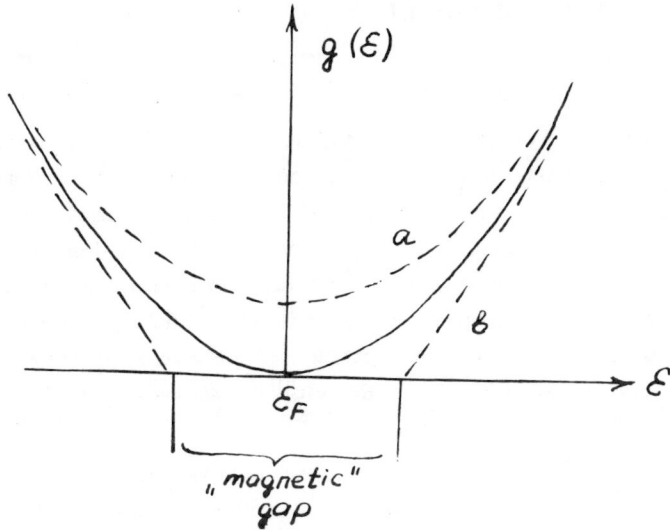

Fig.1   Parabolic Coulomb gap altered as a result of screening (a) and exchange interaction (b).

2) When taking into account spin-spin interactions, an additional term associated with exchange energy will appear in Eq. (2). When this term has the same sign as the Coulomb term then it leads to an increase of the inequality (2) and the appearance of a real gap (Fig. 1b), which could be called "magnetic gap". Therefore the $T^{-1/2}$ exponent in Eq. (1) will change to T^{-1}, i.e., a conductivity with a constant activation energy.

This suggestion of a limit of the $T^{-1/2}$ law is somewhat speculative, but we argue that in order to retain the $T^{-1/2}$ law as $T \to 0$ it is necessary that the strong correlation appearing in Eq. (2) persists at longer and longer distances, which we feel is unreasonable. In contrast, for the case of the $T^{-1/4}$ law there is no such low temperature limit because no correlation between energy and separation of localized states was assumed or is involved.

2. EXPERIMENTAL RESULTS

Let us consider now some experimental data that indicate deviations from the $T^{-1/2}$ law, especially in cases where the existence of the $T^{-1/2}$ law is beyond doubt.

First a few words about the change from $T^{-1/2}$ to $T^{-1/4}$. This is a deviation in the direction of lower resistivities. Unfortunately, there are other reasons for such a deviation. A short bridge between the electrical contacts due to a metal-like conductivity along dislocations and surfaces are examples. A decrease in resistivity may also occur due to selfheating which occurs particularly at superlow temperatures or due to a non-ohmic influence of the electric field on the hopping conductivity, which is the greater, the longer the mean distance of hopping is. Since there are several causes for decreasing the resistivity, the analysis of experimental data has to be done with special care.

Nevertheless as an example I wish to refer to the data by Tokumoto, Mansfield and Lea[2] concerning VRH in n-InSb at superlow temperatures shown in Fig. 2. One finds that the $T^{-1/2}$ law is observed reliably in strong magnetic fields in a wide

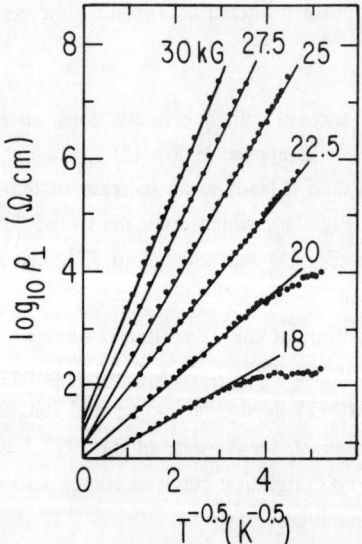

Fig.2.   Transverse magnetoresistivity of n-InSb as a function of $T^{-1/2}$ according to Ref. 2.

interval of resistivities and temperatures. However in weak magnetic fields, close to the MIT, deviations from the $T^{-1/2}$ law to lower resistance are observed. The authors suggest that this may be caused by selfheating. It is very difficult to critically assess data obtained by someone else, but it is interesting to notice that in measurements along the magnetic field no deviations from the $T^{-1/2}$ law were observed in the same interval of temperatures and resistivities. Unfortunately, the range of the deviation was too small to determine whether a $T^{-1/4}$ law followed the $T^{-1/2}$ law or not. Moreover, one has to eliminate other reasons for a decrease in resistivity, such as dislocations and surface effects.

Deviations from the $T^{-1/2}$ law to T^{-1} behavior, i.e., to higher resistivities, are therefore much more interesting. It is clear that such deviations can not be explained by other reasons but changes in DOS spectra. I wish to present here experimental data concerning deviations of this sort which have been observed in Ioffe Physico-Technical Institute in Leningrad. The first observations have been made on samples of specially prepared amorphous germanium-chromium film.[3]

2.1 Amorphous $Ge_{1-x}Cr_x$ Films

The motivation of this work was to study the influence of spin-spin interaction on the hopping conductivity. Therefore chromium impurities were chosen. Because of the strong localization of electrons on chromium atoms, it was necessary to introduce a relatively high concentration, several atomic percent, in order to reach the MIT and to measure the conductivity at low and superlow temperatures. That is why amorphous films of $Ge_{1-x}Cr_x$ were used for this research.

Fig. 3 shows the temperature dependence of the conductivity for films with different x. At $x>0.12$ a metallic-like conductivity is observed, but at $x \approx 0.08$–0.1 VRH-conductivity following the T^{-12} law exists. Fig. 3b shows the logarithmic slopes $d\ln\sigma/d\ln(1/T)$ of two of the curves of Fig. 3a as a function of T on a double logarithmic plot. The slope of these lines is equal to the index m. Deviations from $T^{-1/2}$ to T^{-1} dependence are observed near $T \leq 1K$.

Our explanation of this phenomenon is based on the model discussed above: at $T<1K$ spin-spin interaction leads to the appearance of a magnetic order-like "spin-glass" because of the random distribution of interatomic distances. The exchange energy varies from centre to centre and as a result the correlation expressed by Eq. (2) between energetic position and space distribution is destroyed

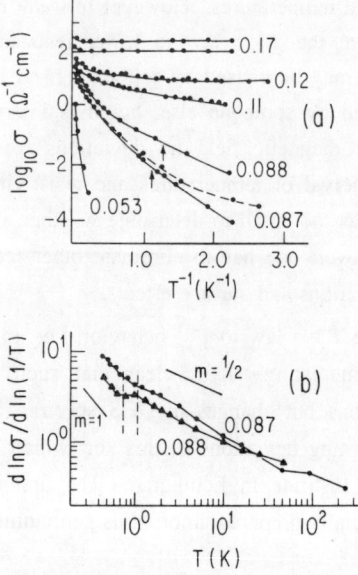

Fig.3 Conductivity (a) and relative activation energy (b) of amorphous films $Ge_{1-x}Cr_x$ as a function of temperature (Ref. 3).

and the Coulomb gap disappears. Instead of that a real "magnetic" gap associated with the necessity of overcoming the exchange energy appears.

If our model is correct, then this deviation from the $T^{-1/2}$ law should disappear in strong magnetic fields. Indeed, in strong magnetic fields all spins will be oriented along the field, so the energy change will be equal for all states and thus should not be taken into account. Therefore we should expect to find a considerable increase of the conductivity with increasing magnetic field. This effect should be larger at lower temperatures because the temperature dependence of the conductivity reverts back from the T^{-1} to the weaker $T^{-1/2}$ dependence. So we expect to find a giant negative magnetoresistance effect. The experimental results support our hypothesis. One can see from Fig. 4 that at $T>1.5K$ the usual positive magnetoresistance is observed in all samples in agreement with the VRH theory. But at $T<1K$ the sign of the magnetoresistance changes to negative with its magnitude increasing rapidly while the temperature is decreased. This is the result of the conductivity reverting back to the $T^{-1/2}$ law as shown for $x=0.087$ by the dashed line in Fig. 3.

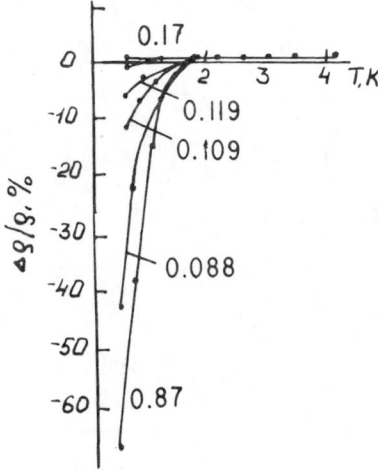

Fig.4 Temperature dependence of the magnetoresistance of $Ge_{1-x}Cr_x$ for different x at H = 10 kOe (Ref. 3).

2.2 Irradiated Polymer Films

Recently results have been obtained on samples of polymer films, in which irradiation with energetic ions led to metallic-like conductivity.[4] The results are shown in Fig. 5. In these samples the observation of a deviation from the $T^{-1/2}$ law to a T^{-1} dependence was unexpected, because no impurities with uncompensated magnetic moments were introduced. It is possible that in this case a high value of the exchange interaction is provided by a specific space distribution of the electron wave functions. Electrons are localized on the broken bonds of polymer molecules. In any case, the relatively high temperature $T \sim 4K$ for the appearance of the magnetic ordering effect allowed us, besides observing the giant negative magnetoresistance (Fig. 6), to estimate the electron spin susceptibility from electron-spin resonance measurements. The result is presented in Fig. 7. At relatively high temperature, where the $T^{-1/2}$ law is observed, the susceptibility is paramagnetic following a Curie law. At $T<4K$ a deviation from this law is observed, which is a direct evidence for magnetic ordering with ferromagnetic character. Moreover, our estimate of the value of the exchange integral agrees with the conductivity activation energy. So measurements of the magnetic susceptibility confirm directly the model for a low-temperature limit of VRH conductivity in the presence of a Coulomb gap as discussed above.

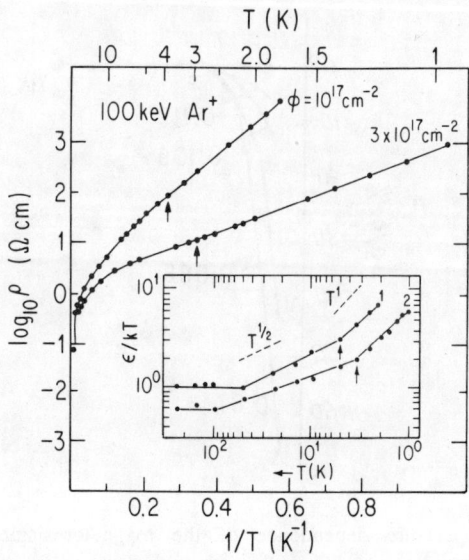

Fig.5　Temperature dependence of the conductivity of polyimide films, irradiated by 100 keV Ar^+ ions with different fluences Φ. Inset shows $d\ln\sigma/d\ln(1/T)$ as a function of T on double logarithmic scales to reveal magnitude of index m (Ref. 4).

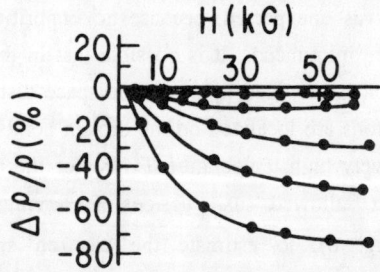

Fig.6　Magnetoresistance of irradiated polyimide films at different temperatures (Ref. 4).

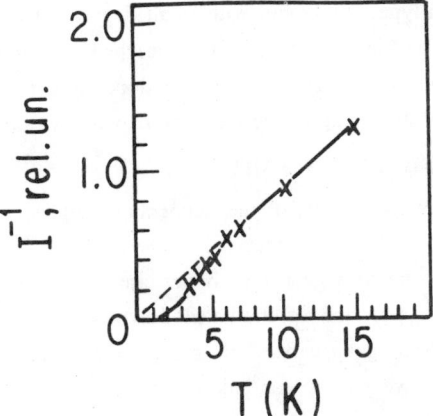

Fig.7   Inverse magnetic susceptibility of one irradiated polyimide film as a function of temperature (Ref. 4).

If our model is correct then this effect should be observed not only in such "exotic" systems as amorphous films and irradiated polymers, but also in "classical" semiconductor where the $T^{-1/2}$ law has been observed reliably. For this reason we began to study doped germanium near the metal-insulator transition at very low temperatures. Preliminary results will be described in the following.

2.3. Transmutation-Doped n-Type Ge

Low-temperature hopping conductivity has been measured in samples of neutron-transmutation doped germanium with specially changed isotropic composition. Neutron transmutation doping is essential for studying hopping conduction close to the MIT, since it enables one to introduce dopants with high accuracy and homogeneity. This is important because in the presence of inhomogeneities, part of the material will be in the insulating and others in the metallic state at low temperatures.

Neutron transmutation doping is especially convenient for germanium because the abundances of the respective isotopes and their cross-sections for capture of thermal neutrons permit doping over a wide range of concentrations, overlapping both the insulating and metallic states. Unfortunately, neutron irradiation of natural germanium produces only p-type material with a compensation ratio $K = 0.4$.

In order to obtain n-type germanium with small degrees of compensation, the germanium crystal used in our experiment was enriched with the isotope ^{74}Ge, which transforms after neutron capture into an ^{75}As donor impurity. After neutron irradiation and subsequent annealing of the radiation damage we obtained n-Ga(As) samples with concentrations close to the MIT.

The results of our measurements of the temperature dependence of the resistance are shown on Fig. 8. One finds that at $T>100mK$ the $T^{-1/2}$ law holds well, but at $T<100mK$ deviations to higher resistivities are noticeable, similar to the cases of amorphous $Ge_{1-x}Cr_x$ and of irradiated polymers. Unfortunately at the lowest temperatures selfheating appears. This prevented us from determining the new conductivity law, but we plan to continue our studies in order to measure the conductivity at the lowest temperatures with higher accuracy as well as the magnetoresistance at $T<100mK$.

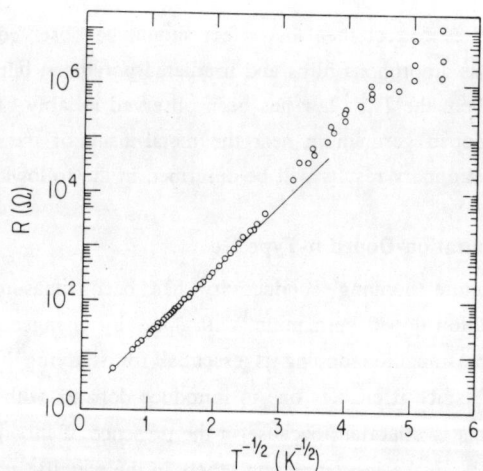

Fig.8 Temperature dependence of the resistance of sample ^{74}Ge(As) doped by the nuclear transmutation method. $N_{As}=2.8\times10^{17}cm^{-3}$.

3. CONCLUSIONS

There are a number of experimental difficulties associated with establishing the low temperature limit of the temperature regime in which the Coulomb gap governs variable range hopping. The temperature of magnetic ordering is usually very low and the resistance accordingly very high for samples with impurity concentrations relatively far from the MIT. On the other hand, as the impurity concentration approaches the MIT, the spin-spin interaction decreases due to the effects of screening and delocalization.

The measurements carried out on homogeneous samples at very low temperatures suggest that one will be able to explore the low temperature limit of the $T^{-1/2}$ range in classical semiconductors such as crystalline germanium doped with shallow impurities. It is very important to observe deviations from the $T^{-1/2}$ law in other semiconducting materials. It is one of the aims of this report to draw the attention of experimentalists to the need of making these observations and to challenge theorists to explain these deviations.

ACKNOWLEDGEMENTS

I wish to thank my colleagues of the Millikelvin Laboratory, Royal Holoway and Bedford New College of the University of London for their help and hospitality during my measurements of transmutation-doped Ge.

REFERENCES

1. Efros, A.L. and Shklovskii, B.I., J. Phys. C. 8, L49 (1975).
2. Tokumoto, H., Mansfield, R. and Lea, M.J., Phil. Mag. B. 46, 93 (1982).
3. Aleshin, A.N., Ionovo, A.N., Parfenjev, R.V., Shlimak, I.S., Heinrich, A., Schumann, I., and Elefant, D., Fiz. tverd, tela, 30, 696 (1988) (Sov. Phys.-Solid State).
4. Aleshin, A.N., Gribanov, A.V., Dobrodumov, A.V., Suvorov, A.V., and Shlimak, I.S., Fiz. tverd. tela, 31, 12 (1989) (Sov. Phys.-Solid State).

HOPPING CONDUCTIVITY AND MAGNETIC TRANSITIONS OF THE Cu^{2+} SPINS IN SINGLE-CRYSTAL La_2CuO_{4+y}

Tineke Thio,[*] R.J. Birgeneau,[*,†] C.Y.Chen,[*] B.S. Freer,[*,†] D.R. Gabbe, H.P. Jenssen, M.A. Kastner,[*] P.J. Picone, and N.W. Preyer,[*]
Center for Materials Science and Engineering,
MIT, Cambridge MA02139.

ABSTRACT

Measurements are reported of the magnetoresistance (MR) for fields up to 23T in La_2CuO_{4+y} single crystals in which the Cu^{2+} spins order antiferromagnetically at $T_N \sim 240K$, and in which the conductivity at low temperature is characterised by hopping between localised states. Using the MR, we map out the phase diagram of the spin flop transition, observed when the magnetic field is applied parallel to the zero-field staggered magnetisation, and that of the weak-ferromagnetic transition, observed with the field perpendicular to the CuO planes. In both transitions the antiferromagnetic propagation vector changes from the $\vec{a}$ direction at zero field to the $\vec{c}$ direction at the highest fields. This rather subtle change of the Cu spin ordering is accompanied by a large increase in the interlayer hopping conductivity: up to a factor 2. We show that the magnetoconductance is proportional to the three-dimensional staggered moment with propagation vector in the orthorhombic $\vec{c}$ direction. The origin of this unusual behaviour is an important unsolved problem.

I. INTRODUCTION

Lamellar cuprates exhibit high temperature superconductivity when doping introduces a sufficiently high density of excess holes or electrons into the CuO_2 layers. In many theoretical models, the pairing necessary for superconductivity involves the coupling of the charge carriers to the Cu^{2+} spins. The study of the lightly doped, insulating antiferromagnetic state is important, because, in the latter, the density of charge carriers can be sufficiently low that the interaction between them is small relative to their interaction with the Cu^{2+} spins.

Materials in the $La_{2-x}(Sr,Ba)_xCuO_4$ system, in which high T_c superconductivity was first discovered,[1] have the simplest structure, so they are ideal for studying the physics of the CuO_2 layers. Undoped La_2CuO_4 is a model two-dimensional (2D) S=1/2 Heisenberg antiferromagnet.[2,3] The in-plane nearest neighbour interaction between the Cu^{2+} spins in La_2CuO_4 is described by the Heisenberg Hamiltonian to an unusually high degree of accuracy: the anisotropies are several orders of magnitude smaller than the nearest neighbour exchange.[4,5] Because of weak interlayer coupling, there is a transition into the Néel state at temperatures close to room temperature.

In this paper we review magnetoresistance (MR) measurements on pure single crystals of La_2CuO_{4+y} for magnetic fields up to 23T. The crystals have transport properties characteristic of doped semiconductors on the insulating side of the insulator-to-metal transition.[6,7] At high T the transport is characterised by thermal activation to a band of highly anisotropic states; at low T the conductivity is dominated by hopping between localised states. These states, introduced by excess oxygen in the crystal, are anisotropic: Chen et al.[7] find an in-plane localisation length $\xi_a=8\pm1$ Å, and an upper bound for the out-of-plane localisation length $\xi_b<3$ Å. Samples with the same Néel temperature as those used for the MR were found from Hall effect measurements[6] to contain a density of $\sim 4\times 10^{19}$ cm^{-3} acceptors.

With the magnetic field applied perpendicular to the CuO_2 planes ($\vec{H} \| \vec{b}$ in orthorhombic notation, space group $Cmca$), a weak-ferromagnetic transition is observed.[8,9] When the external field is applied in the direction of the staggered magnetisation ($\vec{M}^\dagger \| \vec{c}$), features in the MR indicate a spin flop transition.[10] From an extrapolation of the observed critical fields to T=0 we find values for the anisotropies in the nearest-neighbour exchange; theses values confirm earlier measurements[4,5] of the spin Hamiltonian.

The MR provides new insight into the coupling between the excess holes and the background magnetism. We find that the observed MR arises from the interlayer hopping conductivity, and that the latter is proportional to the total staggered moment with propagation vector $\vec{\tau}$ in the $\vec{c}$ direction, both for the spin flop and for the weak ferromagnetic (WF) transitions.

However, the very large size of the MR, and the fact that it is proportional to the staggered rather than the uniform moment present a major unsolved problem.

II. EXPERIMENTAL DETAILS

The samples used for this experiment are large single crystals of pure La_2CuO_{4+y} grown by the top-seeded solution growth method using CuO flux.[11] The crystals have twin domains; we therefore label a magnetic field applied in an orthorhombic in-plane direction $\vec{H} \| \vec{a}, \vec{c}$.

To measure the magnetoresistance, the sample is mounted on a variable-temperature cold finger and placed in the centre of a Bitter magnet. Using a conventional four-probe geometry with the current perpendicular to the CuO_2 planes, the resistance is measured as a function of magnetic field for fields up to 23T at typical sweep rates of 0.08T/sec.

Some of the samples used in this experiment appear to have small cracks, which are visible by eye or under an optical microscope, and which make it difficult to measure the absolute conductivity. In high-quality samples with no observable cracks the conductivity was highly anisotropic.[6] We used the Montgomery method[12] to measure the anisotropy and to determine the change in the conductivity tensor at the WF transition.

III. RESULTS

Figure 1 shows the unusual behaviour of the magnetic moment and the magnetoresistance (MR) for $\vec{H} \| \vec{b}$. At a critical field, there is a jump in the magnetic moment and a corresponding decrease in the resistance, The transition moves to lower field and broadens as T approaches T_N. The critical fields, found from the MR and the magnetisation data, as well as from neutron scattering,[9] are plotted *versus* temperature to generate the phase diagram shown in Figure 2.

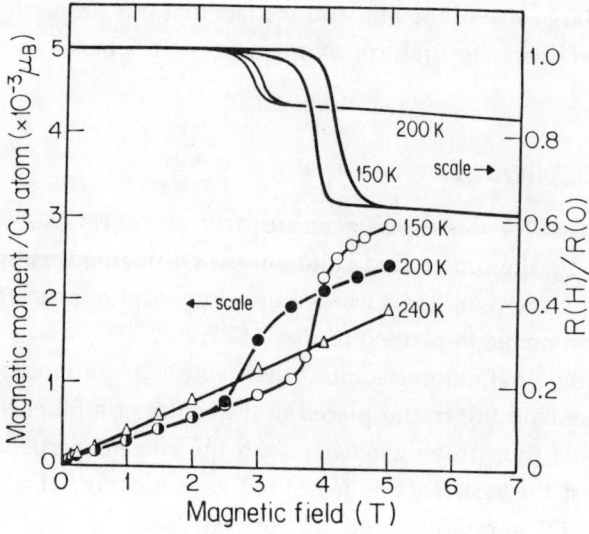

Figure 1. Resistance and magnetic moment as a function of magnetic field $\vec{H} \parallel \vec{b}$.

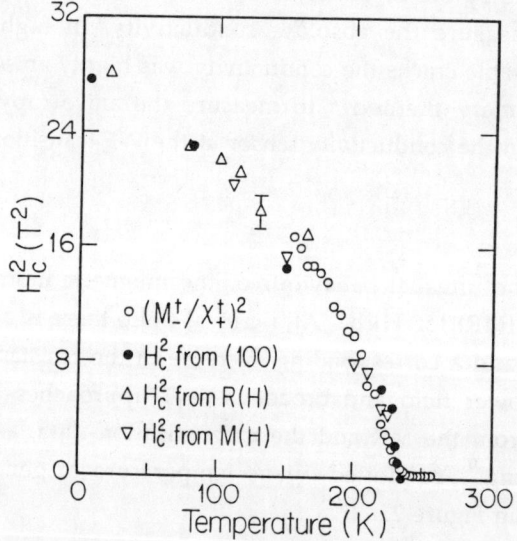

Figure 2. Temperature dependence of the critical fields for $\vec{H} \parallel \vec{b}$ from the MR (triangles), from magnetisation (inverted triangles) and from neutron scattering (filled circles). Also included in the figure is the three-dimensional order parameter (open circles), which is found in mean-field theory[8] to be proportional to the critical field.

For $\vec{H} \| \vec{b}$ there is a single phase boundary which intersects the field axis at $H_c(0)=5.3\pm0.3$ T. The critical field vanishes at $T_N=240$K, where also the three-dimensional antiferromagnetic order parameter $M_-^\dagger$ goes to zero. The zero-temperature extrapolation of the jump in the magnetic moment (not shown) is $M^F(0)=(2.1\pm0.2) \times 10^{-3}$ μ_B/Cu atom.[8]

Figure 3(a) shows the MR normalised to the zero-field resistance R_0 for $\vec{H} \| \vec{a}, \vec{c}$. At all temperatures the resistance decreases monotonically with increasing field. As in the case $\vec{H} \| \vec{b}$, the overall change in resistance is as large as a factor of ~2 at T=24K. At low T the MR has knees at two distinct critical fields; at the highest temperatures only one transition is observed. These features appear more clearly in the derivative dR/dH, plotted as a function of magnetic field in Figure 3(b); the features are free of hysteresis.

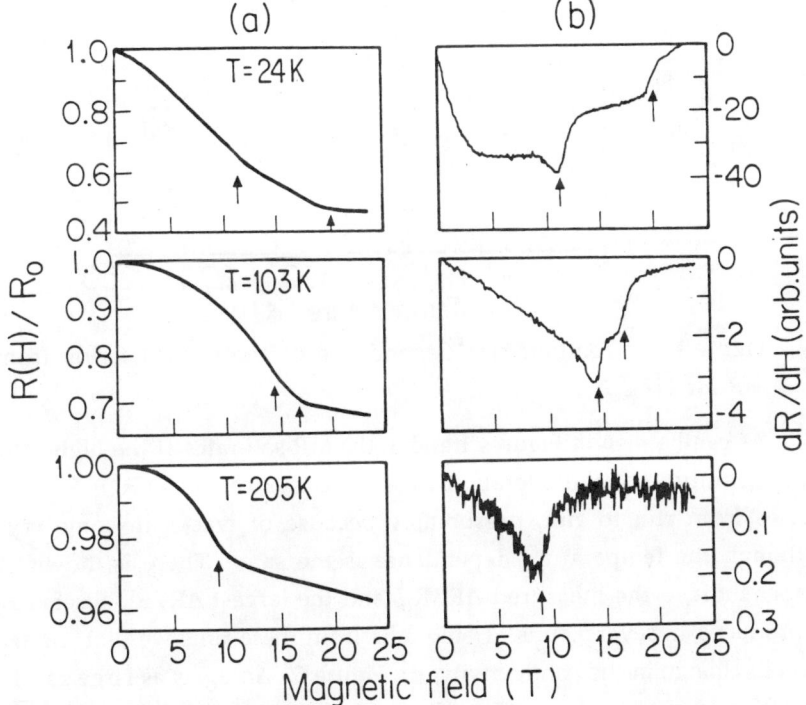

Figure 3. (a) Magnetoresistance normalised to zero-field resistance R_0 for $\vec{H} \| \vec{a}, \vec{c}$; (b) derivative of the MR as a function of field. Arrows indicate critical fields.

When we plot the critical fields as a function of T for $\vec{H} \| \vec{a}, \vec{c}$ (Figure 4), two distinct phase boundaries are seen at low T. The zero-temperature extrapolations of the critical fields are $H_1(0)=10.5\pm1.0$ T and $H_2(0)=20.5\pm1.0$ T. The two boundaries merge at a multicritical point at $T_{mc} \sim 120$K. Above T_{mc} there is only one phase boundary which tends to zero field at $T_N=240$K.

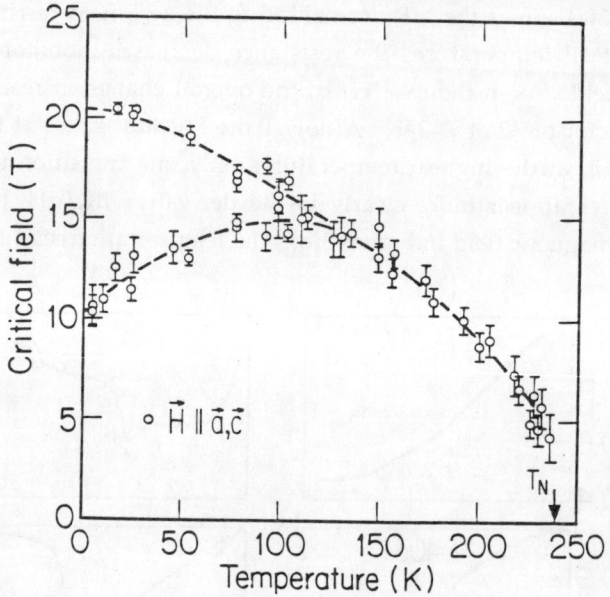

Figure 4. Temperature dependence of the critical fields from the MR for $\vec{H} \| \vec{a}, \vec{c}$.

As can be seen in Figures 1 and 3, the MR saturates at the highest fields. The magnitude of the overall change in resistance, $(R(H)-R_0)/R_0 = \Delta R/R_0$, varies from run to run, presumably because of cracks in some crystals, although the temperature dependence is the same. There is, however, an upper limit to the measured $\Delta R/R_0$, and the largest $\Delta R/R_0$ observed was reproduced in several runs. Using data from these runs we plot $\Delta\sigma/\sigma_0$, the overall change in the conductivity, in Figure 5. $\Delta\sigma/\sigma_0$ is as large as ~1.25 at T=24K, and goes to zero at T_N. The overall change in resistance for $\vec{H} \| \vec{a}, \vec{c}$ is the same as that found in the MR of the same sample for $\vec{H} \| \vec{b}$. This similarity strongly suggests that the MR in both transitions arises from the same coupling mechanism of the holes to the Cu^{2+} spins.

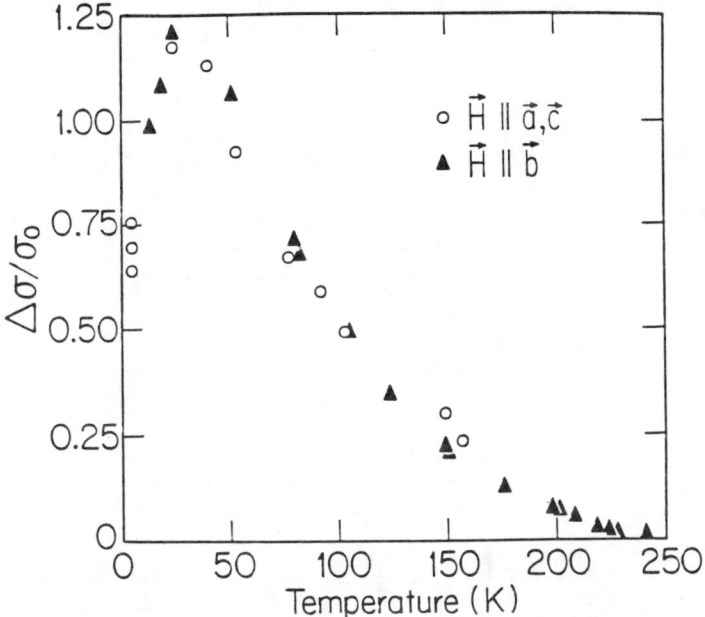

Figure 5. Temperature dependence of the overall change of the conductivity for $\vec{H} \| \vec{a}, \vec{c}$ (circles) and $\vec{H} \| \vec{b}$ (filled triangles).

In Figure 6 are plotted the temperature dependence of the anisotropy of the conductivity at zero field and of the MR for $\vec{H} \| \vec{b}$, measured with the Montgomery method[12] on a crystal which has no observable cracks. Both σ_a/σ_b, the anisotropy of the conductivity at H=0 (Figure 6(a)), and $(\Delta\sigma_b/\sigma_b)/(\Delta\sigma_a/\sigma_a)$, the anisotropy of the *relative* MR (Figure 6(b)), appear to be weakly temperature dependent at low T and increase rapidly with T for T>50K, whereas $\Delta\sigma_b/\Delta\sigma_a$, the anisotropy in the *absolute* MR, is only weakly dependent on T over the entire temperature range (Figure 6(c)).

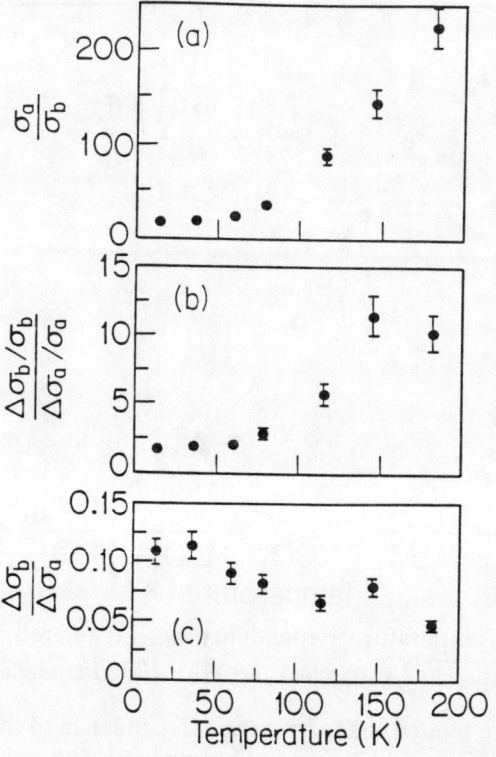

Figure 6. Temperature dependence of the anisotropy of (a) the conductivity at zero field; (b) the relative overall magnetoconductance; (c) the absolute overall magnetoconductance. $\vec{H} \| \vec{b}$.

IV. DISCUSSION

A. Magnetic Transitions

We briefly review the relevant magnetic parameters in La_2CuO_4. Since both the nearest-neighbour and interlayer couplings are antiferromagnetic, the magnetic unit cell contains four spins in two CuO_2 layers (see Figure 7(a)). For two nearest-neighbour spins in the ith layer, one can define a staggered moment $\vec{M}_i^{\dagger} = (\vec{M}_{i\delta} - \vec{M}_{i\epsilon})/2$ and a ferromagnetic moment $\vec{M}_i^F = (\vec{M}_{i\delta} + \vec{M}_{i\epsilon})/2$. To find the field dependence of the Cu spins at T=0, the total magnetic Hamiltonian is minimised, using a mean-field approximation, with respect to $\vec{M}_i^F$ and $\vec{M}_i^\dagger$. Details of this calculation are given elsewhere;[10] here we briefly summarise the results in a heuristic way.

In orthorhombic notation, the in-plane nearest-neighbour exchange is written $H_{ij} = S_i \cdot \overleftrightarrow{J} \cdot S_j$, where

$$\overleftrightarrow{J} = \begin{bmatrix} J^{aa} & 0 & 0 \\ 0 & J^{bb} & J^{bc} \\ 0 & -J^{bc} & J^{cc} \end{bmatrix} \quad (1)$$

Here $J_{nn} = \frac{1}{3}(J^{aa} + J^{bb} + J^{cc}) \sim 128 \text{meV}$.[13] Because $J^{cc} = J^{aa} > J^{bb}$, there is an out-of-plane gap in the magnon spectrum $E_A^b = zS\sqrt{2J_{nn}(J^{cc} - J^{bb})}$, where $z = 4$ is the number of nearest neighbours. The antisymmetric elements J^{bc} arise from a Dzyaloshinsky-Moriya[14,15] interaction which is allowed in the orthorhombically distorted crystal structure, and which contributes a term to the nearest-neighbour Hamiltonian of the form $J^{bc}\hat{a} \cdot (S_i \times S_j)$.[8,16] J^{bc} introduces an in-plane anisotropy gap $E_A^a = zJ^{bc}S$; infra-red absorption and neutron scattering measurements indicate $E_A^a = 1.1 \pm 0.3$ meV and $E_A^b = 2.5 \pm 0.5$ meV.[4,5] These anisotropies cause the staggered moment at zero field to lie in the orthorhombic $\vec{c}$ direction (see Figure 7(a)). The spins cant in the $\vec{b}$ direction, giving rise to weak ferromagnetism which is hidden at zero field because the effective interlayer exchange $J_\perp$ is antiferromagnetic. The interlayer order determines the direction of the AF propagation vector $\vec{\tau}$: for AF interlayer order, $\vec{\tau} \| \vec{a}$, for ferromagnetic interlayer order, $\vec{\tau} \| \vec{c}$.[9]

With the magnetic field applied perpendicular to the CuO$_2$ layers, there is a transition to an overtly weak-ferromagnetic state, when the energy lowering from reorienting the weak-ferromagnetic moments $\vec{M}_i^F$ is larger than the interlayer coupling $J_\perp$. In the high-field state (see Figure 7(b)), the effective interlayer order is ferromagnetic ($\vec{M}_1^\dagger = \vec{M}_2^\dagger$), that is $\vec{\tau} \| \vec{c}$. The critical field for this transition is given by

$$H_c M^F = J_\perp S^2 \quad (2)$$

where $M^F = g\mu_B S J^{bc}/2J_{nn}$ is the size of the canted moment per Cu atom at zero field. From the $T=0$ extrapolated values of H_c and M^F, and using $J_{nn} = 128$ meV,[13] we find $J^{bc} = 0.8 \pm 0.3$ meV and $J_\perp \approx 2\mu$eV.

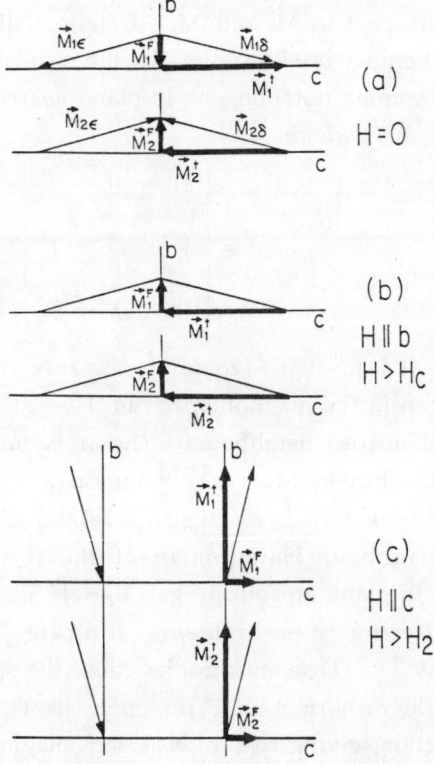

Figure 7. Magnetic moments in the four-sublattice unit cell, projected along the orthorhombic â direction, for (a) H=0, (b) $\vec{H}\|\vec{b}$, $H>H_c$, and (c) $\vec{H}\|\vec{c}$, $H>H_2$.

When the field is applied in the $\vec{c}$ direction, parallel to the staggered moment, there is a spin flop transition when the field is large enough to overcome the in-plane anisotropy J^{bc}. In addition, because of the antisymmetric exchange, there is a continuous out-of-plane motion of the spins, which ends abruptly when the staggered moments are perpendicular to the plane (see Figure 7(c)). In this configuration, the antisymmetric exchange and the external field reinforce one another, at the expense of the out-of-plane anisotropy and the interlayer coupling. The two critical fields which mark the spin flop transition and the completion of the out-of-plane rotation of the spins are, respectively,

$$H_1 = H_M$$

$$H_2 = \frac{1}{H_M}\left[2 H_E H_A^b + 4 H_E H_\perp - H_M^2 \right] \qquad (3)$$

where $g\mu_B H_E = zJ_{nn}S$, $g\mu_B H_A^b = z(J^{cc}-J^{bb})S$, $g\mu_B H_M = zJ^{bc}S$, and $g\mu_B H_\perp = z'J_\perp S$.

From the zero-temperature extrapolation of the lower spin flop phase boundary, $H_1(0)=10.5\pm1.0$ T, we find a spin-wave gap $E_A^a = 1.3 \pm 0.2$ meV. This value is to be compared to that measured in neutron scattering[4] ($E_A^a=1.0\pm0.3$) and infra-red spectroscopy[5] ($E_A^a=1.1\pm0.3$meV). Using the classical expresssion $H_M=zJ^{bc}S$ gives the value $J^{bc} = 0.7\pm0.1$ meV. This value agrees with that measured in the WF transition. To extract the out-of-plane anisotropy, we use $J_{nn}=128$meV, from two-magnon Raman scattering,[13] and $J_\perp=2\mu$eV from the WF transition;[8] with $H_2(0)=20.5\pm1.0$T, we find the out-of-plane gap $E_A^b=g\mu B(2H_E H_A^b)^{1/2} = 1.8\pm0.6$ meV, in agreement with the neutron scattering result ($E_A^b=2.5\pm0.5$ meV). From this we extract the out-of-plane anisotropy $J^{cc}-J^{bb} = 3\pm2$ μeV. We have thus confirmed the conclusion from previous measurements[4,5] that the exchange anisotropies are much smaller than the exchange itself. Since the interlayer coupling $J_\perp$ is also very small, La_2CuO_4 is a model 2D S=1/2 Heisenberg antiferromagnet.

B. Magnetoresistance.

The most intriguing aspect of this work is the high sensitivity of the conductivity to the magnetic order. We demonstrate next that the MR is proportional to M_+^F, the interlayer order parameter with $\vec{\tau}\|\vec{c}$.

The spin flop transition is similar to the WF transition in this respect: in both transitions the direction of the AF propagation vector changes from $\vec{\tau}\|\vec{a}$ at zero field to $\vec{\tau}\|\vec{c}$ at high field. Since the direction of $\vec{\tau}$ is determined by the interlayer magnetic order, we have calculated various interlayer layer order parameters which change at these transitions. The sum $M_+^F=(M_1^F+M_2^F)/2$ and difference $M_-^F=(M_1^F-M_2^F)/2$ of the ferromagnetic moments in two adjacent layers are plotted in Figure 8(a) and 8(b), respectively. $M_-^\dagger$ and $M_+^\dagger$ are the components of the staggered moment per unit cell with propagation vector $\vec{\tau}\|\vec{a}$ and $\vec{\tau}\|\vec{c}$, respectively.[9] Equivalently, $M_-^\dagger$ and $M_+^\dagger$ measure the relative antiferromagnetic and ferromagnetic alignments respectively of the interlayer nearest neighbour spins. They are simply related by $M_+^\dagger = (M_1^\dagger + M_2^\dagger)/2 = g\mu_B S \sin\xi$ and $M_-^\dagger = (M_1^\dagger - M_2^\dagger)/2 = g\mu_B S \cos\xi$, and we therefore plot only $M_+^\dagger$ in Figure 8(c). The experimental data at T=24K are plotted in Figure 8(d) as magnetoconductance.

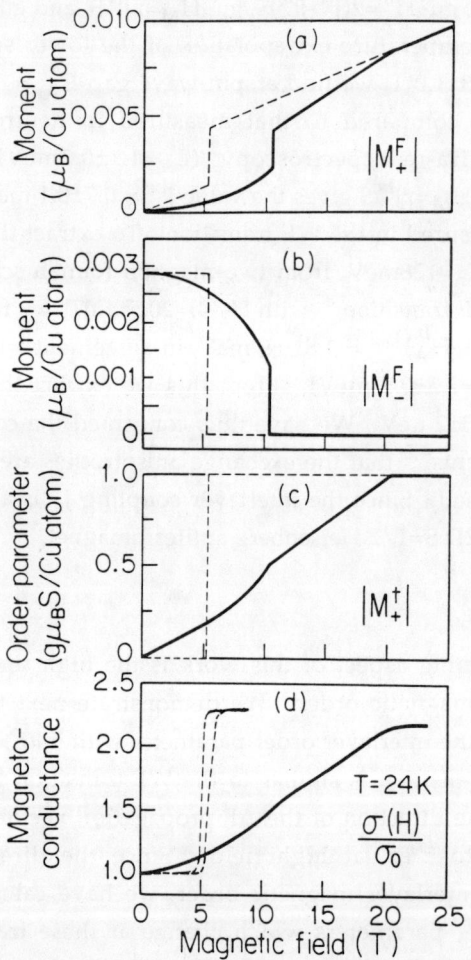

Figure 8. (a) Calculated uniform ferromagnetic moment; (b) calculated difference of the ferromagnetic moments in adjacent layers; (c) calculated interlayer order parameter for $\vec{\tau} \parallel \vec{c}$; (d) magnetoconductance, T=24K. Solid lines indicate the spin flop transition, dashed lines the WF transition.

The MR does not follow the field dependence of the total moment. This is seen clearly in the case $\vec{H} \| \vec{b}$ (Figure 8(a)): $M_+^F(H)$ has a finite slope both above and below the transition, where the MR is independent of field.

For $\vec{H} \| \vec{b}$, M_-^F is independent of field above and below the transition (Figure 8(b)). Its field dependence is consistent with the heuristic idea that the conductivity is enhanced when the canting in adjacent layers becomes identical. However, in the case of the spin flop, the calculated M_-^F indicates that the layers become equivalent at H_1. (For $H>H_1$, $\vec{M}_i^F$ in all layers point in the direction of the applied field). In contrast, the observed MR does not saturate at H_1; in fact, $\sigma(H)$ becomes independent of field only for $H>H_2$.

A comparison of Figures 8(c) and 8(d) shows that the magnetoconductance $\Delta\sigma/\sigma_0$ is proportional to $M_+^\dagger$, for both the spin flop and the WF transitions. Note that at the highest fields, $\vec{M}_i^\dagger \| \vec{c}$ for the WF transition, whereas $\vec{M}_i^\dagger \| \vec{b}$ for the spin flop transition. The fact that the magnitude of the MR is the same in both situations (see Figures 5 and 8(d)) indicates that the holes are not sensitive to the direction of the staggered moment, but only to the relative interlayer *ordering* of the Cu spins.

We show next that the magnetic field has a large effect only on the hopping component of the highly anisotropic conductivity of La_2CuO_4.[6] At low temperature, σ_a/σ_b, the ratio of the dc conductivities in the in-plane and out-of-plane directions, is weakly dependent on temperature (Figure 6(a)). Above $T \approx 50K$ the anisotropy increases strongly. These results, together with measurements of the Hall effect,[6] indicate that at high temperature the conductivity is dominated by thermal activation of carriers into a band of highly anisotropic states. At low temperature the dominant transport mechanism is thermally assisted tunneling between localised states which are anisotropic.[7] In such an anisotropic medium, the in-layer hopping conductivity σ_{ah} is limited by out-of-plane hops and is therefore proportional to the out-of-plane hopping conductivity σ_{bh}. Their ratio is $\gamma = \sigma_{ah}/\sigma_{bh} = (\xi_a/\xi_b)^2$, where ξ_a and ξ_b are the localisation lengths in the two directions.[17] The total conductivities can be written:

$$\sigma_b = \sigma_{bA} + \sigma_{bh}$$

$$\sigma_a = \sigma_{aA} + \gamma \sigma_{bh} \quad (4)$$

where σ_{bA} and σ_{aA} are the activated components of σ_b and σ_a, respectively.

If only the hopping component contributes to the MR, one expects $\Delta\sigma_b/\Delta\sigma_a = 1/\gamma$ independent of temperature, whereas one expects the anisotropy in the *relative* MR, $(\Delta\sigma_b/\sigma_b)/(\Delta\sigma_b/\sigma_a) = \sigma_a/\gamma\sigma_b$, to be strongly T-dependent at high T. If, alternatively, the MR arises from a change in the mobility of the states at the band edge, writing $\sigma_a = ne\mu_a$ and $\sigma_b = ne\mu_b$, one expects both $\Delta\sigma_b/\Delta\sigma_a = \Delta\mu_b/\Delta\mu_a$ and $(\Delta\sigma_b/\sigma_b)/(\Delta\sigma_b/\sigma_a) = \mu_a\Delta\mu_b/\mu_b\Delta\mu_a$ to be weakly T-dependent.

We observe (Figure 6) that $\Delta\sigma_b/\Delta\sigma_a$ is roughly T-independent, whereas $(\Delta\sigma_b/\sigma_b)/(\Delta\sigma_b/\sigma_a)$ is not. Furthermore, the values of γ derived from the low-T value of σ_a/σ_b (~15) and from $\Delta\sigma_a/\Delta\sigma_b$ (~13) agree very well, and is consistent with the lower bound derived[7] from the anisotropy of the localisation lengths $(\xi_a/\xi_b)^2 > 7$. Thus Eq. (4) provides a good description of the conductivity tensor and the magnetic order appears to have influence primarily on the hopping conductivity. Measurements of the ac conductivity and the frequency-dependent dielectric constant at the WF transition confirm this conclusion.[18] Since the in-plane band conductivity σ_{aA} is almost purely 2D, it is not surprising that it is insensitive to the interlayer magnetic ordering. However, it is surprising that this also appears to be the case for σ_{bA}.

Since the low-T magnetoconductance, which is dominated by hopping, is proportional to $M_+^\dagger$ at fixed T, one would expect $\Delta\sigma_{bh}/\sigma_{bh}$, the overall change in the hopping conductivity, to scale with the zero-field interlayer order parameter, $M_-^\dagger$, when T is varied. The overall change of σ_b, plotted in Figure 5, is $\Delta\sigma_b/\sigma_b = \Delta\sigma_{bh}/(\sigma_{bA}+\sigma_{bh})$. This quantity is roughly equal to $\Delta\sigma_{bh}/\sigma_{bA}$ at high T. We find that at high temperature $(\Delta\sigma_b/\sigma_b)(M_-^\dagger)^{-1}$ is indeed proportional to σ_{bA}^{-1}. This explains why $\Delta\sigma_b/\sigma_b$ decreases much faster than $M_-^\dagger$ with increasing temperature.[8]

V. CONCLUSIONS

Using magnetoresistance, we have mapped out the phase diagram for the weak-ferromagnetic ($\vec{H}\|\vec{b}$) and spin flop ($\vec{H}\|\vec{c}$) transitions in insulating La_2CuO_{4+y}. The measurements provide values of the symmetric and antisymmetric anisotropies in the nearest neighbour exchange, and the interlayer exchange. We have confirmed previous results showing that the in-layer nearest neighbour exchange is described, to a high degree of accuracy, by the Heisenberg Hamiltonian. The most important result from this work is the conclusion that the MR arises from the interlayer hopping component of the conductivity, and that it is proportional to $M_+^\dagger$, the staggered moment with $\vec{\tau} \| \vec{c}$.

Despite this thorough characterisation of the magnetoresistance, its microscopic mechanism is still a mystery. The large MR is seen only in the hopping conductivity which varies as $\exp(-R/(\xi_a^2 \xi_b)^{1/3})$ where R is the hopping distance and ξ_a and ξ_b are the localisation lengths. The effect on the conductivity of a change of R or ξ_a, ξ_b with field will therefore be amplified by a factor $R/(\xi_a^2 \xi_b)^{1/3}$. However, estimates based on the theory of Shklovskii[17] for hopping in an anisotropic medium indicate that, even with this amplification factor,[19] $R/(\xi_a^2 \xi_b)^{1/3}$ must change by a factor $\geq 20\%$ to account for the observed change in conductivity. Furthermore, in virtually all models of hopping conductivity,[20] the MR is predicted to be sensitive to the *uniform* magnetic moment because of the overlap of the spin parts of the initial and final eigenstates. The observed proportionality between the MR and $M_+^\dagger$ is unique and shows that the coupling between the charge and spin degrees of freedom is quite unusual.

This work was supported at MIT by NSF Grants DMR84-15336 and DMR87-19217; at Brookhaven by the Division of Materials Science, U.S. Department of Energy, under Contract No. DE-AC02-CH00016. Experiments were performed at the Francis Bitter National Magnet Laboratory, which is supported at MIT by the National Science Foundation. We gratefully acknowledge helpful discussions with A.G. Swanson and B.L. Brandt.

REFERENCES.

*Also Department of Physics, MIT.
‡Also Brookhaven National Laboratory, Upton, NY 11973.
†Present Address: AT&T Bell Laboratories, Murray Hill, NJ.

1. J.G. Bednorz and K.A. Müller, Z. Phys. B **64**, 189 (1986).
2. G. Shirane, Y. Endoh, R.J. Birgeneau, M.A. Kastner, Y. Hidaka, M. Oda, M. Suzuki, T. Murakami, Phys. Rev. Lett. **59**, 1613 (1987); Y. Endoh, K. Yamada, R.J. Birgeneau, D.R. Gabbe, H.P. Jenssen, M.A. Kastner, J.M.

Tranquada, G. Shirane, Y. Hidaka, M. Oda, Y. Enomoto, M. Suzuki, T. Murakami, Phys. Rev. B 37, 7443 (1988).
3. S. Chakravarty, B.I. Halperin and D.R. Nelson, Phys. Rev. Lett. 60, 1057 (1988).
4. C.J. Peters, R.J. Birgeneau, M.A. Kastner, H. Yoshizawa, Y. Endoh, J. Tranquada, G. Shirane, Y. Hidaka, M. Oda, M. Suzuki, T. Murakami, Phys. Rev. B 37, 9761 (1988).
5. R.T. Collins, Z. Schlesinger, M.W. Shafer, T.R. McGuire, Phys. Rev. B 37, 5817 (1988).
6. N.W. Preyer, R.J. Birgeneau, C.Y. Chen, D.R. Gabbe, H.P. Jenssen, M.A. Kastner, P.J. Picone, Tineke Thio, Phys. Rev. B 39, 11563 (1989).
7. C.Y. Chen, N.W. Preyer, P.J. Picone, M.A. Kastner, H.P. Jenssen, D.R. Gabbe, A. Cassanho, R.J. Birgeneau, submitted to Phys Rev. Lett. (1989).
8. Tineke Thio, T.R. Thurston, N.W. Preyer, P.J. Picone, M.A. Kastner, H.P. Jenssen, D.R. Gabbe, C.Y. Chen, R.J. Birgeneau, Amnon Aharony, Phys. Rev. B 38, 905 (1988).
9. M.A. Kastner, R.J. Birgeneau, T.R. Thurston, P.J. Picone, H.P. Jenssen, D.R. Gabbe, M. Sato, K. Fukuda, S. Shamoto, Y. Endoh, K. Yamada, G. Shirane, Phys. Rev. B 38, 6636 (1988).
10. Tineke Thio, C.Y. Chen, B.S. Freer, D.R. Gabbe, H.P. Jenssen, M.A. Kastner, P.J. Picone, N.W. Preyer, and R.J. Birgeneau, submitted to Phys. Rev. B (1989).
11. P.J. Picone, H.P. Jenssen and D.R. Gabbe, J. Cryst. Growth 85, 576 (1987) and J. Cryst. Growth 91, 463 (1988).
12. H.C. Montgomery, J. Appl. Phys. 42, 2971 (1971).
13. R.R.P. Singh, P.A. Fleury, K.B. Lyons, P.E. Sulewski, Phys. Rev. Lett. 62, 2736 (1989).
14. I. Dzyaloshinskii, J. Phys. Chem. Sol. 4, 241 (1958).
15. T. Moriya, Phys. Rev. 120, 91 (1960).
16. A.S. Borovik-Romanov, A.I. Buzdin, S.S. Crotov, N.M. Kreines, Sov. Phys. JETP 47, 600 (in Russian) (1988).
17. B.I. Shklovskii, Soviet Phys. Semicond. 11, 1253 (1977); B.I. Shklovskii, Phys. Stat. Sol. B 85, K111 (1978).
18. C.Y. Chen et al., unpublished.
19. M.A. Kastner, R.J. Birgeneau, C.Y. Chen, Y.M. Chiang, D.R. Gabbe, H.P. Jenssen, T. Junk, C.J. Peters, P.J. Picone, Tineke Thio, T.R. Thurston, H.L. Tuller, Phys. Rev. B 37, 111 (1988).
20. N.F. Mott and E.A. Davis, *Electronic Processes in Non-Crystalline Semiconductors*, Oxford 1977.

VARIABLE RANGE HOPPING IN THE a-Si:H:Au SYSTEM.

A.R.Long and L.Hansmann
Department of Physics and Astronomy,
University of Glasgow, Glasgow G12 8QQ, Scotland, U.K.

ABSTRACT

Measurements of a.c. and d.c. hopping conduction in a range of a-Si:H:Au samples are described and interpreted using the extended pair approximation of Summerfield and Butcher.

1. INTRODUCTION

In this paper investigations of d.c. and a.c. transport in thin films of gold doped hydrogenated amorphous silicon (a-Si_{1-x}:H:Au_x) made between 1K and 100K are reported. At low gold concentrations, such films show variable range hopping characterised by a $T^{1/4}$ law[1]; as the gold concentration is increased, they become steadily more conducting until at a gold concentration x of 14%, a metal-insulator (MI) transition is observed[1,2]. The purpose of the present investigation is firstly to test a recent integrated a.c. and d.c. hopping model (the extended pair approximation (EPA) of Summerfield and Butcher[3]) on a structurally disordered hopping system other than a-Si, which has already been exhaustively investigated[4], and secondly to look at variations of the hopping parameters as x is increased to obtain information about the underlying nature of the changes in the electronic structure of the alloys as the MI transition is approached.

2. SAMPLE PREPARATION AND MEASUREMENT

Samples were prepared by r.f. magnetron sputtering in an atmosphere of Ar containing 33% H_2 from a Si target on which were laid small pieces of Au. The spacing of the gold pieces was used to control the Au concentration in the resultant film. The substates of Corning 7059

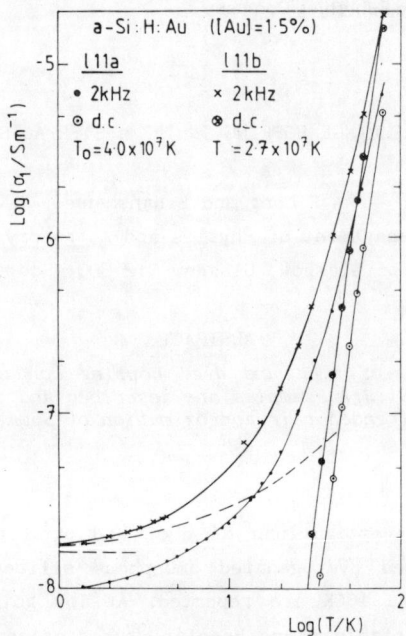

Fig. 1. A.c. and d.c. conductivities of two a-Si:H:Au samples versus T

glass were held at 200°C during sample deposition. Samples were measured in sandwich configuration with Al electrodes; nominal film thicknesses of 1μm and 2μm were prepared during each sputtering run. Films were sputtered under the same conditions onto rock salt substrates and floated off for compositional analysis by the EDX technique in a transmission electron microscope. The X-ray data was analysed using the standardless method of Paterson et.al.[5]. Samples were studied between 1.3K and 20K in an orthodox helium cryostat and between 12K and 100K in a CTI Model 21 cryocooler. A.c. data were recorded using a Hewlett Packard 4274A LCR meter with a typical excitation voltage of 10mV across the sample.

3. RESULTS.
3.1 A Qualitative Discussion

Details of the samples studied are given in the table. All the

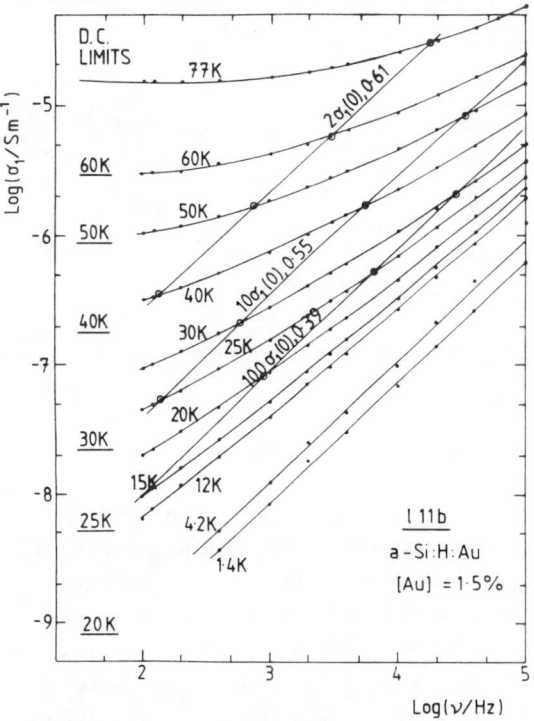

Fig. 2. A.c. conductivity of 111b showing scaling lines with n values.

samples reported followed the Mott $T^{1/4}$ law in their d.c. conductivities over many orders of magnitude. In fig. 1, we plot d.c. and a.c. conductivities of two samples as a function of temperature. These samples were drawn from the same film, but, as the figure and the table show, both the magnitude and the temperature dependence of the conductivity (reflected in the Mott temperature parameter T_0) are significantly different. We believe that this reflects inhomogeneities in doping resulting from the deposition process. In fig. 2 the frequency dependent conductivity is plotted for one of the same samples at different temperatures. As developed by Summerfield[6], the EPA predicts that the reduced conductivities $\sigma_1(\omega)/\sigma_1(0)$ should all scale to a common reduced frequency $\tilde{\omega}$ proportional to $\omega/\sigma_1(0)T^n$. In the original theory, n was equal to 1 to a good approximation, but in

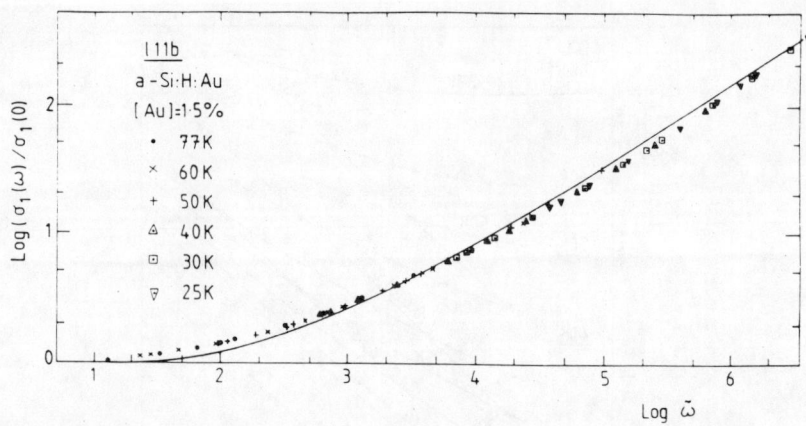

Fig. 3. Scaled data for 111b. Values on abscissa refer to $\tilde{\omega}$ from EPA.

practice rather lower n values have been observed. In the present work, n has been determined by joining points of equal reduced conductivity, as indicated in fig. 2, and is found to be 0.57±0.05 (for $\sigma_1(\omega)/\sigma_1(0) < 10$). Using this scaling relation, the conductivity curves of fig. 2 are reduced to a common frequency scale in fig. 3. The accuracy of the scaling is found to be better than 5% over the full range of the data. Also plotted in this diagram is the curve predicted by the EPA. The fit is generally good, although it deteriorates in the region of the onset of dispersive behaviour. At low temperatures, as fig. 1 shows, the a.c. conductivity tends to a constant value, indicative that intersite correlations are dominant. Such behaviour was not found in the a-Si studies[4] where the low temperature loss was weakly temperature dependent (as $T^{0.4}$). As the temperature rises to become of the order of the intersite correlation energy, then the conductivity will become temperature dependent. The hatched line in the diagram is an attempt to fit the data using the pair model of Efros[7] with an intersite correlation energy of 25K.

3.2 A Detailed Quantitative Analysis

The data for all samples was analysed following the method of ref. 4 to determine the three parameters which govern the hopping current in the simplest EPA model, the Bohr radius a ($\equiv \alpha^{-1}$ in the notation of ref. 4), the density of states at the fermi level $N(E_F)$ (assumed constant) and the hopping rate parameter R_0. One difference from previous practice was that in order to determine a from the conductivity data, the curve for $\sigma_1(\omega)/\sigma_1(0)$ was fitted at a value of 10; this minimised the potential effects of distortions due to low conductance surface layers. Because the scaling is not exactly as the theory predicts, the value of a calculated in this way will vary slowly with temperature. To attempt to minimise any errors introduced by this effect we have evaluated a for each sample as close as possible to the same value of T/T_0; any residual network effects not taken into account by the EPA are likely to depend solely on this reduced temperature. The values for the hopping parameters calculated for all samples are given in the Table. The errors quoted are random errors associated with measurement and with the fitting procedure, and do not take into account systematic effects associated with possible inadequacies of the model. Some plots of the variation of the hopping parameters with the Au concentration x are shown in fig. 4.

Also included in the table are estimates of a derived from the hopping contribution to the low frequency dielectric constant at high temperatures, $\Delta\varepsilon_1(0)$. In these samples ε_1 saturates to a low frequency value which decreases approximately as T^{-1} as the theory demands, unlike the behaviour in the a-Si system[4]. However as the $\Delta\varepsilon_1(0)$ values are likely to be significantly influenced by any residual surface barriers, we use the a values calculated from the fit to the σ_1 data to estimate the density of states.

TABLE

Sample	x	T_0	Bohr radius est. At T	σ_1	a	$N(E_F)$	R_0	Dielectric const. At T	$\Delta\varepsilon_1(0)$	a
19a	6.6	13	20	9.6	2.0	2.6	162	30	66	1.06
19c	12	6.0	15	6.3	2.7	2.3	5.5	30	70	1.0
			12.5	2.2	3.5	1.0	8.1			
110a	3	9.1	20	1.85	2.0	4.0	150	40	43	1.23
110b	2.7	21	40	2.7	.57	71	2.8	40	99	.53
111a	1.5	40	60	.86	1.13	4.7	10	77	51	.54
111b	1.5	27	50	.86	.68	33	1.3	77	90	.30
Unit	%	10^6K	K	10^{-6} Sm^{-1}	nm	10^{18} eV^{-1}cm^{-3}	10^{11} s^{-1}	K		nm
Error	±10%	±2%		±20%	±35%	+250 −60%	±40%		±20%	±35%

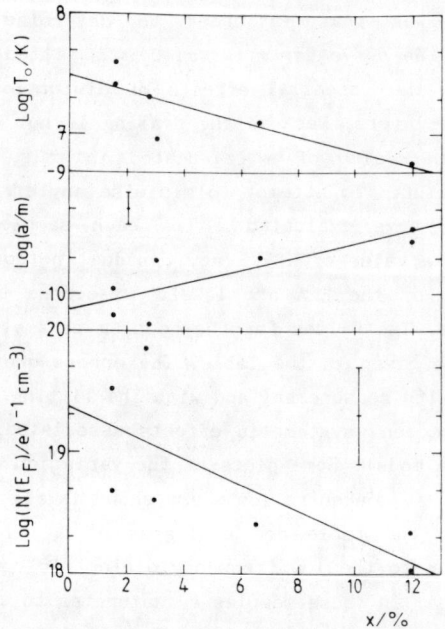

Fig. 4. Hopping parameters versus x. Solid lines are linear fits.

Trends are immediately apparent from this data. As expected and as observed in previous work[1] the Mott temperature parameter declines as the Au concentration increases. Also as expected from the scaling theory[8] and as seen in observations on impurity band conduction[9], a tends to increase as the metal insulator transition is approached. When T_0 and a are put together however to deduce the density of states then, despite the large errors, it can be seen that $N(E_F)$ *decreases* with x rather than increasing as expected. Such an effect is not unknown in this type of measurement. In weakly hydrogenated a-Si[4] the density of states increases with hydrogen content, at variance with physical expectation. This will be discussed again in section 4.

The rate parameter R_0 varies by much more than the error in its evaluation would suggest, implying systematic factors in determining the conductivity which are not taken into account in the analysis. However even allowing for this, there is one striking factor in the data, when it is compared with observations made on a-Si or a-Ge. In this work the range of R_0/T values calculated, from 10^{11} to $10^{13} s^{-1}$, is some four orders less than the equivalent range observed for a-Si[4] and much closer to the value of a phonon frequency as is often assumed. We conclude from this that the structure of the localised states and the mechanism for transfer between them is very different in the two cases.

4. DISCUSSION

This work taken together with the studies of ref. 4 shows that there is a significant difference between the native defects present in a-Si

and those introduced by doping with Au into the hydrogenated material, where the native defects are passivated. Hence the many studies of hopping in a-Si:Au (see refs. 1 and 2 and also ref. 10) are likely to be complicated by the presence of both types of defect. The main question associated with the a-Si:H:Au system, posed by McNeil and Davis[1], is whether the hopping conduction occurs in a band of defects in mid-gap associated with the Au atoms or whether the Au pulls the fermi level down into the valence band tail. On the basis of photoemission data, McNeil and Davis preferred the latter explanation. In the absence of a detailed theory for the rate parameter R_0, these data do not allow us to decide directly between these points of view. The main problem with our analysis is of course associated with the decrease of $N(E_F)$ with x. One possibility for explaining this effect is if the fermi level moves into a band tail of gradually increasing steepness as x increases, as might be expected in the model of McNeil and Davis. A detailed discussion of this model will be given in a future paper.

The possible influence of intersite correlations on the d.c conduction in systems such as this has been considered by Voegele et. al.[10]. There is however no unified theory of the transition from the a.c. (pair) limit, where the sites are close and correlation effects are strong, to the d.c. limit where partial screening of the d.c. current is likely. The potential importance of this may be illustrated by calculating the intersite correlation energy for a pair transition at 2kHz and 10K in sample 111b using the parameters from the Table. The result is equivalent to a temperature of 170K, suggesting that correlations need to be considered at all temperatures of measurement. (Note that the pair fit plotted in fig. 1 used a smaller energy parameter of 25K). None of the samples considered here show any sign of the $T^{-1/2}$ conduction symptomatic of the coulomb gap.

The major problem with these measurements was, we believe, control of sample homogeneity. With better preparation and compositional analysis procedures, the quality of the data could be distinctly improved, and the important trends observed could be clarified.

REFERENCES

1. J.McNeil and E.A.Davis, J. Non-cryst. Solids 59 & 60, 145 (1983).
2. N.Nishida, T.Furubayashi, M.Yamaguchi, M.Shinohara, Y.Miura, Y.Takano, K.Morigaki, H.Ishimoto and K.Ono, J. Non-cryst. Solids 59 & 60 149 (1983).
3. S.Summerfield and P.N.Butcher, J. Phys. C 15, 7003 (1982); ibid. 16, 295 (1983).
4. A.R.Long, J.McMillan, N.Balkan and S.Summerfield, Phil. Mag. B 58, 153 (1988).
5. J.H.Paterson, J.N.Chapman, W.A.P.Nicholson and J.M.Titchmarsh, J. Microscopy 154, 1 (1989).
6. A.L.Efros, Phil. Mag. B 43, 829 (1981).
7. S.Summerfield, Phil. Mag. B 52, 9 (1985).
8. E.Abrahams, P.W.Anderson, D.C.Licciardello and T.V.Ramakrishnan, Phys. Rev. Lett. 42, 673 (1979).
9. A.N.Ionov, I.S.Shlimak and M.N.Matveev, Solid State. Commun. 47, 763 (1983).
10. V.Voegele, S.Kalbitzer and K.Böhringer, Phil. Mag. B 52, 153 (1985).

OBSERVATION OF A CROSSOVER FROM MOTT TO EFROS-SHKLOVSKII VARIABLE RANGE HOPPING IN n-CdSe

Youzhu Zhang, Peihua Dai, Miguel Levy and M. P. Sarachik
City College of the CUNY, New York, New York 10031

ABSTRACT

We report measurements of the resistivity of five samples of insulating compensated n-CdSe which exhibit a crossover as the temperature is decreased from 4.2 to 0.06K from Mott variable range hopping to Efros-Shklovskii hopping in the presence of a gap in the density of states due to Coulomb interactions. The temperature at which the crossover is observed decreases with increasing dopant concentration, indicating that the Coulomb gap narrows as the metal-insulator transition is approached. A determination for each sample of the parameters T_0 and T_0' in the two temperature regimes yields the empirical relation $T_0' \propto T_0^{2/3}$. Possible implications are discussed regarding the behavior of the dielectric constant ε and the localization length ξ as the metal-insulator transition is approached.

INTRODUCTION

The electrical conductivity of materials on the insulating side of the metal-insulator transition is known to be governed by variable range hopping at low temperatures. The form $\ln \sigma \propto -(T_0/T)^{1/4}$ first predicted by Mott[1-2] assumes that the density of states near the Fermi energy is constant or a slowly varying function of energy. More recently, Efros and Shklovskii[3] and Shklovskii and Efros[4] have noted that, as a consequence of the long-range Coulomb interactions between localized electron states, the density of states tends to zero near the Fermi level, giving rise to a "Coulomb gap" and a conductivity which depends on the temperature as $\ln \sigma \propto -(T_0'/T)^{1/2}$. Since activated conduction typically dominates above about 10K or even lower, the range of temperature over which either of these hopping mechanisms is observable is limited to at most two decades, and the observation of both in a single sample is unlikely. To our knowledge, there have been only two reports of observations of both types of hopping in the same material. A crossover from Mott to Efros-Shklovskii(ES) hopping with decreasing temperature has been reported for the granulated materials Sn-Ge and Ag-Ge by Glukhov et.al.[5] Biskupski et.al.[6] have found Mott variable range hopping for InP in magnetic fields below 6T and ES hopping in the same material at higher fields.

In this article we report the observation in five insulating compensated n-CdSe samples of crossover behavior from Mott[1] variable range hopping to ES[3-4] hopping as the temperature is decreased from 4.2 to 0.06K and the hopping energy becomes comparable to and then smaller than the gap. The temperature at which the crossover occurs decreases with increasing donor concentration indicating, as expected, that the Coulomb gap narrows as the metal-insulator transition is approached from the insulating side. The observation of both Mott and ES hopping for the same sample in different temperature ranges allows a simultaneous determination of T_0 and T_0'. By using experimental results for these five samples, we have established an empirical relation between T_0 and T_0' which, if current theory is applicable to this system, has unexpected implications concerning the critical behavior of the localization length ξ and the dielectric constant ϵ as the metal-insulator transition is approached.

PROCEDURES, RESULTS AND DISCUSSION

The indium-doped CdSe crystals used in these studies were provided by T. Dietl and co-workers of the Polish Academy of Sciences except for sample 2 which was grown by Cleveland Crystals and obtained from S. Geschwind. The Table lists the net carrier concentrations deduced from room temperature Hall coefficient measurements; the critical concentration estimated from the Mott criterion is roughly $3\times10^{17}\text{cm}^{-3}$. Earlier studies[7] of the Hall mobility at temperatures above 4.2K yielded estimates for the compensation of these samples of typically 40 to 50%. The resistance was measured in a dilution refrigerator using standard (four-terminal) DC techniques. Additional data for sample 5 were obtained between 4.2 and 1.15K in P. Lindenfeld's laboratory at Rutgers University.

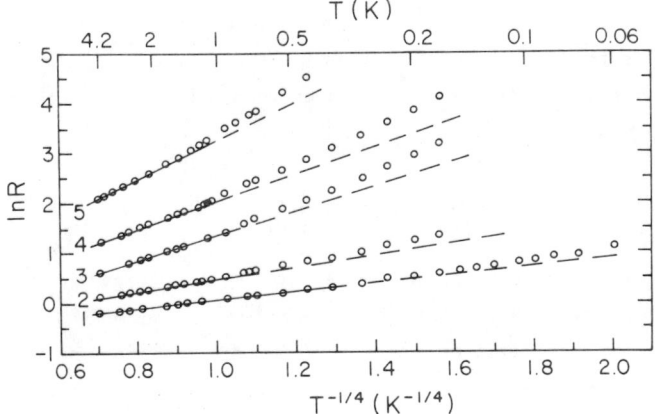

Figure 1: Logarithm of the resistivity of n-CdSe as a function of $T^{-1/4}$.

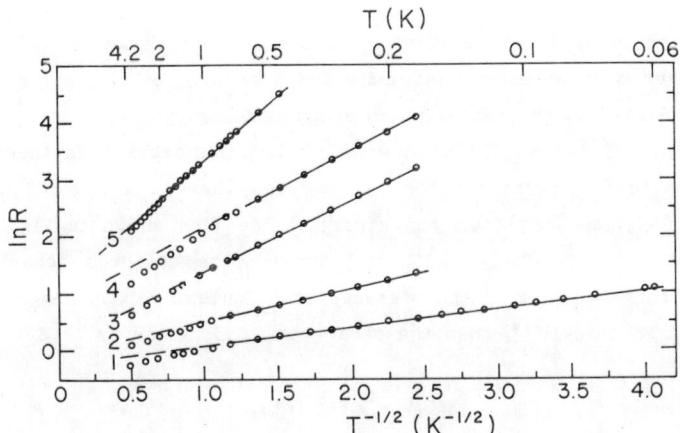

Figure 2: Logarithm of the resistivity of n-CdSe as a function of $T^{-1/2}$.

Figure 1 shows lnR plotted as a function of $T^{-1/4}$. All samples show straight line behavior[8] on this plot over some range at the higher temperatures, and exhibit deviations toward a steeper functional dependence as the temperature is lowered. Mott variable range hopping is thus obeyed in the higher temperature regime, so that $\rho = \rho_0 \exp(T_0/T)^{1/4}$, with

$$k_B T_0 \approx 18/\{N(E_F)\xi^3\}. \qquad (1)$$

Here $N(E_F)$ is the density of states at the Fermi energy and ξ is the localization length. Note that as the samples become more insulating, deviations from the behavior characteristic of Mott hopping occur at increasingly higher temperatures. Values of the parameter T_0 deduced from the slope of the straight line segments of Fig. 1 are listed in the Table. By combining eq. (1) with the expression[2] for the hopping length, $R=\{(8\pi/9)k_B T(N(E_F)/\xi)\}^{-1/4}$, one can obtain the ratio:

$$R/\xi \approx 0.4 \ (T_0/T)^{1/4}. \qquad (2)$$

As we have pointed out earlier[9], in a temperature range where hopping is observed so that experiment indicates that $R>\xi$, eq. (2) gives values for R/ξ (listed in the Table) which are unreasonably low.

A plot of lnR vs $T^{-1/2}$, shown in Fig. 2, indicates that there is a range[8] at low temperatures for all samples where $\rho = \rho_0' \exp(T_0'/T)^{1/2}$. Values of T_0' deduced from the straight line fits shown in Fig. 2 are also listed in the Table. If one ascribes the observed behavior to variable range hopping in the presence of a Coulomb gap as suggested by Efros and Shklovskii[3-4], then the parameter T_0' is given by

$$k_B T_0' \approx 2.8 e^2/(\varepsilon \xi), \qquad (3)$$

where e is the charge of an electron and ε is the dielectric constant of the material. As one can see from the Table, estimates obtained from this equation for $(\varepsilon \xi)$ are quite high. It should be noted, however, that comparable values were obtained for compensated n-Ge:As

TABLE: Sample designations, dopant concentrations and parameters T_0, R/ξ, T_0', $\varepsilon\xi$ and T^* deduced from the data as described in the text.

SAMPLE	$n(10^{17} cm^{-3})$	$T_0(K)$	R/ξ	$T_0'(K)$	$\varepsilon\xi(10^{-6}m)$	T^*
1	≈2.8	0.7	$0.34T^{-1/4}$	0.11	434	0.4
2	2.4	3.0	$0.5T^{-1/4}$	0.33	142	0.9
3	2.2	35	$0.94T^{-1/4}$	1.6	30	1.0
4	2.2	60	$1.1T^{-1/4}$	1.8	26	1.0
5	2.18	220	$1.5T^{-1/4}$	5.2	9	2.3

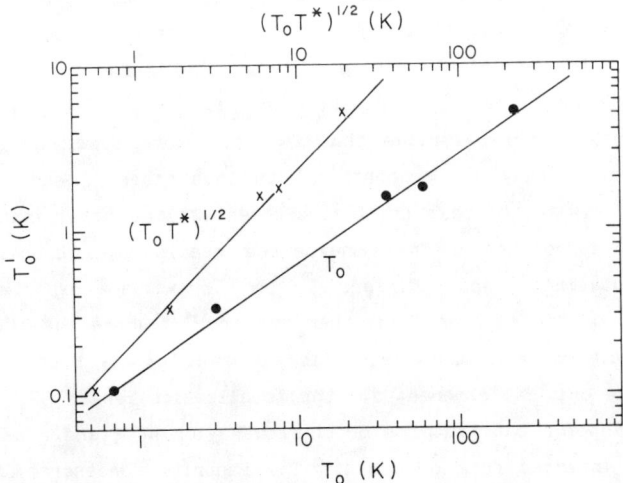

Figure 3: Double logarithmic plot of T_0' versus $(T_0 T^*)^{1/2}$, and of T_0' versus T_0, where the parameters T_0, T_0' and T^* are defined in the text.

by Zabrodskii and Zinov'eva[10] and Ionov et.al.[11]

The major feature of these data is the observation of an apparent crossover from Mott to ES variable range hopping as the temperature is decreased below a crossover temperature T^* which varies with dopant concentration. Approximate experimental values of T^* are given in the table, and from ref. 4 one can estimate that $k_B T^* = e^4 N(E_F) \xi/\epsilon^2$. Combining the latter with eqs. (1) and (3) one obtains that $T_0' \propto (T_0 T^*)^{1/2}$. A plot of T_0' vs. $(T_0 T^*)^{1/2}$ shown on a double logarithmic scale in Fig. 3 demonstrates that the data exhibits the expected linear behavior. Thus, although the above analysis yields values for R/ξ, ξ and ϵ which may seem unreasonably low or high, one should note that the crossover occurs at a temperature T^* which tracks $T_0'^2/T_0$, consistent with theory. Our data suggest that the numerical factors contained in eqs. (1) and (3) require modification and/or that other processes such as multi-electron hopping need to be considered as the transition is approached.

Also shown in Fig. 3 is the parameter T_0 plotted as a function of T_0' on a double logarithmic scale. The data for the five samples closely fit a straight line with slope 2/3 over three decades in T_0, indicating that the empirical relation $T_0' \propto T_0^{2/3}$ holds for these n-CdSe samples. Since $T_0' \propto \xi^{-1} \epsilon^{-1}$ (see Eq. (3)) and $(T_0)^{2/3} \propto \xi^{-2}$ (see Eq. (1)), this experimental finding implies that $\xi \propto \epsilon$. More specifically, it implies that since they are proportional to each other, ξ and ϵ must perforce diverge with the same critical exponent as the metal-insulator transition is approached. This unexpected result goes counter to simple expectations and differs from previous experimental determinations of the exponents in other systems[11-14] where the critical exponent associated with the dielectric constant ϵ was found to be about twice the critical exponent for the localization length ξ.

We should point out that the critical behavior of ξ and ϵ can not be separately inferred from our data. The impurity concentration has not been accurately determined for these samples, and the critical concentration n_c is not well established for n-CdSe. More important is the fact that our samples have different degrees of compensation, and one would therefore expect that they also have different[15] critical concentrations n_c.

CONCLUSIONS

The resistivity was measured between 4.2 and 0.06K for five compensated indium-doped n-CdSe samples with dopant concentrations on the insulating side of the metal-insulator transition. The resistivity of all samples exhibit $\exp(T_0/T)^{1/4}$ behavior characteristic of Mott variable range hopping at the higher end of the temperature range, and $\exp(T_0'/T)^{1/2}$ at lower temperatures. We attribute the latter to Efros-Shklovskii variable range hopping in the presence of a gap in the density of states due to Coulomb interactions. The temperature T* at which the crossover is found to occur increases with decreasing dopant concentration indicating that the Coulomb gap is larger for the more insulating samples, as expected. Although the parameters T_0 and T_0' deduced from the data yield unlikely values of R/ξ and $(\epsilon\xi)$, the theory correctly accounts for the observed variation of the crossover temperature T* in terms of T_0 and T_0'. Using data for the five samples, we find the empirical relation $T_0 \propto T_0'^{2/3}$. If the theory is correct and applicable to this system, this implies that in n-CdSe the localization length ξ and the dielectric constant ϵ diverge the same way as the metal-insulator transition is approached, i. e. that their critical exponents are equal. Work is continuing on these materials to extend measurements to higher and lower temperatures, and to study the possible role of inhomogeneities.[16]

ACKNOWLEDGMENTS

We thank T. Dietl and the Polish Academy of Sciences for contributing four of the samples used in these studies, and to S. Geschwind for sample 2. We are grateful to P. Lindenfeld for the use of his equipment to obtain a thermometer calibration and additional data on sample 5 between 4.2 and 1.15K. We are indebted to R. A. Webb for technical advice and for many useful suggestions. This work was supported by the US Department of Energy under Grant no. DE-FG02-84ER45153. Partial support was also provided by PSC-CUNY.

REFERENCES

1. Mott,N.F., J. Non-Cryst. Solids **1**, 1 (1968).
2. Mott,N.F., <u>Metal-Insulator Transitions</u> (Taylor and Francis, London, 1974).
3. Efros,A.L. and Shklovskii,B.I., J. Phys. C:Solid State Phys. **8**, 249 (1975).
4. Shklovskii,B.I. and Efros,A.L., <u>Electronic Properties of Doped Semiconductors</u>, ed. M.Cardona (Springer, Berlin, 1984).
5. Glukhov,A.M., Fogel,N.Ya. and Shablo,A.A., Sov. Phys. Solid State **28**, 583 (1986).
6. Biskupski,G., Dubois,H. and Briggs,A., J. Phys. C:Solid State Phys. **21**, 333 (1988).
7. Levy,M., Roy,A., Sarachik,M.P. and Isaacs,L.L., Phys. Rev. **B38**, 3323 (1988).
8. The exponents 1/4 and 1/2 were determined in the high and low temperature ranges using a fitting procedure proposed by Finlayson,D.M. and Mason,P.J., J. Phys. C:Solid State Phys. **19**, L299 (1986). Details will be published elsewhere. Also, the exponent 1/4 was shown to hold for these samples between 4.2 and 1.2 in ref. 9.
9. Roy,A., Levy,M.,Guo,X.M., Sarachik,M.P., Ledesma,R. and Isaacs,L.L., Phys. Rev. **B39**, 10185 (1989).
10. Zabrodskii,A.G. and Zinov'eva,K.N., Sov. Phys. JETP **59**, 425 (1984).
11. Ionov,A.N., Shlimak,I.S. and Matseev,M.N., Solid State Commun. **47**, 763 (1983).
12. Hess,H.F., DeConde,K., Rosenbaum,T.F. and Thomas,G.A., Phys. Rev. **B25**, 5578 (1982).
13. Kasiyan,V.A., Nedeoglo,D.D., Simashkevich,A.V. and Timchenko,I.N., Phys. Stat. Sol. B **559**, 559 (1987).
14. Brooks,J.S., Symko,O.G. and Castner,T.G., Jap. Jour. of Appl. Phys. **26**, 721 (1987).
15. Hirsch,M.J. and Holcomb,D.F., in <u>Disordered Semiconductors</u>, ed. by M.A.Kastner, G.A.Thomas and S.R.Ovshinsky (Plenum, New York, 1987).
16. Further experiments are in progress to study the effect of inhomogeneities, which could play an important role at the lower temperatures where the resistivity is a very rapidly varying function of concentration. It is unlikely, however, that this will affect our major findings since similar results are observed in all cases, while the inhomogeneities and the placement of electrical contacts are random.

CONDUCTION IN GRANULAR METALS
FAILURE OF THE COULOMB GAP MODEL

C.J.Adkins

Cavendish Laboratory, Madingley Road, Cambridge CB3 0HE, UK

ABSTRACT

The conductivity of granular metals in the activated regime is found to vary as $\exp(-T_0/T)^x$ where T_0 and x are constants. For cermets, $x \approx 0.5$ while for discontinuous films the parameter is more variable although 0.5 is still typical. Analysis of the classic model for these systems has shown that it cannot explain the above behaviour even when the distributions of structural parameters and the presence of potential disorder are taken into account. In this paper we consider the possibility of conduction being by variable-range hopping in a Coulomb gap. This model would predict the correct temperature dependence, but detailed quantitative analysis of data for three different granular metal systems produces unacceptable values for system parameters. The model therefore fails and we conclude that the problem of the physics of conduction in granular metals remains unsolved. We discuss briefly aspects of conduction in granular metals that remain to be explored in the search for a solution.

1. INTRODUCTION

The effect of electron - electron correlation on transport properties depends on whether the charge carriers are localized or delocalized. On the metallic side of the metal - insulator transition, electron - electron interaction gives rise to weak logarithmic corrections to the temperature dependence of the conductivity.[1] In systems where the charge carriers are localized, it produces a depletion of the density of states in the region of the Fermi level known as the *Coulomb gap*.[2] The physics of this regime was first treated by Efros and Shklovskii[3,4] who showed that the density of single-particle states falls to zero at the chemical potential and rises about the Fermi level parabolically in three dimensions and linearly in two. The width of the depleted region, the Coulomb gap, is set by the density of states in the absence of correlation to which the rising density of states must join smoothly at higher energies. Coulomb gaps in systems of localized carriers have been demonstrated theoretically in computer simulations.[5]

The effect of the Coulomb gap on transport depends on the mechanism of conduction.[2] The simplest processes to consider are those involving the *hopping of individual carriers* to which the densities of states derived by Efros and Shklovskii apply directly. However, correlation necessarily means that hop energies are determined by the current configuration of all particles present, so a mean-field theory cannot contain the whole of the physics and one must consider the possibility of *sequential hopping* processes: single particle transitions, each of which depends on others that have just taken place. The discussion of granular films in this paper assumes the validity of a mean-field approach. We return later to discuss sequential processes. A third conduction mechanism which will not concern us here but which must operate at sufficiently low temperatures involves *many-particle transitions*: because of the correlations, multi-particle transitions can make available processes of lower activation energy than are accessible through single-particle hopping. At higher temperatures, such processes become relatively less important because their probabilities involve products of tunnelling exponents, but at low temperatures the price paid by involving several particles may be less restricting than the requirement for greater activation. The optimization between number of particles participating in a single event and temperature is similar to that in variable-range tunnelling where the optimization is between *distance* and thermal energy (temperature).

Most experimental work on Coulomb interactions has used impurity conduction or inversion layers because these systems are well controlled (see references in [2]), but there is also evidence for the Coulomb gap in photoemission studies of granular palladium films[6] and sodium tungsten bronzes.[7] Evidence for the importance of Coulomb interactions has also been found in studies of hopping transport in indium oxide films.[8] As regards transport in granular metals, the situation is less clear because the observed temperature-dependence of the conductivity, which is *qualitatively* consistent with the presence of a Coulomb gap, has customarily been attributed to structural features of these systems. The models used, however, do not survive critical analysis (see section 2). We therefore discuss in this paper whether the Coulomb gap model of Efros and Shklovskii can provide a satisfactory *quantitative* account of conductivity in granular metals. We shall see that the model does not appear to survive the test of quantitative analysis. The essential elements of the discussion have appeared in an earlier paper.[9]

2. GRANULAR METALS

Granular metals are finely divided two-phase mixtures of metal and non-metal. Three-dimensional composites, generally produced by co-evaporation or co-sputtering of a metal and an insulator, are known as *cermets*.[10] Two-dimensional *discontinuous metal films* are formed during the early stages of film growth by evaporation or sputtering, the deposited metal first forming isolated islands that only later join up to form a continuous film.[11]

The electrical properties of such systems vary continuously as the composition is changed. When the proportion of the metallic phase is large, the metal forms a connected matrix in which the insulator is dispersed in the form of isolated inclusions. In this *metallic regime*, the conductivity is metallic with a positive temperature coefficient of resistance, but, as the proportion of insulator is increased, the reduced volume fraction of metal and the increased scattering from the insulating inclusions result in a progressive reduction of both conductivity and temperature coefficient. When the proportion of insulator is large, the inverse structure is obtained, with isolated metallic inclusions embedded in a matrix of insulator. This is the *dielectric regime* in which the conductivity is small and activated. The extent of the regimes is clearly a percolation problem and, at zero temperature, properties would change sharply at the critical percolation limit.

We are interested here in the dielectric regime where charge is strongly localized in the metallic islands. In this regime, conductivity has generally been discussed in terms of the classic model of Neugebauer and Webb[12] which has two essential features: transfer of electrons between metal islands is by tunnelling, and activation is required to provide the non-negligible electrostatic energy that is associated with placing an electronic charge on an island, creating a 'carrier'. Then, at low temperatures, the conductivity σ should behave as

$$\sigma \propto \exp[-2\alpha s - W/kT] \qquad (1)$$

where α is the tunnelling exponent of electron wavefunctions in the insulator, s the separation of islands, W the island charging energy, k the Boltzmann constant and T temperature. Observed activation energies are of the right order to be consistent with estimated capacitances of islands, but the predicted temperature dependence, simple activation, is not observed. Instead, it is found that

$$\sigma \propto \exp{-(T_0/T)^x} \qquad (2)$$

where T_0 and x are constants. For cermets, $x \approx 0.5$. For discontinuous metal films, various values of x are obtained in the range $0.3 \leq x \leq 1$, although $x \approx 0.5$ is probably still typical. This 'fractional temperature-dependence' has generally been attributed to the distributed nature of s and W in real physical systems. However, all attempts to explain fractional temperature dependence in this way are open to criticism because they involve correlations of these quantities[13,14] that are not normally present, use unrealistic distributions of them,[15] or assume statistical independence of them in individual hopping processes.[16] Another model based on conduction *via* localized states[17] also has to be rejected because it is inconsistent with field-effect experiments[18] and with the observed result of using superconducting metal, which verifies direct tunnelling between grains.[19] Detailed analysis of transport in granular systems by critical path methods (which is similar to effective medium theory in these systems) and incorporating realistic distributions of s and W yields a temperature-dependence which is still very close to simple activation.[20] The same result is obtained if the presence of large random potentials is included in the analysis.[21] (There is experimental evidence for the presence of such random potentials.[18,22])

The best-known mechanism for fractional temperature-dependence of conductivity is variable range hopping.[23] If an electron is tunnelling between localized states which are randomly distributed in energy and space, then, at low temperatures, it will pay for the electron to tunnel further in order to find an empty state more proximate in energy. Optimization of the two terms in equation (1) for a uniform density of states in three dimensions leads to the famous Mott $T^{1/4}$ law (i.e. $x = 0.25$). The argument can be modified to any dimensionality and to non-uniform densities of states. One obtains

$$x = (p+1)/(d+p+1) \tag{3}$$

where d is the dimensionality and p is the index for which the density of states $g(E)$ is assumed to rise with energy E about the Fermi level as

$$g(E) = g_p |E|^p. \tag{4}$$

We note that we obtain $x = 0.5$ for $p = 2$ in three dimensions and for $p = 1$ in two. These are precisely the forms of $g(E)$ predicted to result from electron-electron interaction in disordered systems of localized electrons by Efros and Shklovskii.[3]

3. The Coulomb Gap

The Efros and Shklovskii argument may be summarised as follows: Consider the ground state of a disordered system of localized electrons. States below the Fermi energy will be occupied, those above empty. Let E_i be the site energies *including Coulomb terms* with all other charges. Consider a filled state of energy E_i and an empty state of energy E_j. These are defined as the energies associated with removing the electron from state i or with placing one on state j, *all other charges remaining unchanged*. Now consider the transition of the electron from i to j. The net energy change will be

$$W = E_i + E_j - e^2/4\pi \varepsilon_r \varepsilon_0 r_{ij} \tag{5}$$

where $\varepsilon_r \varepsilon_0$ is the permitivity of the medium and r_{ij} the separation of the sites. The last term is the Coulomb potential between the negative charge added at j and the positive charge added at i by removing the electron. However, for stability of the original (ground) state we require $W \geq 0$. This limits the density of states at low energies. Efros and Shklovskii argue that the charges will adjust their configuration to form a ground state (a *charge glass*) such that $W = 0$ at all r_{ij}. This condition results in three- and two-dimensional densities of states

3-d $\quad g(E) = (3^8 \pi^2 \varepsilon_r^3 \varepsilon_0^3/2^5 e^6)E^2 \equiv g_3 E^2 \tag{6a}$

2-d $\quad g(E) = (2^{11}\pi \varepsilon_r^2 \varepsilon_0^2/3^4 e^4)|E| \equiv g_2 |E|. \tag{6b}$

The numerical constants are obtained by requiring one empty and one filled state within distance r and with mean energy difference equal to the Coulomb term at the mean separation. The values of g_3 and g_2 are given by

3-d
$$g_3 = 8.36 \times 10^{82} \varepsilon_r^3 \ \text{J}^{-3}\text{m}^{-3} \tag{7a}$$

2-d
$$g_2 = 9.49 \times 10^{54} \varepsilon_r^2 \ \text{J}^{-2}\text{m}^{-2}. \tag{7b}$$

It should be noted that equations (6) and (7) contain no material parameters other than the bulk relative permittivity. This remarkable independence from microscopic detail is a result of the densities of states being determined solely by Coulomb's law.

In neither case can the density of states continue to increase, of course, and it is assumed to level off when $g(E)$ rises to g_0, the density of states neglecting Coulomb interactions. This defines the Coulomb gap Δ:

$$g(\Delta/2) = g_0. \tag{8}$$

The possible explanation of $x = 0.5$ in terms of Coulomb interaction therefore involves variable-range hopping in densities of states given by equations 6. However, variable-range hopping has normally been rejected as a possibility for granular metals on the basis of the following argument. Typically, in granular metals, the diameter d of metal islands might be 3 nm and their separation s less than 1 nm. In tunnelling beyond a near neighbour, the tunnelling distance must therefore increase by some 4 nm. In a typical insulator, the tunnelling exponent a would be of order 10^{10} m^{-1} so, it is argued, the relative probability of tunnelling beyond a near neighbour would be negligible. However, this argument is somewhat too simple because it neglects the effect of intervening metal islands on a tunnelling electron. In the region of an intervening island, the energy deficit of a tunnelling electron will become small so that it is only in the insulator that the wavefunction is significantly attenuated. For tunnelling distances greater than d, a simple geometrical argument[24] would then give

$$\alpha_m \approx s \ \alpha_{ins}/(s + d) \tag{9}$$

where α_m is the mean tunnelling exponent and α_{ins} that in the insulator. The reduction of α_{ins} might allow tunnelling beyond near neighbours.

4. Procedure for Analysis of Experiments

It has already been pointed out that variable range tunnelling in a Coulomb gap would correctly account for the *form* of the observed temperature dependence of conductivity in granular metals. However, any valid model must also provide an acceptable *quantitative* analysis of data. We now set out an appropriate approach for analysing *any* transport data in terms of variable range hopping in a Coulomb gap.

4.1 Extraction of Parameters

The essential quantity extracted directly from experiment is T_0 which is obtained by fitting results to equation 2 with $x = 0.5$. From this, one obtains directly W_{opt}, the temperature-dependent optimum hop energy of the variable range tunnelling process:

$$W_{opt} = 0.5k(T_0 T)^{1/2}. \qquad (10)$$

This result, correct for both three- and two-dimensional systems, contains no material parameter. Other quantities depend on the relative permittivity which comes in through its involvement in $g(E)$. The formulae for the tunnelling exponent are

3-d $\qquad \alpha_m = kT_0 (\pi g_3)^{1/3}/10.5 \qquad (11a)$

2-d $\qquad \alpha_m = kT_0 (\pi g_2)^{1/2}/2^{7/2}. \qquad (11b)$

In both cases, for given T_0's, α_m is linearly proportional to ε_r. Numerical expressions for α are

3-d $\qquad \alpha_m/\text{m}^{-1} = 8.42 \times 10^3 \, (T_0/\text{K}) \qquad (12a)$

2-d $\qquad \alpha_m/\text{m}^{-1} = 6.66 \times 10^3 \, (T_0/\text{K}). \qquad (12b)$

The temperature-dependent optimum hop distance R_{opt} is given in both two and three dimensions by

$$R_{opt} = 0.25\, \alpha_m^{-1} (T_0/T)^{1/2} . \qquad (13)$$

R_{opt} is inversely proportional to ε_r.

One needs to be clear about the meaning of ε_r in all these results. The relative permittivity comes in through the Coulomb energy of two charges localized some distance apart on metal islands. The appropriate value will not be exactly the same as that which would be measured by applying a macroscopic electric field to a sample of the material because the field configuration is different. In particular, field lines originate on charged islands, and these are totally surrounded by dielectric so that for the first part of any path integral the permittivity of the dielectric applies. Only if charged islands are relatively far apart does the polarizability of intervening islands contribute to the permittivity, and so only for large distances does ε_r tend to the bulk value. The second point about ε_r is that it diverges as the metal-insulator transition is approached from the insulating side.[25] (Among other consequences is the collapse of the Coulomb gap.) In principle, the *bulk* ε_r can be estimated using effective medium theory[26] and, in some cases, it has been measured. Nevertheless, in view of the uncertainty over the appropriate value to take, we give deduced values of model parameters in Table 1 in terms of ε_r.

Also listed in Table 1 are values of Δ which depend on knowing g_0. The appropriate value to take for granular metals is not obvious. In Table 1 we use the metallic density of states diluted in proportion to the volume (or area) concentration of metal in the system. An alternative, but one we think less soundly based, would be to take the density of *charge* states of the metal islands calculated from their estimated capacitance. This generally gives a value of g_0 about one order of magnitude smaller and a corresponding reduction in the estimate of Δ.

4.2 Limits on Parameters for Validity of Model.

The conditions under which various transport mechanisms may operate in granular metals have been discussed by Entin-Wohlman *et al.*[24]   For the model we are considering to be valid, the model parameters are limited as follows:

T_0 **and** W_{opt} All treatments of activated hopping conduction leading to expressions for the conductivity of the form of equation (1) assume the Boltzmann limit for the phonon statistics. The model set out here can therefore only be valid if deduced values of W_{opt} satisfy

$$W_{opt} \geq kT . \qquad (14)$$

From equation 10 we see this corresponds to

$$T_0 \geq T . \qquad (15)$$

There is also an upper limit on acceptable values of W_{opt} for if activation energies become too large processes involving the constant density of states outside the Coulomb gap will become dominant and one would go over to variable range hopping in a constant density of states with $x = 0.25$ and $x = 1/3$ in three and two dimensions. Calculated values of W_{opt} must also therefore satisfy

$$W_{opt} < \Delta/2 . \qquad (16)$$

Correspondingly, with (14), this implies

$$\Delta \gg kT . \qquad (17)$$

R_{opt} For variable-range tunnelling to be occurring, calculated optimum hop distances must be considerably larger than the distances between neighbouring islands for otherwise the optimization between R and W cannot take place. Thus we also require:

$$R_{opt} \gg d . \qquad (18)$$

As $R_{opt} \to d$ (a possible high temperature limit) the hopping process would go over to near-neighbour tunnelling which would give simple activation.[21]

5. Analysis of Experimental Data

We analyze results for three different kinds of granular metal: two three-dimensional systems, a conventional cermet and a fine grained granular metal film; and one two-dimensional system, a discontinuous metal film. Results of the analysis are given in Table 1.

TABLE I
Analysis of Experimental Data

	Cermet	Granular Film	Discontinuous Film
EXPERIMENTAL DATA			
T_0/K	16000	200	1100
T_r/K	20-300	1.5-200	15-270
d/nm	2.3	3.0	20
s/nm	0.7	0.2	2.0
f	0.24	0.5	0.5
$\varepsilon_{r,ins}$	4	8	4
g_0	2.3×10^{47} J^{-1}m^{-3}	1×10^{47} J^{-1}m^{-3}	5×10^{38} J^{-1}m^{-2}
DEDUCED PARAMETERS			
Δ/eV	$21/\varepsilon_r^{3/2}$	$14/\varepsilon_r^{3/2}$	$660/\varepsilon_r^2$
α_m/m^{-1}	$1.3 \times 10^8 \, \varepsilon_r$	$1.7 \times 10^6 \, \varepsilon_r$	$7.3 \times 10^6 \, \varepsilon_r$
α_{ins}/m^{-1}	$5.8 \times 10^8 \, \varepsilon_r$	$2.7 \times 10^7 \varepsilon_r$	$8.1 \times 10^7 \, \varepsilon_r$
OPTIMUM HOPPING PARAMETERS AT 80 K (kT = 6.9 meV)			
W_{opt}/meV	49	5.5	13
R_{opt}/nm	$26/\varepsilon_r$	$230/\varepsilon_r$	$130/\varepsilon_r$

5.1 Cermet

The data quoted are for a Ni-SiO$_2$ cermet produced by conventional co-sputtering of metal and insulator.[27] We note a high T_0 and that $T^{1/2}$ behaviour is shown over a good range of temperature (T_r). At 80 K the optimum hop energy is considerably greater than kT. In the absence of measurement, we estimate the bulk relative permittivity using effective medium theory.[26] With the quoted volume fraction f of metal, the enhancement factor is 3.6, giving a bulk relative permittivity of about 14. Δ remains adequately large, of order 0.4 eV, α_{ins} becomes very reasonable at 8×10^9 m^{-1} which, for a free electron mass, would correspond to a barrier height of 2.4 eV. There are problems, however, with R_{opt} at 80 K which becomes 1.8 nm, significantly smaller than the particle size of 2.3 nm.

5.2 Granular Film

These data are for a fine-grained granular aluminium film produced by evaporation of aluminium in an oxygen ambient.[28] Although the $T^{1/2}$ law appears to be reasonably well obeyed over a large temperature range, we note immediately that T_0 is low, at the upper end of the temperature range for which the $T^{1/2}$ law appears to be followed. There is a consequent problem with W_{opt} which is less than kT at 80 K. This time, using effective medium theory to estimate the enhancement of the permittivity, we obtain a factor of 10 and a bulk relative permittivity of order 80. This would imply a rather small Δ of order 20 meV but α_{ins} is reasonable at 2.2×10^9 m^{-1} corresponding to a barrier height of 180 meV. Although this would appear small for alumina, it should be pointed out that the tunnelling barriers are believed to be very thin in this system, only a few thenths of a nanometre, so that there would be considerable lowering of the barrier by image forces.[29] R_{opt} again presents problems with a calculated value just under 3 nm, again of the order of the particle size.

5.3 Discontinuous Metal Film

These data are from our own measurements with a typical discontinuous film of gold on glass. The mean thickness of the film was of order 5 nm and this figure was used to calculate g_0. The $T^{1/2}$ law is obeyed over a good temperature range, T_0 is adequately high and W_{opt} is sufficiently greater than kT at 80 K.

In this system there is only a small enhancement of the permittivity by the presence of the metal because, while the film may be approximated to a two-dimensional structure, the electric fields are three-dimensional. For a single isolated island, the effective permitivity would be the average of those of glass and vacuum and that is the figure quoted. Since there will be some concentration of fields towards the plane of the film, there will be some 'effective-medium' enhancement from the presence of other islands and we may reasonably estimate an effective relative permittivity of order 6.

With this value, Δ remains large but there are serious problems with $\alpha_{ins} \approx 5 \times 10^8$ m^{-1}, corresponding to an effective barrier height of only 9 meV. This is unreasonably small; it is smaller than W_{opt} and comparable with kT so that transport would not be by tunnelling but by activation *over* the barrier. (As regards this small value of α_{ins}, it should be remarked that attempts to account for the magnitudes of observed conductivities of discontinuous metal films in terms of their known structures also require small values of α_{ins}.[30,31] No explanation for this has been found.) As serious as the small value of α_{ins} is the small value of R_{opt} which again comes out of the order of the island size.

6. Discussion

In all the cases analyzed, the crucial common result is the small size of the calculated optimum hop distances. In all three systems it comes out of order of the diameter of the metal islands. As pointed out previously, for variable-range hopping to be occurring, the hop range must be sufficient to allow the tunnelling electrons to sample a reasonable number of islands. This would require tunnelling at least beyond near neighbours with $R_{opt} > 2d$.

In the above analysis we have had to estimate ε_r and there must be some uncertainty in the values taken. It is possible that effective medium theory overestimates the enhancement of the permittivity. To see the effect of using a reduced value we consider the figures in the extreme case of no enhancement.

For the cermet film we would have $R_{opt} \approx 3d$ which might just be acceptable; but α_{ins} would then be about 2.3×10^9 m^{-1} corresponding to a barrier height of only 200 meV which would be on the small side even allowing for barrier lowering in the relatively thin tunnel barriers. Clearly the range of ε_r that could yield values for R_{opt} and α_{ins} that might be considered acceptable is very small. Although one might convince oneself that the model could be made viable in this particular case, it seems extremely unlikely that it could account for all cermets. Yet for cermets, the $T^{1/2}$ behaviour is essentially universal in the dielectric regime.

For the fine-grained granular film, with no enhancement of the permittivity, we would obtain $R_{opt} \approx 10d$, which would be satisfactory, but α_{ins} would become unreasonably small at 2.2×10^8 m^{-1}, corresponding to a barrier height of about 2 meV, much smaller than both kT and W_{opt}. For $R_{opt} \approx 3d$, we would require a permittivity that would give $\alpha_{ins} \approx 6.9 \times 10^8$ m^{-1} corresponding to a barrier height of only 18 meV, again unreasonably small. In any case, W_{opt} and T_0 remain unacceptable and one concludes that there is no latitude at all that could allow the possibility of the model being valid in this case.

For the discontinuous film there is little room for adjustment of the numbers and again one concludes that the model cannot apply.

The weight of evidence indicates, therefore, that the conductivity of granular metals cannot be accounted for in terms of electron correlation and variable-range hopping in a Coulomb gap.

Not considered in the earlier discussion is the quantization of electron energy levels in the metal islands. For metal islands smaller than about 3 nm, the level spacing (which varies as d^{-3}) becomes greater than the capacitive charging energy (which varies as d^{-1}). However, it is not apparent how this could significantly affect the mechanism of charge mobility which, in the Coulomb gap scenario, should be independent of the energetics of carrier formation. Similarly, with no carrier correlation, inclusion of energy level quantization in critical path analysis[20] would not alter the prediction of simple activation.

In comparing discontinuous metals with other systems in which the Coulomb gap has been studied, we have to remember an essential difference. In semiconductor systems, carriers of only one sign are present and their concentration is independent of temperature. In granular metals, it is always assumed that 'carriers' are produced by thermal activation from neutral metal islands so that mobile charges of both signs are present in concentrations that are equal and temperature dependent. It is difficult to see how the requirement for activation of carriers could be avoided, and, as long as it is superimposed on the mechanism of mobility, its temperature dependence must be involved in the overall behaviour of the conductivity. In the discussion above, we have treated conduction as if we were simply concerned with the temperature-independent density of states near the Fermi level that is required by the Efros and Shklovskii argument, and we have ignored activation of carriers. At the least, the varying number of mobile charges needs to be incorporated in terms of a temperature-dependent screening length. No attempt to do this seems to have been made.

Perhaps it is more significant to note that much of the direct experimental evidence for Coulomb gaps comes from photoemission studies in which the processes are obviously single-particle in the Efros and Shklovskii sense. It may be that, in transport, the dominance of correlation suppresses single-particle aspects and that forms of correlated motion involving sequential hopping or other many-particle processes would form a sounder basis for theory. An extreme form of carrier correlation was used to explain conductivity and Hall effect data for inversion layers in the regime of activated conduction. Measurements showed a constant carrier concentration and activated mobility rather than the reverse which was expected in terms of the mobility-edge model.[32] Adkins[33] explained this in terms of a model in which correlation dominated carrier energetics so that the carriers behaved as a viscous liquid, the activated mobility then being analogous to the activated fluidity of conventional liquids. More recently, the ideas of the *Coulomb blockade* have been developed in relation to tunnelling between very small conductors.[34] Such a very-short-range approach to Coulomb interaction might be appropriate to granular systems, although the (assumed) presence of carriers of both signs is an added complication that would have to be taken into account. It would be interesting to apply such ideas, based on the assumption that short-range correlation dominates carrier dynamics, to conduction in granular metals and in other systems where Coulomb effects are expected to be important.

All these are aspects of transport needing further investigation. For the moment, we have to conclude that conduction in granular metals remains an unsolved problem.

REFERENCES

1 Altschuler and Aronov in *Electron-Electron Interactions in Disordered Systems* ed. Efros A.L.and Pollak M. (Elsevier Science Publishers B.V.) 1 (1985)

2 Pollak M. in *Electron-Electron Interactions in Disordered Systems* ed. Efros A.L. and Pollak M. (Elsevier Science Publishers B.V.) 287 (1985)

3 Efros A.L.and Shklovskii B.I. *J. Phys. C: Solid State Phys.* **8** L49 (1975)

4 Efros A.L. *J. Phys. C: Solid State Phys.* **9** 2021 (1976)

5 Levin E.I, Nguen V. L, Shklovsii B.I.and Efros A.L. *Sov. Phys. JETP* **65** 842 (1987)

6 Shang-Lin Weng, Moehlecke S.and Strongin M. *Phys. Rev. Lett.* **50** 1795 (1983)

7 Davis J.H.and Franz J.R. *Phys. Rev. Lett.* **57** 475 (1986)

8 Entin-Wohlman O.and Ovadyahu Z. *Phys. Rev. Lett.* **56** 643 (1986)

9 Adkins C.J. *J. Phys.: Condens. Matter* **1** 1253 (1989)

10 Abeles B. *Applied Solid State Science* **6** 1 (1976)

11 Morris J.E.and Coutts T.J. *Thin Solid Films* **47** 3 (1977)

12 Neugebauer C.A.and Webb M.B. *J. Appl. Phys.* **33** 74 (1962)

13 Sheng P, Abeles B.and Arie Y. *Phys. Rev. Lett.* **31** 44 (1973)

14 Heinrichs J, Kumar A.A.and Kumar N. *J. Phys. C: Solid State Phys.* **9** 3249 (1976)

15 Hill R.M.and Coutts T.J. *Thin Solid Films* **42** 201 (1977)

16 Šimánek E. *Solid State Commun.* **40** 1021 (1981)

17 Celasco M, Masoero A, Mazzetti P. and Stepanescu A. *Phys. Rev.* B **17** 2553 (1978)

18 Adkins C.J, Benjamin J.D, Thomas J.M.D, Gardner J.W. and McGeown A.J. *J. Phys. C: Solid State Phys.* **17** 4633 (1984)

19 Adkins C.J, Thomas J.M.D.and Young M.W. *J. Phys. C: Solid State Phys.* **13** 3427 (1980)

20 Adkins C.J. *J. Phys. C: Solid State Phys.* **15** 7143 (1982)

21 Adkins C.J. *J. Phys. C: Solid State Phys.* **20** 235 (1987)

22 Cavicchi R.E.and Silsbee R.H. *Phys. Rev. Lett.* **52** 1453 (1984)

23 Mott N.F. *J. Non-Cryst. Solids* **1** 1 (1968)

24 Entin-Wohlman O. Gefen Y. and Shapira Y. *J. Phys. C: Solid State Phys.* **16** 1161 (1983)

25 Abrahams E, Anderson P.W., Licciadello D.C.and Ramakrishnan T.V. *Phys. Rev. Lett.* **42** 673 (1979)

26 Landauer R. in *Proc. Conf. Electrical Transport and Optical Properties of Inhomogeneous Media* ed. Garland J.C.and Tanner D.B. (AIP, New York) 2 (1978)

27 Abeles B, Ping Sheng, Coutts M.D.and Arie Y. *Advances in Phys.* **24** 407 (1975)

28 Chui T, Deutscher G, Lindenfeld P. and McLean W.L. *Phys. Rev.* **B23** 6172 (1981)

29 Simmons J.G.in *Tunneling Phenomena in Solids* ed. Burstein E and Lundqvist S.

(Plenum, New York) 135 (1969)

30 Simmons J.G. *J. Appl. Phys.* **34** 1793 (1963)

31 Benjamin J.D. *PhD Thesis* University of Cambridge (1981)

32 Mott N.F. and Davis E.A. *Electronic Properties of Non-Crystalline Materials* (Oxford University Press, Oxford) (1979)

33 Adkins C.J. in *The Hall Effect and Its Application*, eds. Chien C.L. and Westgate C.R. (Plenum, New York) 355 (1980)

34 see, Levi B.G. *Physics Today* **41** 19 (1988)

SIMULATION STUDIES OF ELECTRONIC STATES IN COMPENSATED Si:P SYSTEM

Mikio Eto and Hiroshi Kamimura
Department of Physics, University of Tokyo,
Bunkyo-ku, Tokyo 113, Japan

ABSTRACT

Cluster simulation is carried out for both uncompensated and compensated Si:P systems in the intermediate to the low concentration region, by means of Multi-Configuration Self-Consistent Field (MCSCF) method. We determine one-electron orbitals taking into account the electron correlation, and make clear the electronic structures in both systems. Further we predict the characteristic behavior of specific heat and spin-susceptibility of these two systems.

1. INTRODUCTION

The electronic structure of doped semiconductors is one of the most interesting subjects in recent condensed matter physics. In the low concentration region of donor impurities in n-type semiconductors, every electron is localized around each donor. As the donor concentration increases, the wavefunction of an electron becomes extended over some donors (intermediate concentration region) and at a critical concentration n_c a metal-insulator transition takes place. The compensation induces remarkable change in various properties of the doped semiconductor. For example, hopping conduction in the low concentration shows a different feature. At the metal-insulator transition the critical exponent ν of the conductivity, $\sigma(T=0) = \sigma_0 (n/n_c - 1)^\nu$, is 0.5 in the uncompensated Si:P while 1.0 in the compensated Si:P.[1] The other properties such as specific heat, susceptibility, the line width of ESR *etc.*, are also affected by the compensation.[2]

In these systems electron-electron interactions play an important role. In the low concentration region Coulomb gap and multi-electron hopping have been investigated due to the Coulomb repulsion between electrons.[3] In the intermediate region to the metal-insulator transition which is well-known as Anderson localization region, the interaction effect plays an important role together with the random effect. In particular, the spin-dependent interaction is essentially important in this region. To treat this effect, 'spin pair model' has been introduced.[4]

To elucidate the electronic states in the intermediate concentration region of the uncompensated Si:P, we have performed the cluster simulation by means of Multi-Configuration Self-Consistent Field (MCSCF) method with Configuration Interaction (CI).[5] In the MCSCF method one-electron orbitals are determined in the presence of not only coulomb and exchange interactions but also the electron correlation, so that the method is suitable for investigating a strongly correlated random system. In this paper we adopt this method to investigate the electronic states in the compensated Si:P. Then we discuss the difference in features between two systems in the intermediate to low concentration region.

In the next section we briefly explain the MCSCF method. In §3 the formalism of cluster simulation is described and simulation results are presented. In §4 we calculate the specific heat and spin-susceptibility of two systems and discuss their features. Finally, conclusion is given in §5.

2. WHAT IS THE MCSCF METHOD ?

The MCSCF method is a kind of variational method, which has been used in the field of quantum chemistry.[6] In contrast to the Hartree-Fock (HF) method, the trial function is expressed as a linear combination of several Slater determinants;

$$\Phi_{MCSCF} = C_0 \Psi_0 + \sum_{i=1}^{3} \sum_{a=4}^{6} C_{ii}^{aa} \Psi_{ii}^{aa}$$

$$= (C_0 + \sum_{i=1}^{3} \sum_{a=4}^{6} C_{ii}^{aa} a_{a\uparrow}^{\dagger} a_{a\downarrow}^{\dagger} a_{i\downarrow} a_{i\uparrow}) \Psi_0, \qquad (1)$$

where a configuration Ψ_0 is the same as HF trial function, e.g. $|\phi_1 \alpha \phi_1 \beta \phi_2 \alpha \phi_2 \beta \phi_3 \alpha \phi_3 \beta|$ for 6-electron systems, where α and β are spin up and down functions, respectively. Ψ_{ii}^{aa}'s correspond to the two-electron simultaneous excitations from occupied i orbital in Ψ_0 into empty a orbital.

By the variation one-electron orbitals { ϕ_i, ϕ_a } and CI coefficients { C_{ii}^{aa} } in (1) are determined simultaneously. In this procedure the part of the electron correlation represented by matrix elements

$$< aa|\frac{1}{r}|ii > = \int \int \psi_a^*(r_1)\psi_a^*(r_2)\frac{1}{|r_1 - r_2|}\psi_i(r_1)\psi_i(r_2)dr_1 dr_2, \qquad (2)$$

are included in determining the one-electron orbitals. We further perform full Configuration Interaction (CI) calculation, taking MCSCF one-electron orbitals as a basis set. As a result, all the rest of electron-electron interactions are taken into account. Since the pair type of interactions (2) is the largest besides the mean field HF terms, only a small number of configurations are necessary in the result of CI, as will be seen below.

In actual calculations we pay attention to the fact that each many-electron level of a cluster is an eigenstate of the total spin S and its z-component S_z and it makes computation easier. For example, the number of the configurations is reduced to $(_6C_3)^2 = 400$ in the case of the uncompensated systems.

3. CLUSTER SIMULATION

For the uncompensated system we take clusters with 6 electrons, consisting of 6 donors distributed randomly in a sphere. The donor concentrations (n_D) we have selected correspond to $(1.0, 1.7, 2.4, 3.2) \times 10^{18}$ cm^{-3} in the intermediate to the critical region and $(0.25, 0.5) \times 10^{18}$ cm^{-3} in the low concentration region. For the compensated system we choose clusters with 4 electrons. These clusters consist of 8 (or 12) donors and 4 (or 8) acceptors, both of which are distributed randomly in a sphere. The concentrations are

Table.1. The number of electrons, donors and acceptors in the clusters we have selected for the simulation of the compensated Si:P. The concentration to which each cluster corresponds is also listed.

Case	Cluster			Concentration ($\times 10^{18} cm^{-3}$)		
	electron	donor	acceptor	electron	donor	acceptor
a	4	8	4	1.5	3.0	1.5
b	4	12	8	1.0	3.0	2.0
c	4	8	4	1.0	2.0	1.0
d	4	8	4	0.25	0.5	0.25

listed in Table 1. A Gaussian-type hydrogen 1s orbital is attached to each donor site and MCSCF one-electron orbitals are represented by their linear combinations. Acceptors are represented by point charges of -1. The calculations are performed for 30 clusters for each concentration.

Calculated results for the intermediate concentration regions are given below.

<u>Uncompensated System.</u> Most of the clusters have spin-singlet ground state, while 10 to 20 percent of clusters have spin-triplet ground state. An example of spin-singlet clusters and schematic drawing of its MCSCF orbitals are given in Fig.1(a). For the concentration of 1.0×10^{18} cm^{-3} all MCSCF one-electron orbitals are localized in a region of two donors. In higher concentrations, ϕ_1 and ϕ_2 corresponding to lower energy, have almost the same shape as those of the lowest concentration, while the other four MCSCF orbitals become more extended as the concentration increases. The wavefunctions of the ground and first excited states of the spin-singlet cluster shown in Fig.1(a) are, respectively, given as,

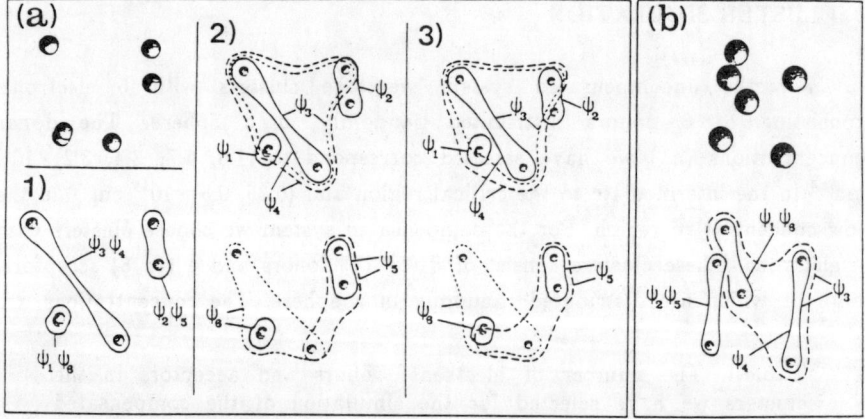

Fig.1. (a) An example of the spin-singlet clusters in the uncompensated system. Schematic drawing of MCSCF one-electron orbitals are shown for 1) $n_D = 1.0 \times 10^{18}$, 2) 1.7×10^{18} and 3) $(2.4 \sim 3.2) \times 10^{18}$ cm^{-3}. The solid and broken lines show the highest and lowest amplitude of orbitals, respectively. (b) An example of spin-triplet cluster in the uncompensated system. MCSCF orbitals are shown for $n_D = 1.7 \times 10^{18}$ cm^{-3}. [After M.Eto and H.Kamimura.[5]]

$$\Phi_g = 0.75|\psi_1\alpha\psi_1\beta\psi_2\alpha\psi_2\beta\psi_3\alpha\psi_3\beta| - 0.65|\psi_1\alpha\psi_1\beta\psi_2\alpha\psi_2\beta\psi_4\alpha\psi_4\beta|$$
$$-0.09|\psi_1\alpha\psi_1\beta\psi_3\alpha\psi_3\beta\psi_5\alpha\psi_5\beta| + 0.08|\psi_1\alpha\psi_1\beta\psi_4\alpha\psi_4\beta\psi_5\alpha\psi_5\beta|$$
$$= a_{1\uparrow}^\dagger a_{1\downarrow}^\dagger(0.99 a_{2\uparrow}^\dagger a_{2\downarrow}^\dagger - 0.12 a_{5\uparrow}^\dagger a_{5\downarrow}^\dagger)(0.75 a_{3\uparrow}^\dagger a_{3\downarrow}^\dagger - 0.66 a_{4\uparrow}^\dagger a_{4\downarrow}^\dagger)|0>, \quad (3)$$

and

$$\Phi_{ex} = 0.70\{|\psi_1\alpha\psi_1\beta\psi_2\alpha\psi_2\beta\psi_3\alpha\psi_4\beta| + |\psi_1\alpha\psi_1\beta\psi_2\alpha\psi_2\beta\psi_3\beta\psi_4\alpha|\}$$
$$+0.08\{|\psi_1\alpha\psi_1\beta\psi_3\alpha\psi_4\beta\psi_5\alpha\psi_5\beta| + |\psi_1\alpha\psi_1\beta\psi_3\beta\psi_4\alpha\psi_5\alpha\psi_5\beta|\}$$
$$= a_{1\uparrow}^\dagger a_{1\downarrow}^\dagger(0.99 a_{2\uparrow}^\dagger a_{2\downarrow}^\dagger - 0.12 a_{5\uparrow}^\dagger a_{5\downarrow}^\dagger)\frac{1}{\sqrt{2}}(a_{3\uparrow}^\dagger a_{4\downarrow}^\dagger - a_{3\downarrow}^\dagger a_{4\uparrow}^\dagger)|0>, \quad (4)$$

for $n_D = 1.7 \times 10^{18}$ cm^{-3}. The ground state can be regarded as consisting of three spin-singlet pairs, while in the first excited state two electrons in ϕ_3 and ϕ_4 form a triplet pair. This corresponds to the situation of 'spin pair model.'[4]

As will be seen in the next section, the existence of the spin-triplet clusters is essentially important for explaining the unusual behavior of the specific heat and spin-susceptibility. Spin-triplet clusters are found to have a common feature; one-electron orbitals in triplet clusters are extended over three or four donors, in which electrons occupy those orbitals satisfying the Hund's rule (*i.e.* lower energy for higher spin state). Fig.2 shows examples of the geometry which gives rise to the Hund's coupling. One spin-triplet cluster is shown in Fig.1(b), whose ground state is,

$$\Phi_{triplet} = 0.56\{|\psi_1\alpha\psi_1\beta\psi_2\alpha\psi_2\beta\psi_3\alpha\psi_4\beta| + |\psi_1\alpha\psi_1\beta\psi_2\alpha\psi_2\beta\psi_3\beta\psi_4\alpha|\}$$
$$+0.30\{|\psi_1\alpha\psi_1\beta\psi_3\alpha\psi_4\beta\psi_5\alpha\psi_5\beta| + |\psi_1\alpha\psi_1\beta\psi_3\beta\psi_4\alpha\psi_5\alpha\psi_5\beta|\}$$
$$+0.18\{|\psi_2\alpha\psi_2\beta\psi_3\alpha\psi_4\beta\psi_6\alpha\psi_6\beta| + |\psi_2\alpha\psi_2\beta\psi_3\beta\psi_4\alpha\psi_6\alpha\psi_6\beta|\}$$
$$= (0.95 a_{1\uparrow}^\dagger a_{1\downarrow}^\dagger - 0.31 a_{6\uparrow}^\dagger a_{6\downarrow}^\dagger)(0.88 a_{2\uparrow}^\dagger a_{2\downarrow}^\dagger - 0.48 a_{5\uparrow}^\dagger a_{5\downarrow}^\dagger)$$
$$\times \frac{1}{\sqrt{2}}(a_{3\uparrow}^\dagger a_{4\downarrow}^\dagger - a_{3\downarrow}^\dagger a_{4\uparrow}^\dagger)|0>, \quad (5)$$

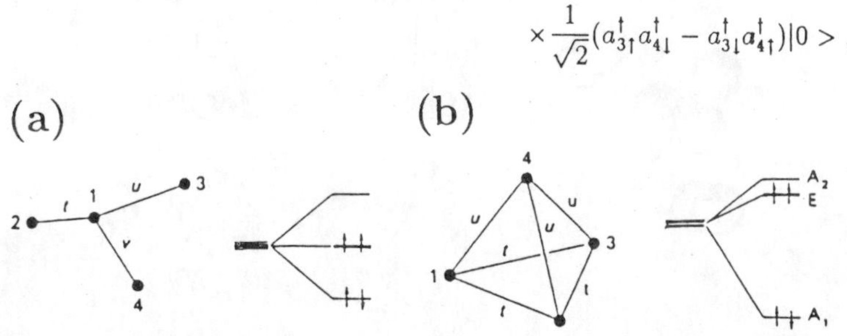

Fig.2. Two types of atomic configurations that have spin-triplet ground states are shown together with their electronic energy levels and the electron occupancy for the half-filled case. (a) A cluster with appreciate transfer energies (t,u,v) only between the central atom and each of surrounding atoms, (b) a cluster with accidental symmetry of C_{3v}.

for $n_D = 1.7 \times 10^{18}$ cm^{-3}, where ϕ_3 and ϕ_4 form a triplet spin pair. They correspond to the wavefunctions of a doubly degenerate level of E symmetry in Fig.2(b).

Compensated System. The ratio between the number of spin-singlet and spin-triplet clusters is almost the same as that of the uncompensated system. In Fig.3 one example of the spin-singlet clusters is shown for the concentration of case a in Table 1. MCSCF orbitals are more extended than those of the uncompensated system at the same electron concentration. The CI wavefunctions of the ground and first excited states are given as,

$$\Phi_g = 0.93|\psi_1\alpha\psi_1\beta\psi_2\alpha\psi_2\beta| - 0.29|\psi_1\alpha\psi_1\beta\psi_3\alpha\psi_3\beta| \\ -0.19|\psi_1\alpha\psi_1\beta\psi_4\alpha\psi_4\beta| \tag{6}$$

and

$$\Phi_{ex} = 0.65\{|\psi_1\alpha\psi_1\beta\psi_2\alpha\psi_3\beta| + |\psi_1\alpha\psi_1\beta\psi_2\beta\psi_3\alpha|\} \\ -0.14\{|\psi_1\alpha\psi_1\beta\psi_2\alpha\psi_4\beta| + |\psi_1\alpha\psi_1\beta\psi_2\beta\psi_4\alpha|\} \\ -0.16\{|\psi_1\alpha\psi_1\beta\psi_2\alpha\psi_5\beta| + |\psi_1\alpha\psi_1\beta\psi_2\beta\psi_5\alpha|\} \tag{7}$$

The ground state wavefunction is much closer to a single Slater determinant, compared with (3) of the uncompensated system. The first excited state is a

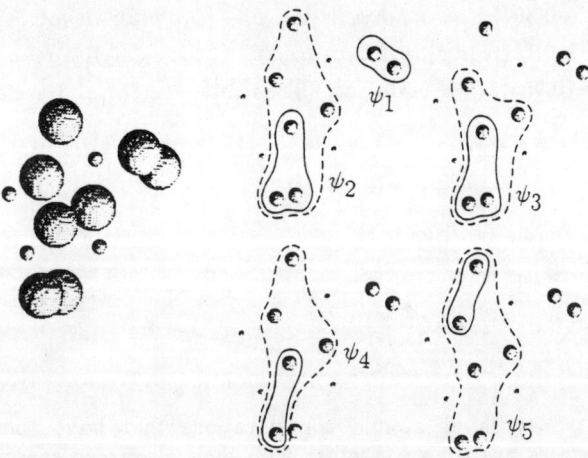

Fig.3. An example of the spin-singlet clusters in the compensated system. The concentration corresponds to the case (a) in Table.1. The large and small spheres represent the donors and acceptors, respectively.

spin-triplet, where first two configurations come from the spin-dependent interaction, just like spin pair model, while the other four terms correspond to the hopping-type excitations, in which the electrons that form the spin pair in the ground state move to other orbitals and as a result, the spin pairing disappears. In this view spin coupling effect is not so strong in the compensated system, compared with that in the uncompensated system, in which spin coupling is very important.

Next, we discuss the low concentration region in both systems. As the concentration decreases, electrons become localized and spin coupling effect diminishes. In the *uncompensated* system the excitation energy from the almost spin-degenerate ground states increases. It corresponds to the intra-state interaction, that is, the Hubbard U repulsion, which becomes larger for more localized orbitals. In the *compensated* system, on the other hand, the hopping-type excitations exist while the ground state is also nearly spin-degenerate. Hence the excitation energy is much smaller.

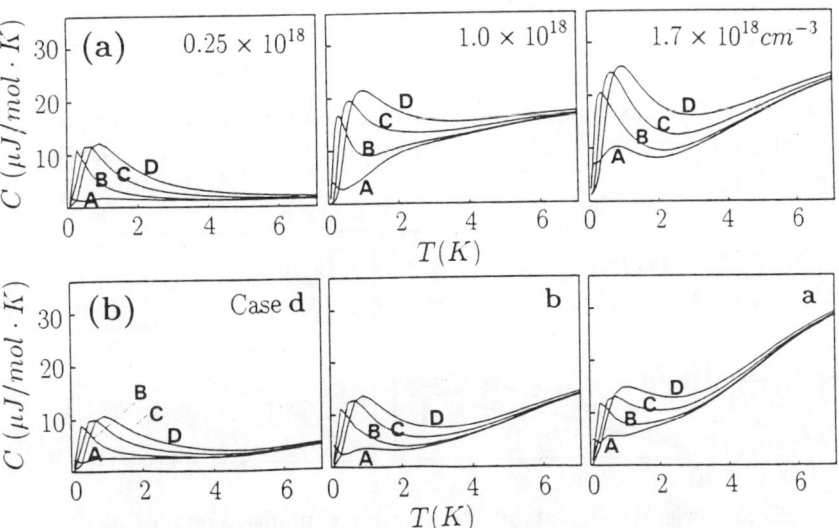

Fig.4. Temperature dependence of the specific heat averaged over 30 clusters, (a) for the uncompensated system and (b) for the compensated system. The concentrations in (b) correspond to case d, b and a in Table.1. The values of the magnetic field are 0 (curve A), 5 (curve B), 10 (curve C) and 15 KOe (curve D).

4. SPECIFIC HEAT AND SPIN-SUSCEPTIBILITY

We first calculate the specific heat in magnetic fields and spin-susceptibility of each cluster which is in a canonical distribution, and then take an ensemble average of them over 30 clusters for each concentration.

Fig.4 shows the T dependence of the specific heat for several magnetic fields. In the *uncompensated* systems it behaves as T-linear in higher temperature region only when n_D is larger than 1.7×10^{18} cm^{-3}. At low temperatures such as about 1 K the hump over T-linear part is seen, which is sensitive to the magnetic fields. These characteristics are in good agreement with experiments.[7] The T-linear part is due to the random distribution of energy spectrum in each cluster. The hump corresponds to the thermal excitation between spin-singlet and spin-triplet, which is the result of the spin-dependent interaction.

In the *compensated* systems the specific heat shows T-linear behavior even for the low concentration region, whose gradient is the increasing function of the concentration. In the low concentration the hopping-type excitations appear in the compensated systems, while they do not appear in the uncompensated systems. At low temperatures, there appears the hump over T-linear part, like the uncompensated system, which is also attributable to the effect of spin-dependent interaction. The hump becomes smaller as the concentration decreases. These results are also in accordance with experiments.[2]

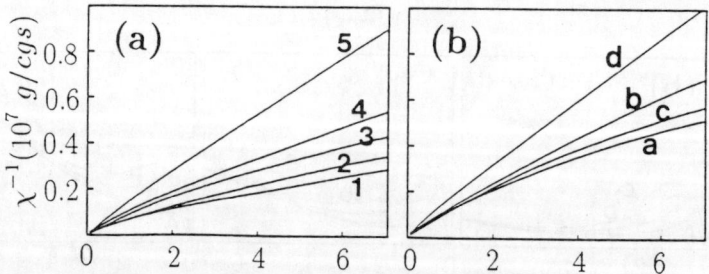

Fig.5. Curie Weiss plots of the spin-susceptibility averaged over 30 clusters, (a) for the uncompensated system and (b) for the compensated system. The concentrations correspond to $n_D = 2.4 \times 10^{18}$ (curve **1**) 1.7×10^{18} (curve **2**), 1.0×10^{18} (curve **3**), 0.5×10^{18} (curve **4**) and 0.25×10^{18} (curve **5**) in (a) and case *a*, *b*, *c* and *d* in Table.1 in (b), respectively.

In Fig.5 the Curie-Weiss plots of spin-susceptibility are shown. T-dependence of the spin-susceptibility is qualitatively the same in both systems; it tends to diverge with decreasing T even in the intermediate concentration. This is due to the existence of spin-triplet clusters. The Curie-Weiss plots are almost T-linear (Curie type) with a downward bending at low temperatures, consistent with experimental results.[8] The downward bending is due to the existence of low-lying triplet excited states in spin-singlet clusters.

5. CONCLUSION

We have performed the cluster simulation for the uncompensated and compensated Si:P, using MCSCF and CI method. We have shown that the MCSCF one-electron orbitals are suitable for investigating the electronic states in both systems, in which electron correlation is important. In the uncompensated system electrons tend to form spin pairing by spin-dependent interaction, while in the compensated system spin dependent interactions are not so strong and hopping-type excitations coexist with spin-excitation states in the intermediate concentration region. In the low concentration the hopping-type excitation is more important in the compensated system. As a result, T-linear behavior of the specific heat becomes dominant. By the present calculation we have clarified the origin of the appearance of the hump in the specific heat and the downward bending T-linear behavior of the spin-susceptibility.

REFERENCES

1. Rosenbaum,T.F., Milligan,R.F., Paalamen,M.A., Thomas,G.A., Bhatt,R.N. and Lin,W., Phys. Rev. B27, 7509 (1983).
2. Nishio,Y., Kajita,K., Iwata,T. and Sasaki,W., Proc. 18th Inter. Conf. Phys. Semicond. (Stockholm, Sweden 1986), 1257; Proc. 18th Int. Conf. Low Temp. Phys. (Kyoto, 1987), 691.
3. Pollak,M. and Ortuno,M., in "Electron-Electron Interaction in Disordered System" eds. M.Pollak and A.L.Efros (Amsterdam: North-Holland, 1985) p.287. Efros,A.L. and B.I.Shklovskii,B.I., *ibid.*, p.409 and related references cited therein.

LONG-RANGE INTERACTIONS IN SYSTEMS WITH LOCALISED STATES

M. Ortuño and R. Chicón
Departamento de Fisica Aplicada, Universidad de Murcia
Murcia 30.071, Spain

ABSTRACT

The density of states of disordered localised systems is studied for both Coulombic and dipole-dipole interactions. The analytic models for obtaining the density of states are analysed. The hardening of the gap, in the case of the Coulomb interaction, when the ground state is made stable with respect to electronic polaron transitions is calculated. A numerical simulation of the density of states for a system whose components interact *via* the dipole-dipole interaction is performed. The results are compatible with the logarithmic behaviour predicted by a self-consistent analytical treatment.

1. INTRODUCTION

Long-range interactions strongly affect the electronic and transport properties of systems with localised wavefunctions due to disorder. In general, the stronger the disorder the more localised the wavefunctions of the system and the more important the effects of interactions. We want to study the effects of Coulomb and dipole-dipole interactions in the limit of strong localisation. A typical example of this situation is impurity conduction for very light doping, well below the critical concentration for the metal insulator transition. We will consider the compensated case because then the interactions play a key role, since the typical carrier-carrier distance is small and the interaction energy large.

The most important consequences of the interactions are the depletion of the density of states (DOS) near the Fermi level, the correlation in the motion of the carriers and the non ergodicity of the system.[1] Here we will concentrate on the first consequence, the depletion of the DOS, for the Coulomb and dipole-dipole interactions.

The depletion of the DOS for a localised system with Coulomb interactions is known as the Coulomb gap, which was first discussed by Pollak.[2] The first analytical model of the DOS in the Coulomb gap was developed by Efros and Shklovskii.[3] They showed that if the ground state is to be stable with respect to one-electron excitations the DOS should go as:

$$N(\epsilon) = c|\epsilon|^{d-1} \qquad (1)$$

d being the dimensionality of the system. A harder gap than this predicted by Eq.(1) is obtained when the ground state is stabilised against many-electron transitions. Efros[4] considered certain type of two electron transitions and obtained and exponential DOS very close to the Fermi level. Davies[5] extended the method to include excitations involving any number of short pairs and one infinite jump, and got a different exponential DOS. We have extended this model, in turn, considering transitions formed by a finite jump and the polarisation (inversion of the occupancy) of short dipoles near the initial and the final sites, i.e., polaron excitations.

In the case of the dipole-dipole interaction, the stabilisation of the ground state with respect to one electron excitations produces no constraint on the energy density of dipoles, at least in its simplest version. However, a self-consistent treatment of the problem predicts a logarithmic singularity at low energies with a vanishing of the density of dipoles at $\epsilon = 0$, a signature of the long range character of the interaction.

In the next section, we will review the previous models for the DOS in the Coulomb gap and describe in more detail our method, showing also the results of our calculations. In section 3, we will do the same thing, but for the dipole-dipole interaction.

2. DENSITY OF STATES IN THE COULOMB GAP

As we are interested in the highly localised regime, we will assume a Hamiltonian with only two contributions: a site energy, which takes random values within an interval $(-W/2, W/2)$, and a direct Coulomb interaction. We minimise the total energy, taking into account the interplay between the disorder energy and the repulsion due to the interaction. Let ϵ_i be the total energy of an electron on site i, equal to the sum of the disorder energy ϕ_i and the interaction energies with all other electrons in their ground state occupancies. It is also the energy which the electron at site i adds to the system. The energy change Δ_{ij} produced by a one-electron transition from site i to site j is therefore

$$\Delta_{ij} = \epsilon_j - \epsilon_i - \frac{e^2}{r_{ij}} \qquad (2)$$

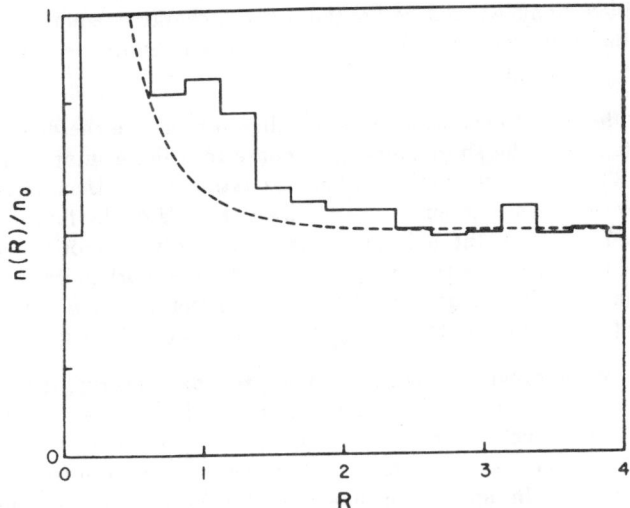

Figure 1 Pair correlation function: numerical results (solid line) and Thomas-Fermi approximation (hatched line).

where r_{ij} is the distance between sites i and j. For the ground state to be stable $\Delta_{ij} > 0$, which establishes a minimum separation between any occupied and any empty site at energy ϵ from the Fermi level: $r_{\min} = e^2/(2|\epsilon|)$. This results in an effective excluded volume around any site, proportional to $|\epsilon|^{-d}$, which produces a DOS as in Eq.(1).

Efros noticed the existence of short pairs with a basically constant density at low energies. The possibility of their polarisation results in a strong constraint on the DOS at very low energies. All the short low energy pairs near a site with energy close to E_F have to be 'correctly' oriented for stability reasons. Poisson distribution tells us that the probabilities of these configurations are proportional to an exponential function, and so it is the DOS.

Our extension of the previous model ensures stability of the ground state against polaron excitations, *i.e.*, an electron jumping over a finite length and short pairs polarising near the initial and final sites. For this calculation we first need a reliable density of short pairs, as a function of energy and distance. Previous workers used a density of excitations obtained from a self convolution of the DOS, assuming no correlation. We get numerically the pair correlation function, and from it we calculate the density of excitations with the help of some relatively weak assumptions. In figure 1 we show the computer simulation results (solid line) for the pair correlation function, together with the Thomas-Fermi approximation[6].

The total average 'induced' charge around a given charge is equal to the value of the latter, but with opposite sign. Thus, there is an almost metallic screening between charges in the ground state.

From the density of excitations it is possible to calculate the distribution of the energy relaxed by the short pairs in response to a finite jump of length R. The short pairs that polarise are those whose excitation energy is smaller than the difference between its interaction energies with the fields of the final and initial sites. The total energy of the polaron excitation is the energy of the long one-electron jump, Eq.(2), minus the energy relaxed by the short pairs, x, which is a random variable with a length dependent distribution function $g(x, R)$. This function can be numerically obtained from the density of short pairs[7].

The effect of the previous extra term in the polaronic energy, the relaxation energy x, is to increase the minimum possible separation between electrons and holes near the Fermi level. There is now a bigger effective excluded volume each state, mainly when they are close to the Fermi level. This results in a smaller DOS, *i.e.*, a hard gap. In figure 2 we show the DOS, in units of the unperturbed DOS (that without interactions), as a function of energy, measured relative to the total width of the gap, $\Delta = (2W^{1/2})^{-1}$. The slashed curve corresponds to the parabolic gap, while the full curve corresponds to the hard gap calculated by us. We can notice that the gap is very strong, but only exists in a small energy range.

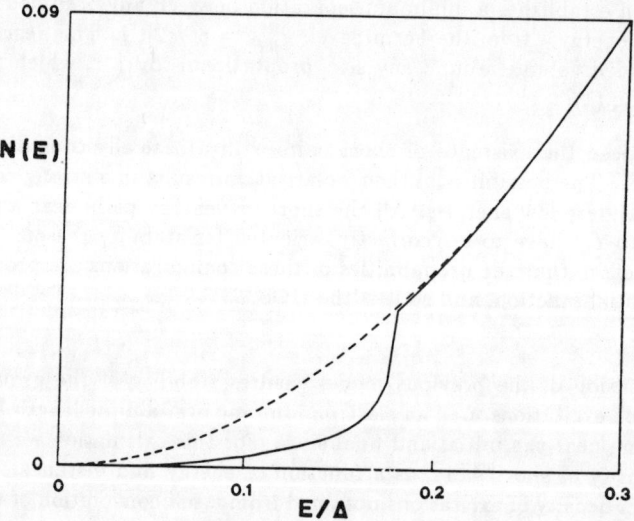

Figure 2 Density of states for the Coulomb gap: hard gap (solid line) and parabolic gap (hatched line).

3. DENSITY OF STATES FOR A DISORDERED SYSTEM OF INTERACTING DIPOLES

As we have mentioned in the previous section, for calculations of the hardening of the Coulomb gap a reliable density of dipolar excitations would be required. This leads to the problem of the calculation of the density of states for a system of interacting dipoles under the influence of a random external electric field. The analysis of such a system is important not only because of its implications on Coulomb gap calculations, but also due to its direct applicability to a variety of systems[8].

Keeping within the assumption of strong localisation, we will consider a Hamiltonian of the form:

$$H = -\sum_i \mathbf{p}_i \cdot \mathbf{E}_i + \frac{1}{2}\sum_{ij}{}' \frac{r^2 \mathbf{p}_i \cdot \mathbf{p}_j - 3(\mathbf{p}_j \cdot \mathbf{r})(\mathbf{p}_j \cdot \mathbf{r})}{r^5} \qquad (3)$$

in which the first sum corresponds to the interaction of the dipoles with the external field and the other to the dipole-dipole interaction. In this equation, $\mathbf{p}_i$ is the dipolar moment of dipole i, $\mathbf{E}_i$ is the random external field acting on this dipole, and r the relative position of dipoles i and j.

For a first approach to the problem, we consider a simple model in which all dipoles have the same length and are located at the sites of a simple cubic lattice. This last assumption should not result in a relevant constraint on the model, in analogy with what happens in the case of the Coulomb interaction. An additional assumption is that, for the time being, we consider dipoles with only one degree of freedom, so that all point in the same direction, which we will take to be the z-axis, with two possible orientations.

The energy of a dipole in this model is defined as

$$\epsilon_i = p_i \phi_i + \sum_j{}' p_j I(\mathbf{r}, p_j) \qquad (4)$$

where $p_i = \pm 1$ depending upon the orientation of the dipole at site i (we take the dipole length equal to unity), so that $\mathbf{p}_i = p_i \hat{\mathbf{z}}$, $\mathbf{E}_i = -\phi_i \hat{\mathbf{z}}$, and $I(\mathbf{r}, p_j)$ is the component along the negative z-axis of the electric field produced at site i by the dipole at site j.

The density of states that we want to obtain is the distribution of the "single-dipole" energies ϵ_i in the ground state. The computer simulation goes in this case along the same paths as for the Coulomb interaction: starting from an initial state with the dipoles randomly oriented, for a given set of $\{\phi_i\}$, we let the system undergo transitions which lower the total energy. First, we consider transitions involving the inversion of just a dipole, say at site i, which lower the total energy by an amount

$$\Delta H = -2\epsilon_i \tag{5}$$

In this way, we change the orientation of a single dipole at a time until all energies ϵ_i are negatives. When this is achieve, no single dipole inversion will lower the total energy of the system, according to Eq.(5). Then we consider transitions involving two dipoles. The energy change associated to these transitions is

$$\Delta H = -2\left(\epsilon_i + \epsilon_j - 2p_i I(\mathbf{r}, p_j)\right) \tag{6}$$

Since ΔH must be positive when the system is in the ground state, Eq.(6) results in a constraint on the relative position of dipoles at sites i and j with energies ϵ_i and ϵ_j respectively. This constraint has to be regarded for the system of dipoles as the analogous to that first obtained by Efros and Shklovskii[3] for the system of charges.

If we calculate, on the basis of Eq.(6), the effective excluded volume around a dipole, it results to be inversely proportional to the energy, ϵ, (remember that for the Coulomb interaction it was proportional to ϵ^3), which leads to a constant density of states. A more refined self-consistent calculation gives a density of states with a logarithmic singularity at zero energy[9]

$$q(\epsilon) = \frac{3}{\pi} \ln^{-1}\left(\frac{CB}{\epsilon}\right) \tag{7}$$

where B is the range of the random energies $\{\phi_i\}$, and C is a constant of integration, which is to be adjusted.

In figure 3 we show the results of our numerical simulations for several values of the range, B, of the random energies. The solid lines correspond to the analytical approximation of the self-consistent treatment, given by Eq.(7). The numerical constant C is only adjusted to one curve and no further fitting is done to reproduce the other curves. The agreement is fairly good and gives strong support to the self-consistent treatment.

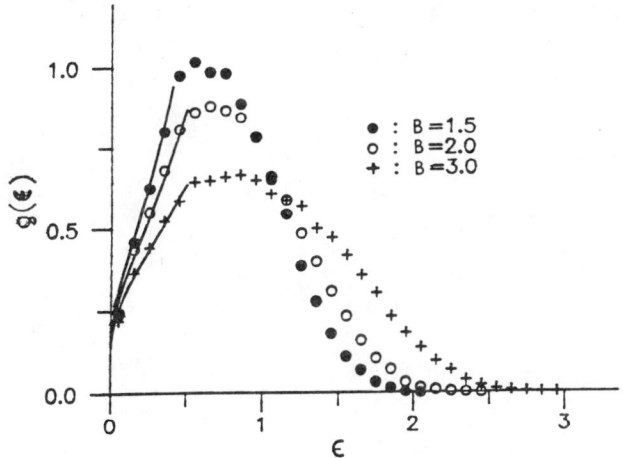

Figure 3 Numerical results for the density of states of a system of interacting dipoles for several values of B. The solid line represents the function $q(\epsilon)$ of Eq.(7).

ACKNOWLEDGEMENTS

We would like to thank M. Pollak and E. Cuevas for collaboration in different stages of this work and J.M.F. Gunn for his help and hospitality in the final stages of the work. We would like to acknowledge financial support from the Dirección General de Política Científica.

REFERENCES

1. M. Pollak and M. Ortuño, in *Electron-Electron Interactions in Disordered Systems*, edited by A.L. Efros and M. Pollak (North-Holland, Amsterdam), 287 (1985).
2. M. Pollak, *Discuss. Faraday Soc.*, **50**, 13 (1970).
3. A.L. Efros and B. Shklovskii *J. Phys. C*, **8** L49 (1975).
4. A.L. Efros *J. Phys. C* **9** 2021 (1976).
5. J.H. Davies, *Phil. Mag. B*, **29**, 511 (1985).
6. M. Ortuño and M. Pollak, *Phil. Mag. B*, **51**, 533 (1985).
7. R. Chicón, M. Ortuño and M. Pollak, *Phys. Rev. B*, **37** 10520 (1988).
8. D. Chowdhury,*Spin Glasses and Other Frustated Systems*, (World Scientific, Singapore), (1986).
9. A.L. Efros and B. Shklovskii, in *Electron-Electron Interactions in Disordered Systems*, edited by A.L. Efros and M. Pollak (North-Holland, Amsterdam), 409 (1985).

STUDIES OF MANY-BODY EFFECTS IN THE COULOMB GAP

M. Mochena and M. Pollak

*Department of Physics, University of California,
Riverside, California 92521, U.S.A.*

ABSTRACT

We describe a new algorithm, devised to study properties of disordered localized systems with Coulomb interactions. In comparison with the algorithm commonly used, the main advantage is that it yields automatically, together with the ground state, also the low lying excited states. In addition, the ground state is not restricted to be stable only with respect to one-particle excitations. The algorithm is expected to be suitable for the study of a variety of phenomena which are currently ill understood. Some of these are specific heat at very low temperatures, transport by correlated many-particle and one-particle hopping excitations, quantum effects and other effects. We present here the computational method, together with some early results on the system density of states, and on the nature of low energy excitations.

1. INTRODUCTION

This study relates to physical properties of disordered localized systems with Coulomb interactions between electrons. When the electronic concentration is small, or the disorder large, electrons are well localized, and the states of the system can be represented by simple configurations, i.e. by specifying which sites are occupied, while allowing at most one electron to occupy a site. The energies of the states are then given by the Hamiltonian

$$H = \sum_i \left[E_i n_i + \sum_j \frac{(q + e\, n_j)(q + e\, n_i)}{R_{ij}} \right] \tag{1}$$

where R is the distance between the two sites, $q = -eK$, $0<K<1$, K being the ratio between the number of electrons and the number of sites in the system, and n is 1 or 0 for sites occupied and unoccupied by an electron, respectively. When quantum effects are important, the Hamiltonian requires the additional term $\sum J_{ij} a_i^\dagger a_j$ (a and $a^\dagger$ destroy and create electrons at the labelled sites and J is the resonance energy between the labelled sites). It produces elastic hopping between configurations. Inelastic hopping is produced by the electron-phonon coupling term, and accounts for the dynamics of the system in the semi-classical limit.

The main aim of the paper is to shed light on the role of correlations on various low temperature properties, a subject long discussed in the literature. Correlations between electrons can be exhibited by phonon induced (inelastic) many-electron hopping excitations (or simultaneous correlations, for brevity), correlations between successive excitations (successive correlations), and, at higher concentrations, quantum effects describable as elastic many-electron hopping. The importance of these correlations can be evaluated when the low energy states of the system, and their energies, are known. For example, if low lying states tend to differ from the ground state by occupation of many sites, many-electron hopping excitations can be expected to affect, or even determine, electronic properties at low temperatures. An evaluation of such properties is possible when the low lying states and their energies are known.

Equilibrium properties can be evaluated from the density of system states alone. Non-equilibrium properties also require the knowledge of transition rates between states, which can be computed when the state configurations are known. The above is true as long as semiclassical conditions can be assumed. But quantum effects can also be evaluated, by computing resonance energies between states.

At present there is no definitive knowledge about the nature of the low lying states, including the ground state, because of formidable analytical and computational difficulties. There are two models of disordered systems with interactions which differ in nature of low energy excitations. According to the Leningrad model[1,2], low energy excitations are dominated by short one-electron transitions; according to our model[3,4], compact many-electron transitions play a major role. In addition, in the Leningrad model, correlations between excitations are not deemed to be particularly important, while in our model they are. These differences between models can have very important implications on low temperature physical properties.

The great difficulty in computational studies of low lying states arises from the existence of a huge number of states, even in relatively small systems. This difficulty is bypassed here by gradually assembling the system from smaller subsystems, beginning with very small ones. As subsystems are combined, the number of states increases very rapidly. But we need to keep at every step only a small fraction of the states, those at very low energy, because the low lying states of the system tend to arise from low lying states of the subsystems. Even though the number of states kept is a minute fraction of the totality of states, the procedure nevertheless requires the use of a supercomputer. In the next section the procedure is explained in more detail.

The so called lattice model has been used here, where sites are placed on a simple lattice, and assigned random energies with a uniform distribution within a range W. The number of electrons is half that of sites. For neutrality of charge, each bare site is assigned a charge $-e/2$. The magnitude of W is made equal to the Coulomb interaction between two electrons on neighboring sites.

2. COMPUTATIONAL METHOD

The main difficulty in a computational study of the nature of low energy excitations and their spectrum, is the huge number of states N even in a relatively small system. A system with N^d (d is the dimensionality) sites and half as many electrons has about $\tilde{N} \approx 2^{N^d}$, so, for N=5 for example, $\tilde{N} \approx 10^{37}$. We circumvent this difficulty by assembling the system gradually from smaller subsystems, beginning with units so small that the energy of all states can be computed easily. The procedure is as follows. We combine two subsystems into a larger system. Specifying the states of the two subsystems specifies the state of the combined system. We can then compute the energies of the states of the combined system, and keep only a certain number of states with low energy. This system is then combined with another system to produce an even larger system, and so forth. It is very important to allow transfer of electrons between two sub-systems when they are combined. This is done by combining subsystems with n_1-i electrons and with n_2+i electrons, for different i's, when the desired number of electrons in the combined system is to be n_1+n_2.

There are, of course, many possibilities how to build up the large unit, and the detailed procedure can be chosen to minimize computer time, or to fit a particular study undertaken. For example, one can build a block from cubes, add monolayer surfaces, or add small clusters of atoms to a surface. Generally, a build up from a smaller units will require fewer states of each step, but more steps. In this particular study we constructed a 4x4x4 cube from 2x2x2 cubes, and then proceeded to a 5x5x5 cube.

In the specific procedure taken, the initial step consisted of 2x2x2 blocks with n ranging from two to six. The energies of all possible configurations were computed and saved in ascending order together with the corresponding configurations. Next, we combined two different 2x2x2 blocks to form 2x4x2 blocks with 6≤n≤10. Blocks with n outside this range were found to have energies outside the needed range. (Our experience has been that the lowest energy states come from subsystems whose n differ by no more than 2 from the K=0.5 value). To make a 4x4x2 we used the 1000 lowest states for each

4x4x2, and for each n. We used also 1000 lowest states for each 4x4x2 and for each n to produce a 4x4x4. We retained 2100 states for the 4x4x4. It is important to note here that as the size of the system gets larger the states become much denser. Therefore, inclusion of more states for large systems is necessary to get properly a ground state, as well as the low lying states. At 4x4x4 we modified the build up process and combined a block with a layer. A 4x4x5 is constructed from a 4x4x4 block and a 4x4 layer. While a large number of states is required for a bigger system, a much smaller number is required for the layer if we wish to include states within a similar energy span. It turns out that 62 states for a 4x4 match 2100 states of a 4x4x4. In the next step we combined a 4x5x4, keeping 2800 states, with a 5x4, keeping 70 states. Finally we combined a 5x5x4, keeping 3500 states, with a 5x5, keeping 70 states. We could keep only the 3500 lowest states of the 5x5x5 due to memory shortage at present.

3. RESULTS

The energies of the lowest 3500 states were binned into bands of .05 W to get the density of states (DOS), fig 3.1. No smoothing procedure has been used in figs 3.1 to 3.3. The system DOS is all the information needed to calculate thermodynamic properties such as the specific heat.

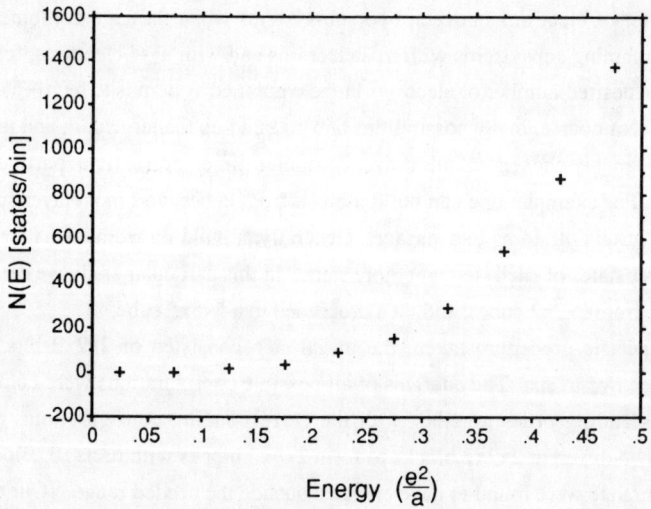

FIG 3.1. The Density of states for a 5x5x5 system. The bin width is .05 e^2/a where a is the lattice constant.

In a microcanonical representation, the energy E of a system state corresponds to the internal energy, while the logarithm of the DOS is proportional to the entropy S. The plot of S vs E in fig. 3.2 is the fundamental thermodynamic equation (see ref. 5) for the system. As a result of assigning the ground state to the first bin, fig 3.2 is unrealistic near E is eqal to zero.

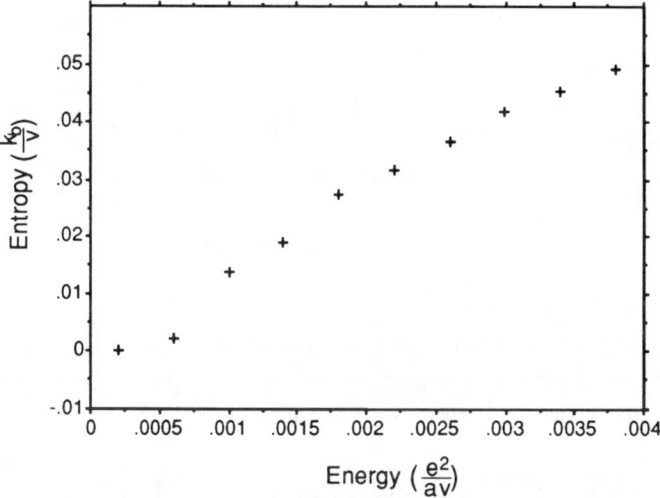

FIG 3.2. The entropy of a system as a function of energy for a 5x5x5. v is the volume of the system and k_b is the Boltzman constant.

The temperature, $T = \partial E/\partial S$, is computed from fig 3.2. In contrast to the first two plots the points are scattered. More statstics is needed to to get a smoother plot. Therefore, a meaningful evaluation of $C_V = \partial E/\partial T$, should await results of DOS on larger systems, but we point out that, in contrast to the work of ref. 4, the method is suitable for computing C_V at very low temperatures.

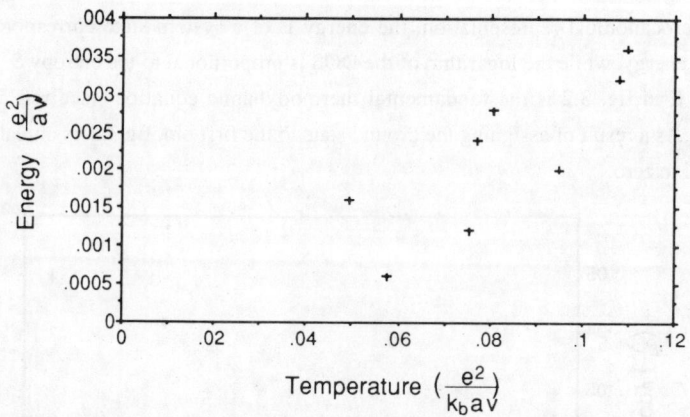

Fig. 3.3. The internal energy as as a function of temperature.

We now turn to the nature of low energy excitations. To eliminate surface effects, we consider first excitations confined to the twenty seven interior sites. The excitations found are all short one-electron and compact two-electron excitations. Furthermore, all the two-electron excitations are uncorrelated or sequential as defined in ref. 4.

EXCITATION ENERGY	TYPE OF EXCITATION	HOPPING DISTANCE
0.0298	1 e	1.4
0.1106	1 e	1.4
0.1176	1 e	1
0.1607	1 e	1.4
0.1853	1 e	1
0.1984	1 e	1
0.2103	1 e	1
0.2143	1 e	1
0.2282	2 e (U)	1, 1.4
0.2298	1 e	1
0.2396	1 e	1.4
0.2650	2 e (U)	1, 1
0.2876	2 e (S)	1, 1
0.3145	2 e (S)	1, 1.4
0.3403	2 e (U)	1, 1.4
0.3486	1 e	1.4
0.4134	2 e (U)	1, 1
0.4715	1 e	1

TABLE 3.1 The low lying excitations confined within 3x3x3. (S) stands for sequentially correlated excitations and (U) for uncorrelated ones. The energy is measured in e^2/a and the distance in units of lattice constant a.

Studies of Many-Body Effects in the Coulomb Gap   135

The low lying excitations not confined to the interior sites show a different character, as shown in table 3.2. The many-electron excitations are of a reasonably complicated nature-consisting of sequentially correlated simultaneous excitations.

EXCITATION ENERGY	TYPE OF EXCITATION	HOPPING DISTANCE
0.0298	1 e	1.4
0.0430	1 e	1
0.0657	3 e	1, 1.4, 1.4
0.0780	2 e	1, 1
0.0817	5 e	1, 1, 1, 1, 1
0.0848	4 e	1, 1, 1, 1
0.1023	3 e	1, 1, 1.4
0.1049	1 e	2.8
0.1051	5 e	1, 1, 1, 1, 1.4
0.1106	1 e	1.4
0.1108	3 e	1, 1, 1.4
0.1148	5 e	1, 1, 1, 1.4, 1.4
0.1176	1 e	1
0.1206	2 e	1, 2.8
0.1213	4 e	1, 1, 1, 1.4
0.1214	3 e	1, 1, 1
0.1219	4 e	1, 1, 1, 1.4
0.1226	4 e	1, 1, 1, 1
0.1267	2 e	1, 1.4
0.1270	4 e	1, 1, 1, 1.4
0.1409	3 e	1.4, 1.4, 1.4
0.1418	1 e	1
0.1425	3 e	1, 1, 1.4
0.1524	1 e	1
0.1570	2 e	1, 1.4
0.1572	7 e	1, 1, 1, 1, 1, 1.4, 1.4
0.1607	1 e	1.4

TABLE 3.2 The low lying excitations of a 5x5x5. The energy is measured in e^2/a and the distance in units of lattice constant a.

4. SUMMARY AND CONCLUSIONS

Though a 5x5x5 is a relatively small system it still sheds some light on the low energy properties of disordered systems with Coulomb interactions. The work for 5x5x5 is done on Cray X-MP/48 which has a memory limit of 4 mega words. This memory limit restricted the number of low lying states we wanted to look at the last stages. Future work will continue on Cray-2 and many more low lying states will be included. The next feasible system within reach is an 8x8x8 block.

One interesting result is that, so far, our ground state was never lower than that obtained with the Leningrad relaxation algorithm (though this relaxation process had to be repeated many times before the lowest ground state was reached). Though it is possible that this result may change when larger systems are explored, we find it remarkable how well the Leningrad algorithm for the ground state works.

The result from the confined excitations is clearly more in keeping with the Leningrad model than with our model. But it is in disagreement with results of ref.4, where, for similar conditions, simultaneous correlations were found to be important. We have no clear cut explanation for this contradiction. A possible reason is that the energy of the excitations explored here is rather lower than the energy of those found in ref.4.

The difference between the confined and unconfined excitations can come from the presence of edge effects in the latter, and from the size of the system. One can expect both the presence of edge effects and the larger size to enhance many-electron excitations. Studies on larger systems should resolve the relative importance of the two effects.

We finally wish to point out that, to our knowledge, this is the first definitive study of low lying excitations in localized disordered systems with Coulomb interactions. The method can, of course, be adopted to other disordered interacting systems, e.g. spin glasses.

5. ACKNOWLEDGEMENTS

One of us (MP) would like to acknowledge helpful discussions on the subject of the paper with Professors A. Efros and B. Shklovskii and with several other members of their group, and their lavish hospitality at the A. F.Yoffe Institute, and also to thank the USA-USSR Academies of Science exchange program for making the visit there possible. We would like to thank Mr. Mark Green for valuble suggestions on the computing aspect of the work.

This work is in part supported by San Diego Supercomputer Center.

6. REFERENCES

[1] A. L. Efros and B. I. Shklovskii, Electron-Electron Interactions in Disordered Solids; A. L. Efros and M. Pollak eds., North Holland, 1985, p. 409
[2] E. I. Levin, V. L. Nguyen, B. I. Shklovskii, and A. L. Efros, Zh. Exp. Teor. Fiz. **92**,1499 (1987) [Sov. Phys. JETP **65**,842,1987]
[3] M. Pollak, M. Ortuño, Electron-Electron Interactions in Disordered Solids; A. L. Efros and M. Pollak eds., North Holland, 1985, p. 2287
[4] B. Hadley, M. Pollak, M. Ortuño, Phys. Rev. **37**, 9006 (1988)
[5] H. B. Callen, Thermodynamics and An Introduction to Thermostatistics, Second Edition, John Wiley & Sons, 1985

Chapter 2

QUANTUM INTERFERENCE EFFECTS

INTERFERENCE PHENOMENA IN VARIABLE RANGE HOPPING CONDUCTIVITY

B.I. Shklovskii

A.F. Ioffe Physical-Technical Institute
194021 Leningrad, USSR*

B.Z. Spivak

Science and Technology Corporation
Academy of Science of USSR, Leningrad

1. INTRODUCTION

In recent years new kinetic interference phenomena have been observed in disordered metals. These include the negative magnetoresistance, the Aharonov-Bohm oscillations with "normal" and "superconducting" magnetic flux quanta, and mesoscopic conductance fluctuations observed in small metallic samples as the magnetic filed increases (Altshuler and Aronov 1985, Aronov and Sharvin 1987, Washburn and Webb 1986). All these phenomena result from interference between different Feynman diffusive paths, which bring an electron to the point of observation. These phenomena are observable only at low temperatures, where the inelastic collision frequency is so low that an electron undergoes many elastic collisions without loss of phase coherence.

In this paper we discuss the occurrence of similar interference phenomena for variable range hopping (VRH) conduction. We show below that the probability of one long distant hop is determined by the interference of many paths of the tunneling electron which include scattering processes. This interference is constructive when all scattering amplitudes are positive or destructive when some of them are negative. In the former case a positive but anomalous magnetoresistance occurs, in the latter, the effect depends on the sample size. In small samples, where only one or a small number of hops determine the

---

*Present address: School of Physics and Astronomy, University of Minnesota, Minneapolis, MN 55455

conductance, it oscillates periodically as the magnetic field increases (mesoscopic oscillations). In macroscopic samples these oscillations average out, and a negative magnetoresistance results.

In a large two-dimensional periodic array of loops a transverse magnetic field leads to Aharonov-Bohm conductance oscillations with "normal" (hc/e) or "superconducting" (hc/2e) flux quanta depending on the concentration of scattering centers that have negative scattering amplitude.

In this paper we outline only briefly the main ideas concerning scattering and interference in VRH conduction. A detailed discussion of the field is given in our forthcoming review article (Shklovskii and Spivak 1990).

2. THE BASIC THEORETICAL CONCEPTS.

Variable range hopping conductivity is observed at low temperatures in systems with localized electron states. The concept of variables range hopping (VRH) was introduced by Mott (1968). His arguments were based on the expression for the conductance G_{12} in the Miller-Abrahams (MA) network which is related to the hop between impurity centers 1 and 2 (see Fig. 1)

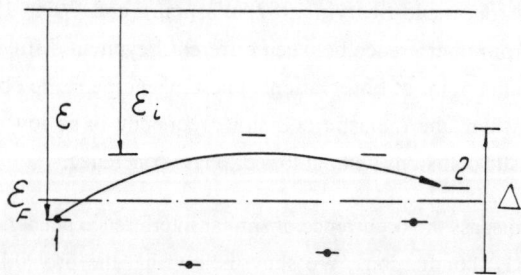

Fig. 1   Levels of impurities located near a straight line connecting centers 1 and 2. The dashed line is the Fermi level ε_F and the solid straight line is the bottom of the conduction band. The transition from center 1 to center 2 is shown by the arrow.

Each term of the series (10) can be interpreted as a contribution ψ_Γ of a zigzag path Γ, which connects points 1 and 2 through the points where scattering occurs. Examples of some of such paths are shown in Fig. 2. An important difference between the series (10) and similar series for free electron multiple scattering in metals is that the spherical scattered waves decay exponentially with increasing distance due to the negative sign of the tunneling electron energy. Therefore only the shortest paths contribute to $\psi_1(\vec{r}_2)$. Along these paths there are no returns, and the electron is scattered only in the forward direction. All such directed paths are concentrated inside the cigar-shaped domain shown in Fig. 2.

We wish to emphasize that all terms of the series (10) are real and in general can have either sign, because the scattering length μ_i can be negative. The wave function $\psi_1(\vec{r}_2)$ obviously depends not only on $\varepsilon_1, \varepsilon_2$ and $r \equiv r_{12}$ but also on the random coordinates and energies of all impurities that are located in the cigar-shaped domain of Fig. 2. One may call these "hidden variables." In this sense, $\psi_1(\vec{r}_2)$ may be considered as random at a given r_{12} and ε_{12}. The wave function ψ can be positive or negative. It was shown by Nguyen et al. (1985 a,b) that with increasing fraction x of scatterers with negative μ_i a second-order phase transition occurs at some value $x=x_c$. This phase transition can be described by saying that at small x the quantity ψ is more often positive than negative (fixed sign phase) whereas at the phase transition both signs have the same probability (random sign phase). One can ask why we should care about the sign of the wave

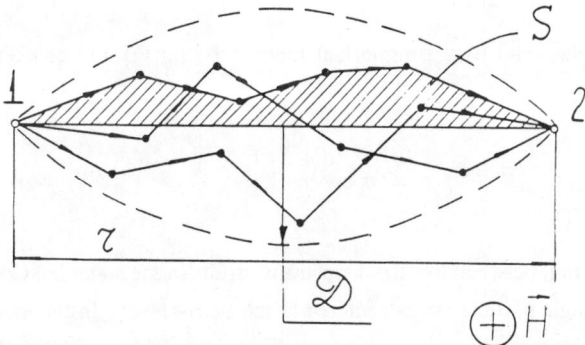

Fig. 2   Three different zigzag paths contributing to the probability amplitude of a hop from center 1 to center 2. The cigar-shaped region containing all important paths is shown by broken lines. Area S attributed to one path is shaded. Scatterers are shown by black circles.

$$G_{12} \propto |\psi_1(r_2)|^2 \exp\left(-\frac{\varepsilon_{12}}{kT}\right) \propto \exp\left(-\frac{2r_{12}}{a} - \frac{\varepsilon_{12}}{kT}\right) \quad (1)$$

where

$$\psi_1(r_2) \propto \exp\left(-\frac{2r_{12}}{a}\right) \quad (2)$$

is the wave function of impurity 1 at point 2. The second exponential factor in Eq. (1) is related to the probability of phonon absorption and to the occupation numbers of impurities 1 and 2. In expression (2) r_{12} is the distance between centers 1 and 2, a is the localization length,

$$\varepsilon_{12} = (1/2)(|\varepsilon_1 - \varepsilon_F| + |\varepsilon_2 - \varepsilon_F| + |\varepsilon_1 - \varepsilon_2|) \quad (3)$$

and ε_1, ε_2 are the energies of impurities 1 and 2, (see for example Shklovskii and Efros 1984). We consider impurities with concentration N and energy levels randomly distributed in a band of width Δ, centered at the Fermi level ε_F. Then the typical value of ε_{12} equals $\Delta_M = \Delta/Nr_{12}^3$. Substituting this value into Eq. (1) one finds that the distance r_{12} and the energy Δ_M that maximize the exponent of the conductivity are equal to

$$r_{12} \approx a(T_0/T)^{1/4} \quad \text{and} \quad \Delta_M \approx T_0^{1/4} T^{3/4} \quad (4)$$

where $T_0 = \beta \Delta / Na^3$ and β is a numerical factor. Using Eq. (1) one obtains for the conductivity

$$\sigma = \sigma_0 \exp\left\{-\left(\frac{T_0}{T}\right)^{1/4}\right\} \quad (5)$$

Coulomb interaction between localized electrons results in the so-called Coulomb gap in the density of single particle states centered at the Fermi level. In the low temperature limit this gap yields a conductivity of the form (Efros and Shklovskii 1975, 1985)

$$\sigma = \sigma_0 \exp\left\{-\left(\frac{T_0}{T}\right)^{1/2}\right\} \quad (6)$$

We now discuss the role of interference in VRH conduction. Each electron hop is accompanied by phonon emission or absorption. Consequently, the electron loses its phase after each hop, and there is no interference between successive hops. But the probability of each hop is determined by scattering and interference effects in the course of electron tunneling. Scattering is due to the fact that in variable range hopping the typical hop distance r_{12} exceeds the average interimpurity distance $N^{-1/3}$. Between impurities 1 and 2 there are many others that affect the probability of electron tunneling from 1 to 2 (see Fig. 1). If there were no other impurities between 1 and 2, the wave function $\psi_1(\vec{r}_2)$ would take the form

$$\psi_1^0(\vec{r}_2) \propto \frac{1}{|\vec{r}_2 - \vec{r}_1|} \exp\left(-\frac{|\vec{r}_2 - \vec{r}_1|}{a}\right) \tag{7}$$

One can say that this wave function describes an electron tunneling from point to point through a potential barrier of height $|\varepsilon_1|$. Each impurity in the barrier produces scattering. Scattering by impurity i results in a spherical wave of the form (Lifshitz and Kirpichenkov 1979)

$$\psi_{scat} = \psi_1^0(\vec{r}_i) \frac{\mu_i}{4\pi |\vec{r} - \vec{r}_i|} \exp\left\{\frac{-|\vec{r} - \vec{r}_i|}{a}\right\} \tag{8}$$

where

$$\mu_i \equiv \frac{8\pi a \, \varepsilon_i}{\varepsilon_1 - \varepsilon_i} \tag{9}$$

is the scattering amplitude or, in other words, the scattering length of impurity i, and $\varepsilon_i < 0$ is the energy of the impurity state i. An electron can be scattered once, twice or many times on its way from 1 to 2. All these waves along with the non-scattered one contribute additionally to the wave function $\psi_1(\vec{r}_2)$ at point $\vec{r}_2$

$$\psi_1(\vec{r}_2) = \psi_1^0(\vec{r}_2) + \sum_i \frac{\mu_i \psi_1^0(\vec{r}_i)}{|\vec{r}_2 - \vec{r}_i|} \exp\left\{-\frac{|\vec{r}_i - \vec{r}_2|}{a}\right\} +$$
$$\sum_{ij} \psi_1^0(\vec{r}_i) \frac{\mu_i}{|\vec{r}_j - \vec{r}_i|} \exp\left\{-\frac{|\vec{r}_j - \vec{r}_i|}{a}\right\} \frac{\mu_j}{|\vec{r}_j - \vec{r}_2|} \exp\left\{-\frac{|\vec{r}_j - \vec{r}_2|}{a}\right\} + \ldots$$
$$= \sum_{\{\Gamma\}} \psi_\Gamma(\vec{r}_2)$$

$$\tag{10}$$

function which is usually considered meaningless. We shall show in section 5 that in a magnetic field where the most interesting effects of scattering arise the distribution of the sign of the wave function plays an important role.

3. MESOSCOPIC OSCILLATIONS

Assume that a uniform magnetic field $\vec{H}$ is applied in the direction normal to $\vec{r}_{12}$. A magnetic field weak enough, that its effect on the single spherical wave between two scattering centers can be neglected, results only in a phase factor $e^{i\varphi_r}$ for each path Γ

$$\psi = \sum_{\{\Gamma\}} \psi_\Gamma e^{i\varphi_\Gamma} \tag{11}$$

where ψ stands for $\psi_1(r_2)$, $\varphi_\Gamma = 2\pi HS_\Gamma/\Phi_0$, $\Phi_0 = hc/e$ is the flux quantum and S_Γ is the area of the figure formed by the zigzag path Γ and the straight line connecting points 1 and 2. The transverse dimension D of the cigar-shape region of Fig. 2 can be estimated as $\sqrt{ra}$ so that the typical S_Γ is of the order of $S = r\sqrt{ra}$ and the typical phase φ_Γ is $\varphi \approx Hr\sqrt{ra}$. When φ reaches a value of the order of π there will be a substantial change in each term and the entire sum will change in random direction by the order of its own value. Therefore the probability of the hop 1-2 will oscillate with H aperiodically with a relative amplitude of the order of 100 percent. The characteristic "period" of oscillations along the H scale is of the order of

$$H_c = \frac{\pi ch}{r^{3/2} d^{1/2} e} \tag{12}$$

In macroscopic samples such oscillations should be averaged out. But in small samples exhibiting variable range hopping, the conductance can be determined by one or a small number of hops (Fowler et al. 1982, Lee 1984). Then the amplitude of conductance oscillations may be substantial. Such samples and oscillations are called mesoscopic. For variable range hopping the mesoscopic oscillations were predicted by Nguyen et al. (1986) and observed by Orlov and Savchenko (1986) and Laiko et al. (1987) and Ovadyahu (this volume) (see also Wainer et al. 1988).

4. NEGATIVE MAGNETORESISTANCE IN VRH CONDUCTION

As we mentioned above, in large macroscopic samples random oscillations of individual conductances are averaged out. Then the question arises whether some monotonic dependence of the conductivity σ on the magnetic field remains. It was shown by Nguyen et al. ((1985C), Sivan et al. (1988) and Entin-Wohlman et al. (1989) that a monotonic effect exists and that it corresponds to a negative magnetoresistance. To obtain this result we shall discuss the averaging procedure of the elements of the MA network which yield the magnetoresistance. Let us discuss the effective conductance G_{eff} which should replace all conductances of the MA network to obtain the same macroscopic conductivity (σ obviously is proportional to G_{eff}). If the MA network were a parallel curcuit, G_{eff} would be equal to $<G>$. On the other hand for a series circuit $G_{eff}=<G^{-1}>^{-1}$. The latter averaging obviously stresses the role of very small G elements. Actually the MA network consists of parallel and series conductances. So some intermediate kind of averaging should be valid for it. Shklovskii and Efros (1984) have shown that for a strongly inhomogenous MA network whose conductivity can be found by a percolation approach one should use logarithmic averaging over "hidden variables" to obtain the magnetoresistance

$$\frac{\sigma(H)}{\sigma(0)} = \frac{G_{eff}(H)}{G_{eff}(0)} = \exp\left\{\left\langle \ln \frac{G(H)}{G(0)} \right\rangle\right\}$$

$$= \exp\left\{\left\langle \ln \left|\frac{\sum_r \psi_r e^{i\varphi_r}}{\sum_r \psi_r}\right|^2 \right\rangle\right\} = \exp\left\{\left\langle \ln \frac{\left(\sum_r \psi_r \cos \varphi_r\right)^2 + \left(\sum_r \psi_r \sin \varphi_r\right)^2}{\left(\sum_r \psi_r\right)^2} \right\rangle\right\}$$

(13)

It can be seen from Eq. (13) that a small magnetic field (φ<<1) affects significantly only cases with small $\sum_r \psi_r$ values. One can say that in a magnetic field it is much less probable to find small G values than without field because both values of $\sum_r \psi_r \cos \varphi_r$ and $\sum_r \psi_r \sin \varphi_r$ must be small for G to be small. Small values of G(0), $(G \leq \varphi^2 \langle G \rangle)$ for which $\ln G(H)/G(0) \approx 1$ contribute to Eq. (13) more than typical values of G, and the sign of this contribution corresponds to a negative magnetoresistance. We shall not discuss the magnetoresistance in further detail (see Shklovskii and Spivak 1990) but we would like to stress two causes for a negative magnetoresistance:

1) Magnetic field induced change of the symmetry of wave functions which become complex instead of real.

2) The presence of series circuits in a strongly inhomogeneous MA network which leads to nonlinear averaging in place of the linear averaging procedure $G_{eff}=<G>$, appropriate for parallel circuits or almost homogeneous systems.

A negative magnetoresistance has recently been observed by several groups (Ovadyahu and Imry 1983, Ovadyahu 1986, Faran and Ovadyahu 1988, Orlov and Savchenko 1986 Laiko et al. 1987, Qiu-Yi Ye et al. 1989, Tremblay et al. 1989). Papers of some of these groups are presented in this volume.

5. AHARONOV-BOHM OSCILLATIONS

It is well known that the resistance of a hollow thin-walled cylinder of a pure metal oscillates as a function of the magnetic flux Φ penetrating the cylinder with a period Φ_0=ch/e called the normal flux quantum (Kulik 1967). It was shown by Altshuler et al. (1981) that the resistance of such a cylinder made of a dirty normal metal should oscillate with a period equal to the "superconducting" flux quantum $\Phi_0/2$. Such oscillations were first observed by Sharvin and Sharvin (1981) and later studied by many other groups mainly using arrays of loops (Aronov and Sharvin 1987, Washburn and Webb 1986) Nguyen et al. (1985 a,b) showed that Aharonov-Bohm oscillations with a period of either Φ_0 or $\Phi_0/2$ can also occur in a cylinder or array of loops made from a material exhibiting variable range hopping conduction. The transition between the two periods, as the concentration of scatterers is changed, is a consequence of the sign phase transition. Aharonov-Bohm oscillations in variable range hopping conduction were discovered by Poyarkov et al. (1986). They studied the resistance of a two-dimensional array of 35 x 6 loops in a transverse magnetic field H. A small part of the array is shown in Fig. 3. This array was made of a thin film of disordered PbTe which exhibits VRH. The temperature dependence of the conductivity has the form $\sigma = \sigma_0 \exp\{-(T_0/T)^{1/3}\}$ in agreement with Mott's predictions for two dimensional systems. The resistance of the array was found to oscillate with increasing magnetic field as shown in Fig. 4. The period of oscillations H_0 = 7 Oe is in rough agreement with $H_0 = \Phi_0/2S$ where S is the area of one loop. In other words the oscillations have the superconducting period $\Phi_0/2$. It is seen in Fig. 4 that the magnetoresistance is negative. Considering that the resistances are as high as about

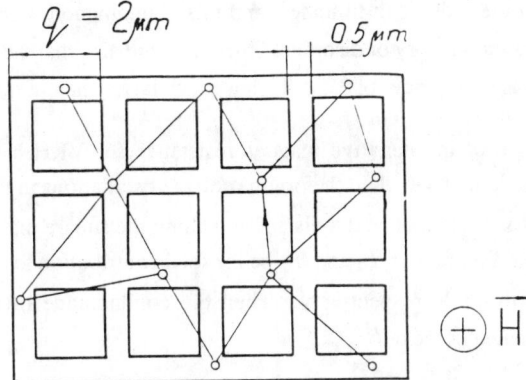

Fig. 3 Part of the two-dimensional array of loops used by Poyarkov et al. (1986) to study Aharonov-Bohm effect. Localized states contributing to VRH at a given temperature are shown by open circles.

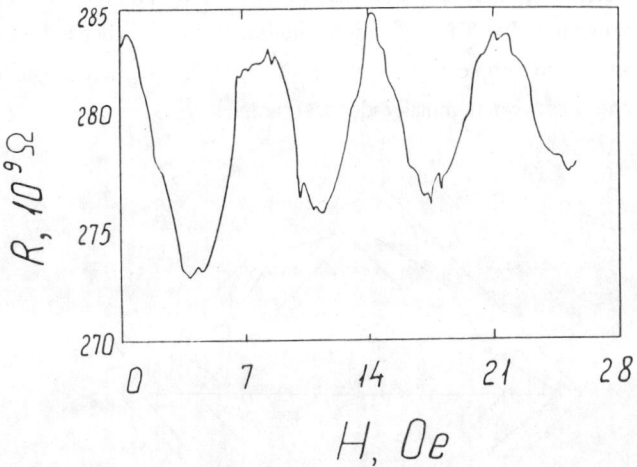

Fig. 4. Array of loop resistance R as a function of transverse magnetic field H at T=1.12K. (From Poyarkov et al. 1986).

3×10^{11} Ohm it is difficult to understand this phenomenon within the framework of the weak localization theory of disordered metals. Note also that the relative amplitude of the oscillations is of the order of 3%, which is much larger than in metallic samples.

The origin of the negative magnetoresistance for VRH has been discussed in the previous section. To explain the origin of $\Phi_0/2$ oscillations in the random sign phase let us consider a single loop and a distant hop along the square diagonal from the point 1 to the point 2 in Fig. 5. Let Ψ_1 and Ψ_2 be the total contribution to Ψ of paths passing above and below the hole respectively. Then the conductance of this loop becomes in a transverse magnetic field $\vec{H}$

$$G_{12}(H) \propto |\psi_1 + \psi_2 e^{i\phi}|^2 = \psi_1^2 + \psi_2^2 + 2\psi_1\psi_2 \cos\phi, \qquad (14)$$

where $\varphi = 2\pi HS/\Phi_0$ is the phase difference between the two "beams" caused by the magnetic flux. Here we neglect the finite width of the strip and therefore the phase difference between two Feynman paths on the same side of the hole. It is seen from Eq. (14) that the difference $G_{12}(H) - G_{12}(0)$ is periodic in the magnetic flux, the period being the normal flux quantum Φ_0. The oscillation amplitude and phase depend on ψ_1 and ψ_2 and vary from sample to sample because of the difference in the microscopic location of the scattering centers between nominally identical samples.

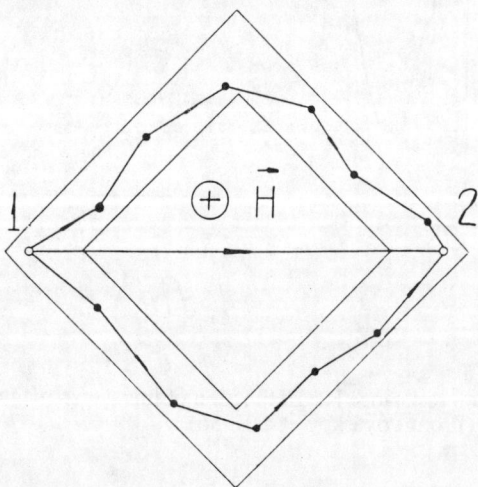

Fig. 5   Single loop with hop along its diagonal. Two paths which contribute to ψ_1 and ψ_2 are shown.

Let us turn now to a macroscopic two-dimensional array of loops and consider the simplest case $\varphi = \pi$ that is $e^{i\varphi}=-1$. Then the effect of a magnetic field is equivalent to a change of sign of ψ_2. In the random sign phase where the wave function ψ_2 can be positive or negative with the same probability 1/2, a change of sign of ψ_2 is equivalent to the replacement of one random realization of the array by another. This should not alter the conductivity of a macroscopic array. So the conductivity at $\varphi=0$ and $\varphi=\pi$ should have the same value. This means that the period of the Aharonov-Bohm oscillations should be equal to $\Phi_0/2$ in agreement with the experimental data of Poyarkov et al. (1986).

1. Al'tshuler B.L., A.G. Aronov, B.Z. Spivak 1981, Pis'ma v Zh. Eksp. Teor. Fiz. 33, 101 - (Sov. Phys. -JETP Lett. 34, 272).

2. Aronov A.G., Ju.V. Sharvin, 1987, Rev. Mod. Phys. 59, 755.

3. Efros A.L., B.I. Shklovskii, 1975, J. Phys. C8, L49.

4. Efros A.L., B.I. Shklovskii, 1985, in: Electron-electron interaction in disordered systems, ed. A.L. Efros fand M. Pollak, (North Holland, Amsterdam).

5. Entin-Wohlman O., Y. Imry, U. Sivan, 1989, in press.

6. Faran O., Z. Ovadyahu, 1988, Phys. Rev. B38, 5457.

7. Fowler A.B., A. Harstein, R.A. Webb, 1982, Phys. Rev. Lett. 48, 196.

8. Kulik I.O. 1967, Pis'ma v Zh. Eksp. Teor, Fiz. 5, 423; Sov. Phys. JETP Lett. 5, 345.

9. Laiko E.I., A.O. Orlov, A.K. Savchenko, E.A. Ilyichev, E.A. Poltoratsky 1987, Zh. Eksp. Teor. Fiz. 93, 2204, Sov. Phys.-JETP, 66, 1264.

10. Lee P.A., 1984, Phys. Rev. Lett., 53, 2042.

11. Lifshitz I.M. and V.Ya. Kirpichenkov 1979, Zh. Eksp. Teor. Fiz. 77, 989; Sov. Phys.-JETP 50, 499.

12. Mott N.F. 1968, J. Non-Crystalline Solids 1, 1.

13. Nguyen V.L., B.Z. Spivak and B.I. Shklovskii, 1985 a, Pis'ma v Zh. Eksp. Teor. Fiz. 41, 35, Sov. Phys.-JETP Lett. 41, 43.

14. Nguyen V.L., B.Z.Spivak and B.I. Shklovskii 1985 b, Zh. Eksp. Teor. Fiz. 89, 1770; Sov. Phys.-JETP 62, 1021.

15. Nguyen V. L., B.Z. Spivak and B.I. Shklovskii 1986, Pis'ma v Zh. Eksp. Teor. Fiz. 43, 35; Sov. Phys.-JETP Lett. 43, 44.

16. Orlov A.O. and A.K. Savchenko 1986, Pis'ma v Zh. Eksp. Teor. Fiz. 44, 34. Ovadyahu Z and Y. Imry 1983. J. Phys. C. 16 L 471.

17. Ovadyahu Z. 1986, Phys. Rev. B 33, 6552.

18. Poyarkov Ya.B., V.Ya. Kontarev, I.P, Krylov, Yu. V. Sharvin 1986, Pis'ma Zh. Eksp. Teor. Fiz. 44, 291; Sov. Phys.-JETP Letters 44, 373.

19. Qiu-Yi Ye, A. Zrenner, F. Koch, K. Ploog 1989, Semicond. Science and Technology 4, 500.

20. Shklovskii B.I. and A.L. Efros 1984, Electronic Properties of Doped Semiconductors. (Springer, Heidelberg).

21. Shklovskii B.I. and B.Z. Spivak 1990 in: Hopping conduction in Semiconductors, M. Pollak, B. Shklovskii, ed. (North Holland, Amsterdam).

22. Sivan, U.,C. Entin-Wohlman and Y. Imry 1988, Phys. Rev. Lett. 60, 1566.

23. Tremblay F., M. Pepper, R. Newbury, D.A. Ritchie, D.C. Peacock, J.E.F. Frost, G.A.C. Jones, 1989, Phys. Rev. B40, 2297 and 5131.

24. Wainer J.J., A.B. Fowler and R.A. Webb 1988, Proc. VII Conference on Electrons Properties of 2-d Systems.

25. Washburn S. and R.A. Webb 1986, Adv. Phys. 35, 375.

ORBITAL MAGNETOCONDUCTANCE IN THE VARIABLE RANGE HOPPING REGIME - PERCOLATION APPROACH

U. Sivan[a,b], O. Entin - Wohlman[b],
and Y. Imry[c]

Abstract

The orbital magnetoconductance (MC) in the variable-range-hopping (VRH) regime is evaluated using a model proposed by Nguyen, Spivak and Shklovskii (NSS), which approximately takes into account the interference among random paths in the hopping process. The MC is obtained using the critical percolating resistor method which is proven to be equivalent to a modified logarithmic averaging. The behavior of the MC is analyzed in detail neglecting backscattering. The small field MC is quadratic in H, it is positive deep in the VRH regime and changes sign when the zero field conductivity is high enough. Very deep in the VRH regime a quasi-linear intermediate field dependence develops. The calculated MC is always positive for strong fields and is predicted to saturate at sufficiently large fields. This behavior and the relevant magnetic field scale are in agreement with recent experiments. The results are argued to be far more general than the model used. They follow from the symmetry crossover of the Hamiltonian from the random orthogonal to the random unitary ensemble. Specific predictions concerning the role of spin - orbit coupling are given.

(a) IBM Thomas J. Watson Research Center, Yorktown Heights, NY 10598, U.S.A.
(b) School of Physics and Astronomy, Raymond and Beverley Sackler Faculty of Exact Sciences, Tel Aviv University, Tel Aviv 69978, Israel.
(c) Dept. of Physics, Weizmann Institute of Science, Rehovot 76100, Israel.

1. INTRODUCTION

The study of the magnetoconductance (MC) in disordered metals in the weak localization regime has given valuable insights on the interference processes in such systems. Various electronic relaxation times have also been determined using this method. In systems of mesoscopic sizes the sample-specific interference is very important, but is expected to average-out in a macroscopic network of such systems. There is no detailed understanding of the magnetotransport in the strongly localized regime. Recent studies [1,2,3] have focused on the magnetotransport in the mesoscopic range where the finite size of the sample is relevant. It is however also of interest to study the MC in a large, macroscopic, sample in the regime where thermal hopping dominates the transport. It turns out that the analysis does produce a definite macroscopic effect due to the fluctuations in the bond conductances.

A recent experimental study[4] of transport properties of indium oxide samples in the variable range hopping regime (VRH) reveals a positive MC. In the absence of a magnetic field, the conductance of these specimens[4] obeys Mott's VRH law, $\sigma = \sigma_0 exp[-(T_0/T)^{1/(d+1)}]$, in two and three dimensions (d=2,3) with $200 \leq T_0/T \leq 1000$. The hopping distance R_M extracted from the data is typically of the order of several ξ, where ξ is the localization length. The behavior of the MC at low fields is as follows: after an initial, fast dependence (perhaps H^4) at extremely small fields, the MC becomes quadratic in the magnetic field for a rather large range of the latter. Studies of the dependence of the MC on the magnetic field orientation relative to that of the film strongly indicate that it results from an orbital, rather than a spin effect. The change in the conductance due to the magnetic field is characterized by the flux $\Phi_M \equiv H R_M^{3/2} \xi^{1/2}$ through an effective area of the order of $R_M^{3/2} \xi^{1/2}$. More recently[5] large, mostly positive, MC was found by Assadullaev and Ciric in amorphous Si_3N_4, but with large measurement voltages and by Benzaquen et al for GaAs. The positiveness of the MC and the anisotropy with respect to the magnetic field orientation were also observed in earlier measurements on 2D Si-inversion layers[6].

There have been theoretical approaches to the MC in the VRH regime but they could not fully account for these experimental data. Shklovskii and Efros[7] and Suprapto and Butcher[8] predict a negative MC, due to the shrinkage of the wavefunctions in the presence of a magnetic field. Nguyen, Spivak and Shklovskii (NSS)[9,10] are the first to consider in this connection the effect of the interference among the various paths associated with the hopping between two sites at a distance R_M apart and a small energy separation of the order of $k_B T^{3/4} T_0^{1/4}$ in three dimensions. They find that the interference between all possible paths within a cigar shaped domain[11] of length R_M and width $\sqrt{R_M \xi}$ might change considerably the hopping probability between two sites. Averaging numerically the logarithm of the conductivity over many random impurity realizations, in the presence of a magnetic field, NSS obtain under certain conditions a positive MC which is *linear* in the field in the whole relevant field range.

A theory for the orbital MC in the VRH regime, based on the NSS model, but employing the critical percolation path picture[7,12,13,14] was presented in Ref.15 (SEI). It yields the sign of the MC and the *quadratic* field dependence in the weak field range, where the field scale was determined by the parameter Φ_M/Φ_o ($\Phi_o = hc/e$ being the quantum flux unit). Furthermore, that model predicted a saturation of the MC for $\Phi_M/\Phi_o \gg 1$. However, the situation very deep in the VRH regime ($R_M/\xi > 10$) was not considered in detail.

In this paper we present a comprehensive discussion of the orbital MC in the VRH regime and argue that the results hold quite generally. We find the regimes in which the various behaviors found by NSS and SEI hold and discuss the various ways to obtain the macroscopic MC from that of the elementary bonds. In section 3 we relate the results obtained for the simplified model of NSS to more general considerations involving the symmetry of the Hamiltonian.

2. ANALYSIS OF THE MC WITHIN THE PERCOLATION MODEL

The conductivity of a sample in the hopping regime may be analyzed in terms of an equivalent resistor network[7,12,13,14]. Any two sites between which the electron hops are taken to be connected by a conductance σ_{ij},

$$\sigma_{ij} \approx \sigma_o |I_{ij}|^2 e^{-\beta \epsilon_{ij}} \qquad (1)$$

where σ_0 is a constant having dimensions of conductance and $\epsilon_{ij} = (|\epsilon_i| + |\epsilon_j| + |\epsilon_i - \epsilon_j|)/2$, where ϵ_i and ϵ_j are the initial and final site energies measured from the Fermi energy. In (1), I_{ij} is the effective overlap integral between the initial and final sites. $|I_{ij}|^2$ depends exponentially on the distance r_{ij} between the two sites, $|I_{ij}|^2 \propto exp\,(-r_{ij}\alpha)$ where α is of the order of the inverse localization length.

Quite generally, σ_{ij} is a random variable, depending upon the site energies, the distance between them, r_{ij}, and the strength of the overlap integral. The conductivity of the macroscopic sample is determined by the percolation threshold condition[7,12]. Given the probability distribution $P(\sigma/\sigma_o, r)$ for the dimensionless conductivity σ/σ_o at a spatial separation r, the sample conductivity σ_c is given by the requirement that the conductors of $\sigma > \sigma_c$ occupy a certain finite fraction

Z_c, of the system volume Ω

$$\int_{\sigma_c}^{\infty} d\sigma \int_0^{\infty} d(r^d) r^d P\left(\frac{\sigma}{\sigma_0}, r\right) = Z_c \Omega , \qquad (2)$$

where d is the system dimensionality. Thus, knowledge of the distribution function will yield an implicit equation for the "critical" conductivity σ_c. In the variable-range hopping regime, the critical conductivity obeys Mott's law, $\sigma_c = \sigma_0 exp(-T_0/T)^{1/(d+1)}$, where $k_B T_0$ is of the order of the average energy spacing between levels localized within a localization volume, ξ^d. The critical conductance connects sites whose spatial separation, R_M, is typically at least several ξ's (depending upon the temperature and impurity concentration) and whose site energies are of the order of a few $k_B T$, much smaller than $k_B T_0$. The percolation argument thus assumes that the conducting properties of the system are determined by hops of elementary conductance σ_c which span a critical network throughout the macroscopic system.

The quantum aspect of the elementary hopping is brought about by the overlap integral I_{if}. It results from electron tunneling between the initial and final sites around which the wavefunctions are localized, with localization length ξ. The spatial separation r_{if} between two sites belonging to the critical network is larger than ξ. There are therefore many different paths connecting the initial and final sites, along which the electron traverses other sites, of site energies in general lying far away from the Fermi level. The effective overlap integral I_{if} thus contains the effects of interference among the various possible paths.

We adopt here the picture of Nguyen, Spivak and Shklovskii (NSS)[9,10,15] according to which the important contribution to I_{if} comes from all oriented paths within a cigar-shaped region of length r_{if} and width $\sqrt{r_{if}\xi}$. One thus may visualize the electron to perform a random walk of step ξ perpendicular to the hopping distance r_{if}. In the specific geometrical model employed by NSS the random walk has an elementary step equal to the microscopic length. The appearance of the ξ in the more general case follows from the discussion of the statistics of the paths contributing to I_{if} given by Shklovskii and Spivak[11]. The resulting effective overlap integral is the sum over the contributions from the various paths

$$I_{if} = e^{-\alpha r_{if}/2} \sum_{\gamma} J_{\gamma} . \qquad (3)$$

As the number of oriented paths is exponential in the path length, a re-definition of the localization length in the expression for the elementary conductance (Eq.(1))

yields
$$\sigma_{ij} = \sigma_o y_{ij}^2 e^{-\alpha r_{ij} - \beta \epsilon_{ij}}, \quad y_{ij}^2 = \frac{1}{n}|\sum_\gamma J_\gamma|^2 . \tag{4}$$

where n is the number of oriented paths and y_{ij}^2 is a random variable of order unity.

In the presence of a constant magnetic field the individual path overlap integral is multiplied by a phase factor $e^{i\phi_\gamma}$, ϕ_γ being the phase acquired from the magnetic field along the γ-th path. This is the only modification due to the magnetic field at small enough fields such that the flux through an elementary area ξ^2 is much smaller than the quantum flux unit Φ_o.[16] (Spin effects which change the site energies are ignored in our discussion of the orbital effects). Eq.(3) is thus modified to read

$$I_{ij} = e^{-\alpha r_{ij}/2} \sum_\gamma J_\gamma e^{i\phi_\gamma} . \tag{5}$$

The elementary conductance of a bond (Eq.(1)) then becomes

$$\sigma_{ij} = \sigma_o y_{ij}^2 e^{-\alpha r_{ij} - \beta \epsilon_{ij}}, \quad y_{ij}^2 = \frac{1}{n}\left|\sum_\gamma J_\gamma e^{i\phi_\gamma}\right|^2 , \tag{6}$$

where y_{ij}^2 represents the quantum interference effect upon the conductance. It depends upon the hopping distance r_{ij} and , in particular, upon the magnetic field.

The interference factor y^2 is a random variable. It is governed by a distribution function which varies under the effect of the magnetic field. Consequently, the percolation condition (2) yields a field-dependent critical conductivity $\sigma_c(H)$ and, in turn, the magnetoconductivity of the sample. It turns out that most features of the MC can be deduced from some general properties of the distribution function for the interference term. We discuss this matter in the next section. For the rest of the present section we adopt the simplified model of NSS.

Consider first the probability distribution of y in the absence of the magnetic field. The individual path overlap integrals may be assumed to be real. Then, for mutually independent path contributions, $|y|$ is distributed normally[17], i.e.,

$$F_o(y^2)dy^2 = \sqrt{\frac{1}{2\pi}} e^{-y^2/2} d|y| , \tag{7}$$

where for simplicity it was assumed that $< J_\gamma^2 > = 1$. Under the assumptions above, the y^2-distribution is independent of the hopping distance and is peaked around $y = 0$. Correlations among the paths were recently considered in Ref.18.

In the presence of the magnetic field, it is convenient to construct the y^2-distribution from the probability distributions of $(J')^2 = \frac{1}{n}(\sum_\gamma J_\gamma \cos\phi_\gamma)^2$ and $J''^2 = \frac{1}{n}(\sum_\gamma J_\gamma \sin\phi_\gamma)^2$ where $|J|^2 = (J')^2 + (J'')^2$. Assuming that for any path with a phase ϕ_γ there is also a symmetric path with a phase $-\phi_\gamma$ (an assumption which is valid in the NSS model), the central limit theorem[17] for both J' and J'' yields (see appendix A)

$$P(J',J'') = \left(\frac{1}{a\sqrt{2\pi}}e^{-(J')^2/2a^2}\right)\left(\frac{1}{b\sqrt{2\pi}}e^{-(J'')^2/2b^2}\right), \tag{8}$$

where

$$a^2 = \frac{1}{n}\sum_\gamma \cos^2\phi_\gamma, \quad b^2 = \frac{1}{n}\sum_\gamma \sin^2\phi_\gamma. \tag{9}$$

Thus $P(J', J'')$ is a product of two normal distributions, for the real (J') and imaginary (J'') parts of the field-dependent effective overlap, with standard deviations a and b, respectively. The resulting y^2-distribution is

$$\begin{aligned}F(y^2)dy^2 &= \int dJ'dJ''\delta(y^2 - (J')^2 - (J'')^2)P(J',J'')dy^2 \\ &= \frac{1}{2ab}e^{-\frac{y^2}{4a^2b^2}}I_o\left(\frac{y^2(a^2-b^2)}{4a^2b^2}\right)dy^2,\end{aligned} \tag{10}$$

where I_o is the modified Bessel function of order zero.

Eqs. (8) and (9) are most easily obtained in the simple case where the J_γ are independent random variables having a variance of unity. Assuming a symmetric distribution for $sin2\phi_\gamma$ one immediately finds that J' and J'' are independent Gaussian random variables distributed according to Eq.(8). This is really the central limit theorem. A fuller analysis leading to Eqs.(8) and (9) under much more general conditions is presented in appendix A.

The effect of the field upon the y^2-distribution is clearly seen from (8) and (10): As the field tends to zero, $a \to 1$ and $b \to 0$. As a result, the second factor in (8) tends to $\delta(J'')$ and the y^2-distribution becomes a normal one, i.e. it is peaked at $y = 0$ (Eq.(7)). However, as the magnetic field is switched on, the weight of the distribution shifts towards finite values of y. The distribution is zero at $y = 0$, increases linearly with y and is peaked at y of the order of b. Thus the effect of the magnetic field is to narrow the distribution of y^2 by favoring intermediate values of y, of order b, instead of those $< b$, and reducing the probability for larger y, as compared to the zero field distribution. This behavior is depicted in Fig.1.

The magnetic field enters the distribution through the phase factors a and b (Eq.(9)). In the strong field limit, such that the magnetic flux through the hopping "area" (of length of the typical hopping distance R_M and width of the order of $\sqrt{R_M \xi}$) becomes larger than the quantum flux unit, a^2 becomes comparable to b^2, $a^2 \sim b^2 \sim \frac{1}{2}$, and the distribution saturates to the form

$$F(y^2)dy^2 \sim e^{-y^2}|y|d|y| , \qquad (11)$$

with a broad peak located at $y \sim \sqrt{1/2}$, and vanishing linearly at $y = 0$. We note that the way the $b \to 0$ limit is achieved is by shrinking of the range (of order b in y) where F vanishes at $y \ll b$ and has a peak for $y \sim b$ (see Fig.1).

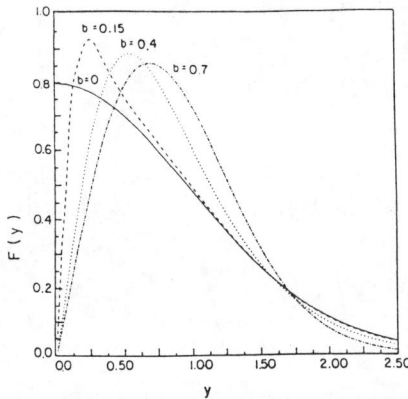

Fig.1 The distribution of $|y|$ (Eq.(10)) for various values of the magnetic field parameter b.

In the presence of a magnetic field, the distribution function of y^2 depends upon the hopping distance r through the phase factors a and b. This complicates the calculation of $\sigma_c(H)$ by the percolation condition (Eq.(2)), [15] but is not expected to be a major effect since everything is governed by the critical bonds whose length does not vary by orders of magnitude. For the sake of simplicity it will be assumed now that the relevant area for the phase factors is characterized by the typical hopping distance R_M. The elementary conductance (Eq.(6)) is then a product of the "quantum" factor y^2 and the "classical" exponential factor, $exp(-\alpha r - \beta \epsilon)$, each obeying its independent distribution function. Assuming the site energies to be distributed uniformly in a band of typical width W, the distribution function of the classical factor of the conductance is (see Eq.(2)).

$$P(z)dz = \int d(r^d)r^d \int d\epsilon P(\epsilon)\delta\left(z - e^{\beta\epsilon + \alpha r}\right)dz$$
$$= \frac{1}{2\beta^2 W^2}\xi^{2d}\frac{1}{2d(2d+1)}\frac{1}{z}(lnz)^{2d+1}dz, \quad 1 \le z \ . \tag{12}$$

This form for $P(z)$ is valid for ϵ smaller than the bandwidth (the relevant range for the variable-range hopping regime). The upper bound on z is determined in fact by the convolution with y^2-distribution (see below) and hence is of no importance.

With the explicit expressions for the distributions, Eqs. (10) and (12), the percolation condition for $\sigma_c(H)$ reads[15]

$$\int_{\sigma_c(H)/\sigma_o}^{\infty} d(\sigma/\sigma_o) \int_1^{\infty} dz P(z) \int dy^2 F(y^2)\delta(\sigma/\sigma_o - y^2/z) = Z_c \ . \tag{13}$$

The zero field conductivity, $\sigma_c(0)$, is given by the same expression with F replaced by its field-free counterpart, F_0 (Eq.(7)). The relative magnetoconductivity (MC) is therefore

$$\frac{\sigma_c(H) - \sigma_c(0)}{\sigma_c(0)} \simeq \int_1^{\infty} dz P(z) \int_{z\sigma_c(0)/\sigma_o}^{\infty} dy^2 \left(F(y^2) - F_o(y^2)\right)$$
$$\left[\int_1^{\infty} dz P(z) z \frac{\sigma_c(0)}{\sigma_o} F_o\left(z\sigma_c(0)/\sigma_o\right)\right]^{-1} \ . \tag{14}$$

Using the explicit expressions for $P(z)$ and F_o, one finds that deep in the VRH regime the denominator on the right-hand-side of Eq.(14) is of the order of $(ln\sigma_0/\sigma_c(0))^{2d+1}$, and in particular is independent of the zero field conductivity $\sigma_c(0)$. The sign and magnitude of the MC will thus be determined by the difference $F - F_o$ of the y^2-distributions, appearing in the numerator on the right-hand-side of (14).

Integration by parts and re-arrangement lead to the following form for the numerator of the MC

$$\int_{\sigma_c(0)/\sigma_o}^{\infty} (lnz\sigma_o/\sigma_c(0))^{2d+2}(F(z) - F_o(z))dz \ . \tag{15}$$

which may be evaluated as follows. For very small magnetic fields such that $b^2\sigma_o/\sigma_c(0) \ll 1$ the argument of the Bessel function appearing in $F(z)$ (see Eq.(10)) is very large. One may then use its asymptotic form to obtain

$$\frac{\sigma_c(H) - \sigma_c(0)}{\sigma_c(0)} \sim \frac{b^2}{(ln\sigma_o/\sigma_c(0))^{2d+1}}$$

$$\int_{\sigma_c(0)/\sigma_o}^{\infty} dz e^{-z/2} \left(z^{-3/2} - z^{-1/2}\right) \left(ln z\sigma_o/\sigma_c(0)\right)^{2d+1}, \quad (16)$$

$b^2 \ll \sigma_c(0)/\sigma_o$.

Hence at very low fields the MC is quadratic in the magnetic field[15]. For high enough values of $\sigma_c(0)/\sigma_o$ ($\sigma_c(0)/\sigma_o > \sigma_l/\sigma_o \sim 10^{-4}$, see Ref.15) the low field MC is negative, as a result of the reduction of F below F_o at large values of z. At lower values of $\sigma_c(0)/\sigma_o$ the small field MC is positive,

$$\frac{\sigma_c(H) - \sigma_c(0)}{\sigma_c(0)} \sim b^2 \sqrt{\sigma_o/\sigma_c(0)}/(ln\sigma_o/\sigma_c(0))^{2d+1}, \quad (17)$$

$$b^2 \ll \sigma_c(0)/\sigma_o \ll 1.$$

We emphasize that the negative MC at $\sigma > \sigma_l$ while being a correct result of the model, is a weaker prediction for real systems. For such large σ's, the neglect of winding paths is not clearly justified. Their inclusion should lead for $\sigma \sim \sigma_o$ to the usual weak localization MC. The magnitude of the relative MC decreases with increasing temperature ($[\sigma_o/\sigma_c(0)]^{1/2} \simeq exp[(T_0/T)^{1/(d+1)}/2]$).

We now turn to the case $\sigma \ll \sigma_l$, not treated in Ref.15. Here, at intermediate values of the magnetic fields such that $1 \gg b^2 > \sigma_c(0)/\sigma_o$ the main contribution to the integration in (15) comes from the region $z > b^2$ and is positive, dominating the negative contribution from the region $\sigma_c(0)/\sigma_o \leq z \leq b^2$. Using again the asymptotic form for the Bessel function one finds for d=2 with $A \equiv ln b^2 \sigma_0/\sigma_c(0)$ (a related expression applies at d=3):

$$\frac{\sigma_c(H) - \sigma_c(0)}{\sigma_c(0)} \sim b(A^5 + 2 \cdot 5A^4 + 2^2 \cdot 5 \cdot 4A^3 + 2^3 \cdot 5 \cdot 4 \cdot 3A^2 + 2^4 \cdot 5!A + 2^5 \cdot 5!)$$

$$/(ln\sigma_o/\sigma_c(0))^{2d+1},$$

$$\sigma_c(0)/\sigma_o < b^2 \ll 1.$$

(18)

In this regime the MC is quasi-linear in the field with important logarithmic corrections. These cause the MC to deviate from linearity (Fig.2) and to appear quadratic over much broader ranges of b. For b^2 sufficiently larger than $\sigma_c(0)/\sigma_o$ the MC has an approximately linear behavior (Fig.2).

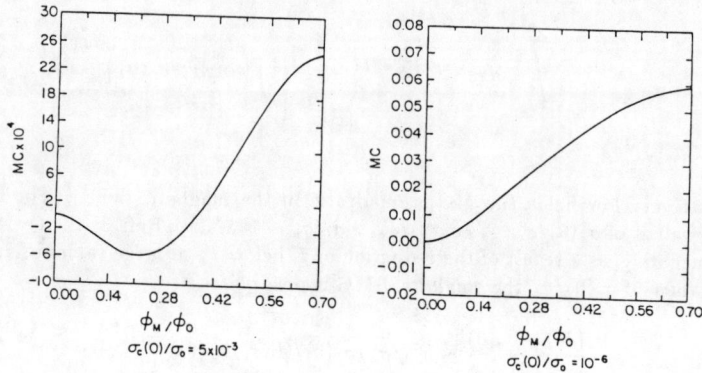

Fig.2   The behaviour of the relative MC as a function of the magnetic field (parametrized by Φ_M/Φ_o) in the various regimes (from Eq.(14)) for two values of $\sigma_c(0)/\sigma_o$.

The temperature dependence of the MC arises from the effective hopping area which determines the phase factor b. This area is proportional to $R_M^{3/2}$, i.e., to $(T_0/T)^{3/2(d+1)}$, again yielding a decrease of the MC with temperature. Finally, when the magnetic field is strong enough such that the flux through the hopping domain is larger than the quantum flux unit (i.e., $b^2 \approx 1/2$), the MC saturates[15] to a relative value of order 0.1, independent of the magnetic field. The behavior of the MC in the various regimes as computed from Eq.(14) is shown in Fig.2. One notes the quadratic-type behavior for moderate values of $\sigma_c(0)/\sigma_o$ which crosses over to quasilinearity as $\sigma_c(0)/\sigma_o$ decreases. In appendix B we show that the present analysis is equivalent to a restricted logarithmic averaging.

The theory presented here is based on critical path analysis and should therefore not regarded as a quantitative one (unlike weak localization theory for example).

A quantitative comparison between experimental results on In_2O_3 and our theory for the same amount of disorder (characterized by T_0/T) is given in Fig. 3 (no other adjustable parameters). The main features qualitatively agree though the magnitude predicted theoretically is 3-4 times smaller than the experimental one.

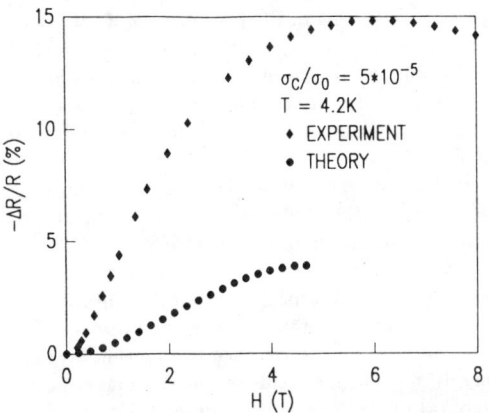

Fig.3   Theory vs. experiment

3. GENERAL PROPERTIES OF F(y) AND HEURISTIC DERIVATION OF THE MC

In the previous section we derived the probability distribution $F(y)$ for the simplified model of NSS (see Fig. 1). In particular, the distribution was shown to display two main features
1) As the magnetic field is increased, $F(y)$ narrows down and becomes peeked around some field dependent value.
2) When the magnetic field is increased even further (more than one flux quantum in the relevant area) the distribution saturates to some limiting form and becomes field independent.

We argue that these features are far more general than the model used. Noticing that y^2 is proportional to the amplitude squared of the tail of the wave function localized at site i, in site f, the above two features of $F(y)$ imply that the fluctuations in the wave function amplitude are reduced as the magnetic field is increased and that they saturate to some limiting value for strong enough magnetic field. This result in terms of transmission coefficients is a well known result in the theory of conductance fluctuations[20,21,22] and is also closely related to random matrix theory.[23,24] In the absence of magnetic field, all states are real and the Hamiltonian pertains to the orthogonal ensemble.[23,24] Upon applying strong enough magnetic field, the symmetry of the Hamiltonian is reduced to the unitary one [23,24] and consequently the fluctuations in the wave function amplitude squared are halved.[21,22,25] The narrowing of F(y) is thus manifestation of the reduced fluctuations characteristic to the unitary ensemble compared with the orthogonal one, and the saturation occurs when the relevant part of the Hamiltonian fully crosses over to the unitary symmetry. A straight forward calculation of the fluctuations in y^2 (using Eq. 10), $<y^4> - <y^2>^2$, for zero and strong magnetic fields show indeed that the fluctuations are twice as large in the first case compared with the second one. In the intermediate field range, $0 < \Phi_M < \Phi_0$, the relevant part of the Hamiltonian neither belongs to the random unitary ensemble nor to the orthogonal one. In the simplified model of NSS the transition is reflected in a gradual cross over from a 1D random walk to a 2D one.

By the same token, the NSS model supplies an intuitive view of the reduced fluctuations in the wave function amplitude in the presence of magnetic field. The reduced probability to have an interference term either larger or smaller than its average value merely reflects the tendency of the magnetic field to improve the interference in realizations with zero field destructive interference and to reduce the interference in cases where the zero field interference is constructive.

The present, general point of view, is also powerful in predicting the role of spin-orbit (SO) coupling. Such an interaction reduces the symmetry of the Hamiltonian even further to the symplectic one.[22,24,25,26]. Consequently, the fluctuations in the wave function amplitude squared are reduced by a factor of two [26,27] relative to the unitary case. Strong enough magnetic field will restore the unitary symmetry (broader distribution) resulting thus in a negative magnetoconductance. In the specific model of NSS, the introduction of SO coupling should result in $F(y) \approx y^3 exp(-2y^2)$ (random walk in 4D) and probably a cross over back to a random walk in 2D in the presence of strong enough magnetic field.

In the rest of the section we employ the general properties of F(y) discussed above and a percolation argument to rederive the main features of the MC in the VRH. Following Eq. 6 we define the classical conductivity $\sigma_{cl} = \sigma_0 exp(-\alpha r - \beta \epsilon)$. The bond conductivity is then given by $\sigma = y^2 \sigma_{cl}$. Since the conductance, and hence the magnetoconductance, are determined by the critical resistors in the equivalent resistor network, it is enough to consider this group alone. Let us divide those resistors into two sub groups. Group A consisting of resistors with $\sigma_{cl} > \sigma_c$ and hence $y^2 < 1$ (such that their conductance is σ_c) and group B consisting of resistors of $\sigma_{cl} < \sigma_c$ and $y^2 > 1$. By the narrowing of the distribution in the presence of magnetic field it readily follows that group A would be characterized by positive MC while group B would display a negative MC. The overall magnetoconductance of the critical resistors would depend on the relative sizes of the two sub groups. For small enough zero field conductance, group A would be the dominant one leading to an overall positive MC while for high enough conductivities the majority of critical resistors would consist of group B resulting in a negative MC. This is the origin of the positive MC deep in the VRH regime and the sign transition depicted in Fig. 2a. The saturation of the MC for strong enough magnetic fields follows simply from the saturation of F(y) to some limiting form. We find then, that in the strong disorder limit where winding paths can be neglected, the orbital magnetoconductance in the VRH regime should generally be positive and should saturate when the flux through a typical critical bond approaches one quantum flux unit.

4. CONCLUSIONS

The conductance of each bond in the VRH model depends on a sum over paths and is thus sensitive to interference, and depends on an applied magnetic field. Each bond, like a mesoscopic system, has a MC of a random sign. The surprising result is that the resultant low field orbital MC of the macroscopic system is not averaged out but instead shows a well defined, substantial, value. This follows from both the critical path analysis and from a modified logarithmic averaging[19] on the critical network.

The MC is governed by a flux in a typical area given by $R_M^{3/2} \xi^{1/2}$ (9,11). At (relatively) large fields, such that the above flux is much larger than a flux quantum, Φ_o, the relative MC tends to a positive constant of the order of 0.1 which depends on the zero field conductivity σ_c. The extremely low field MC depends on σ_c as well. For $\sigma_c > \sigma_l \sim 10^{-4}\sigma_o$, the MC starts negative and crosses over to positive. For $\sigma_c \sim \sigma_l$, it is positive, proportional to H^2 for $\Phi_M \ll \Phi_o$ and saturates at $\Phi_M \gg \Phi_o$. For $\sigma_c \ll \sigma_l$, there is an additional cross-over of the (positive) MC from H^2 to "quasi-linear" behaviour for $(\Phi_M/\Phi_o)^2 \sim \sigma_c/\sigma_o$. In the latter, there is a wide range where the logarithmic corrections cause the MC to be roughly quadratic with an effective linear behaviour at much larger Φ_M/Φ_o values. These features are in agreement with experiment.

We have argued that these results are due to a symmetry crossover from the orthogonal to the unitary ensembles and should therefore be general. Based on this general point of view we also expect a further reduction in the symmetry of the Hamiltonian in the presence of strong spin - orbit coupling and as a result a negative MC. An interesting, still open question is how our results cross-over to the usual weak localization MC at $\sigma_c \sim \sigma_0$ when backscattering becomes important. This cross-over may also influence the negative MC found in our model for the largest σ_c's. For very strong disorder it is possible that the direct path for hopping will become dominant and the MC will vanish. These questions deserve further study.

ACKNOWLEDGEMENTS

The authors are especially grateful to B.I. Shklovskii and Z. Ovadyahu for many discussions. We also thank D. Berman and Y. Meir for useful discussions. This research was partially supported at Tel Aviv University and at the Weizmann Institute by the fund for basic research administrated by the Israeli Academy for Sciences and Humanities, and at the Weizmann Institute by the Minerva Foundation Munich, Germany. One of us (US) was also supported by the Weizmann fellowship.

REFERENCES

1. R.A. Webb, A.B. Fowler, A. Hartstein and J.J. Wainer, Surf. Sci. **170**, 14 (1986); A.B. Fowler, G.L. Timp, J.J. Wainer and R.A. Webb, Phys. Rev. Lett. **57**, 138 (1986); A.B. Fowler, J.J. Wainer and R.A. Webb, IBM J. Res. Dev. **32**, 372 (1988); J.J. Wainer, A.B. Fowler and R.A. Webb, in the Proceedings of EP2DS-7, Santa Fe (1987), Surf. Sci. **196**, 134 (1988).
2. R.K. Kalia, W. Xue and P.A. Lee, Phys. Rev. Lett. **57**, 1615 (1986).
3. Ya. B. Poyarkov, V. Ya. Kontarev, I.P. Krylov and Yu.V. Sharvin, Pis'ma Zh. Eksp. Teor. Fiz. **44**, 291 (1986) [JETP Lett. **44**, 375 (1986)].
4. Z. Ovadyahu, Phys. Rev. B **33**, 6552 (1986); O. Faran and Z. Ovadyahu, Phys. Rev. **B38**, 5457 (1988).
5. N.A. Assadullaev and I. Ciric, Sov. Phys. Solid State **30**, 685 (1988); M. Benzaquen, D. Walsh K. Mazurok, Phys. Rev. **B38**, 10933 (1988).
6. A. Hartstein, A.B. Fowler and K.C. Woo, Physica B **117-118**, 655 (1983).
7. B.I. Shklovskii and A.L. Efros, "Electronic Properties of Doped Semiconductors" pp.210-216, Vol. 45 of Springer Series in Solid-State Sciences (1984).
8. B.B. Suprapto and P.N. Butcher, J. Phys. C**8**, L517 (1975).
9. V.I. Nguyen, B.Z. Spivak and B.I. Shklovskii, Pis'ma Zh. Eksp. Teor. Fiz. **41**, 35 (1985) [JETP Lett. **41**, 42 (1985)].
10. V.I. Nguyen, B.Z. Spivak and B.I. Shklovskii, Zh. Eksp. Teor. Fiz. **89**, 1770 (1985) [Sov. Phys. JETP **62**, 1021 (1985)].
11. B.I. Shklovskii and B.Z. Spivak, J. Stat. Phys. **38**, 267 (1988).
12. V. Ambegaokar, B.I. Halperin and J.S. Langer, Phys. Rev. B**4**, 2612 (1971).
13. A. Miller and E. Abrahams, Phys. Rev. **120**, 745 (1960).

14. M. Pollak, J. Non Cryst. Solids **11**, 1 (1972); B.I. Shklovskii and A.L. Efros Zh Exp. Teor. Fiz. **60**, 867 (1971) [Sov. Phys. JETP **33**, 469 (1971)].
15. U. Sivan, O. Entin-Wohlman and Y. Imry, Phys. Rev. Lett. **60**, 1566 (1988).
16. T. Holstein, Phys. Rev. **124**, 1329 (1961).
17. S.-k Ma, Statistical Mechanics, (World Scientific, Singapore (1985)), Chapter 12.
18. E. Medina, M. Kardar, Y. Shapir and X.R. Wang, Phys. Rev. Lett. **62**, 941 (1989).
19. B.I. Shklovskii and B.Z. Spivak in Hopping Conduction in Semiconductors (editors M. Pollak and B.I. Shklovskii) North Holland, in press.
20. B.L Altshuler, JETP Lett. **41**, 648 (1985).
21. P.A. Lee and D.A. Stone, Phys. Rev. Lett. **55**, 622 (1985).
22. P.A. Lee, D.A. Stone, and H. Fukuyama, Phys. Rev. B **35**, 1039 (1987).
23. E.P. Wigner, Ann. Math. **62**, 548 (1955); ibid. **65**, (1957).
24. F.J. Dyson, J. Math Phys. **3**, 140 (1962); ibid. **3**, 1191 (1962).
25. N. Zanon and J.L. Pichard, J. de Physique **49**, 907 (1988).
26. Y. Meir, Y. Gefen, and O. Entin - Wohlman, Phys. Rev. Lett. **63**, 798 (1989).

A-1
APPENDIX A - DERIVATION OF THE OVERLAP PROBABILITY DISTRIBUTION.

Here we outline the derivation of the probability distribution $P(J', J'')$, (Eq. (8)).

Given the probability density $P(J_1, J_2, \ldots, J_n)dJ_1 \ldots dJ_n$ that the $i-th$ path contributes J_i to the effective overlap ($i = 1, \ldots n$), the explicit form of $P(J', J'')$ is

$$P(J', J'') = \int dJ_1 \ldots dJ_n P(J_1, \ldots, J_n) \delta(J' - \frac{1}{\sqrt{n}} \sum_\gamma J_\gamma \cos\phi_\gamma) \\ \delta(J'' - \frac{1}{\sqrt{n}} \sum_\gamma J_\gamma \sin\phi_\gamma) \ . \quad (A-1)$$

Expressing the δ-functions in the form of Fourier integrals, Eq.(A-1) takes the form

$$P(J', J'') = \frac{1}{(2\pi)^2} \int_{-\infty}^{\infty} dx_1 dx_2 e^{ix_1 J' + ix_2 J''} \\ \prod_\gamma \int dJ_1 \ldots dJ_n P(J_1, \ldots, J_n) e^{-ix_1 \frac{J_\gamma}{\sqrt{n}} \cos\phi_\gamma - ix_2 \frac{J_\gamma}{\sqrt{n}} \sin\phi_\gamma} \ . \quad (A-2)$$

One then expands the exponentials appearing in the second factor of (A-2). Assuming a symmetric distribution of the J_i's and that each path characterized by ϕ_γ has its counterpart with $-\phi_\gamma$, the second factor in (A-2) yields

$$\prod_\gamma (1 - \frac{1}{2} \frac{<J_\gamma^2>}{n} (x_1^2 \cos^2\phi_\gamma + x_2^2 \sin^2\phi_\gamma) + \cdots) \ . \quad (A-3)$$

The higher order terms include factors of order n^{-2}, n^{-3}, etc. For large[17] n, Eq.(A-3) becomes $exp(-\frac{1}{2}(x_1^2 a^2 + x_2^2 b^2))$, to leading order, where $a^2 = \sum_\gamma <J_\gamma^2> \cos^2\phi_\gamma/n$, $b^2 = 1 - a^2$. For simplicity, and without substantial loss of generality, we choose $<J_\gamma^2> = 1$. Inserting this form into (A-2), and performing the x_1 and x_2 integrations, we obtain Eq.(8).

B-1
APPENDIX B - EQUIVALENCE OF RESTRICTED LOGARITHMIC AVERAGING WITH THE PERCOLATION ANALYSIS FOR THE MC

In this section we consider the generic problem of a broad distribution of the conductance, e^{-x}, on top of which the individual bond values of x are randomly changed each by a small Δx. This is evidently a simplified picture, in the VRH model the bonds have varying lengths and energies. Δx can be induced, for example, by a small magnetic field. We denote the distribution of x by $P_o(x)$ and the distribution of Δx by $P_\Delta(x, \Delta x)$ (clearly, moments such as $<\Delta x>$ depend on x). The distribution of $u \equiv x + \Delta x$, is given by

$$P(u) = \int dx \int d\Delta x P_o(x) P_\Delta(x, \Delta x) \delta(u - x - \Delta x) . \qquad (B-1)$$

The critical value of u is given by the percolation condition

$$\int_{-\infty}^{u_c} P(u) du = p_c . \qquad (B-2)$$

We substitute (B-1) into (B-2) and perform the integration over u

$$p_c = \int dx \int d\Delta x P_o(x) P_\Delta(x, \Delta x) \theta(u_c - x - \Delta x) . \qquad (B-3)$$

Expanding the θ function (this is justified for P_Δ which is smooth on the scale of Δx) and writing $u_c = u_c^o + \Delta u_c$ where u_c^o is the critical value before Δx was introduced, we obtain after a further expansion in Δu_c

$$\Delta u_c = \int d(\Delta x) \int dx P_o(x) \delta(u_c^o - x) \Delta x P_\Delta(x, \Delta x) \bigg/ \int dx P_o(x) \delta(x - u_c^o)$$
$$\equiv <\Delta x>_{x=u_c^o}$$
$$(B-4)$$

i.e., the change Δu is given by the average of Δx over the critical bonds before the change. In terms of the bond conductances e^{-x}, this is a (restricted) logarithmic averaging over the critical network (we emphasize that this is not averaging over the whole network which may give undue weight to, e.g., very small conductances). This demonstration is straightforwardly generalizable to the case where $P(x)$ is governed by a joint distribution of several variables (such as r, ϵ, y in our VRH problem). These issues are discussed in Ref.19.

Quantum Interference Effects in the Hopping Conductivity

Y. Shapir and X. R. Wang
Department of Physics and Astronomy
University of Rochester
Rochester, NY 14627 USA

E. Medina and M. Kardar
Department of Physics
Massachusetts Institute of Technology
Cambridge, MA 02139 USA

ABSTRACT

The quantum effects due to the interference between forward scattering paths contributing to the probability amplitude of the critical hop are considered. The correlations between these paths play a crucial role in the behavior of the conductance. Upon averaging only the $h/2e$ periodicity of the Aharonov-Bohm oscillations survives as $R \to \infty$ (R is the hopping distance), for any amount of disorder. The scaling behavior of the normal distribution of the logconductance is computed. While its average decreases linearly with R, its r.m.s. fluctuations increase as R^ω with $\omega = 0.33 \pm 0.05$. The typical extent of the wave function in the direction transverse to the hopping direction behaves as R^ν with $\nu = 0.68 \pm 0.05$. Without disorder there is a very rich structure in the dependence of the conductance upon the enclosed flux which is controlled by the commensurability of the flux per plaquette with the flux quantum. With disorder we find a positive magneto-logconductance in general agreement with other approaches. This behavior is compared with that of the model with random phases which may be relevant in the presence of magnetic impurities or spin-orbit interaction.

I. INTRODUCTION

Novel physical phenomena were revealed recently in weakly localized mesoscopic systems. They include universal conductance fluctuations, Aharonov–Bohm (AB) oscillations of periods h/e and $h/2e$, anomalous magnetoconductance, *etc.* These phenomena take place in systems the size of which is smaller than the phase coherence length and are due to purely quantum effects.[1]

More recently similar effects have been observed in the strongly localized regime at low temperatures where variable range hopping (VRH) is the major transport mechanism.[2-9] In this regime the conductance of *macroscopic samples* is dominated by critical hops of mesoscopic extent (typically 10^3-10^4Å) and the above mentioned effects are expected to be observable. So mesoscopic physics governs the macroscopic transport behavior of strongly localized systems!

A few years ago Nguyen, Spivak, and Shklovskii (NSS) introduced a relatively simple, yet highly non-trivial, model to study the quantum interference effects in this regime.[10,11] This model focuses solely on the interference between forward scattered paths. The model and some of its consequences are described in detail in the articles of Prof. Shklovskii[12] and Dr. Sivan[13] in this volume. A comprehensive review is contained in Ref. 14. Here we briefly recall the essentials of the model and introduce the notations to be used later. The NSS model[10] is based on a tight-binding Anderson model deep in the localized regime whose Hamiltonian is:

$$H = \sum_i \epsilon_i a_i^\dagger a_i + V \sum_{\langle ij \rangle}(a_i^\dagger a_j + c.c.) \qquad (1)$$

The on-site energies ϵ_i are $+W$ with the probability $1-x$ and $-W$ with the probability x, and $V/W \ll 1$. In this limit only the shortest directed paths make important contributions. The quantity of interest is the probability amplitude along the diagonal of the square lattice which in the energy representation is given by a sum over all paths Γ_α:

$$I = \sum_{\Gamma_\alpha} \Pi_{i_\alpha} \frac{V}{\epsilon_{i_\alpha} - \epsilon}. \qquad (2)$$

The energy of the electron may be chosen to be at the Fermi level at $\epsilon = \epsilon_F = 0$. It is useful to factorize out the overall contribution of (V/W) raised to the power of the length of the paths L:

$$I = (V/W)^L J, \qquad (3)$$

with

$$J = \sum_\alpha J_\alpha \tag{4}$$

and

$$J_\alpha = \Pi_{i_\alpha} \eta_{i_\alpha}. \tag{5}$$

The variables $\eta_{i_\alpha} = W/\epsilon_{i_\alpha}$ take values 1 (with prob. $1 - x$) and -1 (with prob. x). In presence of magnetic fields each of the η_{i_α} also acquires a phase according to the vector potential on the bond going between sites i and j. So the quantity J comprises all the quantum coherence effects and is the main subject of our investigation.

In the past few years we have been engaged in a systematic study of the quantum interference effects which dominate the behavior of $\log|J|$ and how they are reflected in the logconductance ($\mathrm{Log} C$) of the system.[15,16] In particular we have considered the subtle and non-trivial effects which originate from the correlations between the different Feynman paths.[14] We have also found it necessary to invoke scaling concepts and renormalization-group ideas to describe the dependence of the different properties on the hopping distance R.[15]

In the following chapters we try to give simple explanations and insight to our major results. The technical details may be found in papers already published[15,16] or to be published soon.[17] In the next chapter (II) we consider the Aharonov–Bohm oscillations. Chapter III is devoted to the universal probability distribution of $\log C$. The magneto-logconductance and its potential relation to the model with random phases are discussed in chapter IV. In chapter V we briefly summarize the effects of magnetic fields on electron tunneling in a perfect lattice with no disorder. The last chapter (VI) is devoted to our conclusions and possible extensions of the NSS model.

II. THE AHARONOV–BOHM OSCILLATIONS

The Aharonov–Bohm (AB) oscillations of the conductivity as a function of the enclosed magnetic flux is one of the most striking manifestations of quantum interference without a classical analog.[18]

We begin with a reminder of results on the periodicity of the oscillations in mesoscopic rings within the weak localization regime[19] (with ring size smaller than the phase-coherence length). While for each ring separately oscillations of both periods $\Phi_0 = h/e$ (flux quantum) and $\Phi_0/2$ were observed, in an ensemble of N rings the amplitude of the h/e oscillations is reduced by $1/\sqrt{N}$ while the $h/2e$ is unaffected. More generally, any averaging (either over realization of the randomness, over energy interval, momentum directions, *etc.*) will eliminate the h/e frequency because of its incoherence. The $h/2e$ oscillations, due to backscattering of time- reversed paths, are all coherent and survive the averaging.

A very similar situation arises for the hopping conductivity: In a macroscopic sample with many critical hops the $h/2e$ will be observed[8] while a mesoscopic system governed by a single critical hop will exhibit the h/e period as well.[20] The origin of the phenomena is drastically different since backscattering is not very likely to affect the behavior on a hopping distance $R >> \xi$ especially if the randomness is so strong the ξ is of the order of the Bohr radius of one impurity. So in the strongly disordered hopping regime one may assume that the electron is scattered only in the forward direction. Then it is the existence of many virtual paths which should lead to the AB effect due to the flux enclosed. NSS have first suggested[10] that AB oscillations may change their period due to the change in the form of the probability distribution of J. For strong randomness (relatively large x) it will be symmetric around zero and only the even moments will survive in the non-linear (Log) averaging. Even moments depend on the field through functions of $\cos^2(2\pi\Phi/\Phi_0)$ and hence the period of $\Phi_0/2$ emerges! This and the following discussions apply to large enough systems with self-averaging.

NSS have convincingly argued[10] that no magnetic field is thus necessary to conclude which period will dominate. It is enough to check the relative values of the first (average) and the second (r.m.s. fluctuations) moments of J in zero magnetic field. If the average decays slower than the fluctuations , as R becomes large, than the h/e period survives the averaging and will be observed. In the opposite case the distribution becomes symmetric and $h/2e$ will be the basic period.

So the question of interest is how the behavior of the ratio $D(x,R) = \langle J \rangle^2 / \langle \Delta J^2 \rangle$ as $R \to \infty$ depends on the disorder parameter x?

Clearly for $x = 0$ the value of J is unique and positive $\Delta J = 0$ and this δ-distribution yields Φ_0 oscillations. On the other hand at $x = 1/2$ J has equal probability to be positive or negative, its average is zero, with $\Phi_0/2$ oscillations.

What will determine the behavior for intermediate values of x? Is there a phase transition[10] from one period to another at a critical value of x? Looking at the statistics of the virtual paths within the NSS model we have found[15] that the correlations between them are essential to answer this question. These correlations are due to the intersections between the different Feynman paths. Upon averaging these intersections induce an effective attraction between the paths.[15] This effective attraction not only depends on x but actually changes with the scale of observation as the different configurations of the paths on smaller scales are integrated out.[16] The essential question is, then, whether this attraction always increases as the scale is increased in which case we say that the randomness is relevant. No matter how small the actual (bare) value of x is, the large scale behavior will be that of a strongly random system. In that case no phase transition is possible. If on the other hand the initial "flow" for small x is back to $x = 0$, randomness is irrelevant with the possibility of a sharp transition at a critical value of x.

It turns out that the answer to this question for small x depends on the number of crossings between the paths which changes with dimension.[14,15]

1. For $d = 2$: the number of crossings increases with R as $R^{1/2}$ and the effective attraction increases exponentially with the scale and randomness is relevant.

2. For $d = 3$: the number of crossings increases as $\text{Log} R$ and the effective attraction and randomness is only marginally relevant with a slow (power-law?) increase of the attraction for small x and a crossover to an exponential increase on larger scales.

3. For $d \geq 4$: the number of crossings is finite and small attraction will decrease. Weak randomness is irrelevant. A phase transition will take place at finite x.

So in the physical dimensions two and three we predict that the $h/2e$ period will always eventually dominate over h/e for large enough R (no matter how small x). Of course for intermediate R both periods may be observed and the precise way the $\Phi_0/2$ takes over will depend on x and on the dimensionality.

There are some similarities between the behavior of $D(x, R)$ at different dimensions d with the behavior of small values of the dimensionless conductance g in weak localizations at dimensions $d - 1$. This is not surprising since the number of intersections of two directed random walks in d-dimensions scales in the same way as the number of returns to the origin of a regular random walk in $d - 1$ dimensions. The latter is known to determine the relative importance of backscattering for weak localization in $d - 1$ dimensions and the corrections to $g \gtrsim 0$ due to interference effects.

III. THE UNIVERSAL DISTRIBUTION OF THE LOG CONDUCTANCE

All the important information on the critical hop is contained in the large-scale ("universal") distribution of Log C and its dependence on other parameters such as external magnetic field, local magnetic impurities, spin-orbit interaction,[14] *etc.* We concentrate here on the simple NSS model in 2D and the effects of these other parameters will be mentioned later.

Even in the simplest situation we would like to be able to answer questions such as the dependence of the average Log C and its fluctuations on the degree of randomness or temperature. While the latter may be simply related to the hopping length R which will be the large distance cut-off in all our subsequent discussions (again we assume $\xi \approx a_o = 1$) the effect of the disorder (or of the parameter x) is difficult to account for. However, based on our conclusions in the previous chapter we do not expect the functional form of the distributions, as well as the scaling with R of their parameters, to actually depend on x. Of course prefactors and coefficients are x-dependent and their estimate is necessary for complete comparison with experiments. Here, however, we relate only to the "universal" properties. These will be extracted for the most random model with $x = 1/2$ (also denoted by $+/-$) which should be a good representative for the whole universality class since in $2d$ randomness becomes more and more important as the length scale on which phenomena are observed becomes larger.

The simplifying feature of $x = 1/2$ is that all paths must always be paired and once this is done no effective randomness is manifested (since $\eta^2 = 1$). Thus all properties will follow from the statistical mechanics of pairs of paths[16]. To compute any moment $\langle J^{2k} \rangle$ one needs to introduce $2k$ paths which are sometimes called replicas and are denoted by indices $\alpha, \beta, \gamma, \ldots$ The averaging procedure couples them into pairs (or larger groups with an even number of paths). All the non-trivial scaling behavior than arises from the effective interaction between two pairs (no randomness anymore!). It turns out[16] that two pairs are attracted to each other due to a subtle "exchange" interaction: when two pairs, say $\alpha\beta$ and $\gamma\delta$, meet at a point they have 3 possible pairings as they leave the intersection: $(\alpha\beta; \gamma\delta)$, $(\alpha\gamma; \beta\delta)$ or $(\alpha\delta; \beta\gamma)$. Hence every intersection of pairs will give a factor of 3 to that diagram. This may be thought of as a Boltzmann factor with energy $-\epsilon/kT = \ln 3$ associated with the crossing. The weight of a diagram will thus increase as 3^I where I is the number of pairwise intersections of pairs. The weights will be modified if a larger number of pairs (say 3,4... etc.) meet at a point,[21] but it is plausible that these higher "vertices" are irrelevant and that the two-pairs (or four-particles) interaction will dominate the asymptotic behavior.

The equivalent model of particles with pairwise δ-function interactions has been solved[22] previously and its consequences are carried-over to the present problem as well. The limit $k \to 0$ of $\langle J^{2k} \rangle$ yields the information on the average of $\text{Log} C \approx \langle \log |J| \rangle = \lim_{k \to 0} [\langle J^{2k} \rangle - 1]/2k$.

We summarize below the important conclusions:
1. The average in linear in R:

$$\langle \log C \rangle \sim -AR.$$

2. Fluctuations have a non-trivial scaling behavior:

$$[\langle \log^2 C \rangle - \langle \log C \rangle^2]^{1/2} = L^x \text{ with } x = 1/3.$$

3. The transverse fluctuations of the wavefunction scale as R^ζ with $\zeta \approx 2/3$. These results were confirmed by numerical simulations with the figures quoted in the abstract.

All this reasoning is applicable to a model with random phases on each bond for which $\eta_i = \exp\{i\phi_i\}$ and the phase ϕ_i is uniformly distributed in the interval $[0, 2\pi)$. Such a phase may arise in presence of magnetic impurities, due to spin-orbit interaction[9] or other mechanisms. It has been also argued that the effect of an external magnetic field is to essentially introduce random phases to the $+/-$ ($x = 1/2$) model[10,23,24]. This possibility is discussed in the next chapter devoted to the magneto-log conductance ($M \text{Log} C$).

IV. MAGNETO-LOGCONDUCTANCE AND RANDOM PHASES

The common conclusions reached, regarding the $M \text{Log} C$, by NSS[10,11], by Sivan et al.[23] (see also Entin-Wohlman et al.[24]), and by our group[17] are as follows:

(i) A positive $M \text{Log} C$ will be seen if a magnetic field is applied to the $+/-$ model.

(ii) No change in the Log C will occur if the field is added upon a system in which random phases are present to begin with.

The understanding of the field effects is therefore intimately related to that of differences between the $+/-$ and the random phases cases. One may argue, based on these conclusions, that the major effect of the field on the $+/-$ model will be to change it to that with random phases. This will certainly be the case for very strong field of the order of flux quantum per elementary plaquette. The situation is far less clear for weak fields of the order of flux quantum throughout the whole area associated with the paths contributing to the critical hop. The phases introduced by a uniform field are <u>not</u> random (unlike these due to random magnetic impurities or to spin-orbit scattering). This may easily be shown[17] by the non-vanishing averages of non-time- reversal invariant quantities like $\langle J^2 \rangle$. Here we concentrate on the regime of strong field or on systems in which other mechanisms do introduce random phases (magnetic impurities and, possibly, spin-orbit interactions[9]). In particular we present the explanation of why the random phases model has a larger Log C than the $+/-$ model.

As explained above for the $+/-$ model, the properties of the Log C distribution are determined by the effective attraction between path-pairs which yield a factor of 3 each time two of them cross each other. What is the essential difference if random phase on each bond replaces the $+/-$ ($x = 0.5$) distribution? While time-reversal symmetry is preserved for the latter case it is broken locally if random phases are present. In particular $J \neq J^*$ and more care is required in calculating moments such as $\langle |J|^{2k} \rangle$. To evaluate J^k we introduce as before k replica $\alpha, \beta, \gamma \ldots$ (which we denote as "particles"), but to evaluate $(J^*)^k$, k different replica $\alpha', \beta', \gamma' \ldots$ (the "holes") are introduced. If a particle acquires a phase ϕ_{ke}, on a bond $\langle ke \rangle$, a hole will obtain a phase $-\phi_{ke}$. It is then clear that only "particle-hole" $(p-h)$ pairs (for which the total random phase on each band cancels out) will survive the averaging.

Pairing such as $p-p$ and $h-h$ which were permitted in the $+/-$ model (with the particle-hole symmetry) will vanish upon averaging the random phases.

Let us explore what happens when two $p-h$ pairs cross each other. Suppose they come in paired as $(\alpha\alpha'; \beta\beta')$, they may go out either in the same configuration or they may exchange partners to $(\alpha\beta'; \beta\alpha')$. Hence, only a factor of 2 (instead of 3 in the $+/-$ case) will be associated with each intersection of pairs. This yields a weaker attraction between pairs with the following immediate consequences:

(i) The correlation length and Log C will increase relative to the $+/-$ case.

(ii) The fluctuations in Log C will still scale as $R^{1/3}$ but with a smaller coefficient.

From these findings we may also conclude that the magnetic field effects will be in these same two general directions, in agreement with previous works.[10-14,23,24] It remains to understand what is the effect of weak fields and the scaling behavior of $\Delta Log C$ upon H.[10-14,17,23,24]

V. MAGNETIC FIELD EFFECTS AT $x = 0$ ("PURE" NSS MODEL):

Some insight as to what may change upon application of the magnetic field may be gained by exploring the "pure" limit $x = 0$ of the NSS model. Since all interferences are constructive in this limit, the magnetic field may only reduce the conductance with respect to its $H = 0$ value and this negative MLog C is to be expected at $x = 0$ and $H = 0$. It turns out, however, that the behavior at any finite, albeit small, field is extremely rich in structure.[25] The behavior of J (and I) may be solved exactly using the transfer matrix approach in a new "diagonal staggered" gauge.

The behavior at any point depends, before anything else, on the commonsurability of ϕ (the flux per plaquette) and the quantum flux ϕ_0.

(i) Commensurate case $\phi/\phi_0 = p/q$:

J has a structure which is a convolution of a periodic structure with period q and its $H = 0$ behavior (which is obtained on all sites on a superlattice with a unit cell q times larger). Upon changing H to bH where b is an integer the period is reduced from q to q/b. This scaling behavior implies a positive MLog C linear in H. However, if the field is changed from one commensurate value to another J may increase or decrease depending on the relative values of the commensurabilities.

(ii) The incommensurate case, ϕ/ϕ_0 is irrational:

The behavior of J looks random-like but is really "quasiperiodic" and deterministic. In this case the average J remains of order 1 and does not increase exponentially as in all previous cases. This implies a much smaller conductance.

VI. CONCLUSIONS AND FUTURE DIRECTIONS

The study of quantum interference effects in the hopping conductivity is only in its early stage. We have presented here the conclusions of our systematic studies of the NSS model which complement and extend upon other works.[10-14,23,24]

We stressed the importance of correlations between the paths which are responsible for the elimination of the h/e period of the AB oscillations at small randomness. They are also crucial to the determination of the scaling properties of the critical hop: The universal lognormal distribution of the conductance and the anomalous spread of the wavefunction in the direction transverse to the hopping. The importance of the effective "attraction" between strongly bounded pairs, which is due to a subtle exchange in replica space, was outlined. In particular we showed how it leads to a larger log C and smaller r.m.s. fluctuations if random phases are present. These may be due to magnetic impurities or spin-orbit interactions.[9] The effect of a weak magnetic field is not to simply induce random phases but its action may have the same tendency since, in accordance with the previous works, we find no change if the field is applied to a model which contains random phases to begin with.

All our investigations were performed within the NSS model which neglects backscattering altogether. It may be adequate in the limit of very strong randomness for which the localization length ξ is of the order of one Bohr radius of the donor. Most experiments[2-9] are performed in more conductive samples for which the localization length may be of one or two order of magnitude larger. In such samples returning loops on a scale of R are still negligible but these an a scale of ξ need to be accounted for.[9] The NSS model which will still apply for distances larger then ξ will have its parameters (V and W) renormalized by the behavior on scales smaller then ξ. These renormalized values (and ξ itself) will be affected by the external field[9]. A more comprehensive model to account for the intricate convolution of backscattering effects with the forward scattering mechanism will be required in order to fully explain all the interference effects in the VRH regime.

ACKNOWLEDGEMENTS

We have benefitted enormously from ongoing interactions and discussions with Zvi Ovadyahu. Fruitful correspondence with Boris Shklovskii and conversations with Ted Castner are acknowledged (YS). Work at the University of Rochester was partially supported by the Eastman Kodak Company (X.R.W), work at M.I.T. was supported by the NSF (grant No. DMR-86-20386), by the Sloan Foundation (M.K.) and by the "Centro de Investigacion y Desarrollo" INTEVEP S. A. Venezuela (E.M.).

REFERENCES

1. For a general overview see *e.g.* R. A. Webb and S. Washburn, Physics Today 4, 46 (1988).
2. A. Hartstein, A. B. Fowler, and K. C. Woo, Physica B 117-118, 655 (1983).
3. Z. Ovadyahu and Y. Imry, J. Phys. C16, L741 (1983).
4. Z. Ovadyahu, Phys. Rev. B33, 6552 (1986).
5. E. I. Laiko, A. O. Orlov, A. K. Sarchenko, E. A. Ilyichev, and E. A. Poltoratsky, Zh. Eksp. Fiz., 93, 2204 (1987). [JETP 66, 1258 (1988)].
6. O. Faran and Z. Ovadyahu, Phys. Rev. B38, 5457 (1988).
7. F. Tremblay, M. Pepper, D. Ritchie, D. C. Peacock, J. E. F. Frost, and G. A. C. Jones, Phys. Rev. B39, 8059 (1989).
8. Ya. B. Poyarkov, V. Ya. Kontarev, I. P. Krylov, and Yu. V. Sharvin, Pis'ma Z. Eksp. Teor. Fiz. 44, 291 (1986) [JETP Lett. 44, 373 (1986)].
9. Y. Shapir and Z. Ovadyahu, "Effects of Spin-Orbit Scattering on Hopping Magnetroconductivity". Submitted to Phys. Rev. B; Z. Ovadyahu, these proceedings.
10. V. L. Nguyen, B. Z. Spivak, and B. I. Shklovskii, Pis'ma Z. Eksp. Teor. Fiz. 41, 35 (1985) [JETP Lett. 41, 42 (1985)] and Z. Eksp. Fiz. 89, 1770 (1985) [JETP 62, 1021 (1985)].
11. V. L. Nguyen, B. Z. Spivak and B. I. Shklovskii, Pis'ma Z. Eksp. Teor. Fiz. 43, 35 (1986) [JETP Lett. 43, 44 (1986)].
12. B. I. Shklovskii, these proceedings.
13. U. Sivan, these proceedings.
14. For a recent comprehensive review see *e.g.* B. I. Shklovskii and B. Z. Spivak

"Scattering and Interference Effects in Variable Range Hopping Conduction" to be published in "Hopping Conduction in Semiconductors" (M. Pollak and B. I. Shklovskii, Editors).
15. Y. Shapir and X. R. Wang, Europhysics Lett. $\underline{4}$., 10 (1987).
16. E. Medina, M. Kardar, Y. Shapir, and X. R. Wang, Phys. Rev. Lett. $\underline{62}$, 941 (1989).
17. E. Medina, M. Kardar, Y. Shapir and X. R. Wang, in preparation.
18. Y. Aharonov and D. Bohm, Phys. Rev. $\underline{115}$, 484 (1959).
19. S. Washburn and R. A. Webb, Adv. Phys. $\underline{35}$, 317 (1986).
20. R. A. Webb and Z. Ovadyahu, private communication.
21. Y. C. Zhang, Phys. Rev. Lett. $\underline{62}$, 979 (1989) and Europhys. Lett. $\underline{9}$, 113 (1989).
22. M. Kardar, Phys. Rev. Lett. $\underline{55}$, 2235 (1985); Nucl. Phys. $\underline{B290}$ [FS20], 582 (1987).
23. U. Sivan, O. Entin-Wohlman, and Y. Imry, Phys. Rev. Lett. $\underline{60}$, 1566 (1988).
24. O. Entin-Wohlman, Y. Imry, and V. Sivan, submitted to Phys. Rev. $\underline{B}$.
25. D. R. Hofstadter, Phys. Rev. $\underline{B14}$, 2239 (1976).

INTERACTIONS AND QUANTUM INTERFERENCE IN THE VARIABLE-RANGE-HOPPING REGIME IN n-TYPE GaAs

F. Tremblay, M. Pepper, R. Newbury, D. A. Ritchie, D. C. Peacock,[*] J. E. F. Frost and G. A. C. Jones

Cavendish Laboratory, University of Cambridge, Madingley Road, Cambridge, CB3 0HE, United Kingdom

G Hill

Department of Electronic and Electrical Engineering, University of Sheffield, Sheffield, S1 3JD, United Kingdom.

ABSTRACT

We report results of low temperature transport measurements made on n-type bulk GaAs with net donor concentration below the metal-insulator transition and exhibiting variable-range-hopping. It is shown that in this material electron-electron interactions are important in both the ohmic regime and under conditions of high electric field in the activationless regime. A low field negative magnetoresistance is observed and is associated to the suppression of destructive quantum interference. Results of a quantitative analysis of the effect are compared with the theoretical predictions and with experimental data obtained in $In_2 O_{3-x}$ in a recent study.

Quantum interference effects in the variable-range-hopping (VRH) regime have been the object of recent interest both theoretically and experimentally. In this mode of conduction, the optimum hopping length R_M exceeds the average distance χ between impurities and electrons can be scattered many times on several possible trajectories in the course of a single hopping event. Because the scattered waves decay exponentially in space, only the shortest possible paths for tunneling contribute substantially to interference. For a sufficiently lightly doped semiconductor with typically $n \leq n_c/2$ where n_c is the critical concentration, contributions by paths with returns and loops can be neglected and only forward going paths are important. Nguen, Spivak and Shklovskii[1] first showed that the interference among the various forward going paths associated with

---

[*] also at GEC Hirst Research Centre, Wembley, Middlesex, HA9 7PP, United Kingdom

the hopping between two sites can be predominantly destructive. Sivan, Entin-Wohlman and Imry[2] (SEI) used a different approach to reach a similar conclusion. The destructive interference is suppressed by a low magnetic field B and consequently the magnetoresistance (MR) is negative. SEI found a MR quadratic in field proportional to a flux square through a coherent area square. Assuming that an electron loses completely its phase after a single optimum hop and that the electron executes a random walk along the direction perpendicular to the hopping direction, this gives

$$\Delta R/R \sim R_M^3 \chi B^2 \tag{1}$$

Possible effects related to electron-electron interactions were not discussed by these authors.

The ohmic VRH conductivity depends on temperature according to

$$\sigma = \sigma_0 \exp(-R_M/\xi) = \sigma_0 \exp[-(T_0/T)^x] \tag{2}$$

where $X = 1/(d+1)$ in d dimensions ($d = 2,3$) for a non-interacting system with a constant density of states at the Fermi level[3], $X = 1/2$ for an interacting system with a parabolic Coulomb gap,[4] and ξ is the localization length. The parameter σ_0 is generally assumed to be temperature independent to a good approximation. Beyond the ohmic regime, in a high electric field E exceeding a critical value E_c, electrons no longer absorb phonons and gain all their energy from the electric field. The conductivity becomes temperature-independent and in this activationless regime, the current is expressed in the form[5,6]

$$I = I_0 \exp[-(E_0/E)^s] \tag{3}$$

where $S = X$,

$$E_0 \sim 1.4k\, T_0/e\xi \tag{4}$$

and T_0, X are defined in Eq. (2). The critical field depends on temperature according to

$$E_c \sim kT/e\xi \tag{5}$$

irrespective of the value of X.

In this paper, we report results of conductivity and low field magnetoresistance measurements performed on bulk n-type GaAs doped below the critical concentration for the metal-insulator transition, $n_c = 1.6 \times 10^{16}$ cm^{-3}, and taken in both the ohmic and activationless VRH regime.

The samples used were molecular-beam-epitaxy grown GaAs. Their concentration and compensation ratio are listed in table I. The sample numbers are denoted W_{xy} where y identifies a particular sample cut from a wafer number x. Details on the samples and techniques of measurements can be found elsewhere.[7,8]

TABLE I. Room temperature electronic concentration n_{300k}, compensation K, best T_0 and X values [Eq. (2)], and best values of the parameters defined by Eq. (6).

Sample	n_{300k} (10^{16}cm^{-3})	$K=N_A/N_D$	T_0 (K)	X	α	$yB^2(K^\alpha)$ ($\alpha=1.22$)
W1a	1.4	0.2	10.5* 1.3	0.17±0.09		
W1b	1.4	0.2	16.5*	0.25±0.12	1.23±0.03	1.64
W2	1.6	0.2	39.7*	0.23±0.15	1.25±0.03	1.21
W3a	0.8	0.4	9.8	0.59±0.08	1.19±0.04	1.191
W3b	0.8	0.4	9.4	0.49±0.06		
W3c	0.8	0.4	9.6	0.51±0.07		
W4	0.7†	0.3	10.5	0.47±0.12	1.22±0.05	1.277
W5	0.44	0.3	27.2	0.50±0.07	1.15±0.07	1.043
W6	0.28†	0.2	59.1	0.60±0.15		

* $T_0(X = 1/4)$ [$T_0 (X = 1/2)$ otherwise]
† n (T=77K)

The temperature dependence of conductivity of the samples was measured in the ohmic regime and in the temperature range $1.3 < T < 4.2K$ with a standard four-terminal low frequency technique. Figure 1 shows a representative result obtained for sample W5. The best value of X in Eq. (2) was evaluated using a least square fit procedure and the error ΔX on X defined by $SD(X_{min} + \Delta X) = 3 SD(X_{min})$ where $SD = \Sigma [\ln \sigma_{expt} - \ln \sigma_{theor}]^2$. The best values of X and T_0 are listed in table I. For $n \leq 0.8 \times 10^{16}$ cm^{-3} the data are in agreement with the Coulomb gap theory, indicating that electron-electron interactions are important in this system. This is in agreement with results obtained by others.[9,10] However, for $n \geq 1.4 \times 10^{16}$ m^{-3} the exponent appeared lower in this temperature range, with a value close to what is expected from a non-interacting system. Interactions are important in an insulator only if the bandwidth of energy Γ providing conduction is less than the width Δ of the Coulomb gap and if the Coulomb gap is not smeared out by the temperature, that is $\Delta > \Gamma > kT$. We define the density of states D/D_{free} = y, $n/n_c = \phi$, α a factor of order unity, $\xi = \Omega \xi_{iso}$ and the dielectric constant $\varepsilon_r = \eta \, \varepsilon_{iso}$,

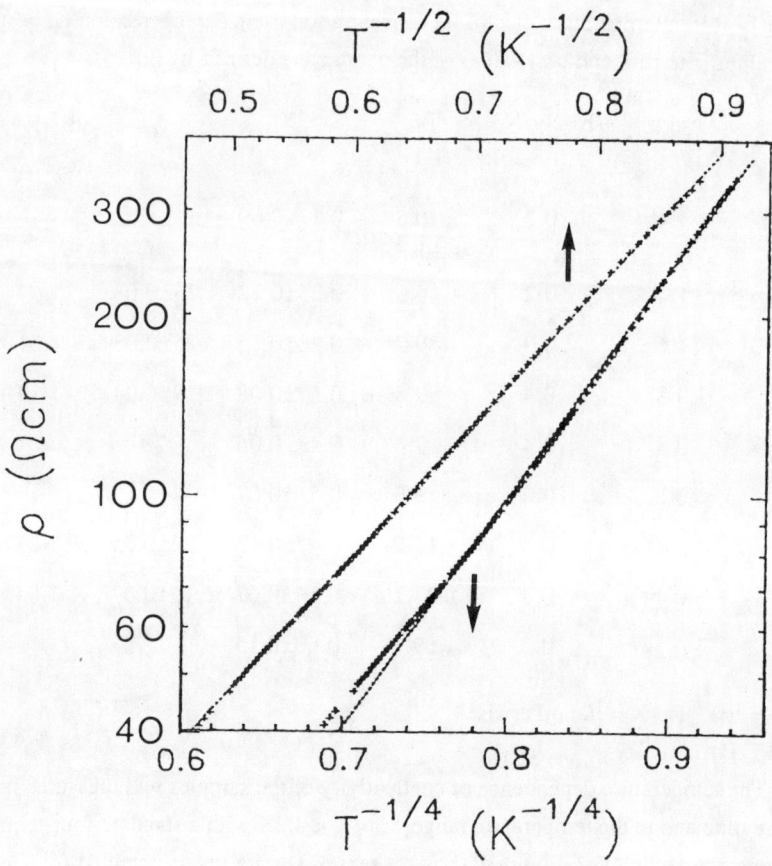

Figure 1. Resistivity as a function of temperature for sample W5.

where ξ_{iso} is the radius of an isolated impurity and ε_{iso} the corresponding dielectric constant. For GaAs, the first condition implies $T_c = 3.7 \, \alpha y \phi^{1/3} \Omega/\eta^2$ and the second $T_c = \alpha T_0(X = {}^1\!/_2) = 360\alpha / \Omega\eta$ where T_c is the critical temperature above which interactions are no longer effective. As the carrier concentration is increased, both Ω and η increase to diverge at the metal-insulator transition. The increase is accompanied by a decrease in the value of T_c. Sample W1a was investigated in the temperature range $0.05 < T < 0.8 K$. Under these conditions the best power obtained is $X = 0.51 \pm 0.06$ with the low value

$T_0(X=1/2) = 0.99$. A smearing out of the Coulomb gap could possibly explain the lower value of X observed in the temperature range 1.3<T<4.2K for this sample.

There are at present a number of reports on various materials showing a decrease of the best value of X in the vicinity of the transition.[11-14] Katsumoto et al[14] also evaluated the magnitude of the dielectric constant in the critical regime in $Al_{0.3}Ga_{0.7}As$:Si. This allowed the calculation of the ratio Δ/Γ. It was found that the inequality $\Gamma<\Delta$ holds up except in the vicinity of the transition where X decreases. On the other hand, in the immediate vicinity of the transition, where the conductivity is more weakly temperature dependent, the evaluated best values of X and T_0 could be affected by a temperature dependence of the pre-exponential factor.[15] Also, the use of the Boltzmann limit for the description of VRH is justified only for $T_0 \geq T$. This can render difficult the analysis of transport in the critical regime $n \sim n_c$.

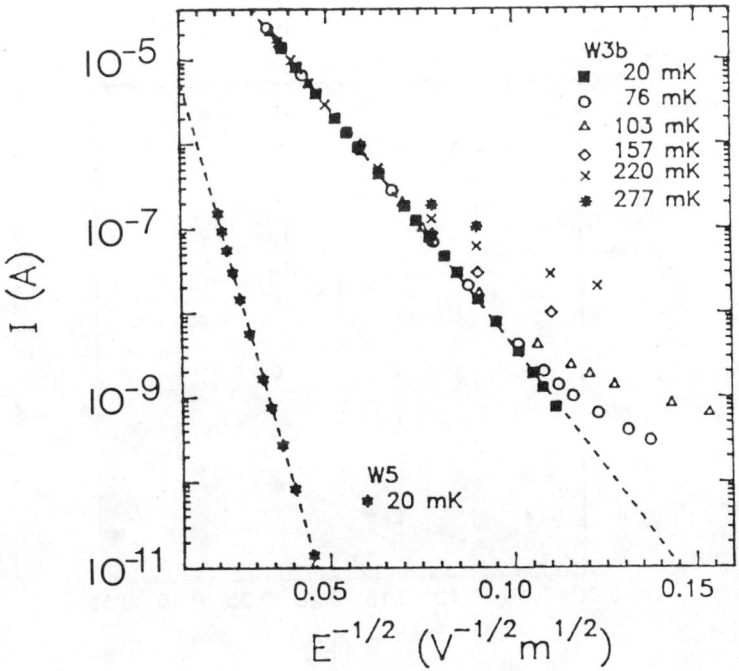

Figure 2. Current as a function of electric field and temperature for samples W3b and W5.

Samples W3b, W3c and W5 were investigated under conditions of high DC-electric field and low temperature T<300mK. Figure 2 shows the electric field dependence of the measured current for samples W3b and W5. It is seen that the curves corresponding to different temperatures merge at a temperature dependent critical field E_c. For reasons of clarity, only a subset of points is shown for each temperature T≥77mK for sample W3. The best fit to Eq. (3) of the data obtained in the activationless regime at T_{bath} = 20mK gives the values of E_o and S. These are listed in table II. The value S ~0.5 is obtained for the 3 samples, indicating that interactions remain important in this regime. The presence of a gap in both regimes of conduction was also recently demonstrated in amorphous silicon.[16] Figure 3 shows the temperature dependence of E_c for W3b and W5. The agreement with Eq. (5) is satisfactory. The evaluation of ξ using Eqs. (4) and (5) gives the values presented in table II. The values $\xi(T_o,E_o)$ and $\xi(E_c)$ are of the same order of magnitude and not too much greater than the radius of an isolated impurity in this material (ξ_{iso}~100Å). The values of $\varepsilon \sim \beta e^2/4\pi\varepsilon_o k_B \xi T_o$ where $\beta = 2.8$

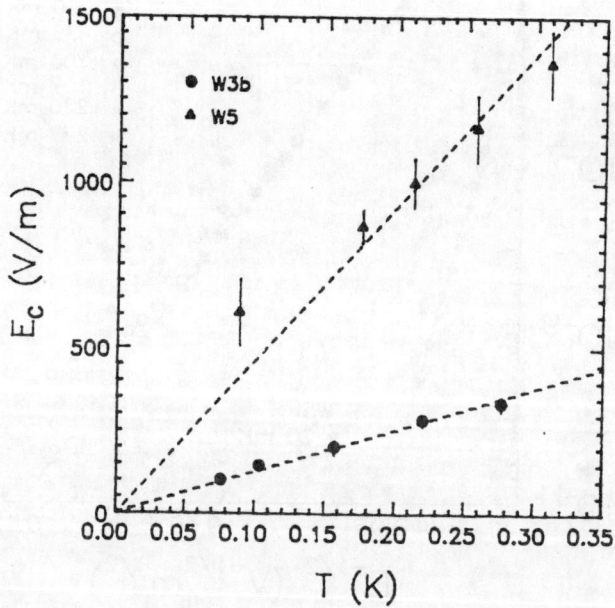

Figure 3. Temperature dependence of the critical electric field for samples W3b and W5.

TABLE II. Best values of S and E_0 obtained from Eq. (3), and calculated values of the localization length ξ and relative permittivity ε.

Sample	E_0 (S=1/2) V/cm	S	$\xi(T_0,E_0)$ Å	$\xi(E_c)$ Å	ε
W3b	175.6	0.54±0.07	647	680	77
W3c	186.3	0.49±0.04	622		78
W5	1344.5	0.43±0.09	245	189	70

are also listed. These are somewhat greater than $\varepsilon_{iso} = 13$, but as they scale with an unknown factor of order unity and the samples are doped not too far from the transition, the results may be considered reasonable.

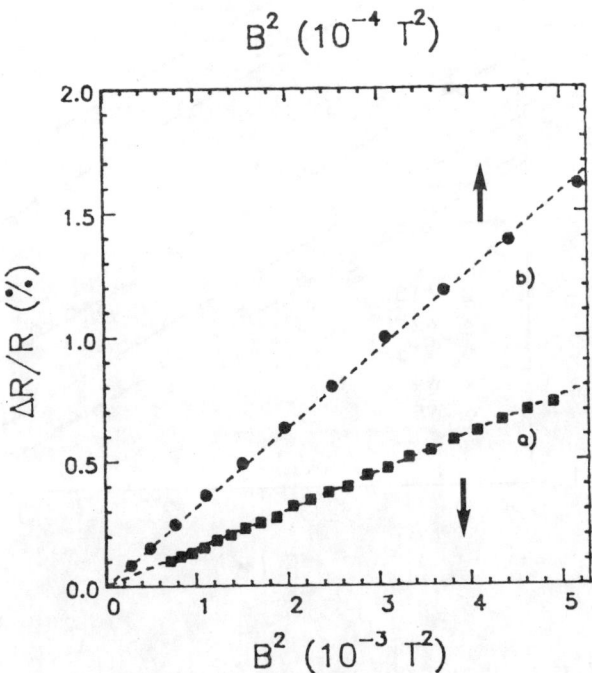

Figure 4. $\Delta R/R$ as a function of magnetic field a) sample W3a : T = 1.4K and ohmic regime b) sample W3b : T_{bath} ~ 20mK, E = 154 V/m in the activationless regime.

The low field MR data in the ohmic regime and temperature range 1.3<T<4.2K is now discussed. It is found that the MR is negative with a variation ΔR/R = (R(B) − R(O))/R(O) initially quadratic in a magnetic field range typically of the order of ~600G at 1.3K to ~1500G at 4.2K (Figure (4a)). The temperature dependence of ΔR/R for a field of 720G is shown on a log-log scale in figure 5. A best fit of the data to an equation of the form

$$\Delta R/R = yT^{-\alpha} B^2 \qquad (6)$$

gives $\alpha \sim 1.2$ for all samples investigated. The values of α and yB^2 are listed in table I. According to Eq. (1), the scaling of the MR with temperature must depend on the shape

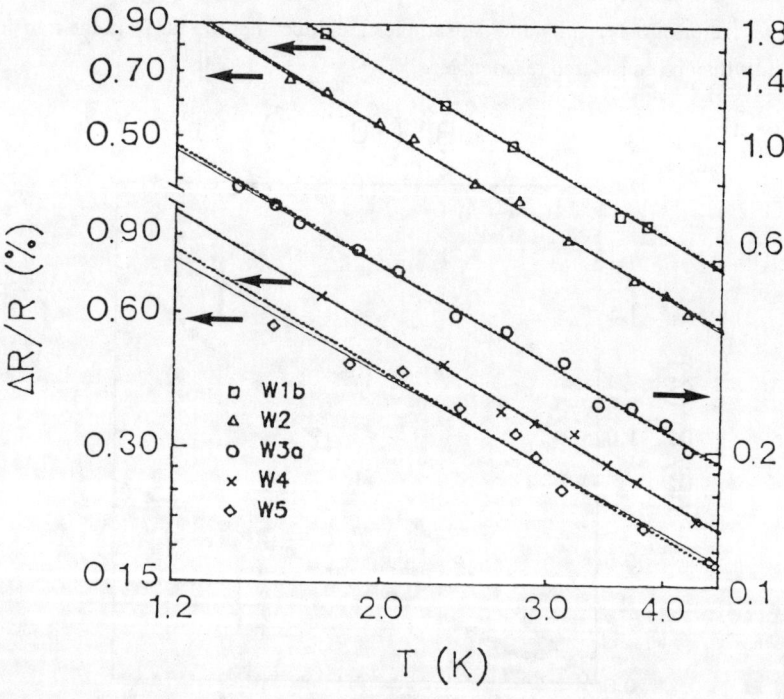

Figure 5. ΔR/R as a function of temperature in the ohmic regime for samples W1b,W2,W3a,W4 and W5. The applied magnetic field is equal to 720G. The solid lines are least-squares fits to the data, and the dotted lines are the least-squares fits to the data for $\alpha=1.22$ in Eq.(6).

of the density of state spectrum. For a non-interacting system it is expected that $\alpha = 0.75$. Assuming that Eq. (1) is also valid for a system with a Coulomb gap, in this case $\alpha \sim 1.5$ should be found. Results in In_2O_{3-x} under conditions where interactions were negligible, showed a low field negative MR nearly quadratic and scaling such that $\alpha \sim 0.75$ in agreement with the SEI model.[17] The scaling in GaAs is stronger but still weaker than the expected value $\alpha \sim 1.5$. Furthermore, it is independent of disorder up to very near the transition where the best value of X drops and the neglect of paths with returns and loops is no longer justified. It should be noted that the use of $\Delta\sigma$ versus temperature rather than $\Delta\sigma/\sigma$ (or equivalently $\Delta R/R$ for small relative changes) would give a disorder dependent value of α and which changes sign as the disorder is increased ($\alpha = 0.85$ to -0.5). This indicates that contrary to the weak localization regime where $\overline{\Delta\sigma}$ is a relevant parameter for quantum interference studies, in the insulating phase the only relevant parameter is the ratio $\Delta\sigma/\sigma$. This was emphasized by Ganor et al.[18]

The present data do not show clear evidence of scaling with the hopping length. This could be because the true shape of the coherent volume is not known or because the assumption that an electron loses completely its phase after a single inelastic hop is not valid under our experimental conditions. This last possibility was expressed by Xie and Das Sarma.[19] Some experiments also suggest that the phase-breaking length $L\phi$ might be greater than R_M.[20]

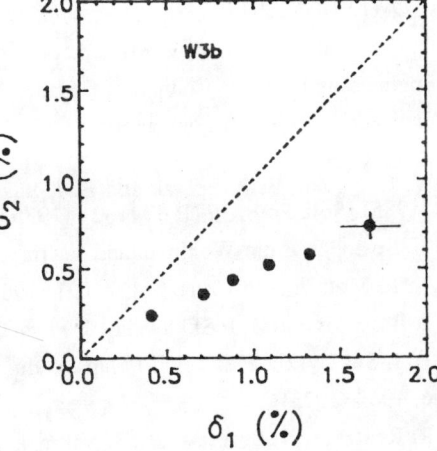

Figure 6. δ_2 versus δ_1 for sample W3b. The dashed line corresponds to the equality $\delta_1 = \delta_2$.

In In_2O_{3-x}, it was further observed that both the conductivity and the MR become temperature independent above E_c, and that if δ_1 (E_o,B,R) is the fractional change of the resistance due to the application of a perpendicular field B while $E_o > E_c$, and δ_2(T,B,R) is the value taken at a temperature T in the ohmic regime such that $R(T) = R(E_o)$, in general $\delta_1 > \delta_2$.[17] A similar study in GaAs reveals that above E_c, the MR becomes activationless together with the conductivity. Above E_c, the MR remains a quadratic function of B (figure 4(b)) and the inequality $\delta_1 > \delta_2$ has been checked to be also true (figure 6). There are several similarities in the behaviour of the MR in these two materials. Further experiments in various materials will be necessary before such features are shown to be general and before the major distinctions between effectively interacting and non-interacting systems are established. Further theoretical work on the effect of interactions, backward steps and cross-over to weak localization is also needed.

CONCLUSION

It is shown that in insulating n-type GaAs electron-electron interactions are important and create a Coulomb gap in both the ohmic VRH regime and under conditions of high electric field in the activationless regime. A low field negative magnetoresistance was observed and a quantitative analysis of the effect was compared with recent theories on quantum interference. Only partial agreement was found. The results were also compared with those obtained in In_2O_{3-x} in a recent study. The similarities and differences were underlined. The latter may well be related to the different degree of strength of interaction in these two systems.

ACKNOWLEDGMENT

One of us (F.T.) acknowledges partial financial support from the Natural Science and Engineering Research Council of Canada. This work was supported by the Science and Engineering Research Council of the United Kingdom.

REFERENCES

1. V. L. Nguen, B. Z. Spivak and B. I. Shklovskii, Zh. Eksp. Teor. Fiz. **89**, 1770 (1985) [Sov. Phys. JETP **62**, 1021 (1985)].
2. U. Sivan, O. Entin-Wohlman and Y. Imry, Phys. Rev. Lett. **60**, 1566 (1988).
3. N. F. Mott, J. Non-Cryst. Solids **1**, 1 (1968).
4. A. L. Efros and B. I. Shklovskii, J. Phys. C **8**, L49 (1975).
5. B. I. Shklovskii, Fiz. Tekh. Polaprovodn. **6**, 2335 (1972) [Sov. Phys. Semicond. **6**, 1964 (1973)].
6. R. Rentzsch, H. Berger and I. S. Shlimak, Phys. Stat. Sol. A**54**, 487 (1979).
7. F. Tremblay et al., Phys. Rev. B**39**, 8059 (1989).

8. F. Tremblay et al., Phys. Rev. B**40**, (1989).
9. R. Rentzsch et al., Phys. Status Solidi (b) **137**, 691 (1986).
10. D. Redfield, Phys. Rev. Lett. **30**, 1319 (1973).
11. D. M. Finlayson and P. J. Mason, J. Phys. C**19**, L299 (1986).
12. D. M. Finlayson, P. J. Mason and I. F. Mohammed, J. Phys. C**20**, L607 (1987).
13. A. G. Zabrodskii and K. N. Zinov'eva, Pis'ma Zh. Eksp. Teor. Fiz. **37**, 369 (1983) [JETP Lett. **37**, 436 (1983)].
14. S. Katsumoto, F. Komori, N. Sano and S. Kobayashi, J. Phys. Soc. Japan (1989).
15. D. M. Finlayson, J. Phys. C**21**, 2792 (1988).
16. A. V. Dvurechenskii, V. A. Dravin and A. I. Yakimov, Pis'ma Z. Eksp. Teor. Fiz. **48**, 144 (1988) [JETP Lett. **48**, 155 (1988)].
17. O. Faran and Z. Ovadyahu, Phys. Rev. B**38**,5457 (1988).
18. O. Ganor, Y. Lereah and R. L. Rosenbaum, Phys. Rev. B**39**, 8764 (1989).
19. X. C. Xie and S. Das Sarma, Phys. Rev. B**36**, 9326 (1987).
20. Ya. B. Poyarkov, V. Ya. Kontarev, I. P. Krylov and Yu. V. Sharvin, Pis'ma Zh. Eksp. Teor. Fiz. **44**, 291 (1986) [JETP Lett. **44**, 373 (1986)].

Orbital Magnetoresistance in the Variable Range Hopping of Indium-oxide samples

Z. Ovadyahu*
IBM Research Division
T. J. Watson Research Center
Yorktown Heights, NY 10598, USA

Abstract

Some basic features of the magnetoresistance observed in the hopping regime in In_2O_{3-x} samples are described. Experimental results are given as a function of disorder, temperature, magnetic and electric fields. The anisotropy of the magnetoresistance for reduced effective dimensionalities, establish the orbital nature of the effect and allows a semi-quantitative estimates of the coherence length involved. The latter turns out to be the temperature (or electric field) dependent hopping length.

The presence of spin orbit scattering is shown to have a non-trivial effect on the field dependence of the magnetoresistance which, qualitatively, looks similar to that observed in the diffusive regime. It is argued that, as long as the localization length exceeds the Bohr radius, the magnetoresistance may be modulated by significant contributions from "backscattering".

Introduction

The possible relevance of quantum interference to the magneto transport properties of systems with hopping conductivity has been first considered almost thirty years ago. Shortly after the basic Aharonov and Bohm (AB) paper appeared, Holstein[1] suggested that the AB effect may be responsible for the Hall effect observed in samples where conduction proceeds via hopping. Interestingly, it was only in this decade that an extensive attention has been given to quantum interference effects in disordered conductors and then, it was the metallic (diffusive) rather than the hopping regime that was heavily investigated. There were two main reasons for that: The seminal paper by Abrahams et al[2] that challenged some previously held views of the metal-insulator transition and the relative ease of fabricating and measuring disordered metallic samples. Thin films, in particular can be readily made from

---

* Permanent address: The Racah Institute of Physics, The Hebrew University, Jerusalem.

virtually any metal or alloy and their resistance can be controlled by a variety of techniques. Many experimental groups turned their effort towards the elucidation of the new theoretical ideas. The most frequently studied transport property was the magnetoresistance (MR) which yielded a wealth of information and it is fair to say that our understanding of the diffusive regime has greatly expanded as a result of these studies. It is now generally accepted that a dominant mechanism for the MR observed in disorderd metals is an orbital AB effect associated with suppression of, so called, backscattering. The sign of the MR may be negative or positive depending on the strength of spin-orbit scattering[3]. The occurrence of similar effects in the variable range hopping (VRH) regime is less widely recognized. Prior to 1985, most researchers in the field held to the view that the main mechanism for MR in VRH systems are due to spin effects or, for sufficiently high fields, shrinkage of the wave functions. The anisotropic, negative MR reported for insulating Si-inversion layers[4] and In_2O_{3-x} films[5] was then, a somewhat surprising a result. Later work confirmed the orbital origin of the effect even deep into the VRH regime. To date, such orbital effects have been observed on at least three different systems: In_2O_{3-x}, GaAs[6] and Si-inversion layers. In all three systems, orbital negative MR exists in the diffusive regime where it is well established that the MR is due to "delocalization". Recently, several theoretical models were offered suggesting another MR mechanism of an orbital nature that seems more appropriate for VRH systems. Thses models are detailed elsewhere in these proceedings (see the papers by Shklovskii[7], Sivan et al[8] and Shapir et al[9]).

In this paper, we describe some of the empirical features of the MR observed in In_2O_{3-x} samples. We demonstrate the orbital nature of the effect in the hopping regime and review the evidence in favor of the conjecture that the hopping length plays the role of the coherence length associated with the phenomenon.

Main Features of the MR

Like many other disordered metals and semiconductors, In_2O_{3-x} exhibit negative MR at low temperatures. This phenomenon is easily observable in indium oxide films at temperatures as high as 77K due to the high resistivity of the material on one hand and its weak electron-phonon interaction on the other. In thin films (thickness, d, of the order of 100Å), the fractional change of the resistance, $\Delta R/R$ is typically, 1% for $H \simeq 0.5T$ at $T \simeq 4K$ over a wide range of sheet-resistances, $R_\square$ as seen in figure 1:

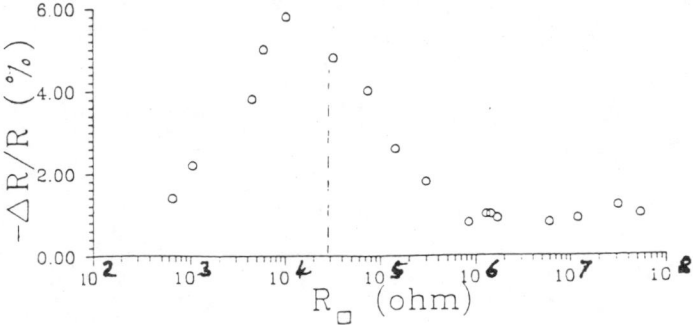

Figure 1. The dependence of the MR on disorder (characterized by the sheet resistance at T=4K). The series of samples shown is for a single batch of In_2O_{3-x} film with d=100Å (where heat treatment was employed to vary $R_\square$). The data show the fractional change of the resistance due to the application of a fixed magnetic field. (H=6kOe, T=4K).

Judging just by the size (and sign!) of the MR, it is hardly possible to distinguish between a film with, say, $R_\square \simeq 1G\Omega$ and that of one with $R_\square \simeq 1k\Omega$. What is perhaps intriguing is that samples with $R_\square > 30k\Omega$ exhibit exponential R(T) while samples with $R_\square < 30k\Omega$ show much weaker, logarithmic temperature dependence which is characteristic of the diffusive regime[3]. To illustrate, In_2O_{3-x} with $R_\square = 1k\Omega$ at T=4K has typically, $R_\square = 1.02k\Omega$ at T=1K while a sample with $R_\square = 10M\Omega$ at T=4K attains resistances of the order of $10G\Omega$ at T=1K. Both films, however, show negative MR at this range of temperatures and $\Delta R/R$ for a fixed magnetic field, has the **same** temperature dependence (namely, $\Delta R/R$ inversely proportional to T) and a similar magnitude! In simple words, while the R(T) of these films unambiguously reflect the expected change in the type of conduction (diffusion versus hopping, the two being "seperated" in figure 1 by the dashed vertical line), the MR does not look all that different. The only **qualitative** difference in the MR of "insulating" vis-a-vis that of "conducting" samples appears to be the functional dependence of $\Delta R/R$ on H (and even that applies for a limited range of disorder). Characteristic MR curves for two VRH samples are shown in figure 2:

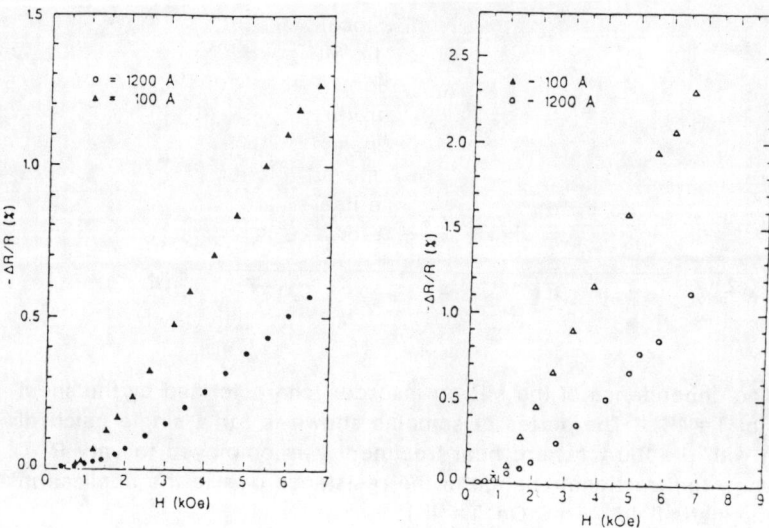

Figure 2. The MR as a function of magnetic field for a typical 2D sample (d = 100Å) and 3D sample (d = 1200Å). Left-T = 4.11K, right-T = 2.17K.

These curves resemble a parabola for small fields which is the behavior seen in diffusive samples. But actually, for films that are not too deeply in the VRH regime, the MR curves do not conform to any simple power law as the next figure illustrates (the dashed line depicts the $\Delta R/R \simeq H^2$ law for the sake of illustration):

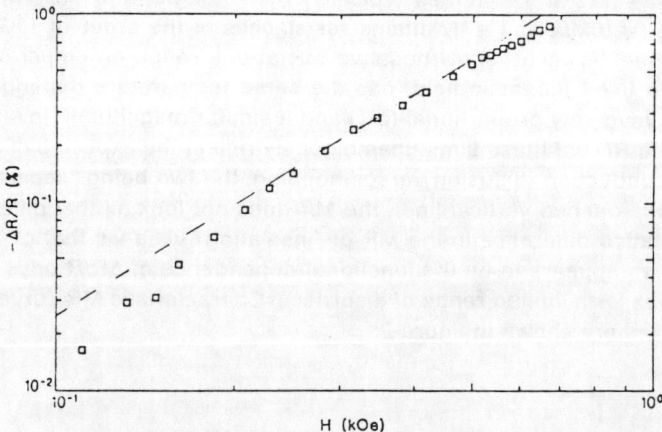

Figure 3. MR as a function of field demonstrating the deviation from a power-law. The sample has d = 100Å, ξ = 25Å, and was measured at T = 4.1K.(see also references 5, 10 and 11).

However, samples that are more strongly localized tend to show again a parabolic MR. Thus, the faster than quadratic MR seems to be a feature peculiar to the immediate vicinity of the transition. This curious field depedence has been discussed several times in the literature[5,10,11] and is still a mystery. A less surprising difference concerns the sensitivity of the MR to the electric fields used in the measurement which is quite pronounced in VRH samples. The size of the MR (as well as the resistance itself), decreases monotonically with electric field as shown in the next figure for a typical case:

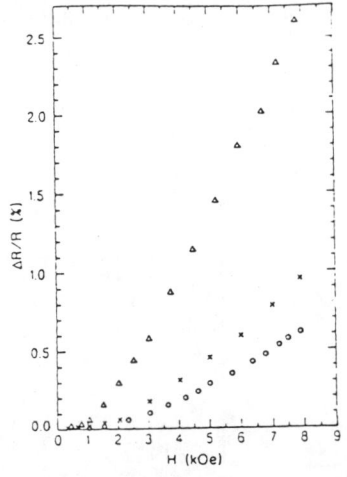

Figure 4a. Dependence of the MR on the longitudunal electric field, F. Sample parameters: d=100Å, ξ=50Å measured at T=4.11K at a perpendicular field of 6kOe. Triangles- F=0.3 volt/cm. Crosses- F=50 volt/cm. Circles- F=200 volt/cm.

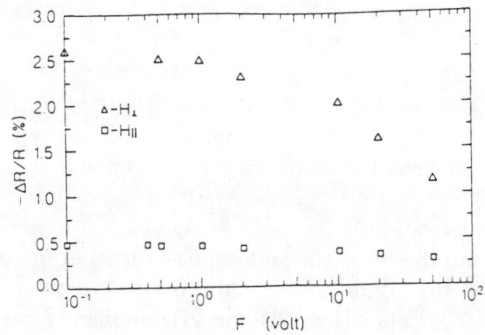

Figure 4b. Dependence of the MR on F for perpendicular versus parallel H of the same magnitude (H=6kOe) at T=4.11K. Samples parameters: d=100Å, ξ=50Å, distance between voltage probes was 0.8cm. Note the strong anisotropy of the effect and the reduced size and anisotropy as F becomes larger.

This, again, is not a qualitatively different behavior and it may merely reflect the inherent sensitivity of the hopping process to electron heating[11].

One interesting feature of the MR that is common to VRH and diffusive samples is the anisotropy[10,11]. For sufficiently thin films, $\Delta R/R$ is bigger when the magnetic field is oriented perpendicularly to the sample plane than when it is parallel to it. This fact is suggestive of an orbital MR mechanism (as opposed to, e.g., effects due to spin alignment). Faran and Ovadyahu[11] argued that the MR anisotropy can be used to estimate the cut-off length associated with the coherent AB effect responsible for the phenomenon. These authors have found that the MR anisotropy, quantified by $\beta = \{\Delta R/R(H_\perp)\}/\{\Delta R/R(H_\parallel)\}$, scales with the hopping-length, r. Some of these results are summarized in the following figure:

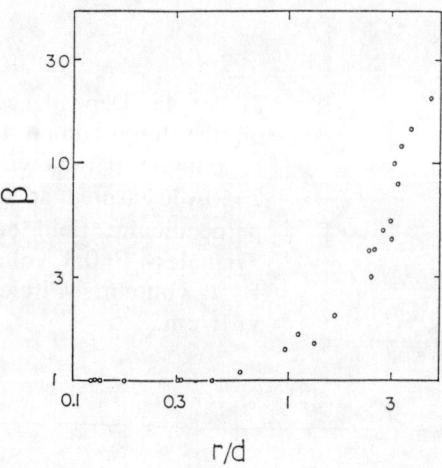

Figure 5. The dependence of the anisotropy parameter (see text) on the ratio of hopping-length/film-thikcness (c.f., reference 11 for details).

The hopping length, r, and the localization length, ξ, used in the above figure were estimated from the R(T) data for the films used. Typical such data are depicted in figure 6 and 7 for effectively 2D and 3D samples respectively.

The values of ξ's attached to the various curves were calculated using the standard expressions for Mott's VRH.

A natural way to explain all these findings is to assume that the hopping length, whether controlled by temperature or electric field, is the coherence

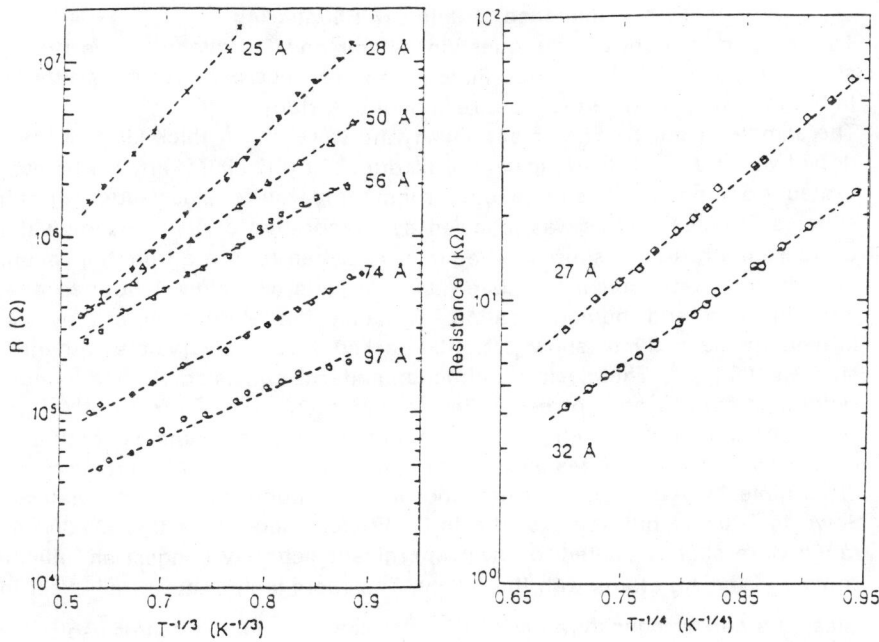

Figure 6. Temperature dependence of the resistance for thin (100Å) In_2O_{3-x}. The localization length associated with each curve was calculated from the logarithmic slopes using Mott's VRH formulae for the 2D case.

Figure 7. Same as in figure 6 but for 3D In_2O_{3-x} films (d = 1200Å).

length that determines the cut-off for the quantum interference effect associated with the MR. One is then led to consider the conjecture that the **only** difference between VRH and diffusive samples as far as the orbital MR is concerned, is that r replaces the inelastic diffusion length, L_{in} which, in turn, is known to set the scale for quantum interference in the latter regime. In other words, the question is how deeply insulating should a system be for the backscattering mechanism to be insignificant.

Influence of Spin-Orbit Scattering

To shed further light on this question, Shapir and Ovadyahu[12] have recently studied the MR of indium oxide films in the presence of spin-orbit scattering. In the following we give the main results of this study.

The samples used by Shapir and Ovadyahu were 150Å thick films of In_2O_{3-x} doped with Au using the following procedure: Pure (99.997%)In_2O_3 was evaporated from an e-gun source onto room-temperature glass-slides through suitable SS-masks. This was followed by evaporation of 5Å mass-equivalent of Au from a Knudsen source. The films were then removed from the vacuum system and placed on a hot-plate (200c) for approximately 1 hour to affect crystallization and homogenization. Transmission elctron-microscopy performed on such films, showed tightly-packed In_2O_3 polycrystals with grain sizes of 100-300Å. The electron diffraction patterns consisted of the previously reported[13] set of rings corresponding to bcc In_2O_{3-x} with no trace of Au precipitation. Specrophotometry revealed that the Au-doped samples are 10-20% less transmissive in the visible then the pure In_2O_{3-x} films (a difference easily discernible by the eye). No extra specific absorption modes were detected down to $200cm^{-1}$ but some of the In_2O_3 Frolich modes[14] in the $600-300cm^{-1}$ range were slightly shifted to lower energies which may suggest an intimate contact of the Au atoms with the lattice. The room-temperature Hall-effect indicated a carrier concentration of 10^{20} electrons/cm^3 that is quite close to the value usually found[14] in In_2O_{3-x} samples.

Five different batches of Au-doped In_2O_{3-x} were studied. Within each batch, several samples (in the form of 5x8mm strips) were measured, having $R_\Box$ (samples are identified by their $R_\Box$ at $T=4.11K$), ranging from $1.5k\Omega$ to $100M\Omega$. The different $R_\Box$ values were generated by heat-treatment as described before[14] in detail for undoped In_2O_{3-x} samples.

Resistance and MR data were taken by a standard 4-probe dc technique employing the high-impedance Keithley's current-source (K220) and electrometer (K617) controlled by a PC. The MR data points represent computer averaged results of 2 to 10 bipolar-readings, depending on noise level. Measurements were made in a ^{4}He immersion cryostat mounted in the air gap of a split-coil electromagnet. The magnetic field was applied perpendicularly to the samples plane except where otherwise noted. Temperature was measured by means of a calibrated Ge Thermometer. In the following, we describe results for one particular batch for which all the diagnostic procedures described above were made simultaneously with the transport measurements.

MR data for several $R_\Box$ values are shown in figures 8, 9 and 11 and some aspects of these are compared with data for undoped films in figure 10.

Roughly speaking, the results fall into two qualitatively different groups: Au-doped samples with $R_\square < 1M\Omega$ exhibit a pronounced positive component remeniscent of that usually found in metal films in the presence of moderately strong spin-orbit scattering[15]. In fact, up to a certain field, H*, the MR is **positive**. H* is well defined experimentally in samples with $R_\square < 500k\Omega$ (figure 9). This feature is not observed in undoped films[11], where, independent of $R_\square$, the MR is negative. Au-doped samples with $R_\square > 2M\Omega$, on the other hand, show only **negative** MR within the range of measurements. Au-doped samples in this group could not be told apart from undoped films with comparable $R_\square$ (except for their optical characteristics mentioned above). In particular, such films exhibited only negative MR which also showed field, temperature and disorder dependences in close similarity with those of undoped films (c.f., figures 8, 9 and 10 and reference 11).

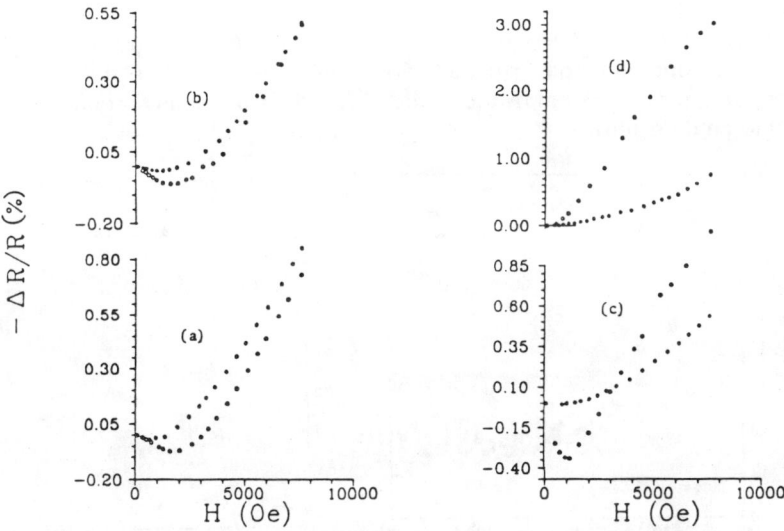

Figure 8. MR curves for Au-doped samples for various $R_\square$ values (full-circles; $T \simeq 4.1K$, empty-circles; $T \simeq 1.4K$).
(a)-A weakly localized sample, $R_\square = 4.5k\Omega$. The other data sets are for samples exhibiting VRH conductivity with the following parameters:
(b)-$R_\square = 45k\Omega$, $T_0 = 32K$, $\xi = 200Å$.
(c)-$R_\square = 300k\Omega$, $T_0 = 1100K$, $\xi = 35Å$.
(d)-$R_\square = 4.5M\Omega$, $T_0 = 11000K$, $\xi = 12Å$.

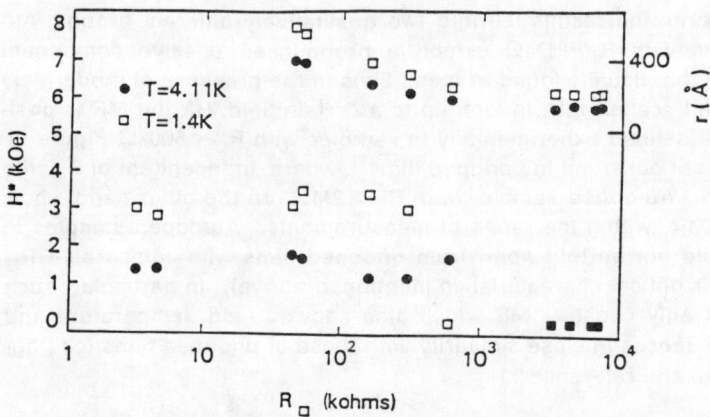

Figure 9. The dependence of the "zero-crossing" field, H* (lower set of data), and the hopping-length, r(upper set) on the disorder. Note that most of the variation in r occurs **below** the "critical" $R_\square$. (H* = 0 means that no positive MR could be detected).

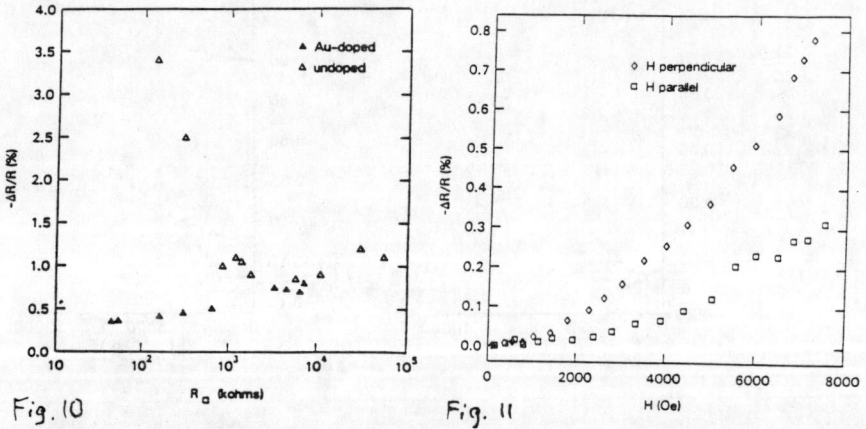

Fig. 10   Fig. 11

Figure 10. MR for H = 6.5kOe and T≃4K as a function of disorder for doped and undoped In_2O_{3-x} samples. Note the similarity above $R_\square > 1M\Omega$ (and compare with figure 2).

Figure 11. MR (at T = 4.11K) for a doped sample with $R_\square = 6.4M\Omega$ illustrating the orbital origin of the effect (c.f., reference 3).

The "history" of measurements in this batch was as follows: The as-prepared film had $R_\square = 4.5k\Omega$ and was measured first. Heat-treatment was then used to generate the samples with $R_\square$ of 300, 4500, 650, 7100, 54, 3300, 3.5 and 6400kΩ respectively, measured by that order. Clearly, the salient features of the transition observed in figures 9 and 10 are independent of "history". That, along with the diagnostic tests alluded to above, suggests that this transition is "disorder-driven" rather than an artifact due to, e.g., Au expulsion or segragation to grain-boundaries which could hardly be expected to be reversible. To better understand the role played by the Au impurities, several additional batches with increasing amounts of Au were made and measured. For batches with Au doping exceeding 15(atomic)% , the above mentioned reproducibility did not hold. Heat treatment cycles resulted in Au precipitation and the optical transmission of such films increased upon rinsing in acetone. This apparently suggests a Au miscibility limit in In_2O_{3-x} of the order of 10%. Since In_2O_{3-x} has a 10% oxygen defficiency[13] it seems plausible to assume that most of the Au atoms in our studied films reside on oxygen voids. It was also found that the Au inclusion allows less lattitude in the range of $R_\square$ realizable with heat treatment: The lowest $R_\square$ achievable within a given batch increased with the Au-concentration. The choice of the particular ($\simeq 2\%$) Au-concentration, was largely dictated by the desire to be able to include diffusive samples in the batch to serve as a useful reference.

Shapir and Ovadyahu have argued that these results indicate that, up to a certain degree of disorder, the main mechanism for the MR is backscattering or, in simple terms, just an extension of the behavior seen in the diffusive regime. One expects the direct effect of backscattering to become increasingly less important as r/ξ grows bigger: The "delocalizing" effect of the field is associated with a reduction in the quantum mechanical probability of the electron to "return". In the strongly localized regime this probability is already close to unity due to short range ($\simeq \xi$) scattering and quantum interference on scale r (such that $r > > \xi$), is only a small added correction. The oriented path interference, on scale r, is likewise an exponentially small entity. However, its contribution modulates the probability to **forward-scatter** which, in the VRH regime, is, by itself, exponentially small function of r/ξ. This qualitative argument merely means that as r/ξ becomes larger than unity, the forward scattering interference must become the dominant one as far as **observability** of orbital MR effects is concerned. (It is tacitly assumed, as suggested by the theoretical models[7,8,9], that a non-linear ensemble averaging is inherently involved). But, returning loops, specifically ignored by these models, may still be important even in the $r > > \xi$ case. This should be considered as an indirect effect due to backscattering and it may be operative as long as $\xi > a_0$. (a_0 is the Bohr radius). We refer to the effect of backscattering (important on scales $< \xi$) on the probability to forward scatter (on scales $\simeq r$). The latter is the outcome of compounding many (elastic) tunneling events along each Feynman trajectory. The number of such steps is, typically, larger than r/ξ. Clearly, the probability for the local (scale $<\xi$) events will be affected by backscattering; when a magnetic field is applied, it will be enhanced or suppressed dependent on the local relative strength of spin orbit scattering. Due to the smallness of the relevant area ($\simeq \xi^2$) these local effects are individually small. Nevertheless, they may be amplified to an observable magnitude through the compounded probability associated with the forward scattering since the underlying time reversal symmetry dictates a common trend in the local steps. More formally, the effective tunneling matrix elements, V, which enter in the Nguyen et al[16] model (where even small returning loops are ignored), will be normalized if returning loops **are** present and this renormalization may be sensitive to the details of the local effect of the field. If this is indeed true, then the current versions of the oriented path models are strictly valid only in the $\xi/a_0 \simeq 1$ limit. We note that the formal justification for neglecting "returning-loops" in the oriented path models, rests on a numerical simulation[16] for a system where $\xi \simeq a_0$ and $r/\xi > 20$ which, to our knowledge, may exclude from comparison most of the experimental results currently in print. It is emphasized that, in the strongly localized regime, it is r/ξ that appears to be the physically relevant parameter for quantum interference phenomena: The probability-amplitudes for the quantum interference, of either kind, as well as their relative significance depend on r/ξ. That does not mean that r/ξ is the only relevant parameter for the problem at hand even for the non-interacting system. In fact, the arguments raised above suggest that ξ/a_0 should also be considered. It is rather that $R_\square$ is certainly not expected to be the "universal" single parameter for the strongly localized system. Inasmuch as the contribution of backscattering to $\Delta R/R$ vis-a-vis that of the oriented-path interference is concerned, r/ξ appears to be a natural measure of disorder. This point should be borne in mind when comparing results obtained on systems with highly disparate carrier densities. To illustrate; the GaAs specimens reported by Laiko et al[6] had $R_\square \simeq 10^7 \Omega$ for r/ξ of 5-6. This should be compared with a In_2O_{3-x} film with $R_\square$ that is 4 orders of magnitude smaller for the **same** r/ξ. Similarly, the sample of PbTe studied by Poyarkov et al[17], had r/ξ of only[16] 3-4 and it is dubious whether it can be treated as being in the limit of very strong disorder despite the huge ($>10^{11}\Omega$) value of the resistance measured.

In summary, some of the evidence for the existence of orbital MR in the VRH regime of indium oxide samples has been reviewed. We have presented experimental results pertaining to the influence of spin-orbit scattering on this intriguing phenomenon. The insensitivity of the MR to spin-orbit at sufficiently strong disorder, seems to suggest the relevance of the oriented-path mechanism in this limit. At the same time, attention is called to the possible role of backscattering for intermediate degree of disorder. Our experiments may be interpreted as indicating that returning loops may significantly influence the magneto-transport in the VRH regime.

Many questions remain to be answered. Perhaps the most important of which is the generality of the phenomenon, or rather, the apparent lack of it. It is realized that in most VRH systems the MR, if at all present, has been usually attributed to anything **but** a quantum interference mechanism. We believe that, in part, the reason for that is the very limited study of VRH systems of low dimensionalities where the field anisotropy (if present), might have helped to identify the MR mechanism. It is emphasized that it is the anisotropy, rather than the sign of the low field MR that is the characteristic feature of the effect. It would then, be of interest to re-examine some of the "traditional" VRH system to determine what part of their MR is due to a quantum interference mechanism of a kind described above. It would be very surprising if it turns out that quantum interference effects are less general in the hopping regime than they are in the diffusive one.

This research has been partially supported by a grant adminisrered by the Israel-US Binational Science Foundation.

1. L. Friedman and T. Holstein, Ann. Phys., **21**, 494 (1963).
2. E. Abrahams, P. W. Anderson, D. C. Licciardello and T. V. Ramakrishnan, Phys. Rev. Lett., **43** (1979).
3. G. Bergmann, Phys. Rep. **107**, 1 (1984); S. Kobayashi and F. Komori, Prog. Theor. Phys. **84**, 224 (1985).
4. A. Hartstein, A. B. Fowler, and K. C. Woo, Physica B **117-118**, 655 (1983).
5. Z. Ovadyahu and Y. Imry, J. Phys. C, **16**, L471 (1983).
6. E. I. Laiko, A. O. Orlov, A. K. Savchenko, E. A. Ilyichev, and E. A. Poltoratsky, Zh. Eksp. Theor. Fiz., **93**, 2204 (1987); F. Tremblay, M. Pepper, D. Ritchie, D. C. Peacock, J. E. F. Frost, and G. A. C.Jones, Phys. Rev. B **39** 8059 (1989).
7. B. I. Shklovskii, these proceedings and V. I. Nguyen, B. Z. Spivak, and B. I. Shklovskii, Pis'ma Zh. Eksp. Theor. Fiz., **89**, 1770 (1985).
8. U. Sivan, O. Entin-Wohlman, and Y. Imry, Phys. Rev. Lett. **60**, 1566 (1988) and these proceedings.
9. Y. Shapir and X. R. Wang, Europhys. Lett., **4**, 1165 (1987); E. Medina, M. Kardar, Y. Shapir and X. R. Wang, Phys. Rev. Lett., **62**, 941 (1989). and these proceedings.
10. Z. Ovadyahu, Phys. Rev. B **33**, 6552 (1986);
11. O. Faran and Z. Ovadyahu, Phys. Rev. B **38**, 5457 (1988).
12. Y. Shapir and Z. Ovadyahu, Phys. Rev.B (to be published).
13. Z. Ovadyahu, B. Ovryn, and H. W. Kraner, J. Electrochem. Soc., **130**, 917 (1983).
14. Z. Ovadyahu, J. Phys. C, **19**, 5187 (1986).
15. D. Abraham and R. Rosenbaum, Phys. Rev. B **27**, 1409 (1983).
16. B. I. Shklovskii and B. Z. Spivak, **"Scattering and Interference Effects in Variable Range Hopping Conduction"** (preprint) and private communication.
17. Ya. B. Poyarkov, V. Ya. Kontarev, I. P. Krylov and Yu. V. Sharvin, JETP Lett. **44** 373 (1985).

HOPPING PROCESSES IN INDIUM OXIDE FILMS

Meir Nissim and Ralph Rosenbaum
Tel-Aviv University, Raymond and Beverly Sackler Faculty of Exact Sciences, School of Physics and Astronomy, Ramat-Aviv, 69978, Israel

ABSTRACT

Strongly insulating thin films of granular In_2O_{3-x} exhibited a Mott VRH temperature dependence in the R vs T data; the observed small negative magnetoresistance (MR) could be explained using the quantum interference theory of Sivan, et al. However for less strongly insulating films such that $T_{Mott} < 100$ K, this theory failed to account for the almost linear dependence of the MR upon magnetic field. There is also no explanation for the temperature independent behavior of the Hall constant. In very thin amorphous In_xO_y films, a Coulomb VRH dependence of R vs T was observed. The MR was positive and small and could be explained by the wave function "shrinkage" theory. The temperature dependent Hall constant data could be fitted to an expression suggested by Nemeth and Muhlschlegel. Below 4 K, experimental results are presented on a new hopping process that dominated over the Coulomb interaction process.

I. INTRODUCTION

Below the metal-insulator transition, the charge carriers become localized to impurity sites. When the thermal energy k_BT is much smaller than the energy difference between adjacent localized states, the localized wave function has to sample an increasingly larger volume to find a state whose energy is close enough to be accessible. Charge transport takes place by a hopping process, known as variable-range-hopping (VRH). The process is a tunneling one in which thermal activation is supplied by the phonons. For a constant density of states (DOS), the temperature dependence of the resistance follows Mott's famous law: $R(T) = R_o \exp(T_{Mott}/T)^{1/4}$ for a 3D film and $R(T) = R_o \exp(T_{Mott}/T)^{1/3}$ for a thin 2D film. [1] For a long time it had been assumed that VRH should be accompanied by a strong positive magnetoresistance (MR). The reason for the positive MR arises from the "shrinkage" of the wave function in directions perpendicular to the applied magnetic field, thus making the hopping process less probable. Indeed, many doped semiconductors exhibit gigantic positive MR's.[2]

II. EXPERIMENTAL RESULTS ON GRANULAR In_2O_{3-x} FILMS - MOTT VRH REGIME

However, there are some experimental investigations in which negative MR's are observed in the Mott VRH regime.[3],[4],[5] The negative MR's reported by Ovadyahu in granular In_2O_{3-x} films motivated Sivan, et al. to develop a theory based upon interference among random paths in the hopping process.[3],[6]

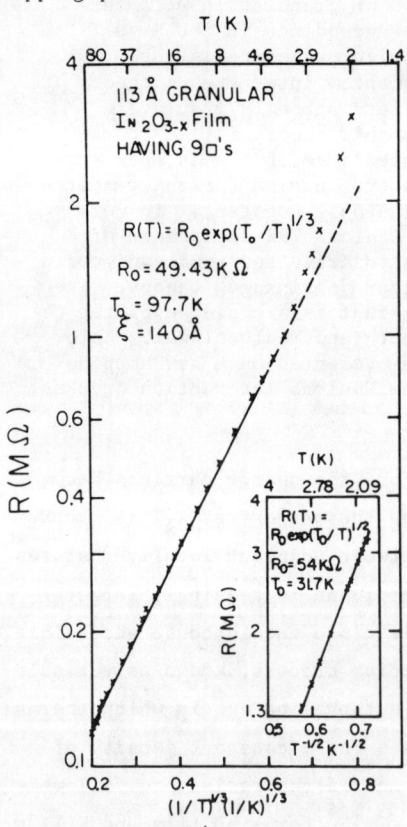

Fig. 1: R vs $T^{-1/3}$ exhibiting the Mott law. Note the small value for $T_0 = T_{Mott} = 97.7K$. Below 4 K, the data followed the Coulomb VRH law.

We have recently observed the Mott law in granular transparent In_2O_{3-x} films. The films were fabricated by evaporating In_2O_3 powder in an O_2 atmosphere of 5×10^{-4} mmHg onto heated glass substrates at 220 C. Provided that the film is strongly insulating, namely $T_{Mott} > 250$ K, many aspects of the Sivan, et al. theory are confirmed including the negative sign of the MR, the quadratic field dependence of the MR in small fields, and the linear field dependence in moderate fields. The reader is referred to Ref. 7] for details.

Granular In_2O_{3-x} films that are not so strongly insulating and localized still exhibit the Mott VRH dependence in R vs T as shown in Fig. 1. For 10 K < T_{Mott} < 100 K, the magnetoconductance (MC) no longer follows a quadratic B^2 dependence but almost a linear dependence as shown in Fig. 2. Simi-

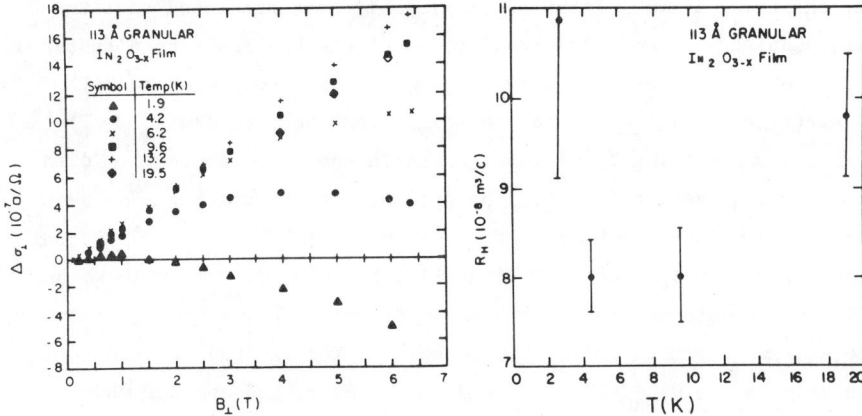

Fig. 2: MC, Δσ, vs B data exhibiting the surprising linear dependence upon B for small fields.

Fig. 3: Temperature independent property of the Hall constant R_H. Theory predicts an $\exp(0.15 T_{Mott}/T)^{1/4}$ law.[8]

lar behaviors have been observed by Faran and Ovadyahu.[8] This surprising result needs theoretical clarification.

Moreover, the Hall constant $R_H = V_H d/IB$ appears to be independent of temperature as illustrated in Fig. 3. Tousson and Ovadyahu have also observed the temperature independent behavior.[9] This is surprising in view of the prediction by Gruenewald, et al. in which an easily observable increase in the Hall constant should be detected with decreasing temperatures according to the relation $R_H \alpha \exp(0.15 T_{Mott}/T)^{1/4}$. Such an dependence of R_H has been observed in SiAs samples by Koon and Castner;[10] but we emphasize that their samples exhibited positive MR effects characteristic of the wave function "shrinkage" process. There has been no theoretical prediction yet for the temperature dependence of R_H using the interference model of Sivan, et al.[6]

III. EXPERIMENTAL RESULTS ON THIN AMORPHOUS $In_x O_y$ FILMS - COULOMB VRH REGIME ARISING FROM ELECTRON INTERACTIONS

At low temperatures such that $T \ll T_{Mott}$, the Coulomb interaction between hopping sites leads to a gap in the density of states (DOS) at the Fermi level. Efros and Shklovskii have predicted that the DOS will vanish at E_F.[11] This depletion of the DOS leads to a faster diverge of the resistance, namely - the Coulomb VRH law where $R(T) = R_o \exp(T_{ES}/T)^{1/2}$.[2] Such a crossover from the Mott $\exp(T_{Mott}/T)^{1/4}$ to the Coulomb $\exp(T_{ES}/T)^{1/2}$ laws occurs in the granular In_2O_{3-x} data near 4 K as seen in Fig. 1. The crossover temperature T_c has been predicted by Entin-Wohlman, et al. to be $T_c = T_{ES}^2/T_{Mott}$;[12] using the values of T_{ES} and T_{Mott} from Fig. 1, we predict $T_c \simeq 10$ K. Below 2K, the MR data in the Coulomb regime takes on positive signs; owing to equipment limitations, we have not studied the Coulomb VRH regime in granular In_2O_{3-x} films in detail. However, Schoepe has made an excellent study on a doped Ge thermometer in the Coulomb VRH regime.[13]

Fortunately, the amorphous opaque form of indium oxide, namely - In_xO_y exhibits the $\exp(T_{ES}/T)^{1/2}$ law in R vs T over a wide temperature range between 90 K to 9 K. This Coulomb VRH dependence is observed only in relatively thin In_xO_y films having thicknesses d less than 500 Å. The positive MR data can be fitted nicely using one of two theories. One expression is based upon shrinkage of the wave function in directions perpendicular to the applied field. This expression appears as Eq. 9.2.12 in Shklovskii and Efros's book.[2] The second expression comes from the spin-splitting of the electron energies upon application of the magnetic field and appears as Eq. 3.43b in Lee and Ramakrishnan review paper.[14] However, this expression has to be modified by a multiplicative prefactor - $1/\exp(T_{ES}/T)^{1/2}$ to

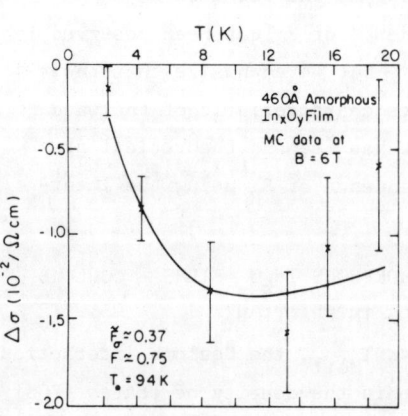

Fig. 4: Comparison of the modified Lee and Ramakrishnan expression for the MC (solid line) to the MC data taken at different temperatures in the fixed field of 6 Tesla. At these temperatures, the R vs T data were described by the Coulomb VRH law with $T_{ES} = T_o = 94$ K.

account for the strong temperature dependence of the conductivity in these insulating films. Fig. 4 shows the fit using the Lee and Ramakrishnan expression to the MC data. The reader is referred to Ref. 15] for details.

Recently, Nemeth and Muhlschlegel have predicted a strong temperature dependence of the Hall constant, namely $-R_H \propto \exp(0.602 T_{ES}/T)^{1/2}$ in the Coulomb VRH regime.[16] We have compared our poor quality Hall constant data to their prediction as shown in Fig. 5. Agreement is satisfactory; but there is great need for much better quality Hall constant data on highly insulating films.

The above amorphous $In_x O_y$ films were fabricated by evaporating In_2O_3 powder in an O_2 atmosphere of 5×10^{-5} mmHg onto RT cooled substrates. the evaporation rate was similar to that of the granular films - 1/4 Å/s.

IV. EXPERIMENTAL RESULTS ON AMORPHOUS $In_x O_y$ FILMS AT VERY LOW TEMPERATURES - A NEW HOPPING PROCESS?

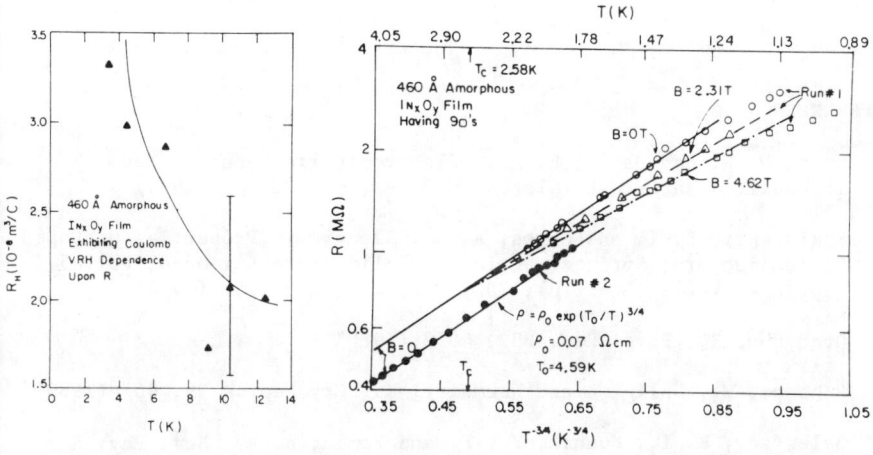

Fig. 5: Comparison of the Nemeth and Muhlschlegel expression (solid line) to the Hall constant, R_H. For this film, $T_{ES} = 13.9$ K.

Fig. 6: R vs T data in three different magnetic fields showing the crossover temperature T_c of 3 K. Above 3K, the MR is positive, and below it, the MR is negative.

There is some experimental evidence now for a new hopping process that dominates at very low temperatures such that $T \ll T_{ES}$. We observe a crossover regime near 3 K in the amorphous In_xO_y films where the R vs T data deviate from an $\exp(T_{ES}/T)^{1/2}$ law and **follow** an $\exp(T_o/T)^{3/4}$ dependence as shown in Fig. 6. Below 2 K the MR becomes negative as shown in Fig. 7 (a relatively strong effect as seen in Fig. 6), and the MR takes on a linear B dependence for small fields. We have no Hall voltage data in this regime. Clearly much more data are needed to clarify the experimental properties of this process.

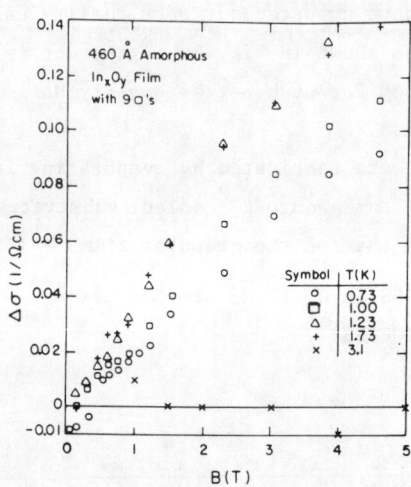

Fig. 7: The MC data vs B illustrating the linear B dependence at low fields and the negative sign of the MR data.

REFERENCES

1. Mott, N. F. and Davis, E. A., Electronic Processes in Non-Crystalline Materials (Clarendon Press, Oxford), 1 (1979).

2. Shklovskii, B. I. and Efros, A. L., Electronic Properties of Doped Semiconductors, Springer Series in Solid State Sciences, vol.45 (Springer-Verlag, Berlin), 202, 212, 228, 240, 357 (1984).

3. Ovadyahu, Z., Phys. Rev. B33, 6552 (1986).

4. Schoepe, W., Uhlig, K and Neumaier, K., Cryogenics 29, 467 (1989).

5. Belevtsev, B. I., Komnik, Yu. F. and Fomin, A. V., Sov. Phy. Solid State 30, 1598 (1988).

6. Sivan, U., Entin-Wohlman, O. and Imry, Y., Phys. Rev. Lett. 60, 1566 (1988)

7. Ganor, O., Lereah, Y. and Rosenbaum, R. L., Phys. Rev. $\underline{B39}$, 8764 (1989).

8. Faran, O. and Ovadyahu, Z., Phys. Rev. $\underline{B38}$, 5457 (1988).

9. Tousson, E. and Ovadyahu, Z., Solid State Commun. $\underline{60}$, 407 (1986); Phys. Rev. $\underline{B38}$, 12290 (1988).

10. Koon, D. W. and Castner, T. G., Solid State Commun. $\underline{64}$, 11 (1987); Phys. Rev. Lett. $\underline{60}$, 1755 (1988).

11. Efros, A. L. and Shklovskii, B. I., J. Phys. $\underline{C8}$, L49 (1975).

12. Entin-Wohlman, O., Gefen Y. and Shapira, Y., J. Phys. $\underline{C16}$, 1161 (1983).

13. Schoepe, W., Z. Phys. $\underline{B71}$, 455 (1988).

14. Lee, P. A. and Ramakrishnan, T. V., Rev. Mod. Phys. $\underline{57}$, 287 and 308, (1985).

15. Nissim, M., Lereah, Y. and Rosenbaum, R., preprint entitled "Dominating Coulomb Interaction Effects in Amorphous In_xO_y Films"; accepted for publication in Phys. Rev. B.

16. Nemeth, R. and Muhlschlegel, B., Solid State Commun. $\underline{66}$, 999 (1988).

Chapter 3

MESOSCOPIC SYSTEMS

DISTRIBUTION FUNCTION OF CONDUCTANCE OF FINITE SIZE INHOMOGENEOUS BARRIER STRUCTURES

M.E.Raikh and I.M.Ruzin
A.F.Ioffe Physical Technical Institute
Leningrad, 194021, USSR

ABSTRACT

The conductance distribution function of a finite size barrier with exponential spread of local conductivity is studied. This function is shown to be universal and independent of concrete origin of disorder. The shape of distribution function is different in three exponentially wide intervals of the sample area.

1. INTRODUCTION

Consider a plane sample representing a barrier for the tunneling of an electron. We assume the barrier parameters to undergo spatial fluctuations originating, for instance, from the roughness of the boundaries or random potential fluctuations of the impurities present in the barrier region. Since the transmittancy of a barrier depends exponentially on its parameters, even small fluctuations in the latter can result in an exponentially large spread of the local transmittancy. If the barrier area is sufficiently large, the major contribution to the conductance of the barrier structure will come from "punctures", namely sparse regions with an exponentially large compared to conventional transmittancy. The total conductance of the sufficiently large sample will be dominated by the optimal punctures i.e. by the punctures with the maximum product of transmittancy by the probability of formation. In this case, the conductivity should be area-independent and a self-averaging quantity.

Conductance calculation based on ideology of optimal punctures was carried out for a great number of specific barrier structures [1-13]. In Refs 2-8 the structures with current transport determined by electron hops over impurity sites were studied: hopping conduction in 2D - case[2-6] and 1D-case[7,8]. The barriers with current dominated by direct tunneling were considered in Refs 9-13.

In actual fact, the area of a barrier structure is always finite. Since the distances between punctures are exponentially large, a situation may occur where no optimum puncture will be present in a sample. Under these conditions the conductivity of a typical sample will be determined by a few highest-transmittancy punctures of those existing in the sample. As a result, the conductivity of a sample will, first, depend strongly on sample area, and, second, will vary in a random way from sample to sample by an amount on the order of itself, the only possibility being to use the conductivity distribution function (DF) over the samples. This was first pointed out in Ref 1, where a simple model of a random system of independent filaments with traps was analyzed.

The form of the DF is a nontrivial problem even for an ensemble of large area samples containing a large number of optimum punctures. It might seem that the relative spread of the conductance determining the width of the DF would in this case be on the order of $M_{opt}^{-1/2}$, where

M_{opt} is the number of optimum punctures in the sample. As we will see later on, this is not so, namely, the conductance spread is dominated by more sparse rather than optimum punctures and exceeds substantially the value $M_{opt}^{-1/2}$. The derivation and analysis of the distribution function can be carried out in a very general way without invoking any concrete model of the nonuniform barrier [14]. This is what we are going to do in the present work.

2. WIDTH AND PEAK POSITION OF THE DISTRIBUTION FUNCTION

In our analysis of the conductance DF over samples we will conveniently classify all possible punctures according to the value of the equivalent cross section of transmission, $A = A_o \exp(-u)$[*]. We introduce now the puncture density $\rho(u)$, a characteristic of a disordered barrier significant for our further consideration and defines as

$$\rho(u) = \lim_{S \to \infty} \frac{1}{S} \sum_i \delta(u - u_i) \qquad (1)$$

where S is the barrier area. The summation in eq.(1) is performed over all possible punctures. Thus $\rho(u)du$ is the mean concentration of punctures with log transparency lying in the interval from $-u$ to $-u + du$. We will be interested here in sparse punctures, i.e. in such values of u for which the dependence $\rho(u)$ is exponential:

---
[*]
In the region of a puncture, the local transmittancy drops exponentially as one moves away from its center. The quantity $\exp(-u)$ is the local transmittancy at the puncture centre, and $A_o^{1/2}$ is in order of magnitude the distance from the centre in which the local transmittancy drops e-fold.

$$\rho(u) = \frac{1}{S_o}\exp(-\Omega(u)) \qquad (2)$$

where $S_o^{1/2}$ is the characteristic puncture size which is small compared with the spacing between punctures, $\Omega(u) \gg 1$ and decreases with increasing u (We will also assume that $d^2\Omega/du^2 > 0$).

Using eq.(2), the sample-averaged conductivity (we will not distinguish conductivity and area-averaged transmittancy of the sample) can be written in the form

$$\langle\sigma\rangle = A_o \int_0^\infty du\, e^{-u}\rho(u) = \frac{A_o}{S_o}\int_0^\infty du\, \exp(-u-\Omega(u)) \qquad (3)$$

The integrand has a sharp maximum at $u = u_{opt}$ which can be found from the equation

$$\Omega'(u_{opt}) + 1 = 0 \qquad (4)$$

For the $\ln\langle\sigma\rangle$ we thus obtain

$$\ln\langle\sigma\rangle \simeq \ln(\frac{S_o}{A_o}\langle\sigma\rangle) = -u_{opt} - \Omega(u_{opt}) \qquad (5)$$

Eq. 5 defines the conductivity of a typical sample only in the case where the sample area S is sufficiently large, in other words, where sufficiently large is the number of optimum punctures present in the sample. This condition can be rewritten

$$S\rho(u_{opt}) \gg 1 \qquad (6)$$

Since $|\ln(S_o \rho(u_{opt}))| = \Omega(u_{opt}) \gg 1$ the condition (6) reduces to $\nu > 1$, where the parameter ν is defined by the expression

$$\nu = \frac{1}{\Omega(u_{opt})} \ln \frac{S}{S_o} \qquad (7)$$

If condition (6) is not met, i.e. if $\nu < 1$, then optimum punctures will be present only in exponentially rare samples which, thus, will determine the ensemble-averaged conductivity (5). However the conductivity of a typical sample will be dominated by a few of its punctures possessing the highest transmittancy. Quantitatively, this means that when calculating the conductivity of a typical sample one should take for the lower limit in eq.(3) $u = u_f > u_{opt}$, where u_f is found from the condition $S\rho(u_f) \sim 1$. This condition can be rewritten

$$\Omega(u_f) = \nu\Omega(u_{opt}) \qquad (8)$$

For a typical sample this yields

$$\ln(\frac{S_o}{A_o}\sigma(\nu)) = -\nu\Omega(u_{opt}) - u_f(\nu), \quad \nu < 1 \qquad (9)$$

$\ln \sigma(\nu)$ has the meaning of the position of the maximum in the DF of the log conductivity for $\nu < 1$. Naturally, $\ln \sigma(\nu) < \ln \langle\sigma\rangle$; for $\nu \to 1$, eq.(9) transforms into eq.(5).

Let us analyze now the dependence of the width of the DF on parameter ν. To do this, we first calculate the variance of the conductivity over an ensemble of samples of a given area S. Since all punctures are distributed in a random and independent way, the fluctuations in the number of punctures of any type in the sample are described by Poisson statistics. Thus we have

$$\langle(\delta\sigma)^2\rangle = \langle\sigma^2\rangle - \langle\sigma\rangle^2 = \frac{A_o^2}{S}\int_0^\infty du\, \rho(u)e^{-2u} \quad (10)$$

Substituting eq.(2) into (10) we find that the intergrand passes through a sharp maximum at $u = u_d$, such that

$$\Omega'(u_d) + 2 = 0 \quad (11)$$

A comparison of eqs.(11) and (4) reveals that $u_d < u_{opt}$, i.e. the major contribution to the variance comes from the punctures with much higher transmittancy than from the ones determining the mean conductivity. Thus

$$\langle(\delta\sigma)^2\rangle \sim \frac{A_o^2}{SS_o}\exp(-2u_d - \Omega(u_d)) \quad (12)$$

Eq.(12) is valid for an ensemble of samples of any area. However it defines the width of the DF only in the cases where the mean number of punctures with $u = u_d$ within the area S is large, i.e. where $S\rho(u_d) \gg 1$. The latter condition may be conveniently presented in the form $\nu > \nu_d$, where ν_d is the solution of the equation

$$\varphi(\nu_d) = 2 \quad (13)$$

and the function φ is defined in the following way

$$\varphi(\nu) = -\Omega'(u_f(\nu)) \quad (14)$$

with $u_f(\nu)$ found from eq.(8). The function φ is positive and grows with increasing ν, and $\varphi(1) = 1$, so that $\nu_d > 1$.

Thus for $\nu > \nu_d$ the width Δ_o of the DF of $\ln\sigma$ is on the order of

$$\Delta_o \sim \frac{(\langle(\delta G)^2\rangle)^{1/2}}{\langle G \rangle} \sim \exp\left[u_{opt} + (1-\frac{\nu}{2})\Omega(u_{opt}) - u_d - \frac{\Omega(u_d)}{2}\right]$$
(15)

This quantity is exponentially small and inversely proportional to $S^{1/2}$. It is, however, exponentially greater than $M_{opt} \sim \exp((\nu-1)\Omega(u_{opt})/2)$.

For $\nu < \nu_d$ the dominant contribution to the variance (12) is due to exponentially rare samples containing punctures with $u = u_d$ whereas the width of the DF is determined by typical samples so that in its calculation one should include only the punctures which can be found in a typical sample. This is equivalent to replacing the lower limit in the integral (10) by $u_f(\nu)$:

$$\Delta_o \sim \frac{1}{G(\nu)}\left[\frac{A_o^2}{SS_o}\int_{u_f}^{\infty} du\, e^{-2u-\Omega(u)}\right]^{1/2} = \frac{A_o}{S_o G(\nu)} e^{-u_f(\nu) - \nu\Omega(u_{opt})}$$
(16)

For $1 < \nu < \nu_d$ we have $G(\nu) = \langle G \rangle$ (see eq.(5)), while for $\nu < 1$, $G(\nu)$ is defined by eq.(9). Substituting the corresponding expressions into (16) yields

$$\Delta_o \sim \exp\left[-u_f(\nu) + u_{opt} + (1-\nu)\Omega(u_{opt})\right], \quad 1 < \nu < \nu_d$$
(17a)

$$\Delta_o \sim 1, \quad \nu < 1$$
(17b)

Since for $\nu = \nu_d$, $u_f(\nu_d) = u_d$, eqs.(15) and (17a) are joined at $\nu = \nu_d$.

This consideration is invalid for $\nu \ll 1$, since in this case fluctuations in the transmittancy of the highest-transmittancy puncture close to the value $\exp(-u_f)$ become essential where u_f is defined by eq.(8). The scale of these fluctuations is related with the uncertainty in the value of u_f: $S\rho(u_f) \sim 1\div 2$. Hence the uncertainty δu_f is of the order of $(S \rho'(u_f))^{-1} \sim 1/\varphi(\nu)$. Thus for $\nu \ll 1$ we have $\Delta_o \sim \delta u_f \sim 1/\varphi(\nu) \gg 1$.

We have thus shown that in the model of a randomly nonuniform barrier there exist three exponentially broad intervals of area variation: $0 < \nu < 1$, $1 < \nu < \nu_d$ and $\nu > \nu_d$ within which the DF has different forms. In the first interval the position of its maximum and width are given by eqs.(9) and (17b), in the second, by eqs.(5) and (17a), and in the third, by eqs.(5) and (15). The exact expression for the distribution function is derived in the next section.

3. DERIVATION OF EXPRESSIONS FOR THE DISTRIBUTION FUNCTION

We write the conductivity of a given sample in the form

$$\sigma = \frac{A_o}{S} \sum_i n_i e^{-u_i} \qquad (18)$$

where n_i is the number of punctures of ith species within the sample area (the puncture species is characterized, for instance, by a certain impurity configuration in the barrier region); $\exp(-u_i)$ is the transmittancy of punctures of ith species. The probability of finding n_i punctures within an area S is defined by Poisson distribution

$$p(n_i) = \frac{\exp(-\bar{n}_i)}{n_i!} \bar{n}_i^{n_i} \qquad (19)$$

where $\bar{n}_i$ is the mean number of punctures of ith species whithin the area S.

We introduce the distribution function of the quantity $Q = -\ln(S_o \sigma /A_o)$:

$$f(Q) = \langle \delta \{ Q + \ln [\frac{S_o}{S} \sum_i n_i e^{-u_i}] \} \rangle \qquad (20)$$

The averaging in eq.(20) is carried out by means of the distribution (19)

$$f(Q) = e^{-Q} \sum_{n_i=0}^{\infty} \delta [e^{-Q} - \frac{S_o}{S} \sum_i n_i e^{-u_i}] \prod_k p(n_k) \qquad (21)$$

Substituting eq.(19) into (21) and replacing the δ-function by its Fourier transform we obtain

$$f(Q) = \frac{e^{-Q}}{2\pi} \int_{-\infty}^{\infty} dt \, e^{ite^{-Q}} \sum_{n_i=0} \prod_i \frac{e^{-\bar{n}_i}}{n_i!} [\bar{n}_i \exp(-\frac{itS_o}{S} e^{-u_i})]^{n_i} \qquad (22)$$

The summation over n_i can be easily performed for each i separately

$$f(Q) = \frac{e^{-Q}}{2\pi} \int_{-\infty}^{\infty} dt \, e^{ite^{-Q}} \prod_i \exp \{ \bar{n}_i [\exp[-\frac{itS_o}{S} e^{-u_i}] - 1] \} \qquad (23)$$

The final answer is obtained after replacing the product over i by an interval in the exponent. It can be readily expressed in terms of the puncture density $\rho(u)$(1):

$$f(Q) = \frac{e^{-Q}}{2\pi} \int_{-\infty}^{\infty} dt \, \exp \{ ite^{-Q} + S \int_0^{\infty} du \, \rho(u) [e^{-\frac{itS_o}{S} e^{-u_i}} - 1] \} \qquad (24)$$

An analysis of this expression permits us to reproduce all the results for the width and position of the maximum of the DF derived in the preceding Section. Apart from this, one can find from eq.(24) the actual form of the distribution function for each of the three intervals of variation of the area (or parameter ν)[14].

3.1 Case (a): $\nu<1$

Substituting expression (2) for $\rho(u)$ into (24) and using eq.(7) for ν we can rewrite the distribution function as

$$f(Q) = \frac{e^{-Q}}{2\pi} \int_{-\infty}^{\infty} dt \, \exp\left[ite^{-Q} + I_\nu(t)\right], \qquad (25)$$

$$I_\nu(t) = \int_0^\infty du \, \exp\left[\nu\Omega(u_{opt}) - \Omega(u)\right]\left[\exp(-ite^{-u-\nu\Omega(u_{opt})}) - 1\right] \qquad (26)$$

According to results of Sec.2, Q and u are nearly equal to $u_f(\nu) + \nu\Omega(u_{opt})$ and $u_f(\nu)$, respectively. It is therefore convenient to make the following change of variables in the integrals in (25) and (26)

$$t = \mathbf{v} \exp\left[\nu\Omega(u_{opt}) + u_f(\nu)\right] \qquad (27a)$$

$$u = u_f(\nu) + u_1 \qquad (27b)$$

Since u_1 is typically much less than u_f the argument of the exponential in the first factor in (26) can be expanded to second order in the small parameter u_1/u_f. Using eq.(14) for $\varphi(\nu)$ we get

$$I_\nu = \int_{-\infty}^{\infty} du_1 \, \exp\left[\varphi(\nu)u_1 - \tfrac{1}{2}\Omega''(u_f)u_1^2\right]\left[e^{-i\nu e^{-u_1}} - 1\right] \qquad (28)$$

The evaluation of (28) depends on the range over ν varies:

$$I_\nu = -\frac{\Gamma(1-\varphi)}{\varphi} e^{i\pi\varphi/2} \nu^\varphi, \quad 1-\varphi(\nu) \gg (\Omega_0'')^{1/2}, \qquad (29a)$$

$$I = -\frac{\pi\nu}{2} + i\nu \left\{ \ln\nu - \frac{2\pi}{\Omega_0''} \exp\left[\frac{(\varphi-1)^2}{2\Omega_0''}\right] \Phi\left[\frac{1-\varphi}{(2\Omega_0'')^{1/2}}\right] \right\}, \qquad (29b)$$

where

$$\Phi(x) = \pi^{-1/2} \int_x^\infty dw\, e^{-w^2} \qquad (30)$$

is the error integral and $\Omega_0'' \equiv d^2\Omega/d^2u\big|_{u=u_{opt}}$. Formula (29a) follows by neglecting the second term in the argument of the exponential in (28) so that the integral reduces to the Γ-function. The real part of (29b) follows in precisely the same way. The imaginary part of (29b) can be found by expanding the exponential in the second factor in (28) and replacing the lower limit of integration by $u_1 = \ln\nu$.

Formulas (29) assume that $\nu > 0$, which is no restriction since it is clear from (28) that $I_\nu(-\nu) = I_\nu^*(\nu)$. The final expression for the DF follows upon inserting (29a) and (29b) into (25).

For the case $1-\varphi(\nu) \gg (\Omega_0'')^{1/2}$ it is convenient to center the DF near $Q = \nu\Omega(u_{opt}) + u_f(\nu)$ by introducing the quantity

$$\Delta = Q - \nu\Omega(u_{opt}) - u_f(\nu) + \frac{1}{\varphi}\ln\frac{\Gamma(1-\varphi)}{\varphi} \qquad (31)$$

Making the change of variable $x = v(\Gamma(1-\varphi)/\varphi)^{1/\varphi}$ we obtain

$$f(Q) = \frac{e^{-\Delta}}{\pi}\int_0^\infty dx\, \exp(-x^\varphi\cos\frac{\pi\varphi}{2})\cos(xe^{-\Delta} - x^\varphi\sin\frac{\pi\varphi}{2}) \qquad (32)$$

Figure 1 shows the function f calculated numerically for several values of φ. For large positive and negative Δ we have the asymptotic formulae

$$f(Q) = \left[\frac{\gamma\beta(\beta+1)}{2\pi}\right]^{1/2}\exp\left[\frac{\beta\Delta}{2} - \gamma e^{\beta\Delta}\right],\quad \Delta \gg 1 \qquad (33a)$$

$$f(Q) = \frac{\sin\pi\varphi}{\pi}\Gamma(\varphi+1)\, e^{\varphi\Delta},\quad \Delta < 0,\ |\Delta| \gg 1 \qquad (33b)$$

where

$$\beta = \varphi/(1-\varphi),\quad \gamma = (1-\varphi)\varphi^\beta \qquad (34)$$

The derivation of eqs (33) is given in Appendix 1.

As shown in Appendix 1, for $\varphi \ll 1$ the expression for the distribution function simplifies to

$$f(Q) = \varphi\exp(\varphi\Delta - e^{\varphi\Delta}) \qquad (35)$$

We see, that the width of the DF increases as $1/\varphi(\nu)$ as ν decreases, in agreement with the result of our qualitative analysis.

For the case $|1 - \varphi(\nu)| \ll 1$ (i.e. at $|1-\nu| \ll 1$)

we find upon inserting (29b) in (25) and writing

$$\Delta_1 = Q - u_{opt} - \Omega(u_{opt}) + \ln\left\{\left[\frac{2\pi}{\Omega_o''}\right]^{1/2} \Phi\left[\frac{1-\varphi}{(2\Omega_o'')^{1/2}}\right]\right\} \quad (36)$$

$$w_1^{-1} = \left[\frac{2\pi}{\Omega_o''}\right]^{1/2} \exp\left[\frac{(1-\varphi)^2}{2\Omega_o''}\right] \Phi\left[\frac{1-\varphi}{(2\Omega_o'')^{1/2}}\right] \quad (37)$$

that the distribution function is

$$f(Q) = \frac{1}{\pi w_1} \int_0^\infty dv\, e^{-\pi v/2} \cos(v\frac{\Delta_1}{w_1} - v\ln v), \quad |1-\varphi| \ll 1 \quad (38)$$

which is plotted in fig.1b. For large negative Δ_1:
$|\Delta_1| \gg w_1$, $f(Q)$ decays as w_1/Δ_1^2, while for $\Delta_1 \gg w_1$ it falls off as $\exp(-\exp(\Delta_1/w_1))$.

The DF (38) has width $\sim w_1$, for which we have the asymptotic formulas

$$w_1 = 1 - \varphi, \qquad 1 - \varphi \gg (\Omega_o'')^{1/2} \quad (39a)$$

$$w_1 = \left[\frac{\Omega_o''}{2\pi}\right]^{1/2} \exp\left[-\frac{(1-\varphi)^2}{2\Omega_o''}\right], \quad \varphi - 1 \gg (\Omega_o'')^{1/2} \quad (39b)$$

in the two limiting cases. Expression (37) for w_1 can be used to analyze how the distribution function becomes narrower as one goes from large ($\nu < 1$) to small ($\nu > 1$) fluctuations.

3.2 Case (b): $1<\nu<\nu_d$

For this case, as it was shown in Sec.2, the width of the DF is determined by the values $u \simeq u_f(\nu)$ and the position of the maximum by the values $u \simeq u_{opt}$. Expression (28) for I_ν was derived under the assumption that $|u - u_f(\nu)| \ll u_f(\nu)$. Since this does not hold in the present case, we must separate out the contribution from $u \simeq u_{opt}$ in the integral (26) by recording the latter as a sum of two terms:

$$I = I_\nu^{(1)} + I_\nu^{(2)},$$

$$I_\nu^{(1)} = -i\nu \exp(\nu\Omega(u_{opt})+u_f(\nu)) \int_0^\infty du \exp(-\Omega(u)-u) \quad (40)$$

$$I_\nu^{(2)} = \int_{-\infty}^\infty du_1 \exp(u_1 - \tfrac{1}{2}\Omega''(u_f)u_1^2)\left[e^{-i\nu e^{-u_1}} + i\nu e^{-u_1} - 1\right] \quad (41)$$

where u_1 and v are related to u and t by eqs.(27).

Since the integrand in (40) is sharply peaked at $u = u_{opt}$, the integral is readily evaluated by the method of steepest descent:

$$I_\nu^{(1)} = -i\nu \left[\frac{2\pi}{\Omega_o''}\right]^{1/2} \exp(u_f(\nu) - u_{opt} + (\nu-1)\Omega(u_{opt})) \quad (42)$$

Since the integral for $I_\nu^{(2)}$ converges for $|u_1| \ll u_f$, in deriving (41) we have expanded the argument of the exponential in the first factor in (26) in powers of the parameter u_1/u_f. As in the case of (28), the evaluation of the integral (41) depends on ν :

$$I_\nu^{(2)} = \frac{\Gamma(2-\varphi)}{\varphi(\varphi-1)} e^{i\pi\varphi/2} v^\varphi, \quad 2-\varphi \gg (\Omega_d'')^{1/2} \tag{43a}$$

$$I_\nu^{(2)} = -v^2 \left[\frac{\pi}{2\Omega_d''}\right]^{1/2} \exp\left[\frac{(\varphi-2)^2}{2\Omega_d''}\right] \Phi\left[\frac{2-\varphi}{(2\Omega_d'')^{1/2}}\right], \quad |\varphi-2| \ll 1 \tag{43b}$$

where u_d is defined by (11) and $\Omega_d'' \equiv \Omega''(u_d)$. Expressions (43) follow from (41) in exactly the same way as (29a) and (29b) follow from (28).

Substituting the sum of (42) and (43a) into (25) and making the change of variable

$$x = v\left[\Gamma(2-\varphi)/\varphi(\varphi-1)\right]^{1/\varphi}$$

we obtain

$$f(Q) = \frac{1}{\pi w_2} \int_0^\infty dx \, \exp(x^\varphi \cos\frac{\pi\varphi}{2}) \cos(x\frac{\Delta_2}{w_2} - x^\varphi \sin\frac{\pi\varphi}{2}),$$

$$2-\varphi \gg (\Omega_d'')^{1/2} \tag{44}$$

for the distribution function, where

$$\Delta_2 = Q - u_{opt} - \Omega(u_{opt}) + \ln(2\pi/\Omega_o'')^{1/2} \tag{45}$$

$$w_2 = \left[\frac{\Omega_o''}{2\pi}\right]^{1/2} \left[\frac{\Gamma(2-\varphi)}{\varphi(\varphi-1)}\right]^{1/\varphi} \exp\left[u_{opt} - u_f(\nu) + (1-\nu)\Omega(u_{opt})\right] \tag{46}$$

Equations (45), (46) show that the DF is centered at $Q \simeq u_{opt} + \Omega(u_{opt})$ and has width $\sim w_2$, in agreement with eqs. (5), (17a) from qualitative analysis. Figure 1c plots the DF for several values of φ.

For $|\varphi(\nu) - 2| \ll 1$ the integral in (25) can easily be evaluated by substituting the sum of expressions (42) and (43b). The resulting DF is gaussian:

$$f(Q) = (2\pi)^{-1/2} w_3^{-1} \exp(-\Delta_2^2/2w_3^2) \qquad (47)$$

with width

$$w_3 = (\Omega_o)^{1/2}(2\pi\Omega_d'')^{-1/4} \exp\left[u_{opt}+(1-\frac{\nu}{2})\Omega(u_{opt})-u_d-\frac{\Omega(u_d)}{2}\right]$$

$$\times \phi^{1/2}\left[\frac{2-\varphi}{(2\Omega_d'')^{1/2}}\right] \qquad (48)$$

3.3 Case (c): $\upsilon > \upsilon_d$

In all this region the DF is gaussian as is readily seen by expanding the integrand in (26) to the second order in t, after which the integrals in (26) and (25) are easily evaluated. Expression (48) for the width of DF simplifies for $\varphi(\nu) - 2 \gg (\Omega_d'')^{1/2}$:

$$w_3 = (\Omega_o'')^{1/2}(2\pi\Omega_d'')^{-1/4} \exp\left[u_{opt}+(1-\frac{\nu}{2})\Omega(u_{opt})-u_d-\frac{\Omega(u_d)}{2}\right] \qquad (49)$$

which agrees with eq.(15) in Sec.2 up to a prefactor.

4. CONCLUSION

The main result of this paper is that the log conductance distribution function $f(Q)$ has a universal form which doesn't depend on the concrete model of disordered barrier. The only restriction is the exponentially wide spread of local transmittancy. The functional form $f(Q)$

is controlled by the only dimensionless parameter $\varphi(\nu)$ which is the monotonic function of the logarithm of the sample area.

It is convenient to extract parameter φ directly from experimental data by comparing the moments of DF with their theoretical values. For most interesting case $\varphi < 1$ the moments of DF(32) are equal to

$$M_n = n! \sum_{m=1} \frac{1}{m!} \sum_{k_i=1} a_{k_1} a_{k_2} \cdots a_{k_m} \delta_{k_1+k_2+\ldots+k_m,\, n-m}, \quad (50)$$

$$a_k = \frac{(-1)^{k+1}}{k+1} \zeta(k+1) \left[\frac{1}{\varphi^{k+1}} - 1 \right]$$

where $\zeta(n)$ is Riemann ζ-function. The derivation of eq.(50) is presented in Appendix 2. For small n eq.(50) reduces to

$$M_2 = \frac{\pi^2}{6}\left[\frac{1}{\varphi^2} - 1\right], \qquad M_3 = -2\zeta(3)\left[\frac{1}{\varphi^3} - 1\right],$$

$$M_4 = \frac{\pi^4}{15}\left[\frac{1}{\varphi^4} - 1\right] + \frac{\pi^4}{12}\left[\frac{1}{\varphi^2} - 1\right]^2 \qquad (51)$$

For $\varphi < 1$ the average value of log conductivity depends on the sample area S. Using eqs (7-9) this dependence can be expressed in the form

$$\frac{d \ln \sigma(\nu)}{d \ln S} = \frac{1}{\varphi} - 1 \qquad (52)$$

Thus for $\varphi < 1$ the dependence of total conductance $G = S\sigma$ on area is superlinear and close to power law: $G \propto S^{1/\varphi}$. Eq.(52) gives additional possibility to find

φ from experimental data.

Experimental investigation of the conductance DF involves measurements on a large lot of samples. There is, however, another possibility of studying experimentally the DF, with only one sample needed, if there exists an external factor redistributing the contributions of different punctures to the total conductance. Then the variation of this factor (applied voltage, temperature and so on) will produce random but reproducible fluctuations of conductivity. This "incoherent mesoscopic" effect was observed in Refs 15, 16 and studied theoretically in Refs 14, 17, 18. The DF can be restored from the mesoscopic pattern.

This procedure was carried out in computer experiments [19], in which the dependence of hopping resistance of 1D chain on Fermi level position was studied. The value of the variance of log resistance was found in [19] to be 4.9. Substituting this value into first equation in (51) yields $\varphi = 0.5$. It turns out that for this value of φ the integral (32) can be evaluated exactly

$$f(\Delta) = \frac{1}{2\pi^{1/2}} \exp\left[\frac{\Delta}{2} - \frac{1}{4}e^{\Delta}\right] \qquad (53)$$

Note that in order to apply this expression for 1D chain we must change formally the conductance by the resistance, i.e. Δ by $-\Delta$ in eq.(53). The fig.2 plots the dependence of f upon $\Delta = \ln R - \langle \ln R \rangle$ by solid line. It is seen to fit well to results of numerical simulation[19] shown by circles.

Recently the investigation of the DF of hopping conduction mesoscopic fluctuations in short channel GaAs MESFET was carried out[6]. The results are in satisfactory agreement with theory presented in this paper.

APPENDIX 1

We rewrite the distribution function (32) in the form

$$f(Q) = \frac{e^{-\Delta}}{2\pi i} \int_C dz \, \exp(z \, e^{-\Delta} - z^\varphi) \qquad (A1.1)$$

where the path of integration C in the complex plane is shown in fig.3. The integrand in (A1.1) has a saddle point at $z = z_0 = (\varphi \, e^\Delta)^{1/(1-\varphi)}$ on the real axis. We shift C so that it passes through the point z_0 (contour C_1 in fig.3). For $\Delta \gg 1$ the integral along C_1 can be evaluated by the method of steepest descent, because most of the contribution comes from values z such that $|z - z_0| \ll z_0$. This leads to eq.(33a).

To derive the asymptotic formula (33b), we deform C into the contour C_2 passing along the edges of the cut in fig.6. Expression (A1.1) then becomes

$$f(Q) = \frac{e^{-\Delta}}{\pi} \int_0^\infty dt \, \exp(-te^{-\Delta} - t^\varphi \cos \pi\varphi) \sin(t^\varphi \sin \pi\varphi) \qquad (A1.2)$$

For $\Delta < 0$ and $|\Delta| \gg 1$, most of the contribution to the integral comes from values $t \ll 1$. We can therefore replace the sine by its argument and omit the second term in the exponential; the resulting integral is readily evaluated to yield (33b).

Expression (A1.2) is also useful for finding the distribution function when $\nu \ll 1$. In this case it simplifies to

$$f(Q) = \varphi e^{-\Delta} \int_0^\infty dt\ t^\varphi \exp(-te^{-\Delta} - t^\varphi) \qquad (A1.3)$$

which after the change of variable $u = te^{-\Delta}$ gives

$$f(Q) = \varphi e^{\varphi\Delta} \int_0^\infty du\ u^\varphi \exp(-u - u^\varphi e^{\varphi\Delta}) \qquad (A1.4)$$

Since this integral converges for $u \sim 1$, we can replace u^φ in the integrand by 1 when $\varphi \ll 1$, and eq.(35) follows immediately.

APPENDIX 2

To calculate the moments of DF it is convenient to use expression (A1.2) for the DF. According to definition

$$M_n = \langle(\Delta - \langle\Delta\rangle)^n\rangle = \frac{1}{\pi}\int_0^\infty dt \int_{-\infty}^\infty d\Delta\ \sin(t^\varphi \sin\pi\varphi)(\Delta - \langle\Delta\rangle)^n$$

$$\times \exp(-\Delta - t^\varphi \cos\pi\varphi - te^{-\Delta}) \qquad (A2.1)$$

Changing the variables in both integrals: $\Delta = \ln(t/z)$, $t = u^{1/\varphi}$ yields

$$M_n = \frac{1}{\pi\varphi}\int_0^\infty du\ e^{-u\cos\pi\varphi}\frac{\sin(u\sin\pi\varphi)}{u}\int_0^\infty dz\ e^{-z}(\frac{1}{\varphi}\ln u - \ln z - \langle\Delta\rangle)^n$$

$$(A2.2)$$

Expressing the internal integral as

$$\lim_{\varepsilon \to 0} \frac{d^n}{d\varepsilon^n} \int_0^\infty dz\, e^{-z} \exp\left\{\varepsilon\left[\frac{\ln u}{\varphi} - \ln z - \langle\Delta\rangle\right]\right\} \quad (A2.3)$$

we obtain

$$M_n = \frac{1}{\pi\varphi} \frac{d^n}{d\varepsilon^n} \int_0^\infty dz\, e^{-z} \exp\left[-\varepsilon(\ln z + \langle\Delta\rangle)\right] \int_0^\infty du\, e^{-u\cos\pi\varphi}$$

$$\times \sin(u \sin\pi\varphi)\, u^{\varepsilon/\varphi - 1} \quad (A2.4)$$

We see that the double integral had turned into product of two separate integrals, both readily reducing to the Γ-function

$$M_n = \frac{1}{\pi\varphi} \frac{d^n}{d\varepsilon^n}\left[e^{-\varepsilon\langle\Delta\rangle} \sin(\pi\varepsilon)\, \Gamma(1-\varepsilon)\, \Gamma(\varepsilon/\varphi)\right] \quad (A2.5)$$

Using properties of Γ-function one can simplify eq.(A2.5) to

$$M_n = \frac{d^n}{d\varepsilon^n} \frac{\Gamma(1+\varepsilon/\varphi)}{\Gamma(1+\varepsilon)} e^{-\varepsilon\langle\Delta\rangle} \quad (A2.6)$$

The value of $\langle\Delta\rangle$, as it is seen from definition of M_n, is given by (A2.6) when putting formally in it $\langle\Delta\rangle = 0$, $n = 1$. Thus we get

$$\langle\Delta\rangle = \Gamma'(1)\left[\frac{1}{\varphi} - 1\right] = \psi(1)\left[\frac{1}{\varphi} - 1\right] \quad (A2.7)$$

where $\psi(x) = d\ln\Gamma(x)/dx$. To calculate the higher moments it is convenient to rewrite Γ-function as

$$\Gamma(1+x) = \exp\left[\int_0^x dv\, \psi(1+v)\right] \tag{A2.8}$$

Substituting (A2.7), (A2.8) in eq.(A2.6) yields

$$M_n = \frac{d^n}{d\varepsilon^n} \exp\left\{\int_0^\varepsilon dv\left\{\frac{1}{\varphi}\left[\psi\left[1+\frac{v}{\varphi}\right] - \psi(1)\right] - \psi(1+v) + \psi(1)\right\}\right\} \tag{A2.9}$$

Expanding the integrand in powers of v and performing the integration we obtain

$$M_n = \lim_{\varepsilon \to 0} \frac{d^n}{d\varepsilon^n} \exp\left\{\sum_{k=1}^\infty \frac{\varepsilon^{k+1}}{(k+1)!}\left[\frac{1}{\varphi^{k+1}} - 1\right]\psi^{(k)}(1)\right\} \tag{A2.10}$$

where $\psi^{(k)}(1) = d^k \psi(x)/dx^k\big|_{x=1}$. To get the final result (eq.50)) one should expand (A2.10) in powers of ε, use the relation $\psi^{(k)}(1) = (-1)^{k+1} k!\, \zeta(k+1)$ and take the limit $\varepsilon \to 0$.

REFERENCES

1. Lifshits, I.M., Gredeskul, S.A. and Pastur, L.A., Sov. Phys. JETP 56, 1370 (1982).
2. Pollak, M. and Hauser, J.J., Phys.Rev.Lett. 31, 1304 (1973).
3. Tartakovskii, A.V., Fistul, M.V., Raikh, M.E. and Ruzin, I.M., Sov. Phys. Semicond. 21, 370 (1987).
4. Levin, E.I., Ruzin, I.M. and Shklovskii, B.I., Sov. Phys. Semicond. 22, 101 (1988).
5. Raikh, M.E., Ruzin, I.M. and Shklovskii, B.I. Sov. Phys. Semicond. 22, 1254 (1988).
6. Orlov, A.O., Raikh, M.E., Ruzin, I.M. and Savchenko,

A.K., Sov. Phys. JETP 69 (1989), to be published.
7. Kurkijarvi, J., Phys. Rev. B8, 922 (1973).
8. Raikh, M.E., and Ruzin, I.M., Zh. Eksp. & Teor. Fiz. 95, 1113 (1989, Sov. Phys. JETP 68).
9. Lifshits, I.M. and Kirpichenkov, V.Ya., Zh. Eksp. & Teor. Fiz. 77, 989 (1979, Sov. Phys. JETP 50),
10. Krylov, M.V. and Suris, R.A., Sov. Phys. JETP 61, 1303 (1985).
11. Gusyatnikov, V.N. and Raikh, M.E., Sov.Phys. Semicond. 18, 670 (1984).
12. Raikh, M.E. and Ruzin, I.M., Sov. Phys.Semicond. 19, 745 (1985).
13. Raikh, M.E. and Ruzin, I.M., Sov. Phys. Semicond. 21, 283 (1987).
14. Raikh, M.E. and Ruzin, I.M., Sov.Phys. JETP 65, 1273 (1987).
15. Fowler, A.B., Harstein, A. and Webb, R.A., Phys. Rev. Lett. 48, 196(1982).
16. Orlov, A.O. and Savchenko, A.K., JETP Lett. 44, 41 (1986).
17. Raikh, M.E. and Ruzin, I.M., Sov.Phys. Semicond. 22, 799 (1988).
18. Raikh, M.E. and Ruzin, I.M., Zh.Eksp. & Teor. Fiz. 95, 1113 (1989 , Sov.Phys. JETP 68).
19. Serota, R.A., Kalia, R.K. and Lee, P.A., Phys.Rev. B 33, 8441 (1986)

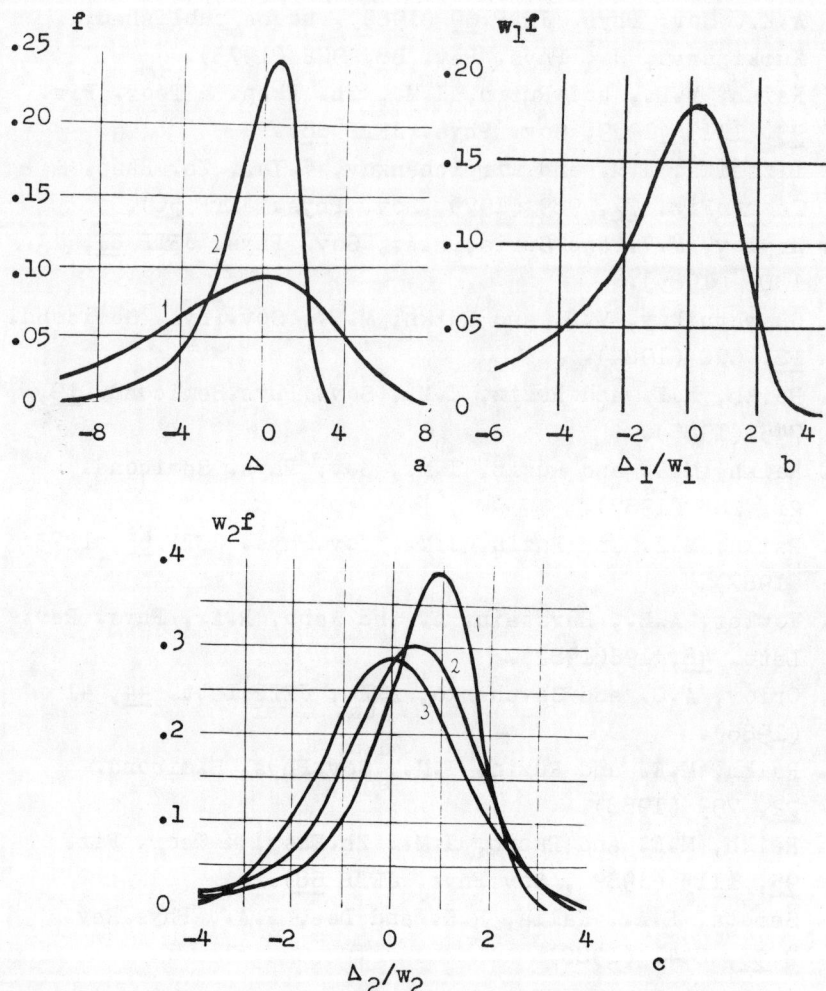

Fig. 1. Distribution function for the logarithm of the conductivity for several values of the parameter φ:
a) 1 − 0.25; 2 − 0.5; b) 1.0; c) 1 − 1.44; 2 − 1.69; 3 − 1.96. The distributions were calculated by eqs. (32), (38) and (44), respectively.

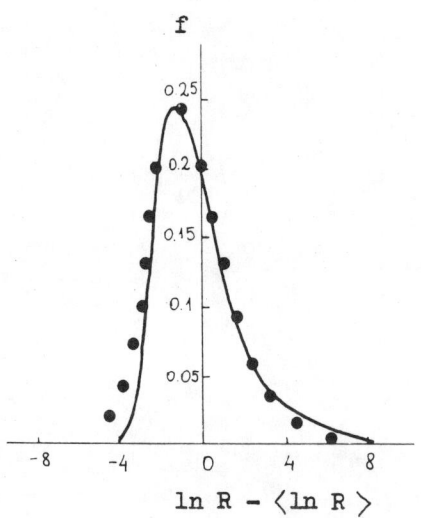

Fig.2. Distribution function of the logarithm of hopping resistance of 1D system obtained by computer simulation[19] The solid line - calculation by eq. (53).

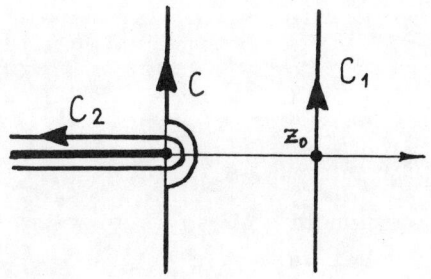

Fig. 3.

1D VARIABLE-RANGE HOPPING IN WIDE MESOSCOPIC MOSFETS

Dragana Popović
Department of Physics, Brown University
Providence, RI 02912, USA

ABSTRACT

Conductance fluctuations have been observed in wide MOSFETs of large area ($\approx 3.3 \times 10^{-6} m^2$). In the region of gate voltages where conduction proceeds via variable-range hopping, the temperature dependence of the averaged conductance logarithm $< \ln G >$ has been found to obey Mott's law for a one-dimensional system. This is believed to be consistent with the model of effectively one-dimensional isolated impurity chains along which electrons move by hopping between impurities. The fluctuations caused by the change of a chemical potential or gate voltage have been studied. $\ln G$ has been found to oscillate on two different scales. Although there is no detailed understanding of this at present, it is most likely due to a competition between different hops that give the most dominant contributions to the conductance. The distribution function of $\ln G$ for a population of specimens (different gate voltages) has been extracted from the data. Its main features are found to agree qualitatively with the theoretical predictions. The fluctuations do not seem to be significantly affected by the applied perpendicular magnetic field (0–1 T). The temperature dependence of $< \ln G >$ does not appear to be changed from the zero-field case within the measurement error. The oscillations of $< \ln G >$ with the field have also been observed. The characteristic "period" of the oscillations suggests that their origin lies in the fluctuations of the resistance associated with the single hop.

I. INTRODUCTION

A lot of recent research effort has concentrated on the physics of small devices (~ 1 μm or less) especially since large reproducible fluctuations in the conductance were first discovered in Si MOSFETs in the strongly localized regime[1]. Similar, sample-specific effects were subsequently observed in metallic devices[2,3,4]. Such systems have been termed "mesoscopic" to signify the fact that their properties cannot be described as a statistical average but are too large to be considered microscopic either. In metallic samples, where the interference of electron waves propagating along different paths provides a significant correction to the classical expression for the conductivity, this lack of averaging may be observed in devices the size of which is comparable to the phase coherence length L_ϕ. Its typical value at low temperatures is of the order of a micron. The amplitude of the fluctuations decays as L/L_ϕ increases, where L is the sample size. In the case of variable-range hopping (VRH), however, the limitations on the length of a sample for the observation of fluctuations are not as obvious as in the case of metallic conduction. As we shall see later, due to the specific nature of the VRH, the existence of conductance fluctuations is possible even in samples of fairly large dimensions (~1 mm), where the total number of hops in the system is large and, therefore, one might expect that the details associated with the microscopic structure of the sample would be lost due to averaging. Even though the fluctuations in large samples have been observed as early as 1965[5] in the transconductance of MOSFET devices, they were either ignored or attributed incorrectly to a host of causes. A study of mesoscopic effects in large, nominally two-dimensional samples has been of major interest in this work. In addition, the effect of magnetic field on interference processes in the VRH regime has been investigated. Before the experimental results are presented and discussed, the major ideas and models relevant to the interpretation and understanding of the data will be briefly reviewed.

II. THEORETICAL BACKGROUND

The problem of calculating the hopping conductivity is commonly replaced by the problem of transport in an equivalent random resistance network proposed by Miller and Abrahams[6]. The hop between the states i and j with energies E_i and E_j is represented by the resistor R_{ij} such that

$$R_{ij} = R_{ij}^0 \exp\left[\frac{2r_{ij}}{\xi} + \frac{1}{2k_BT}(|E_i - E_j| + |E_i - \mu| + |E_j - \mu|)\right], \quad (1)$$

where r_{ij} is the spatial separation of two states, ξ the localization length and μ the chemical potential. It is important to note that this random network has an exponentially wide spectrum of resistances because R_{ij} depends exponentially on the random variables E_i, E_j and r_{ij}. In order to find the conductivity of this network, the percolative approach[7] is often employed. All the resistors are first removed and then introduced into the network one by one, in an increasing order of resistivity until the lattice percolates, thus enabling the current flow. Due to the exponential spread of resistors R_{ij}, the order of magnitude of the conductivity of the sample is determined by the conductivity of the resistor which just makes the sample connected. In other words, the sample resistance is dominated by a single hop at the percolation threshold. In short one-dimensional devices, the number of states within several k_BT of μ that give relevant contributions to the resistance can be quite small, so that this resistance will be a strong random function of the relative values of μ and the energies of the states[8]. In fact, the one-dimensional case is specific in that the fluctuation effects are very weakly dependent on the sample length[9] due to the unavoidability of energetically unfavourable hops.

In wide samples, the current paths, i.e. the lowest resistance paths, represent nearly rectilinear isolated chains of impurities along which electrons move by hopping[10, 11]. The distance between the impurities in chains is much smaller than the hopping length. The sample conductance is equal to the sum of the conductances of the various paths consisting of chains of hops and, following the ideas of percolation theory, it is determined by the conductance of the best-conducting chain whose resistance, in turn, is determined by the largest of the series connection of resistances making up a chain. In this way, the conductance of a sample is actually determined by a very small number of hops even though the total number of hops may be fairly large. The so-called optimal chain, i.e. the chain for which the product of its conductivity and the probability of formation is maximum determines the average conductance. If the sample area is so small that the average number of optimal chains over its area is less than 1, the conductance will fluctuate by up to $\sim 100\%$ from one sample to another and its logarithm will thus fluctuate by an amount of the order of unity. In this case, only the distribution function (DF) for the logarithm of the conductance for a population of samples is meaningful. Raĭkh and Ruzin (RR)[12] have calculated the distribution functions for samples of large, small and intermediate areas and found qualitative differences between them. The DF for large samples is gaussian and its maximum is determined by the optimal chain. The maximum of the DF for the intermediate area case is also determined by the optimal chain but the DF is no longer gaussian, its shape is asymmetric, or skewed. This is even more pronounced in the DF for small samples, whose maximum is determined by the most transparent chains. In general, the width of the DF depends in a complicated way on the area but, of course, it decreases as the area is increased. Raĭkh and Ruzin introduce the parameter

$$\nu = \frac{2}{Q_0} \ln \frac{S}{S_0}, \qquad (2)$$

where Q_0 is the transparency logarithm in an infinitely large sample, S is the cross-sectional area of the sample and S_0 is the area corresponding to a single chain. For the two-dimensional system, S/S_0 has to be replaced by W/W_0, with W being the sample width and W_0 the typical width of a single chain. In the theory of Raĭkh and Ruzin, the condition $\nu < 1$ means that most samples will not contain even a single optimal chain and that the few chains with the largest conductance will give the dominant contributions to the sample conductance. This is the case with the largest fluctuations. The intermediate area case is realized when the parameter ν satisfies the condition $|1 - \nu| \ll Q_0^{-1/2}$, while larger values of ν characterize the large area situation.

Raĭkh and Ruzin also show[12, 13] that the change in an external parameter (e.g. applied field, temperature) alters the properties of the chains, changing their relative contributions to the sample conductance. In this way, the effect of changing an external parameter is equivalent to changing the random configuration of impurities, i.e. to replacing one sample by another. Therefore, it should be possible to observe mesoscopic effects on a single sample, in particular, the oscillations of the conductance logarithm with the variation of an external parameter. These oscillations are characterized by the correlation functions of the conductance logarithm for various types of parameters. Raĭkh and Ruzin[12] find the expression for the correlation function for the case when $\nu < 1$. The knowledge of the correlation function allows one to determine ν and, hence, calculate the corresponding distribution function. According to some estimates of RR, mesoscopic effects should be observable even in "macroscopically" large samples, of size up to 100 mm.

If the conductance of a relatively large sample is determined by a small number of hops, then it should also be possible to observe the oscillations of the conductance logarithm with the magnetic field, proposed by Nguen, Spivak and Shklovskiĭ (NSS)[14]. The origin of these oscillations lies in the fluctuations of the resistance associated with the single hop. An electron, hopping between the states 1 and 2, is scattered on its way by other impurities in a cigar-shaped region of length r (r is the length of the hop) and transverse dimension $(r\xi)^{1/2}$ so that the overlap integral I is formed by summing the contributions I_i from scattering over all possible paths i. In the presence of impurities with a negative scattering amplitude, I will be equal to the sum of terms with random signs. Due to the Aharonov-Bohm effect, the application of a magnetic field perpendicular to the hopping direction gives rise to additional random phase factors $\exp(2\pi i\Phi_i/\Phi_0)$, associated with each of these paths. Here $\Phi_0 = ch/e$ is a magnetic flux quantum and Φ_i is the magnetic flux through an area bounded by the ith path and by a straight line joining the sites 1 and 2. Therefore, when the magnetic flux through the cigar area becomes comparable to $\Phi_0/2$, there will be a substantial change in each term of the sum and the entire sum will change in a random direction by an amount of the order of its own value. The resistance associated with one hop will, consequently, fluctuate in a magnetic field with the characteristic "period"

$$H_c = \frac{hc}{2er^{3/2}\xi^{1/2}}. \qquad (3)$$

Of course, if all the electron paths surround the same area, periodic oscillations of the conductance with the magnetic field should appear. They have been observed experimentally in PbTe films[15].

The oscillations of the conductance with a change in the chemical potential (gate voltage in Si MOSFETs) and magnetic field of the type described above, have been observed[16] in wide but short ($\sim 1\mu$m) samples. Those results appear consistent with the model of conduction via isolated chains of impurities. The data obtained on much longer, but also wide samples will be presented and discussed in the next section.

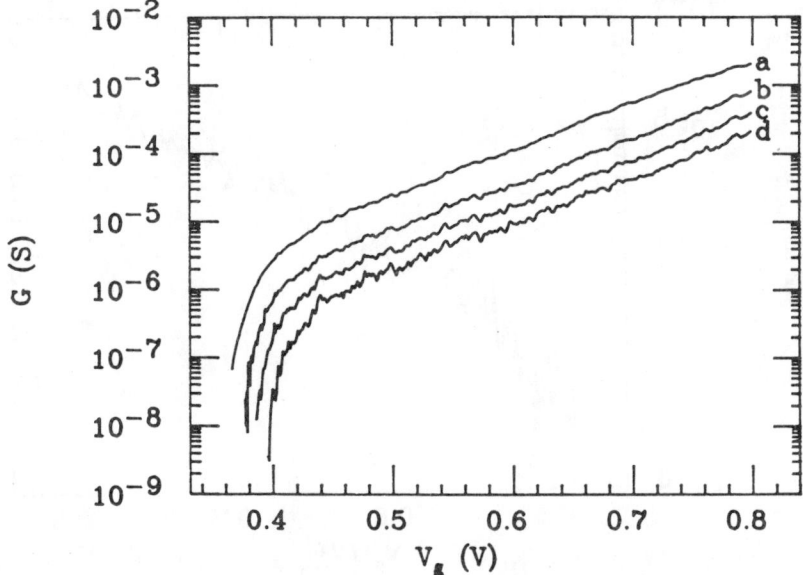

Figure 1: Conductance vs. gate voltage at: a) T=790 mK; b) T=555 mK; c) T=420 mK; d) T=330 mK.

III. EXPERIMENTAL RESULTS AND DISCUSSION

The measurements were carried out on n-channel circular MOSFETs fabricated on the (100) surface of p-type silicon doped to a level of $\approx 8.3 \times 10^{14}$ acceptors/cm^3. The gate oxide thickness was 435 Å. The channel length was 0.4 mm with $W/L = 20$. The amount of oxide charge estimated from the threshold voltage at 77 K was found to be about 3×10^{10}cm^{-2}. The peak mobility of these samples at 4.2 K was not more than 10^4cm^2/Vs. The conductance measurements were performed using a lock-in technique at a frequency of ~10 Hz and keeping the source-drain voltage constant at 10^{-4}V. The samples were placed in a dilution refrigerator. The fluctuations in temperature were not more than 0.1%.

The conductance has been measured at a series of temperatures between 330 mK and 1.2 K. Fig. 1 shows some representative data taken at several different temperatures. The fluctuations in the conductance appear in the whole range of gate voltages measured. The structure is fully reproducible over a period of months as long as the sample is kept at helium temperatures (below 1.2 K). All the data that will be presented, analyzed and discussed have been taken on the same sample. Another sample has also been tested and it exhibited the same type of behaviour, i.e. large, reproducible conductance fluctuations, the detailed structure of which was

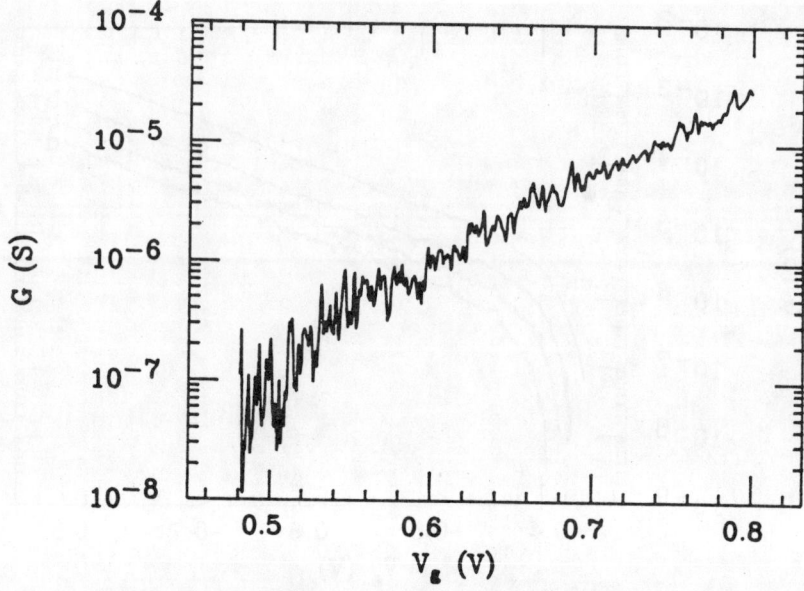

Figure 2: Conductance vs. gate voltage at T=90 mK.

sample-dependent. Some data have been taken at even lower temperatures (Fig. 2), where the size of the fluctuations reached an order of magnitude. The magnetic field measurements were performed by keeping the field constant and sweeping the gate voltage. The field was varied up to 1 T, in steps of 0.1 T, at each value of temperature. The structure in the conductance does not appear to be significantly affected by the magnetic field: the peaks do not seem to shift with the gate voltage and their amplitudes hardly change at all (Fig. 3). Some peaks may be split or slightly reduced. On the other hand, the background conductance is visibly reduced at higher fields.

In the variable-range hopping regime, the average of the conductance logarithm $<\ln G>$ over different impurity realizations (gate voltages) is expected[8] to obey Mott's law

$$<\ln G> \sim \left(\frac{T}{T_0}\right)^n. \qquad (4)$$

The autocorrelation function

$$C(\Delta V_g) = <[\ln G(V_g + \Delta V_g) - <\ln G(V_g + \Delta V_g)>][\ln G(V_g) - <\ln G(V_g)>]>, \qquad (5)$$

where the angular brackets denote averaging with respect to V_g, has been calculated for each set of data (Fig. 4). The ranges of gate voltages over which the averaging of $\ln G$ has been performed contained typically 15–20 correlation lengths. The temperature dependence of $<\ln G>$ obtained in this way, was fit to (4). In order to

1D Variable-Range Hopping in Wide Mesoscopic MOSFETS

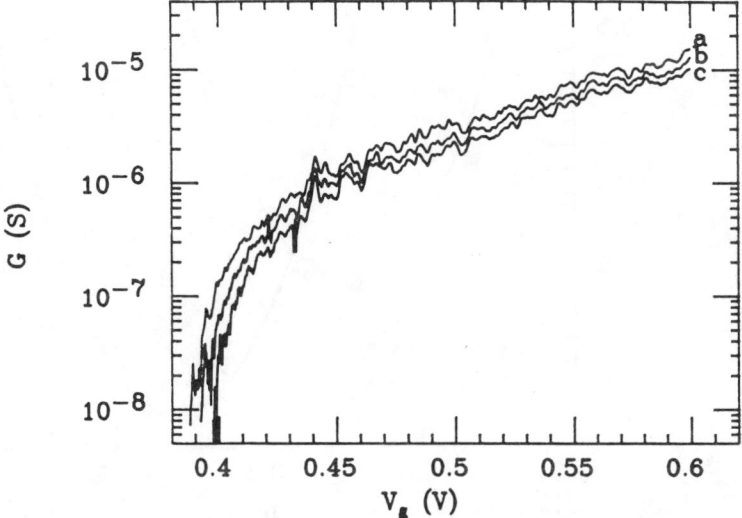

Figure 3: Conductance vs. gate voltage at T=400 mK: a) B=0; b) B=0.7 T; c) B=1.0 T.

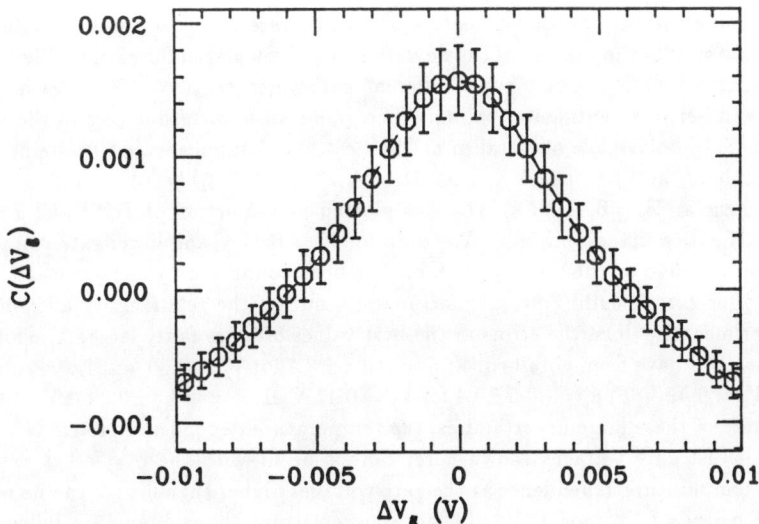

Figure 4: The autocorrelation function calculated on the interval 0.76 V$\leq V_g \leq$ 0.80 V (T= 555 mK). The error bars are also shown.

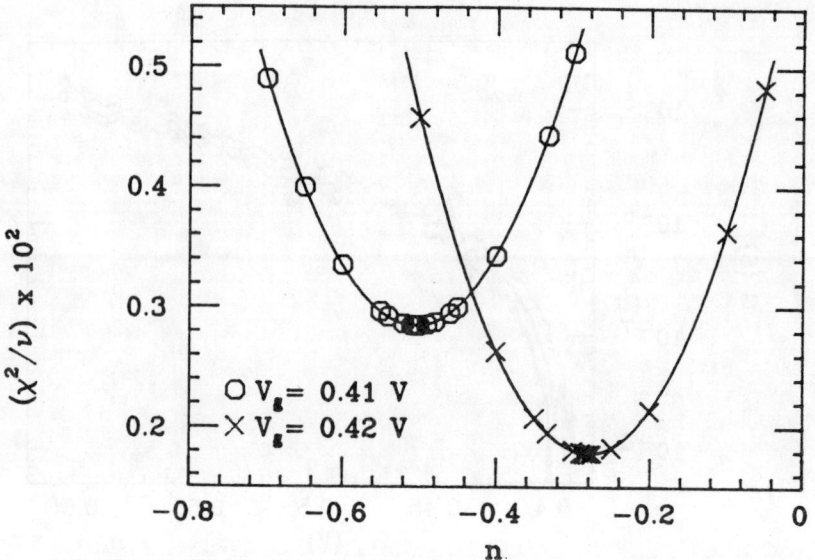

Figure 5: χ^2/ν vs. n for two gate voltages.

determine the best exponent, least-squares fits were made for different values of n and the sum of the squares of the deviations (χ^2) calculated for each n. Fig. 5 shows χ^2/ν as a function of n for two different gate voltages. ($\nu = N - 2$, where N is the number of experimental points.) The values of n corresponding to the minima of the parabolas have been taken as the best fits. It appears that the temperature dependence at $V_g = 0.42$ V is best fit with $n = -1/3$, while $n = -1/2$ is the best exponent at $V_g = 0.41$ V. $< \ln G >$ is plotted as a function of $T^{-1/2}$ and $T^{-1/3}$ for several gate voltages in Fig. 6. The data for $V_g = 0.41$ V and lower gate voltages are better rectified on the $T^{-1/2}$ plot. This has been confirmed by comparing the values of χ^2 for corresponding fits. Unfortunately, due to the relatively small number of experimental points, the errors in the best values of n are fairly large. The following values of n have been obtained: a) $n = -0.8 \pm 0.4$ for $V_g = 0.39$ V; b) $n = -0.7 \pm 0.2$ for $V_g = 0.40$ V; c) $n = -0.5 \pm 0.4$ for $V_g = 0.41$ V; d) $n = -0.3 \pm 0.2$ for $V_g = 0.42$ V. In spite of these large uncertainties, the temperature dependence of $< \ln G >$ at the two lowest gate voltages shown here, cannot be fit with the $n = -1/3$ exponent. The temperature dependence at the gate voltages higher than 0.42 V can be rectified with neither $T^{-1/2}$ nor $T^{-1/3}$. In fact, it seems that the conductance follows some kind of power law dependence in this region. It persists all the way up to $V_g = 0.8$ V, which is the highest value of the gate voltage measured. This region has not been analyzed in any detail so far. It may be concluded, therefore, that the Mott's VRH

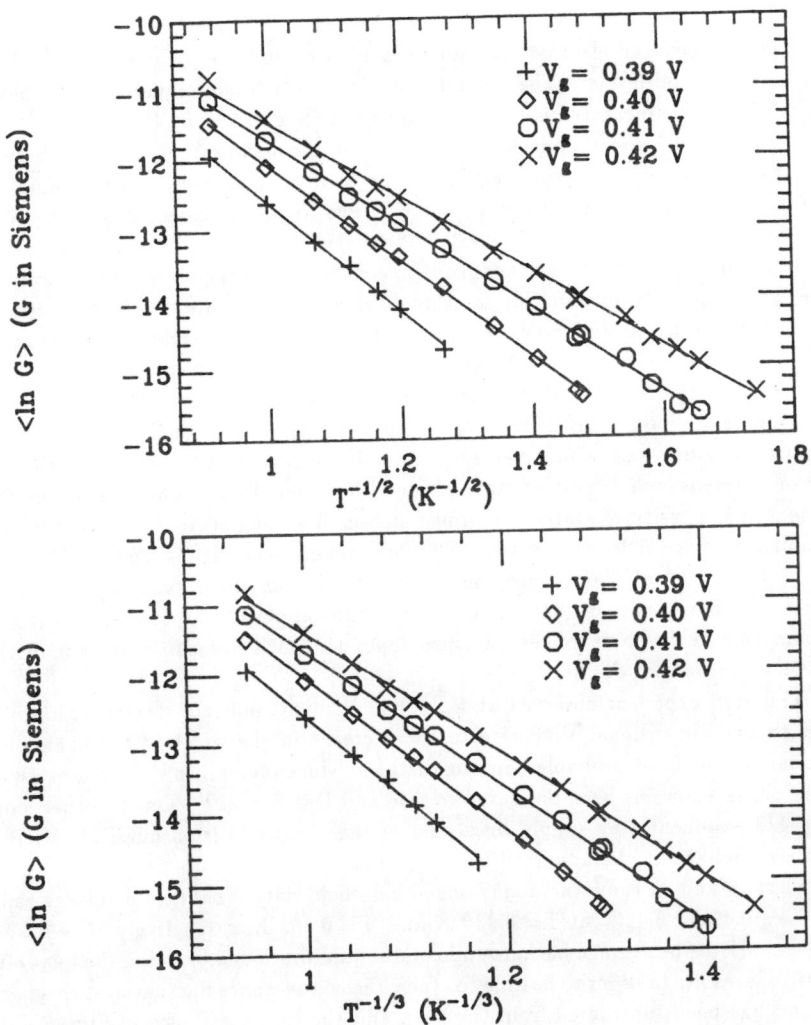

Figure 6: $<\ln G>$ vs. $T^{-1/2}$ and $T^{-1/3}$ for several gate voltages.

conduction dominates the transport for the carrier densities corresponding to gate voltages of less than 0.42 V. In addition, in that region the conductance obeys the law (4) with $n = -1/2$.

The $-1/2$ exponent in (4) is expected in two cases. The first one is when the interactions between electrons cannot be neglected and the density of states has a gap at the Fermi level[17]. The second one is for a one-dimensional sample obeying Mott's law. The measured values of T_0 vary between 30 and 60 K. In the Coulomb gap model, this corresponds to a variation in the localization length ξ between about 2000 and 4500 Å for electrons in a Si inversion layer. These values are unreasonably, by an order of magnitude, larger than the expected ones[18]. If the Coulomb gap argument was to be valid, the dielectric constant of silicon would have to be about ten times larger. Therefore, it seems that this theory is inconsistent with the data. The other explanation for the observed temperature dependence, namely the one-dimensionality of the system, might seem surprising at first since one expects that electrons in such a large MOSFET form a two-dimensional system. Keeping in mind, however, that this structure exhibits large fluctuations, it seems plausible that the $-1/2$ exponent really reflects the effective one-dimensionality of the isolated impurity chains along which electrons hop. If the width of the chain is W_0, then the one-dimensional density of states D_1 will be about $W_0 D_2$, where D_2 is the two-dimensional density of states. W_0 must be small compared to the most probable hopping distance R for the system to be one-dimensional. If $\xi \approx 300 - 500$ Å and $D_2 \leq 1.6 \times 10^{14}$ eV^{-1}cm^{-2} are used, it is found that W_0 is typically 100–300 Å and W_0/R is typically 0.1–0.5. These values thus appear to be consistent with the explanation of the observed temperature dependence which assumes the conduction via one-dimensional chains.

The $-1/3$ exponent observed at $V_g = 0.42$ V might indicate the transition from one- to two-dimensional VRH due to the increase in the width of the chains and decrease of the most probable hopping distance, since increasing V_g increases $D(E)$. Such a transition has been predicted by Xie and Das Sarma [19]. On the other hand, the $-1/3$ exponent may simply be a part of the transition from quasi-1D VRH to localized behaviour characterized by the power law temperature dependence.

Figs. 7a and 8a show the nonaveraged zero-field data for the gate voltage range between 0.4 and 0.42 V at T= 0.330 K and T= 0.365 K, respectively. The dashed lines represent the average background conductance obtained by doing the quadratic least-squares fit. In order to be able to study the conductance fluctuations, this background has been subtracted from the data and the result is shown in Figs. 7b and 8b. It is immediately apparent that the deviation of the conductance logarithm from its average value has two scales: small-scale fluctuations are superimposed on the fluctuations of a larger scale, which seem nearly periodic. The Fourier transform of the data (Fig. 9, where $k = 2\pi/\Delta V_g$) displays two distinct peaks at frequencies corresponding to periods of 5 and 10 mV. Even more convincing is the shape of the autocorrelation function $C(\Delta V_g)$ shown in Fig. 10 for T= 0.330 K. (The higher

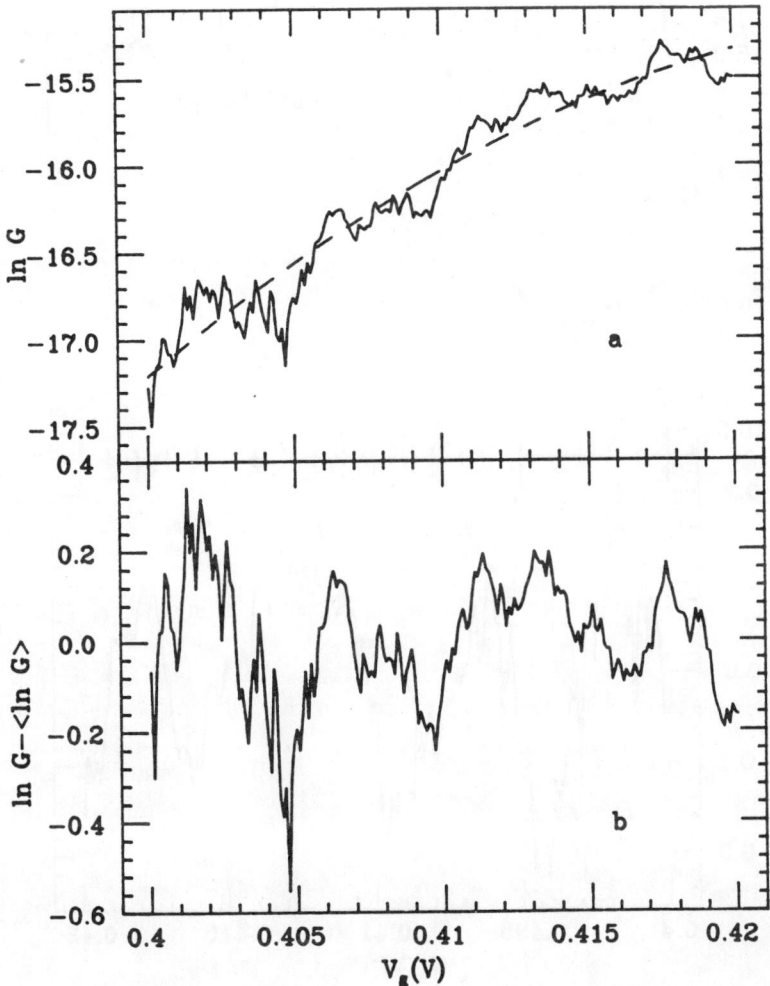

Figure 7: $\ln G$ vs. V_g at T= 0.330 K and in B=0: a) the dashed line is the background; b) the same data with the background subtracted.

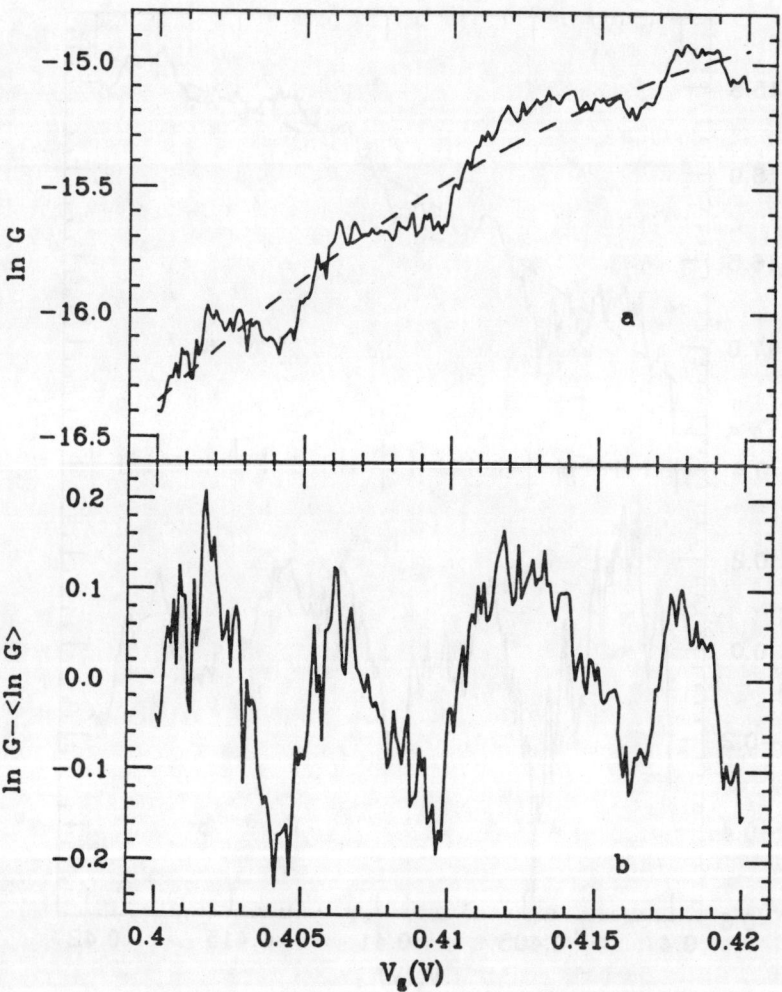

Figure 8: $\ln G$ vs. V_g at T= 0.365 K and in B=0: a) the dashed line is the background; b) the same data with the background subtracted.

1D Variable-Range Hopping in Wide Mesoscopic MOSFETS

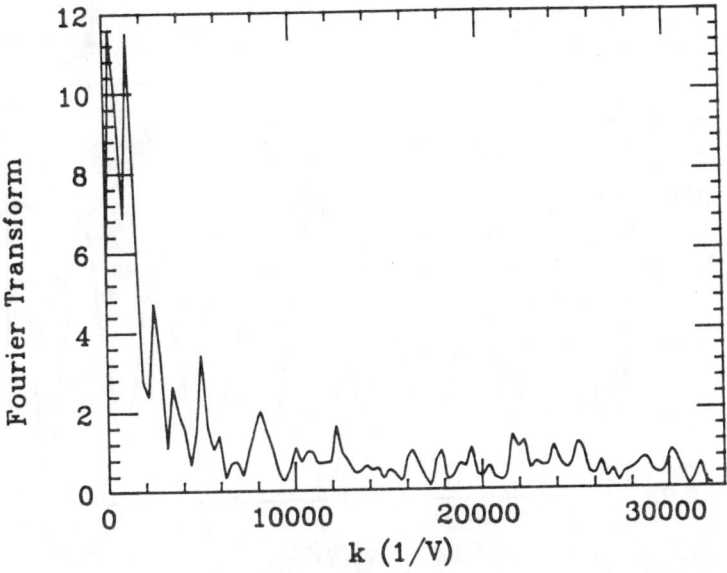

Figure 9: Fourier transform of $\ln G - <\ln G>$ vs. V_g at T= 0.330 K.

temperature results are qualitatively the same.) The amplitudes of the "secondary" peaks, instead of gradually falling off to zero with the increasing distance from the central peak, actually become larger and the oscillations in the autocorrelation function appear to be periodic. While the main peak (centered at $\Delta V_g = 0$) characterizes the fluctuations of the shorter range, the behaviour of the other peaks certainly reflects the almost periodic large-scale fluctuations in the conductance logarithm. The parameter ν, introduced by RR[12], may be determined from $C(\Delta V_g = 0)$. RR provide an expression for the autocorrelation function only for the case of $\nu < 1$, when

$$C(\Delta V_g = 0) = \frac{\pi^2}{6}\left(\frac{1}{\nu^4} - 1\right). \qquad (6)$$

The values of ν obtained using this relation are 0.9972 and 0.9987 for T= 0.330 K and T= 0.365 K, respectively. It is important to note that the equation (6) breaks down when $\nu = 1$ (it gives the incorrect value of zero for $C(\Delta V_g = 0)$, which is a measure of the dispersion of the data) and, by definition, it always gives $\nu < 1$. The distribution functions of the data shown in Figs. 7b and 8b have also been calculated and plotted in Fig. 11. They have been normalized so that the area under each curve equals unity. The values on the abscissa are given by [12, 20]

$$\Delta = \ln G - <\ln G> + 0.577\left(\frac{1}{\nu^2} - 1\right). \qquad (7)$$

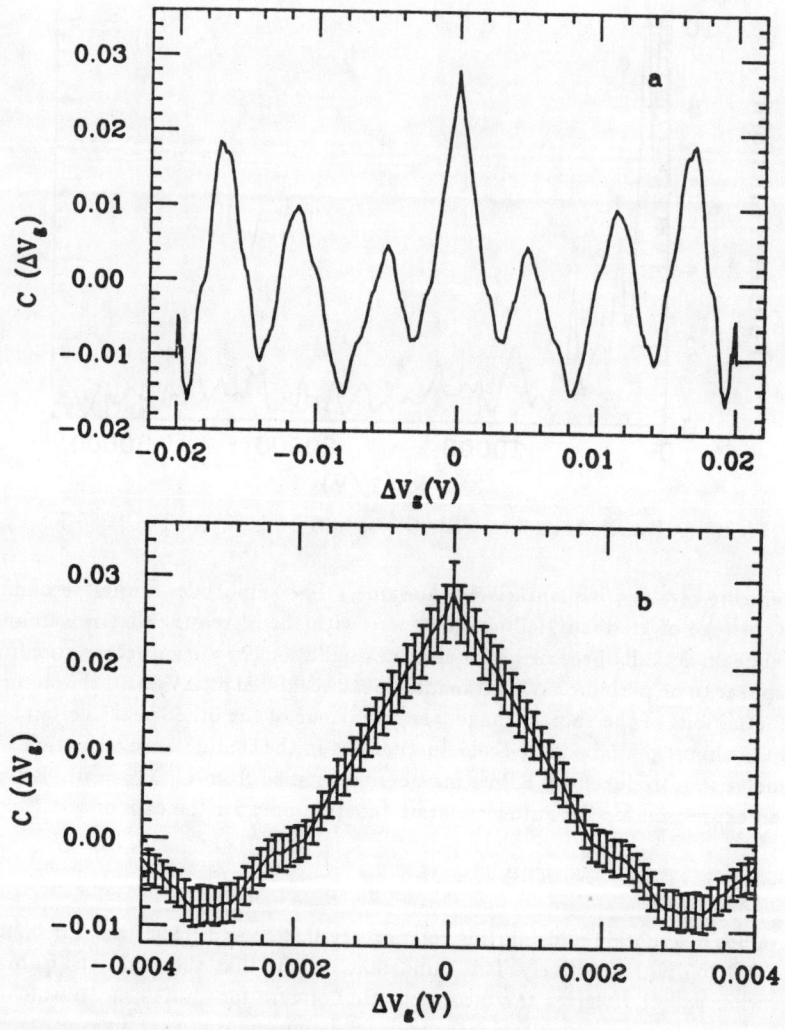

Figure 10: Autocorrelation function $C(\Delta V_g)$ vs. ΔV_g for 0.40 V$\leq V_g \leq$0.42 V at T= 0.330 K: a) calculated on the whole range of gate voltages; b) an expanded view of the same result; the error bars are also shown.

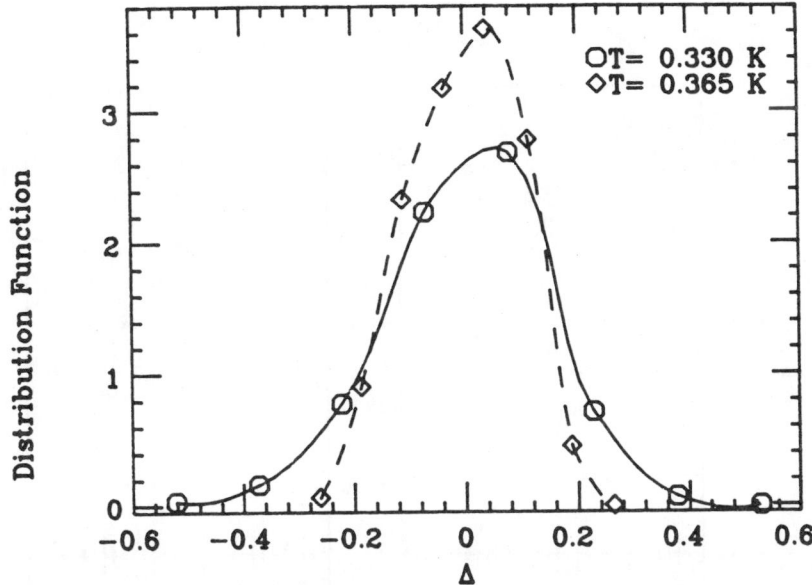

Figure 11: Distribution functions of $\ln G - <\ln G>$ at T= 0.330 K and T= 0.365 K.

These distribution functions are equivalent to those calculated by RR for a population of specimens. They are both asymmetric, just as predicted by theory, with the higher temperature DF becoming narrower and more symmetric. The theoretical expression for the DF, given by RR for the small area case, is not consistent with (6) for the values of ν that are as large as those found here and, therefore, cannot describe the measured DF. It seems very likely that the values of ν for these data are actually greater than unity. Unfortunately, RR have not provided an expression for the autocorrelation function for $\nu > 1$ so that there is no way at present to determine ν and to compare the data with the theory for the large area case. It is also possible that the experimental results correspond to the intermediate area situation in which case the DF is universal and ν determines only the width of the distribution function. The work on fitting the results to the theoretical expression for the DF in this case is currently under way.

The conductance logarithm has also been found to fluctuate with the change in the perpendicular magnetic field. In order to test the ideas of NSS[14], it was of interest to look at the magnetic field dependence of the $<\ln G>$, where the angular brackets still denote the averaging with respect to different gate voltages (ensemble average). Fig. 12 shows some representative data obtained at different temperatures and gate voltages. The magnetoconductance curves are not smooth: they exhibit oscillations but there is no strict periodicity. The existence of these

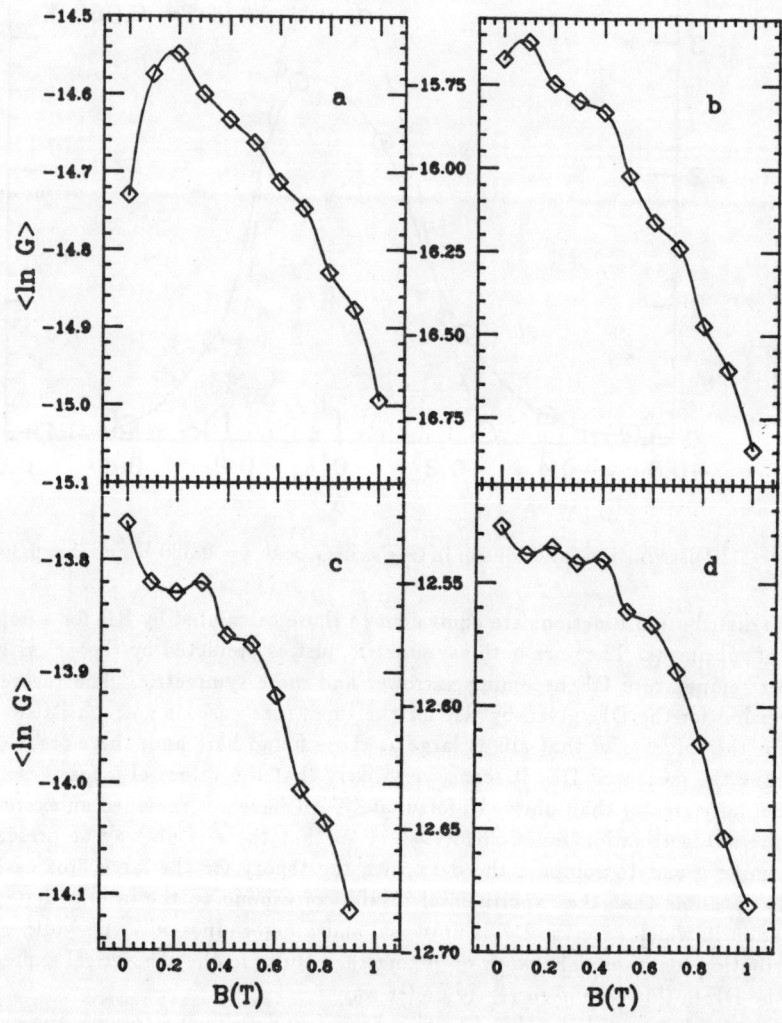

Figure 12: $<\ln G>$ vs. B for: a) $V_g = 0.39$ V and T= 0.625 K; b) $V_g = 0.41$ V and T= 0.365 K; c) $V_g = 0.41$ V and T= 0.555 K; d) $V_g = 0.41$ V and T= 0.790 K.

oscillations in $<\ln G>$ might seem surprising at first since one might expect that this type of behaviour would be already averaged out. It is important to point out, however, that even though the averaging has been performed over different impurity realizations, i.e. over different "sets" of the most dominant hops, it is not very likely that the most probable hopping distance has changed appreciably, by an order of magnitude, in the range of gate voltages over which the averaging has been carried out. In other words, the area of the cigar-shaped region associated with the single hop has remained approximately the same. The typical oscillation "period" that can be inferred from the data is about 0.2–0.5 T. This is in agreement with the period of the conductance oscillations for one jump in a magnetic field proposed by NSS and given by (3). For the localization length $\xi \approx 300$–500 Å, as assumed in the above analysis, the hopping distances of 700–1500 Å are obtained from (3), which is quite reasonable and in agreement with the values found from T_0 in Mott's law (eq.(4)). The observed oscillations are superimposed on the monotonously changing magnetoconductance, which appears to be negative and can be fitted rather well with the quadratic dependence. An example is shown in Fig. 13, where the $<\ln G>$ data from which the monotonous background has been subtracted, are also presented.

It is also of interest to determine whether a magnetic field modifies the temperature dependence of $<\ln G>$. The fits have been made to (4) in the same way as in the zero-field case. $<\ln G>$ at $V_g = 0.41$ V is plotted vs. $T^{-1/2}$ for three different magnetic fields in Fig. 14. The non-zero field data appear slightly curved on this plot since the effect of the magnetic field seems to be stronger at lower temperatures. The fitting shows that the absolute value of the exponent n increases gradually with the magnetic field but the uncertainty in n is again fairly large so that all the data taken in fields up to 1 T may be fitted well with $n = -1/2$ within the error. The best value of n at the highest fields (1 T) reaches approximately unity.

IV. CONCLUSION

The VRH conduction in wide mesoscopic MOSFETs of large area ($\approx 3.3 \times 10^{-6} \mathrm{m}^2$) has been investigated for the first time. The temperature dependence of the averaged conductance logarithm has been found to obey Mott's law for a one-dimensional system. This is believed to be due to the fact that the conduction occurs via hopping of electrons along the widely spaced chains of impurities that connect the source and drain contacts. In this model, the conductance is determined by a very small number of hops even in a fairly large sample, which also provides an explanation for the origin of conductance fluctuations with a change in an external parameter.

The fluctuations caused by the change of a chemical potential or gate voltage have been studied. The conductance logarithm has been found to oscillate on two different scales. This is not quite understood at present but it is most likely a consequence of the competition between different hops that give the most dominant

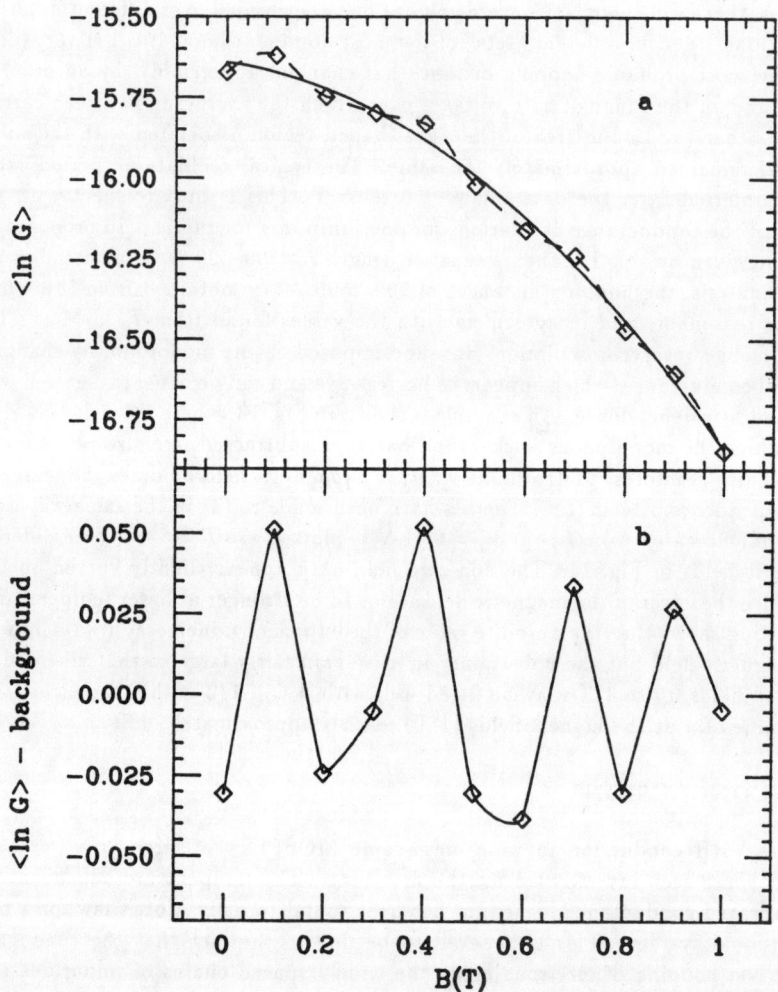

Figure 13: $< \ln G >$ vs. B for $V_g = 0.41$ V at T= 0.365 K: a) the solid line is a quadratic least-squares fit and the dashed line is a guide to the eye; b) the same data with the background (quadratic fit) subtracted; the solid line is a guide to the eye.

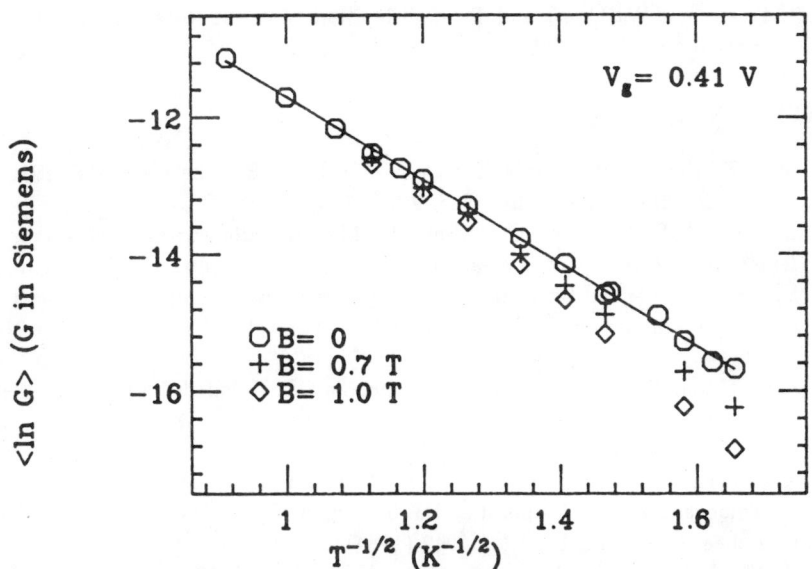

Figure 14: $<\ln G>$ vs. $T^{-1/2}$ at $V_g = 0.41$ V for different magnetic fields.

contributions to the conductance. The distribution function of the conductance logarithm for a population of specimens has been calculated from the data. Its main features are found to agree qualitatively with the theoretical predictions of Raĭkh and Ruzin[12] but the quantitative comparison still needs to be made.

The effect of the magnetic field has also been studied. The structure observed in the conductance does not seem to be significantly affected by the applied field. The temperature dependence of the averaged conductance logarithm does not appear to be changed from the zero-field case within the error of the measurement. Due to relatively large uncertainties, however, the final conclusion cannot be drawn. There may be a tendency for the temperature dependence to become stronger in the magnetic field. The oscillations of $<\ln G>$ with the field have been observed. The characteristic "period" of the oscillations suggests that their origin lies in the fluctuations of the resistance associated with the single hop.

Obviously, more experimental work still needs to be done in order to clarify and understand fully the physics of large mesoscopic systems in the strongly localized regime. Furthermore, the theoretical description of these systems and phenomena is far from complete.

ACKNOWLEDGEMENTS

The research reported in this paper was performed in collaboration with

A. B. Fowler, S. Washburn and P. J. Stiles, and was supported in part by the NSF Grant No. DMR–8717817.

REFERENCES

1. A. B. Fowler, A. Hartstein and R. A. Webb, Phys. Rev. Lett. **48**, 196 (1982)
2. G. Blonder, Bull. Amer. Phys. Soc. **29**, 535 (1984)
3. R. A. Webb, S. Washburn, C. P. Umbach and R. B. Laibowitz in *Localization, Interaction, and Transport Phenomena in Impure Metals*, edited by G. Bergmann, Y. Bruynseraede and B. Kramer (Springer-Verlag, Heidelberg, 1985); C. P. Umbach, S. Washburn, R. B. Laibowitz and R. A. Webb, Phys. Rev. B **30**, 4048 (1984)
4. J. C. Licini, D. J. Bishop, M. A. Kastner and J. Melngailis, Phys. Rev. Lett. **55**, 2987 (1985)
5. W. E. Howard and F. F. Fang, Solid State Electron. **8**, 82 (1965)
6. A. Miller and E. Abrahams, Phys. Rev. **120**, 745 (1960)
7. V. Ambegaokar, B. I. Halperin and J. S. Langer, Phys. Rev. B **4**, 2612 (1971)
8. P. A. Lee, Phys. Rev. Lett. **53**, 2042 (1984)
9. R. A. Serota, R. K. Kalia and P. A. Lee, Phys. Rev. B **33**, 8441 (1986)
10. M. Pollak and J. J. Hauser, Phys. Rev. Lett. **31**, 1304 (1973)
11. A. V. Tartakovskiĭ, M. V. Fistul', M. É. Raĭkh and I. M. Ruzin, Sov. Phys. Semicond. **21**, 370 (1987)
12. M. É. Raĭkh and I. M. Ruzin, Sov. Phys. JETP **65**, 1273 (1987)
13. M. É. Raĭkh and I. M. Ruzin, JETP Lett. **43**, 562 (1986)
14. V. L. Nguen, B. Z. Spivak and B. I. Shklovskiĭ, JETP Lett. **43**, 44 (1986)
15. Ya. B. Poyarkov, V. Ya. Kontarev, I. P. Krylov and Yu. V. Sharvin, JETP Lett. **44**, 373 (1986)
16. E. I. Laĭko, A. O. Orlov, A. K. Savchenko, É. A. Il'ichev and É. A. Poltoratskiĭ, Sov. Phys. JETP **66**, 1258 (1987); A. O. Orlov and A. K. Savchenko, JETP Lett. **47**, 471 (1988); A. O. Orlov and A. K. Savchenko, JETP Lett. **44**, 43 (1986); A. O. Orlov, A. K. Savchenko, E. V. Chenskiĭ, É. A. Il'ichev and É. A. Poltoratskiĭ, JETP Lett. **43**, 542 (1986); M. Pepper and M. J. Uren, J. Phys. C: Solid State Phys. **15**, L617 (1982)
17. A. F. Efros and B. I. Shklovskii, J. Phys. C **8**, L49 (1975)
18. C. J. Adkins, S. Pollitt and M. Pepper, J. Phys. (Paris), Colloq. **37**, C4-343 (1976)
19. X. C. Xie and S. Das Sarma, Phys. Rev. B **36**, 4566 (1987)
20. M. É. Raĭkh, private communication.

Computer Simulations of Elastic Tunneling Through Mezoscopic MOSFETS*

M. Green and M. Pollak

*Department of Physics, University of California,
Riverside, California 92521, U.S.A.*

ABSTRACT

Recent experiments on mesoscopic MOSFETs show sharp peaks in the source-drain current as a function of the gate voltage. The peaks are due to coherent tunneling between the source and drain electrodes, via a localized state near the center. The structure of such peaks at low temperatures is expected to be Lorentzian, but the observed peaks had a non-Lorentzian character, and also exhibited resolvable secondary peaks on the shoulders of the main peak. These features were interpreted as due to Coulomb interactions between the tunneling site and electrons which may enter or exit the localized states during the experiment and excitations internal to the system. At present our results are in partial agreement with this interpretation.

1. INTRODUCTION

This paper reports the results of computer modeling of experiments reported by Fowler et al.[1] on very small MOSFETs. The experiments were done at low temperatures (< 150 mK) so that transport between the source and drain electrodes is predominantly by elastic tunneling. The electron states of concern are thought to be strongly localized. The total number of sites in these systems is on the order of 1000. This makes computer simulations feasible with present day supercomputers.

Reference 1 reported on two different MOSFETs with a width of 1-1.75 µm and length of 0.5 µm. The oxide layer was 10 nm with the electrons in the Si substrate thought to be 4 nm below the Si/SiO$_2$ interface. The temperatures of the samples ranged from 22 mK to greater than 150 mK. The source-drain voltage was kept small. The gate voltage was changed and the source-drain conduction was noted. (For more details see Ref. 1).

Experimental results show conductance peaks as a function of gate voltage and hence of the chemical potential. As the chemical potential is varied, it becomes coincident with different sites in the sample. According to the model of Ref. 1, when the chemical potential is coincident with certain sites, a conductance peak results.

The experimental results pose two questions: none of the conduction peaks is a pure Lorentzian and each exhibits fine structure as a function of the chemical potential. Since at equilibrium the sample can be in any accessible configuration consistent with its temperature, one possible explanation is that transitions between configurations would change the energy of the tunneling site due to the changing Coulomb interactions. If the thermal shifting between configurations is rapid in comparison with experimental times, and slow compared to elastic tunneling times, the conductance measured at a given gate voltage will have contributions from each configuration, weighted by the thermal probabilities of being in that configuration. This will cause the conduction peaks to deviate from a pure Lorentzian by widening them and also providing the fine structure. This paper explores the peaks resulting from this approach and compares them with experiments.

2. COMPUTER MODEL

The salient structural features of the MOSFET are shown in fig. 1. The accepted model for a MOSFET[2] assumes that charged impurities in the oxide layer and imperfections at the Si/SiO_2 interface produce electron trapping centers near the interface. Such randomly distributed imperfections are represented by a random potential, so that there is a disorder energy associated with each center. In this paper, the disorder energy is chosen to have a uniform probability within a certain energy interval.

The computer model randomly places N_s sites within the sample area and randomly occupies them with N_e electrons, where $N_s > N_e$. A site may have at most one electron. We take an unoccupied site to be neutral and an occupied site to have the charge of an electron, here denoted by q. The effect of the ions in the oxide is taken as a random potential, so we do not assign specifically any charges to the oxide layer. Charge neutrality of the system is achieved by a neutralizing charge in the gate electrode. In this model, this is automatically achieved by the technique of image charges, discussed below.

2.1 Interaction Potential

The electron-electron interactions are reduced by the dielectric of the Si and by screening due to the dielectric interface and the metal electrode. To calculate the interaction energy, we calculate the spatial dependence of the potential of a point charge in the Si, as indicated in fig. 1. The screened potential was solved for in the region $d \geq z > a$, where a is the distance from the electrode to the interface and d is the distance from the electrode to the electrons as shown below.

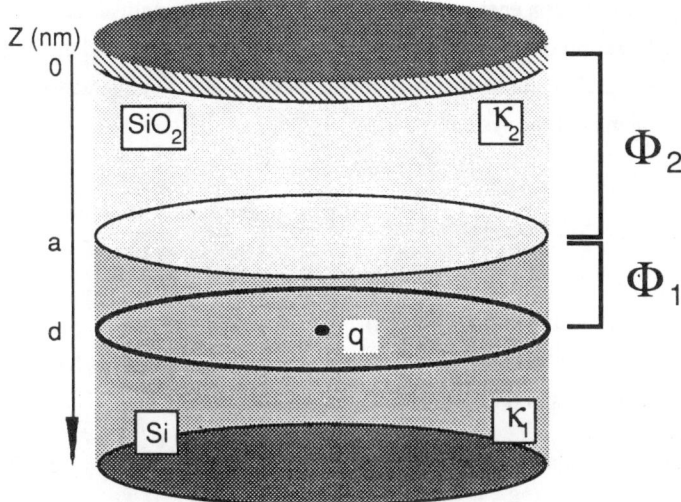

Figure 1. Shown is part of a MOSFET, including the gate electrode, the silicon oxide layer and the silicon. κ_1 and κ_2 are the dielectric constants of the silicon and the silicon oxide, respectively.

The potential, assuming an infinite sample, can be expanded in terms of Bessel functions[3]. The final result in region 1 is:

$$\Phi_1(r,z) = \frac{q}{\kappa_1} \left(\frac{1}{\sqrt{r^2 + (z-d)^2}} + \frac{e^{-2b}}{\sqrt{r^2 + (z-2a+d)^2}} - \left(1 - e^{-4b}\right) \sum_{n=0}^{\infty} \frac{e^{-2nb}}{\sqrt{r^2 + (z+2na+d)^2}} \right)$$

where $b = \tanh^{-1} \frac{\kappa_2}{\kappa_1}$ and $\kappa_1 > \kappa_2$. This potential is the sum of the potential from the original point charge plus an infinite number of image charges.

2.2 Hamiltonian

The total energy at a site occupied by an electron is the sum of the random energy W_m at that site and the interaction energy due to Coulomb interactions with all other electrons in the system. The Hamiltonian is the sum of the random site energies and the interaction energies, with care taken not to double count the electron-electron interactions.

The Hamiltonian H of the system is:

$$H = q \sum_{m>j}^{ele} (1-n_m)(1-n_j)\Phi(R_{mj},d) + \sum_m (1-n_m)W_m$$

where n_m is 0 (1) if occupied (unoccupied).

This can be written in terms of the site energies ε_m, which is the potential at site m due to all other charged sites in the system, plus the disorder energy W_m. This is defined as:

$$\varepsilon_m = W_m + \sum_{j \neq m}^{ele} (1-n_j)\Phi(R_{mj},d) .$$

The final Hamiltonian then becomes

$$H = q \sum_m^{ele} (1-n_m)\left(\frac{\varepsilon_m}{2}\right) + \sum_m (1-n_m)W_m .$$

3. PROCEDURE

For any given N_e, the system is first put into a pseudo ground state, using an algorithm developed by Baranovskii et al[4]. This is a state stable with respect to one-electron excitations, and in addition has a well defined Fermi level, such that all sites with a higher (lower) site energy are empty (occupied).

According to Ref. 1, a peak is observable if the tunneling site lies within a localization length L_0 of the center of the sample. (Reference 1 finds L_0 to be 42 nm). We assume that tunneling occurs through a site within this area, when it is the lowest unoccupied site. One may object to this assumption, as it destroys the natural electron-hole symmetry. Tunneling may in fact occur just as well through the highest occupied site, or indeed through a fractionally occupied site. As the chemical potential is swept through the lifetime broadened tunneling level, the occupation changes between 0 and 1. The resulting change in charge on that site may induce a transition of electrons among the other sites. This can happen when the ground state of the N_e electron system differs from the ground state of the N_e+1 electron system (other than by the change in occupation of the tunneling site). When this is not the case, the tunneling through the lowest unoccupied site of the N_e electron state and through the highest occupied site of the N_e+1 electron state are smoothly connected. In the other cases the situation is more complex. The expected experimental behavior then depends on how the transition rate between the ground state configurations of the N_e and N_e+1 occupations compares with the rate at which the chemical potential is swept through the broadened tunneling state. If the former is larger, one expects a sudden change in the tunneling current. This is not expected when the chemical potential is swept through the tunneling state before the configuration can change. In this case a hysteresis can be expected, as the "downward" sweep of the chemical potential will not reproduce the "upward" sweep. (To our knowledge such an effect has not been reported). Our assumption of tunneling through the lowest unoccupied site then simulates the "upward" sweep.

To test whether the proposed mechanism of thermally induced hopping between configurations can account for the observed lineshape and structure, the following procedure could ideally be adopted.

1. For a given N_e, find the pseudo ground state. For each excited state examine the excitation energy E from the pseudo ground state.
2. For each state with $E < 5\ kT$, calculate the shift Δe in site energy of the tunneling site, and the transition rate ω to reach such a state. Keep those states with ω large enough to allow transitions during experimental times.
3. Superimpose all the peaks, with energy shifts Δe, amplitudes proportional to $\exp(-\Delta E/kT)$, and appropriately lifetime broadened. The resulting plot should correspond to an experimentally observed peak.

We choose a more realistic approach by testing configurations which can be reached from the ground state by one electron excitations. No attempt was made to calculate the transition rate.

There are two types of one electron excitations. The first is an excitation within the sample and the second type are excitations between the sample and the source or drain electrode. Both types of excitations are from the pseudo ground state, as outlined above.

The first type of excitation is simulated by making all one-electron excitations in the system excluding the lowest unoccupied site. E is computed for each excitation. If E < 5 kT, Δe (which can be positive or negative) is computed, and both E and Δe are stored. If E > 5 kT, the excitation is deemed unobservable and discarded. The second type of excitation is simulated by making all transitions of one electron from the system to an electrode, and from an electrode to the system. As above, E is computed for each such excitation, and discarded if larger than 5 kT. For E < 5 kT, Δe is calculated and Δe and E are stored.

4. RESULTS

The computer results given below are for the smaller device mentioned in Ref. 1 with an area of 500 x 1000 nm. The number of sites was 780 and the disorder energy was uniformly distributed within a range of 4 meV. The temperature used was 22 mK for all calculations. The density of states based on the above is 3.9×10^{17} m^{-2}eV^{-1}. A value of 11.7 was used for the dielectric constant in the silicon and 3.95 in the oxide. The oxide layer was 10 nm wide and electron states were 4 nm below the oxide. (These parameters were obtained in part from Ref. 1 and in part from private communications[5]).

Two "samples" were used, which differed in the positions of the sites in the samples and the disorder energy on each site. Initially, one electron is placed in the sample, in the site of lowest random energy, which is clearly the ground state for this case. The probability of tunneling is calculated as outlined in Part 3 and noted. The number of electrons in the sample is then incremented by one, finding for each increment the pseudo ground state, whether tunneling can occur, and, if so, the set of Δe's and E's (for E < 5 kT). The procedure is repeated up to a total of 779 electrons.

Figures 2 and 3 show the un-normalized probability $e^{-\Delta E/kT}$ versus the associated change in energy Δe of the tunneling site. The pseudo ground state, where the tunneling site would have an un-normalized probability of unity, is not included in Figs. 2 and 3.

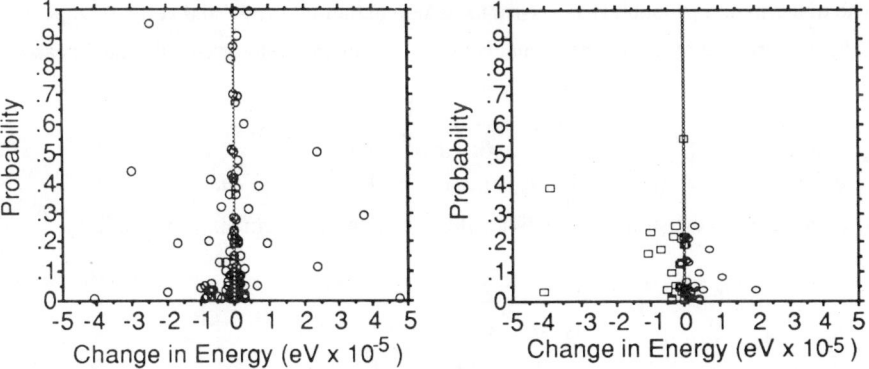

Figure 2 Both graphs are for sample 1 where the lowest unoccupied site becomes the highest occupied site. Left graph shows excitations within the system. Right graph shows transition to or from an electrode.

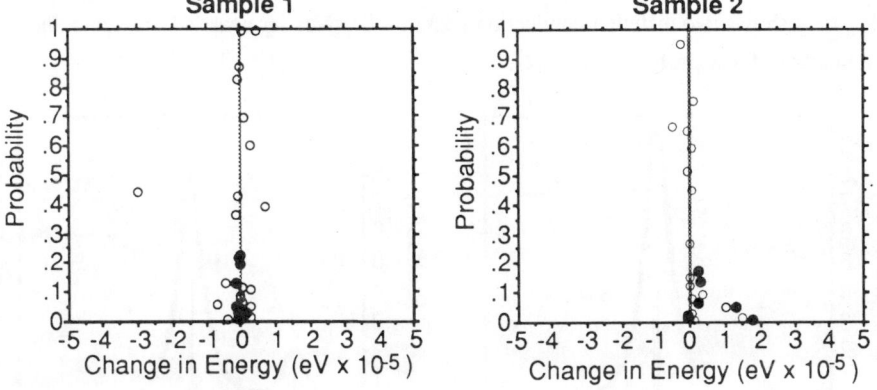

Figure 3 In both graphs lowest unoccupied site becomes highest occupied site. Lowest unoccupied site is within a localization radius of center of sample. Open circles are excitations within system. Filled circles are transition from the source or drain electrode into the system or from the system to either electrode.

As explained in section 3, it is possible to construct a theoretical graph of a conductance peak from the thermal excitation probability, the associated changes in the tunneling site energy given in Fig. 3, and the lifetime broadening. For this purpose, the probabilities of Figs. 3 and the central peak not shown must be included.

The transmission probability is known to be Lorentzian in form[6], so that the final peak will be a superposition of the individual peaks. The transmission for an individual peak is:

$$T_i(E) = \frac{(\tfrac{1}{2}\Gamma)^2}{(E - E_i)^2 + (\tfrac{1}{2}\Gamma)^2}$$

where Γ is the line width of the resonance and E_i is the change in energy.

The final peak is the sum of the individual peaks times the normalized probability.

$$T_f(E) = \frac{\sum_i T_i(E) \, e^{-\Delta E_i / kT}}{\sum_j e^{-\Delta E_j / kT}}$$

According to Ref. 1, the transmission peaks should remain constant in amplitude and peak width if the temperature broadening is less than the line width, i.e. $kT < \Gamma$. Fowler shows that one peak remains constant for temperatures less the 100 mK and therefore estimates that the intrinsic width of the peak is 8.6 µeV.

Figure 4 shows two different conduction peaks $T_f(E)$. Neither peak is Lorentzian due to a number of low probability peaks.

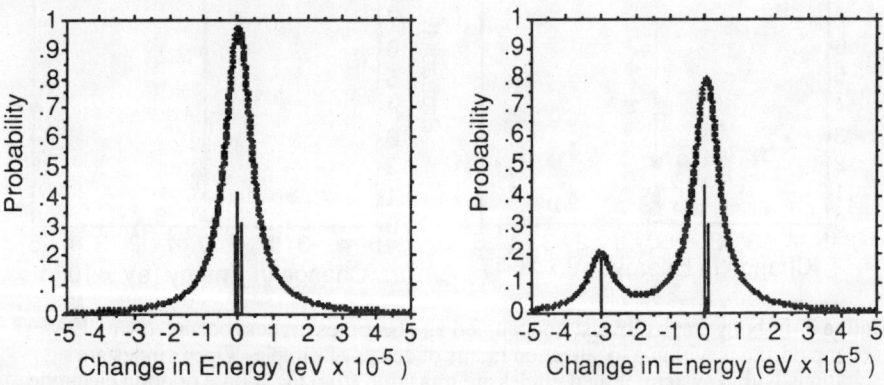

Figure 4 Superposition of peaks produce conduction peak. Lines represent probability and position of individual peaks.

5. CONCLUSION

The experimental value to Γ is 8.6 µeV, while the observed width of a conduction peak is of the order of 100 µeV. If the proposed mechanism of Coulomb effects is to explain the experimental results, the line width would have to be entirely due to the shifts Δe. Thus, Δe should go up to $\approx \pm 50$ μeV, and there should be about $100/8.6 \geq 12$ excitations with such Δe's for each tunneling site. While the first requirement is partially fulfilled, the number of excitations with E < 5kT we observed is much smaller. According to our results, the line rather than being broadened, is split into a small number of peaks, each with a line width ≈ 8.6 μeV. It is possible that this result may be modified by two or three electron excitation, but we have not yet investigated these possibilities.

Although a uniform disorder energy was used above, we have also investigated an exponential distribution of disorder energy, as proposed by Pollitt[7]. Initial results indicate that the nature of the disorder is not important, at least not for the effects reported here.

Since this work was completed, we have learned that the number of states in the system may be half of that used. We do not feel that this can quantitatively effect the conclusions.

The method we use to find the pseudo ground state randomly occupies the sample with N_e electrons, and then applies the algorithm cited in Section 3 for finding a pseudo ground state. This is done repeatedly, each time randomly occupying the sample with N_e electrons and keeping the lowest ground state; the procedure is limited only by the computer time available. Another approach would be to start with the (N_e-1) ground state, add one electron to the tunneling site, and then find the ground state. This would be done only once. The first method simulates a MOSFET where the gate voltage is moving slow enough that the system has time to relax. The second method implies a fast moving gate voltage. Both methods can realistically reflect experimental procedure and should be investigated.

The procedure we used assumes that the system is strongly localized. Using a nearest neighbor distance of 13 nm, a Bohr radius of 42 nm and other values as given in the text, the Anderson criteria (for 3 dimensions) would indicate that the system is borderline localized. While in two dimensions the localization is stronger, we feel that this point requires closer examination.

6. ACKNOWLEDGEMENTS

We would like to thank A. B. Fowler for discussions and for providing information about his experiments.

7. REFERENCES

* Supported in part by the San Diego Supercomputer facility.
[1] A. B. Fowler, G. L. Timp, J. J. Wainer and R. A. Webb, Phys. Rev. Lett. **57**, 138-141 (1986).
[2] T. Ando, A. B. Fowler and F. Stern, Rev. Mod. Phys. **54**, 54 (1982).
[3] W. K. F. Panofsky and M. Phillips, Classical Electricity and Magnetism, Addison-Wesley, Second Edition, pg. 88.
[4] Baranovskii *et al.*, J. Phys. **C12**, 1023-1034 (1979).
[5] Private communications with Dr. Fowler.
[6] A. Douglas Stone and P.A. Lee, Phys. Rev. Lett. **54**, 1196 (1985).
[7] S. Pollitt, Commun. Phys. **1**, 207 (1976).

Probability Distribution Functions and Wavelength Correlations
for
Transmission of Waves Through Random Media:
A New Numerical Method

Itzhak Edrei
Serin Physics Laboratory
Rutgers University
P. O. Box 849
Piscataway, NJ 08855-0849

ABSTRACT

We present a new numerical method for calculating interference phenomena for waves propagating through random media. We apply the model to calculate probability distribution functions for the transmission $P_L(T)$ in two dimensions, T being the transmission coefficient. The model yields new results for two dimensional systems. The distribution function $P_L(T)$ in two dimensions, in the diffusive regime, is found to be close to a Gaussian with a variance proportional to the mean, in agreement with the results of diagrammatic calculations. We also apply the method to calculate the wavelength dependence of the intensity correlation functions in the transmission and find it to be in good agreement with the theoretical results.

Introduction

In the past few years the analogy between light scattering and electron transport in random media was realized[1]. Interference between multiply scattered waves leads to interesting phenomena such as universal conductance fluctuations in weakly disordered metals[2,3]. One expects that in the optical waves something similar will occur. We deal here with systems in which the mean free path ℓ, is larger than the wavelength λ, i.e. we use the diffusion approximation. In this region long multiple scattering paths are important and the interference between those paths will cause coherent backscattering pick, intensity fluctuations (speckle), large transmission fluctuations and intensity correlations.

In this paper I want to present a numerical method for studying interference phenomena in random media, and to use this method to study the full distribution function $P_L(T)$, where T is the transmission coefficient and L is the length of the system. We also apply the method to study the wavelength intensity-intensity correlation function $C(\Delta\lambda)$. $P_L(T)$ and $C(\Delta\lambda)$ were calculated in two-dimensional systems.

The Method and the Model.

The method used for the simulation is a general method. It follows the propagation of the wave in the medium. Each time step the wave is scattered locally (single scattering) at each site, and the multiple scattering effects are built up in time until a steady state is achieved.

Generally we can write the following equation

$$(1) \quad \psi_\alpha(m+1) = \sum_\beta G_{\alpha\beta} \psi_\beta(m)$$

where $\psi_\alpha(m)$ is a complex number that defines the wavefunction (or electric field) at a discrete time m on site (or bond) α. The matrix $G_{\alpha\beta}$ is a unitary matrix that includes the symmetries of the problem (i.e. time reversal and space symmetries if they exist).

The specific model[5] that we used is built from blocks such as illustrated in Fig. 1.

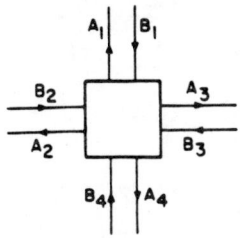

Fig. 1. Two dimensional scatterer A and B are the outgoing and ingoing amplitudes, respectively and related by $A_i = \sum_{j=1}^{4} S_{ij} B_j$ where S_{ij} is 4x4 S matrix for the scatterers.

Each block represents a scatterer and a 2Dx2D S matrix is assigned to it. Each bond carries two waves propagating in opposite directions. At each time step the S matrix transforms the 2D incoming waves to 2D outgoing waves. This process continues until a steady state picture is achieved.

The S matrix is defined by four complex numbers that correspond to the scattering in four directions. The transmission coefficient $t = |t| e^{i\phi_t}$, the reflection coefficient $r = |r| e^{i\phi_r}$ and scattering to right and left $r_R = r_L = |r_L| e^{i\phi_L}$. On the bonds we have random phases $2\phi_i$ ($i=1...N_b$, N_b=number of bonds), which are related to the free propagation on the bond, and are added to the S matrix. (To be more accurate, half of the phase, i.e. ϕ_i is added to each scatterer connected to this bond.)

We choose all scatterers to have the same amplitudes t, r, r_L, and the phases $2\phi_i$ to be random between $0-2\pi$. Because the disorder is weak we expect that the specific choice of the disorder will be of no importance.

The S- matrix attached to the scatterer in Fig. 1 will be the following

(2) $$S = \begin{pmatrix} r\phi_{11} & r_L\phi_{12} & r_L\phi_{13} & t\phi_{14} \\ r_L\phi_{21} & r\phi_{22} & r\phi_{23} & r_L\phi_{24} \\ r_L\phi_{31} & t\phi_{32} & r\phi_{33} & r_L\phi_{34} \\ t\phi_{41} & r_L\phi_{42} & r_L\phi_{43} & r\phi_{44} \end{pmatrix}$$

where $\phi_{ij} = \exp i[\phi_i + \phi_j]$

and we have the following constraint (choosing $\phi_L = 0$)

(3) $$\text{tg}\phi_t = (|t| + |r|\cos\phi_o)/|r|\sin\phi_o$$

where $\phi_o = \phi_r - \phi_t = \cos^{-1}(-|r_L|^2/|r||t|)$

and $|r|^2 + 2|r_L|^2 + |t|^2 = 1$

It is straight forward to show that the unitarity of the individual S matrix ensures the unitarity of the matrix $G_{\alpha\beta}$.

Transmission Probability Distributions.

Applying this method to one dimensional systems[6] gave results for the resistance probability in weak and strong disorder, with very good agreement with theory[7]. In two dimensions we chose symmetric scatterers with $|r|^2 = |t|^2 = |r_L|^2 = 0.25$. The geometry we used is a slab geometry with thickness L and width W, where W>>L. W was illuminated by a plane wave. The diagrammatic calculations[8] gave for the average transmission $\langle T \rangle \simeq \ell/L$ and for the variance $\langle \Delta T^2 \rangle \simeq \frac{\langle T \rangle}{N}$ (where N is the number of channels). In Fig. 2 we plot $\langle \Delta T^2 \rangle$ vs. $\langle T \rangle$ and $\langle T \rangle$ vs. $1/L$ for four lengths and we see that to a good approximation they fall on a straight line. Next in Fig. 3 we plot the full probability $P_L(T)$ as a function of $X = (T - \langle T \rangle)/\langle \Delta T^2 \rangle^{1/2}$ and we draw the Gaussian function of x. There is no

adjustable parameter, and we can see that the bulk of the
distribution can be represented by a Gaussian (not the tails)

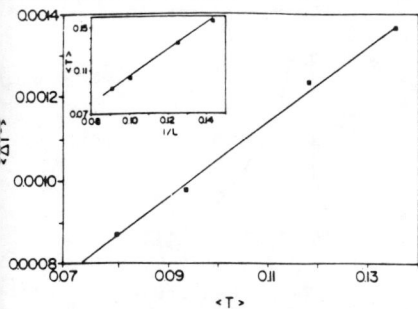

Fig. 2 The squares represent the variance of T vs ⟨T⟩ for a slab with W=50, L=7,8,10,11. Inset: ⟨T⟩ vs. 1/L.

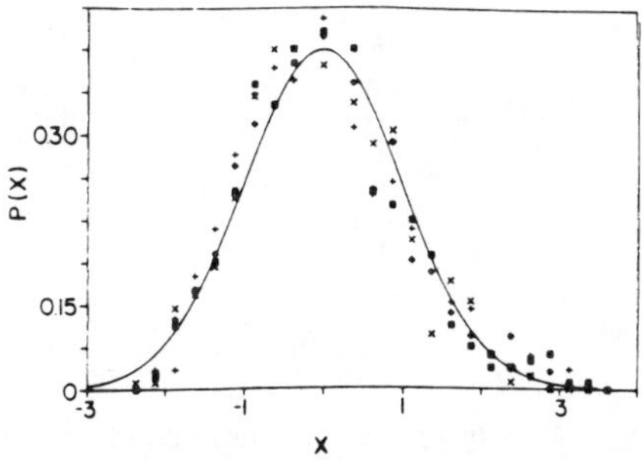

Fig. 3 The four distribution functions for 4 values of L. The symbols ■, ×, ◊, + correspond to L=7,8,10,11. The solid line is the Gaussian $(2\pi)^{-1/2}\exp(-X^2/2.)$

In other geometry that we used (W<L) we saw a crossover to log-normal behavior.[6]

Wavelength Intensity Correlations C(Δλ)

The intensity correlation function is defined as the following cumulant

(4) $$C(r,\Delta\lambda) = \langle |E(r,\lambda)|^2 |E(r,\lambda+\Delta\lambda)|^2 \rangle_c$$

In order to calculate this theoretically, it is enough to calculate $\langle E(r,\lambda)E^*(r,\lambda+\Delta\lambda)\rangle$ and square it (factorization approximation)[8-10] because here we are looking at the correlations at the same point, r.

The geometry used is the same as the above. Here W=200 and L varies between 20 to 60. The scatterers are identical and the phases are random. The transmission coefficients are: $|t|^2 = 0.85$, $|r|^2 = |r_L|^2 = 0.05$. In Fig. 4 we plot the intensity correlation function and the field correlation function square, for the transmitted and reflected waves. We see that up to $\Delta\lambda$ that corresponds to the half of $C(\Delta\lambda)$ the factorization approximation is very good.

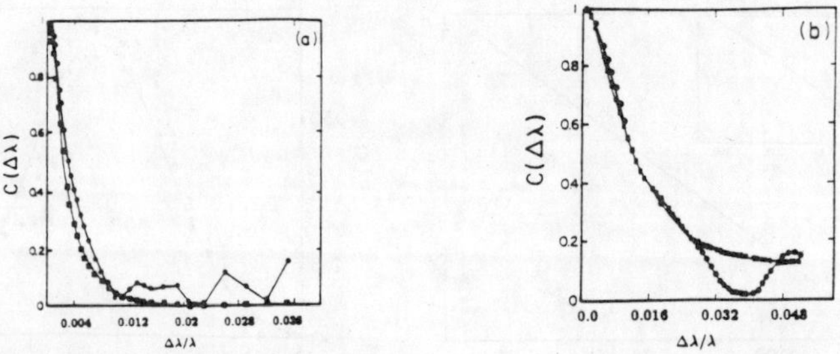

Fig. 4 $C(\Delta\lambda)$ for transmitted waves
• intensity correlations
field correlations (squared)

Reflected waves.

We also see that the intensity correlation has larger fluctuations than the field correlations. We believe that this is due to the finiteness of the system, and in the field correlations the phases average out the oscillations.

The theoretical correlation[10] function is a function of $\Delta\lambda L^2/\lambda^2 \ell$. In order to check the theoretical results we plot in Fig. 5 $C(\Delta\lambda)$ for four different lengths (L=20,30,40,60) and the analytical results (the full curve) vs. $\Delta\lambda L^2/\lambda^2 \ell$ with one adjustable parameter for all four curves. We see that the agreement is good (more details will be published elsewhere[11]).

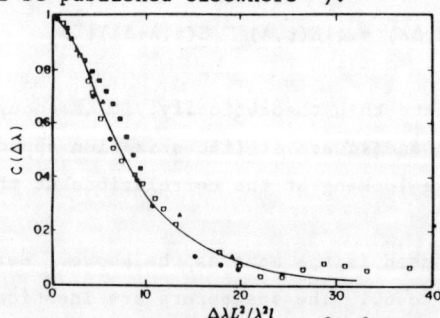

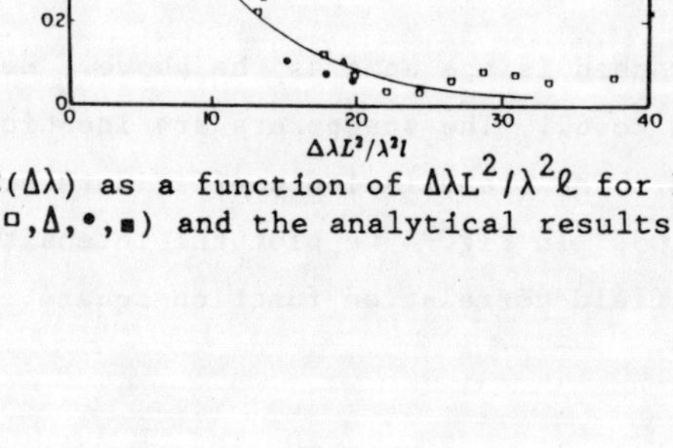

Fig. 5 $C(\Delta\lambda)$ as a function of $\Delta\lambda L^2/\lambda^2 \ell$ for L=20,30,40,60 (□,△,•,■) and the analytical results (full curve).

In summary we developed a numerical method for wave propagation and applied it to calculate probability distributions for transmission and found that it can be represented by a Gaussian. We confirmed the results for $\langle \Delta T^2 \rangle$. We showed that the factorization approximation is a good approximation and that $C(\Delta\lambda)$ really scales like $\Delta\lambda L^2/\lambda^2 \ell$ as the theoretical calculations show.

This work was done in collaboration with Professor M. Kaveh, Bar Ilan University, Israel and Professor B. Shapiro, Technion, Israel.

I would like to thank Professor M. J. Stephen for going over the manuscript.

References

1. P. W. Anderson, Phil. Mag. **B52**, 502 (1985).
2. P. A. Lee and A. D. Stone, Phys. Rev. Lett. **55**, 1622 (1985).
3. B. L. Altshuler and D. E. Khmelnitskii, Pis'ma Zh. Eksp. Teor. Fiz. **42**, 291 (1985) [JEPT Lett. **42**, 359 (1985)].
4. For review see M.J. Stephen, to appear in Mesocopic Phenomena in Solids, Eds. P. A. Lee, R. Webb, B. L. Altshuler, Elsevier Publication, Amsterdam, Netherlands.
5. This model for a stationary case was defined in B. Shapiro, Phys. Rev. Lett. **48**, 823 (1982) as a generalization of the model of P. W. Anderson, D. J. Thouless, E. Abrahams and D. S. Fisher, Phys. Rev. **B22**, 3519 (1980).
6. I. Edrei, M. Kaveh and B. Shapiro, Phys. Rev. Lett. **62**, 2120 (1989).
7. B. Shapiro, Phil. Mag. **56**, 1031 (1987).
8. M. J. Stephen and G. Cwilich, Phys. Rev. Lett. **59**, 285 (1987).
9. B. Shapiro, Phys. Rev. Lett. **57**, 2168 (1986).
10. I. Edrei and M. Kaveh, Phys. Rev. **B38**, 950 (1988).
11. I. Edrei and M. Kaveh, to be published.

Chapter 4

HIGH FIELDS AND FREQUENCY EFFECTS

HIGH FIELD HOPPING AND NEGATIVE DIFFERENTIAL CONDUCTANCE IN WEAKLY COMPENSATED SILICON.

D.I. Aladashvili, Z.A. Adamiya, K.G. Lavdovskii
Tbilisi State University, Tbilisi, USSR, 380028

E.I. Levin, B.I. Shklovskii
A. F. Ioffe Institute, Leningrad, USSR, 194021

ABSTRACT

We investigated nonohmic electron hopping conductivity σ in n-and p-type-Si containing impurity concentrations between 5×10^{16} and $9 \times 10^{17} cm^{-3}$ and compensation ratios $4 \times 10^{-5} \leq K \leq 10^{-2}$. At low temperatures (T<4K) electrons localized on acceptors situated near donors in p-Si must be activated to hop among the majority of acceptors. In this range the number of mobile electrons increases with field E due to a Poole-Frenkel-like lowering of the activation energy. At the high temperature end of the hopping regime (T>10K) the mobile electron concentration saturates. Here σ decreases with E. In two samples with 5.9 and 7.3×10^{16} cm^{-3} acceptors, current oscillations with periods 0.5-4 sec were observed at $E \geq E_c$ where $E_c = 40$ and 80 V/cm, respectively. We believe these oscillations result from the negative differential conductivity predicted by Nguyen and Shklovskii.

1. INTRODUCTION

The effect of high electric fields on the hopping transport of charge carriers among localized states is still poorly understood. The amorphous semiconductors in which the conductivity increases exponentially with field strength a quantitative interpretation is hampered by the fact that one does not know well enough the density of states function and the localization radius. In crystalline semiconductors experimental data over a wide range of doping parameters are missing, in particular those on lightly doped and weakly compensated samples. These are most suitable for theoretical interpretation because they constitute the simplest systems in which hopping conduction is observed at low temperatures. We therefore describe here non-ohmic hopping conduction in a well-characterized material, lightly doped and a very weakly compensated silicon. Low temperature hopping conduction in the low concentration and low compensation regime is well understood[1,2,3] so that the theoretical basis for understanding the non-ohmic effects is quite firm.

We begin by briefly describing the physical picture of the low field hopping conduction. We then discuss the familiar non-ohmic Poole-Frenkel effect in the low temperature thermally activated hopping regime. Finally we describe the occurrence of current oscillations and the observation of a negative differential conductance range.

2. TEMPERATURE DEPENDENCE IN LOW CONCENTRATION HOPPING REGIME

We shall consider a lightly doped and a very weakly compensated semiconductor (specifically with p-type conduction). At sufficiently low temperatures the bulk of the acceptors in such a semiconductor is neutral. The electrons lost by donors are located at those few acceptors whose potential is reduced by the potential of a nearby positive donor. Near one donor there may be one, two, or no negatively charged acceptors, i.e., 1-, 2-, ..or. 0-complexes may be formed.[3] Figure 1 shows not only a giant δ-like peak of the density of states $g(\varepsilon)$ at $\varepsilon = 0$, which is due to neutral acceptors, but also the density of states $g_1(\varepsilon)$ at an acceptor nearest to the donor and the density $g_2(\varepsilon)$ near a second charged acceptor located near a 1-complex. The 0- and 2-complexes are formed when an electron destroying at 1-complex leaves the vicinity of its donor and drops to the lower state of the second electron in the vicinity of the second donor. This "spillover" is identified by an arrow in Fig. 1. The Fermi level position μ is governed by the condition of electrical neutrality) $N_0(\mu)=N_2(\mu)$, which corresponds to[3]

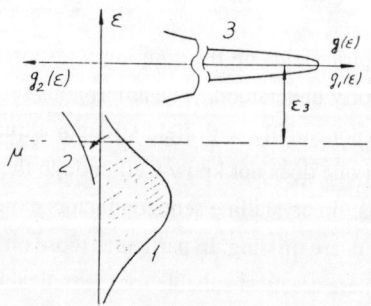

Fig. 1. Densities of states at acceptors and their occupancy. Continuous curves: 1) density of states at an acceptor closest to a donor; 2) density of states of a second charged acceptor near a 1-complex; 3) density-of-states peak of neutral acceptors. The filled states are shown shaded occupancy in zero electric field (the chain line is the Fermi level).

$$\mu = -0.99 \frac{e^2}{\kappa} N_A^{1/3}. \tag{1}$$

Since the concentration of the complexes is low compared with the acceptor concentration, at moderately low temperatures only the electrons activated to a δ-like density-of-states peak ("mobile electrons") contribute to the hopping conduction process. Therefore, the low-temperature hopping conductivity is proportional to $\exp(-\varepsilon_3/kT)$, where $\varepsilon_3 = |\mu|$ is the energy separation between the Fermi level and the δ–like peak. This corresponds to region III of Fig. 2. With increasing temperature the concentration of mobile electrons increases and at

$$T = T_s = \frac{\varepsilon_3}{k} \ln \frac{N_A}{N_D} \tag{2}$$

becomes close to the total concentration N_D of electrons. At higher temperatures $T>T_s$ most electrons are excited to the N_A-N_D acceptors and $n \approx N_D$. This corresponds to the saturation region II of the hopping regime in Fig. 2. The high temperature limit of the saturation range is the onset of excitations of holes from the acceptors to the valence band with the activation energy ε_1, the acceptor binding energy. This region is marked I in Fig. 2.

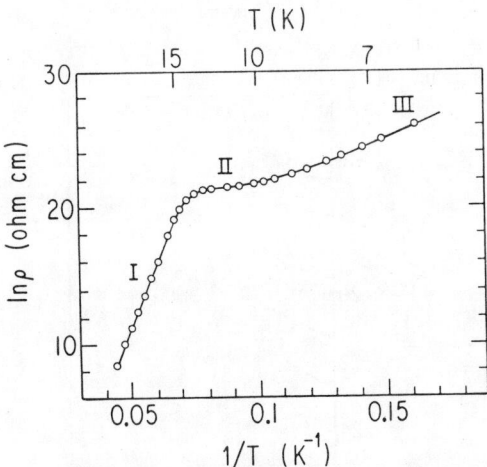

Fig. 2. Temperature dependence of the electrical resistivity of weakly compensated silicon.

In the following we discuss the field-induced changes in the hopping conductivity. These changes have opposite signs in regions II and III. In the ε_3-region the conductivity increases with field strength E because ε_3 is decreased by the Poole- Frenkel effect, whereas σ decreases with field in the saturation region II and shows a negative differential conductance as predicted by Nguyen and Shklovskii [4].

3. LOW TEMPERATURE ε_3 - REGIME

A typical dependence of the hopping conductivity on field strength is shown in Fig. 3 for a p-type Si sample containing $N_A=3,5 \cdot 10^{17}$ cm^{-3} and K=0,01. The conductivity increases with E in the low temperature ε_3-regime in all n and p-type Si samples studied. This increase of σ with E appears to be due to a decrease of the electron detachment energy ε from an acceptor - donor pair to the N_A-N_D more distant acceptors. This effect is equivalent to the Poole-Frenkel effect. The Poole-Frenkel effect[5] is usually defined as an increase in the electron density in the conduction band because of an electric-field-induced reduction in the ionization energy of a Coulomb impurity. It is shown in Fig. 4a how the ionization barrier is reduced in an electric field E. If the potential energy of an electron along the x axis, which is parallel to the field, is described by

$$V(x) = -\frac{e^2}{\kappa |x|} - eEx, \tag{3}$$

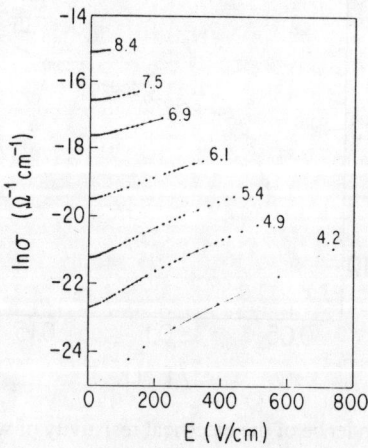

Fig. 3. Dependence of ln[σ(E)] on E for a p-type Si sample containing $N_A=3.5 \times 10^{17}$cm^{-3} at different temperatures.

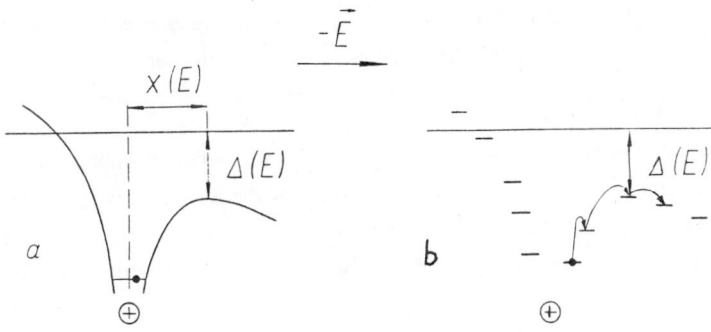

Fig. 4. Potential energy of an electron (black dots) in the field of a positively charged donor (crosses) and in an external electric field. a) Bending of the bottom of the conduction band. The horizontal straight line gives the position of the bottom of the band far from a donor when E = 0. b) Modulation of the energy level of an electron at a neutral acceptor. The continuous straight line is the position of an isolated acceptor level when E= 0.

where e is the electron charge and κ is the permittivity, the separation from the point of bending x(E) and the reduction of the barrier at this point Δ(E) are easily calculated to be

$$x(E) = \sqrt{e/\kappa E} \quad \text{and} \quad \Delta(E) = 2\sqrt{e^3 E/\kappa} \qquad (4)$$

Clearly, the probability W(E) of impurity ionization increases on increase in E in accordance with the law

$$W(E) = W(0)\exp|\Delta(E)kT| \qquad (5)$$

whereas the exponential dependence of the electrical conductivity on E is of the form

$$\sigma(E) = \sigma(o)\exp\left[\alpha\frac{\Delta(E)}{kT}\right] \qquad (6)$$

The coefficient α is 1 or 1/2 in those cases when recombination is of monomolecular and binomolecular nature, respectively. The former case is realized in a compensated semiconductor and the latter in a practically uncompensated material.

We shall show that an effect similar to the Poole-Frenkel effect occurs also in doped and weakly compensated semiconductors under hopping conduction conditions. A similar idea was put forward earlier[6] to account for electronic properties of polymers.

We shall now consider the influence of an electric field on the mobile electron density. We shall concentrate on a 1-complex and see how the field affects the probability of its ionization, i.e., how it influences the detachment of an electron from an acceptor closest to a donor and the transfer to more distant acceptors responsible for the δ-like density-of-states peak.

It is shown in Fig. 4b how the energy levels of an electron are modified by the Coulomb field of the donor and by the field E in the detachment process. We can see that, as in the conventional Poole-Frenkel effect, an electric field increases W(E) in accordance with Eq. (5). However, if only the 1- and 0-complexes exist in a field E = 0, then for $n \ll N_0(\mu)$ (n is the mobile electron density), i.e., at sufficiently low temperatures, the recombination rate should be proportional to $nN_0(\mu)$. This would mean that the conductivity $\sigma(E)$ should then be described by Eq. (6) with $\alpha = 1$.

In reality we have to allow for the existence of the 2-complexes. When an electron is detached from a 2-complex, the complex remains neutral so that the field alters the probability of its ionization much less than predicted by Eq. (5). A careful analysis[7] which includes 2-complexes yields $\alpha = 0.69$.

In Fig. 5 we test the relation (6) by plotting $\ln[\sigma(E)/\sigma(0)]$ against $\Delta(E)/kT$ for an n-type and a p-type Si sample. The data were obtained at different temperatures in the ε_3-region of hopping conduction. We find that Eq. (6) describes the data well for $\Delta(E)/kT > 4$ with $\alpha = 0.35 \pm 0.05$. The reason of discrepancy between the theoretical and experimental values α is not clear. We may assume that it is partly due to the fact that Eq. (6) is derived only with an exponential precision. Allowance for the preexponential dependence dropped from Eq. (6) may have a considerable influence on $\sigma(E)$ in the relatively moderate fields used by us and it may improve the agreement with the experimental results.

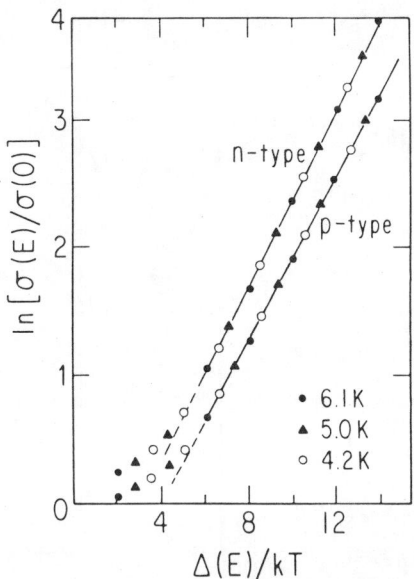

Fig. 5. Dependence of ln [σ(E)/σ(0)] on ΔE/kT for an n-type and a p-type Si sample.

4. SATURATION HOPPING REGIME

The field dependence of conductivity of a p-type Si sample doped with $N_A = 5.5 \times 10^{16}$ cm^{-3} boron acceptors and $K = 5 \times 10^{-3}$ is shown in Fig. 6 for various temperatures. At the two highest temperatures, one is still in the ε_1 - regime and conduction is by holes in the valence band. With decreasing T, hopping conduction becomes predominant, and σ(E) begins to decrease with increasing field E. As the temperature is further decreased one observes a change in sign of d ln σ(E)/dE from negative to positive at the transition from the saturation region to the thermally activated ε_3-region. This is shown in Fig. 7 for another sample having a somewhat larger $N_A = 9 \times 10^{16}$ cm^{-3} in the saturation region of the hopping conductivity one has observed a decrease of σ(E) by as much as a factor two in weakly compensated Si, both n-type and p-type doped with As, P, Sb, B, or Ga. A slight decrease (up to 30%) in σ has been observed previously in weakly compensated Ge.[8] Current saturation in Ga-doped Si was reported by Baron and Young[9].

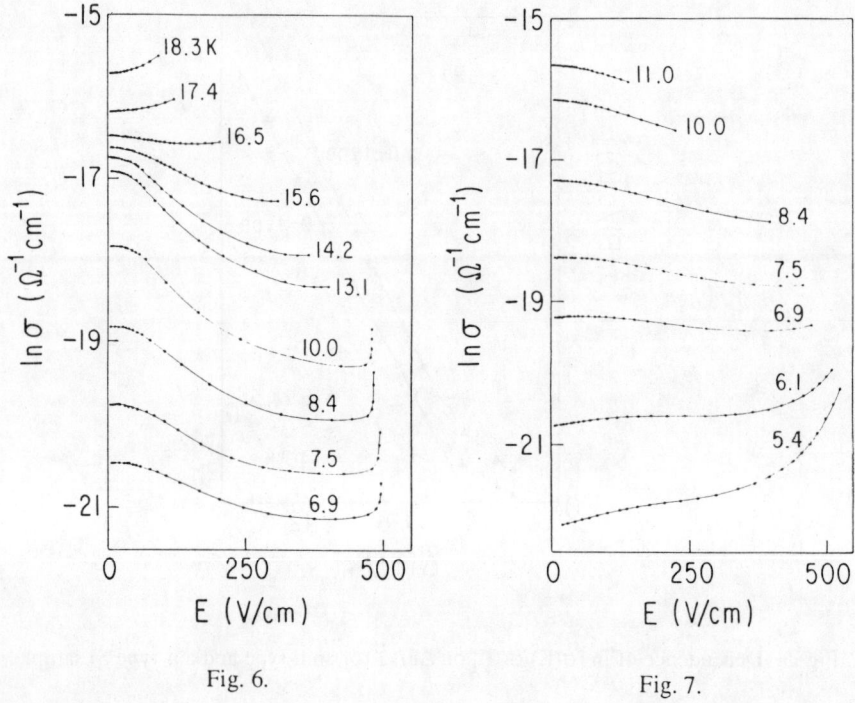

Fig. 6.

Fig. 7.

Fig. 6. Dependence of ln σ on electric field E for a p-type silicon sample doped with $N_A = 5.5 \times 10^{16}$ cm^{-3} acceptors at different temperatures.

Fig. 7. Dependence of ln σ on E for a p-type sample Si containing $N_A = 9 \times 10^{16}$ cm^{-3} at different temperatures.

A negative sign of dln σ(E)/dE occurs only in the saturation region of hopping conduction. This is demonstrated by the following experiment. The extent of the saturation region can be diminished by applying an uniaxial stress. The stress decreases the acceptor ionization energy ε_1 and increases the hopping activation energy ε_3. As a consequence, the saturation range is decreased from the high as well from the low temperature side as shown in Fig.8 . As soon as the saturation range disappears, the negative slope of dlnσ(E)/dE observed at stress $\chi = 0$ is replaced by a positive value which is caused by a Poole-Frenkel lowering of the activation energy as shown in Fig. 9 for a Ga-doped Si sample containing $N_A = 4 \times 10^{17}$ cm^{-3} and $K = 5 \times 10^{-3}$.

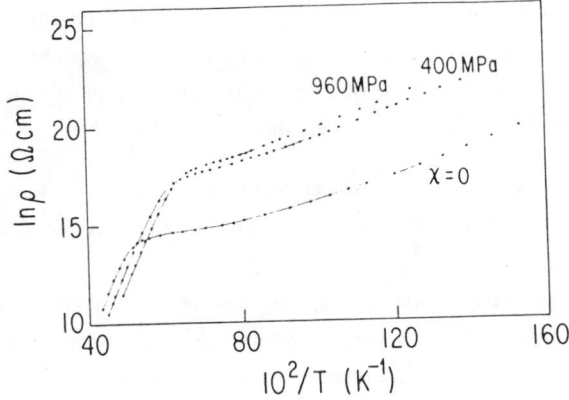

Fig. 8. Temperature dependence of the electrical resistivity of p-type Si sample with $N_A = 9.2 \times 10^{16}$ cm^{-3} acceptors at different stresses.

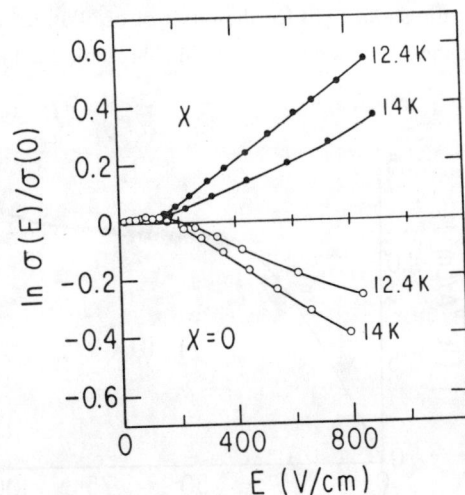

Fig. 9. Dependence of $\ln[\sigma(E)/\sigma(0)]$ on E for a p-type Si sample doped with $N_A = 4 \times 10^{17}$ cm^{-3} gallium acceptors at different temperatures and for zero stress and $\chi = 900$ MPa.

4.1 VERY WEAK COMPENSATION

Interesting new effects such as a negative differential conductance and current oscillations have been observed in B-doped Si samples which were very weakly compensated.[10] We shall describe measurements on two samples: Sample A containing $N_A=5.9\times10^{16}cm^{-3}$, $K=4\times10^{-5}$, and B containing $N_A=7.3\times10^{16}cm^{-3}$, $K=2\times10^{-4}$.

Fig. 10 shows the dependence of the current density j of sample A on electric field E at T=10K. As soon as the j-E characteristics begins to flatten out one observes an onset of current oscillations which suggest a region of negative differential conductivity. The current oscillations have maximum amplitude near the onset voltage. The period of the oscillations lies between 0.5 and 4 sec. The period decreases with increasing electric field until the oscillations cease with the onset of impact ionization.

The current oscillations are limited to the temperature region in which the hopping conduction is saturated. This is illustrated in Fig. 11 which shows for the two samples the temperature dependence of the low field resistivity and the amplitude of the oscillations. The threshold field for the onset of oscillations is nearly independent of temperature while the oscillation period near the threshold field increased with decreasing T.

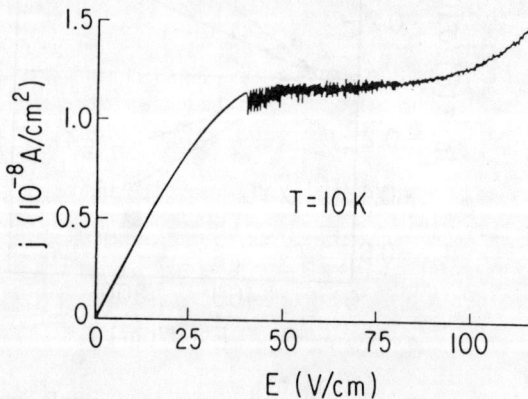

Fig. 10. Behavior of the current density as E increases slowly with time for sample A (total duration of sweep is 10 min).

The field dependence of the oscillation amplitude and period is not smooth in most samples but exhibits certain step-like features. An example observed on the sample B is shown in Fig. 12. These steps occur reproducibly at the same field in the whole temperature range in which the oscillations are observed. The pattern of steps is different though, for the opposite field direction. We have no explanation for this effect but suggest that it is associated with inhomogeneities in the sample or near the contacts. The domain formation which is associated with current oscillations is a nonlinear effect which is easily influenced by local variations in the nonohmic hopping conductance.

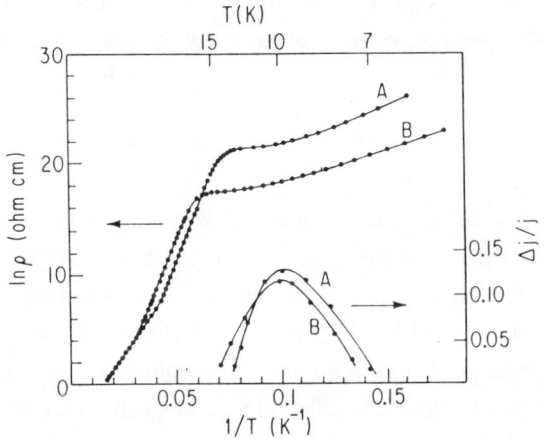

Fig. 11. Resistivity (scale on the left) and relative amplitude of the current oscillations near the threshold (scale on the right), as a function of the reciprocal temperature for samples A and B.

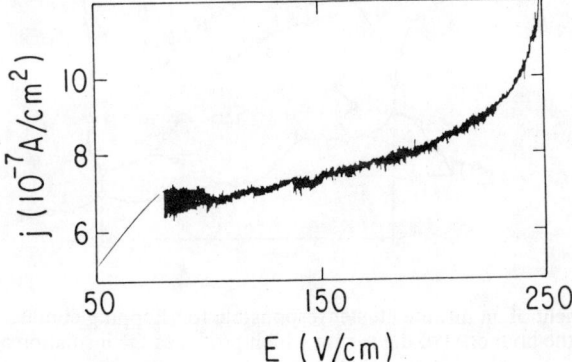

Fig. 12. Dependence of current density on E for sample B in negative differential resistance range.

4.2 MODEL FOR NEGATIVE DIFFERENTIAL CONDUCTIVITY

Nguyen and Shklovskii[4] considered the effect of an electric field on the hopping conduction of a few electrons among a random distribution of neutral impurities. This situation arises in the saturation region of hopping in very weakly compensated semiconductors. The ohmic hopping transport is known to be governed by a percolation network of acceptors which connects the two electrodes. A section of this network, called an infinite cluster, is sketched in Fig. 13. Those acceptors belong to this infinite cluster whose separations do not exceed $r_c+a/2$ where $r_c=0.87\ N_A^{-1/3}$ is the percolation radius and a is the localization length of the acceptors. The typical period or length scale of the cluster is

$$L_0 = \frac{1}{3}(2r_c/a)^v N_A^{-1/3} \qquad (7)$$

where $v=0.88$ is a critical index. The important point to note is that connected to the infinite cluster are always some dead ends. These do not disturb the low field ohmic hopping conduction because an electron can easily hop in and out of such dead ends of length L_0 when $kT \gg eEL_0$. However, at higher fields, where $eEL_0 \gg kT$, the probability for thermal excitation from a dead end properly oriented with respect to the field, such as type 1 in Fig. 13, is reduced by a factor $\exp(-eEL_0/kT)$. Dead ends of this

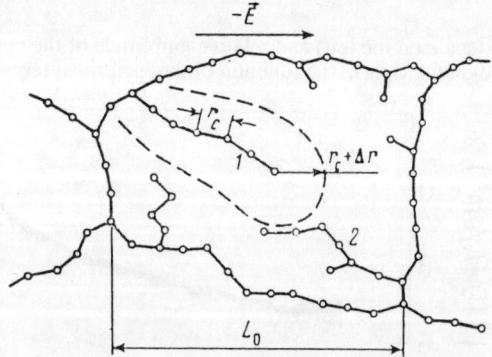

Fig. 13. Fragment of an infinite cluster responsible for hopping conductivity. The dashed line is the boundary of the region which produces the insulation around dead end 1.

kind will then act as electron traps. The number of field-induced traps increases with E. As a consequence, the concentration of freely hopping electrons decreases with increasing field causing a decrease in current and a negative differential conductivity.

In order to estimate the number of such field-induced traps one needs a more precise description of their structure. Since an electron can escape a dead end of length X by thermal excitation with probability exp(-eEX/kT), one requires that each acceptor belonging to the dead-end is isolated from any other acceptor by an extra distance Δr such that the escape probability by tunneling is equal to the thermal excitation probability, or

$$\frac{eEX}{kT} = \frac{2\Delta r}{a} \tag{8}$$

The probability of finding a trap of size X is then

$$W(X) = \exp(-N_A \Delta V) \tag{9}$$

where $\Delta V = 2\pi r_c X \Delta r$ is the extra isolation volume surrounding the dead end. Using the relation (7) with $\nu=1$ and assuming that the concentration of electrons is very small compared to that of acceptors, one obtains[10]

$$\sigma = \sigma(0)\exp(-eEL_0/2kT) \tag{10}$$

which leads to a negative differential conductivity when $E > E_c \equiv 2kT/eL_0$. This critical field E_c agrees fairly well with the experimental results. We observe the onset of oscillations near E_c=80V/cm at 10K for a sample containing N_A=7.3x10^{16}cm^{-3} and for which $2r_c/a \approx 20$ and $L_0 \approx 10^{-5}$ cm.

5. SUMMARY AND CONCLUSIONS

Lightly doped and weakly compensated crystalline semiconductors offer the opportunity to test present theories on the effect of electric fields on hopping conduction. Whenever the conduction process is thermally activated, the field tends to lower the thermal activation energy thereby increasing the conductivity. The theoretical estimate of the field-induced decrease of the activation energy ε_3 based on the Poole-Frenkel effect overestimates the measured decrease by nearly a factor 2.

Theoretical predictions of a negative differential conductivity in the saturation range of the hopping process have been confirmed in very weakly compensated B-doped silicon. The negative differential conductivity manifests itself by the presence of current oscillations having a period between 0.5 and 4 sec. These suggest very low frequency motion of space charge domains. The presence of such domains has not yet been demonstrated directly, and we have no theoretical understanding of the nature of these current oscillations. The theoretical model of the negative differential conductivity suggests that as the field is increased, an increasing fraction of hopping electrons becomes trapped in dead-end sections of the hopping network for longer periods of time. The theoretical prediction for the onset of the differential conductivity agrees reasonably well with experiments on boron-doped silicon.

A detailed theory of negative differential conductivity including finite compensation effects and transition to Poole-Frenkel like current-voltage characteristic has been given recently.[11]

ACKNOWLEDGMENTS

We wish to thank Professor Hellmut Fritzsche for helpful discussions and valuable comments.

REFERENCES

1. H. Fritzsche Phys. Rev. 99, 406 (1955)

2. H. Fritzsche, in "Metal-Nonmetal Transition in Disordered Systems," ed. by L. R. Friedman and D. P. Tunstall, (Scottish Universities School in Physics Publications, Edinburg, 1978) p. 193

3. B. I. Shklovskii and A. L. Efros "Electronic Properties of Doped Semiconductors." (Springer Verlag, Berlin, Heidelberg, New York, Toyko, 1984)

4. V. L. Nguyen and B. I. Shklovskii, Solid State Commun. 38, 99 (1981)

5. J. Frenkel, Phys. Rev. 54, 647 (1938); J. L. Hartke, J. Appl. Phys. 39, 4871 (1968)

6. H. Hirsh, J. Phys. C$\underline{12}$, 321 (1979)

7. D. I. Aladashvili, Z. A. Adamiya, K. G. Lavdovskii, E. I. Levin and B. I. Shklovskii, Sov. Phys. Semicond. $\underline{23}$, 213 (1989)

8. A. G. Zabrodskii and I. S. Shlimak, Sov. Phys. Semicond. $\underline{11}$, 430 (1977)

9. R. Baron and M. H. Young, Solid State Electron. $\underline{28}$, 204 (1985)

10. D. I. Aladashvili, Z. A. Adamiya, K. G. Lavdovskii, E. I. Levin and B. I. Shklovskii, JETP Letters $\underline{47}$, 390 (1988)

11. D.I. Aladashvili, Z.A. Adamiya, K.G. Lavdovskii, E.I. Levin and B.I. Shklovskii, Sov. Phys. Semicond. $\underline{24}$ (1990)

ONSET OF NONLINEAR HOPPING CONDUCTION IN SEMICONDUCTORS WITH SMALL COMPENSATIONS*

J. Talamantes and M. Pollak
Department of Physics
University of California,
Riverside, CA 92521
USA

The problem of nonlinear transport in disordered Anderson insulators is addressed. In previous theoretical works, four nonlinear effects have been proposed, and their relative importance is currently not clear. Eliminating two by an appropriate choice of model, the other two, namely the field-induced charge redistribution, and the backward flow on segments of the percolation path, are studied here in detail near the onset of nonlinear behavior. The model is a reasonable representation of impurity conduction, so results can be compared with experiment.

An expansion of the current density to third order in the field has been developed, and used to construct an algorithm for a computer simulation study using percolation theory. Strong fluctuations of conductance existing on a scale less than a correlation length of the current carrying percolation cluster are accounted for in detail, while the small fluctuations which exist on a larger scale are treated by effective medium theory.

1. Introduction.

The work presented here is a computer simulation of DC hopping conductivity in disordered systems at moderate electric fields (i.e., near the onset of nonohmic effects). This regime is treated by including effects of second and third order in the electric field E. The work uses percolative methods on a scale where fluctuations are very large (i.e. for lengths smaller than the correlation length L of the current-carrying paths or macrobonds) and an effective medium approach (EMA) when fluctuations are small (i.e. on a scale larger than L).

In the recent past, considerable effort has been spent, both theoretically[1-15] and experimentally[16-20], on this problem. The different theoretical works emphasized different aspects and, not surprisingly, they reached different results. Experiments on different systems also yield different results and it turns out that some experiments can be found to fit any theory.

---

* This work was supported in part by the San Diego Supercomputer Center.

The problem at hand, in the absence of Coulomb interactions, is reasonably well understood in the ohmic regime. At electric fields beyond the linear response, the problem becomes more complicated. A serious complication in the nonohmic regime is due to the local chemical potential μ_i at site i, which measures the field-induced change of that site's occupation probability. The problem is that μ_i is not just a property of site i, but depends on some extended neighborhood of i.

Four effects have been proposed which give rise to a nonlinear response: (i) in ref. 3, as E increases, there is a field-induced increase in the density of sites which can participate in transport in the down side of the current, giving rise to a superlinear current. (ii) In ref. 4 the important effect is a rearrangement of charge along the macrobonds which occurs in response to the increase in E, i.e., the field-induced changes of μ_i. This also leads to a superlinear current. (iii) The effect emphasized in ref. 6 is the (field-dependent) length of the segments of the macrobonds where the current flow is against the field. This effect produces a current sublinear in the field. (iv) As the field increases, the number of carriers in the current-carrying cluster may decrease[8]. This is due to the "dead ends": parts of the percolation cluster which act as "carrier traps". This effect also leads to a sublinear behavior of the current. This paper attempts to contribute to the understanding of the interplay between the various effects. We construct a model where only effects (ii) and (iii) are present. Effect (i) is eliminated by choosing equi-energetic sites (simulating impurity conduction), and effect (iv) is eliminated by keeping the global chemical potential constant. We find that effect (iii) dominates over effect (ii) at the moderate fields investigated.

2. The Macrobond Currents.

The starting point to find the current on an isolated macrobond is the equation for the current between two sites i and j,

$$I'_{i,j} = q\left[\, \gamma_{i,j}\, f_i\, (1-f_j) - \gamma_{j,i}\, f_j\, (1-f_i)\,\right], \tag{1}$$

where q is the charge of the carriers, $\gamma_{i,j}$ is the transition rate from site i to site j, and f_i is the occupation probability of site i.

The local chemical potential μ_i is defined by $f_i = \{ \exp [(E_i - \mu_i)/kT] + 1\}^{-1}$, where E_i is the site energy, T is the temperature, and $\gamma_{i,j} = \gamma \, (E_i - E_j)/kT \exp(-2r_{i,j}/a)$ $\times \{\exp[-(E_i - E_j)/kT] - 1\}^{-1}$, where $r_{i,j} = |\mathbf{r}_{i,j}| \equiv |\mathbf{r}_i - \mathbf{r}_j|$, $\mathbf{r}$ being the radius vector of site i, γ is a constant, and a is the localization radius. We introduce the following reduced variables: the electric field $F = qEa/kT$, the intersite separation $\rho_{i,j} = 2r_{i,j}/a$, its projection along the field $\rho'_{i,j} = \rho_{i,j} \cos \theta_{i,j}$ ($\theta_{i,j}$ is the angle between $\mathbf{E}$ and $\mathbf{r}_{i,j}$), the site potential $\varepsilon_i = \tilde{E}_i/kT + F\rho'_{i,j}$, and the local electrochemical potential μ_i/kT $= \delta\eta_i - q\mathbf{E}\cdot\mathbf{r}_{i,j}/kT$ (the global chemical potential has been set equal to zero). Eq. 1 can be shown to be

$$I_{i,j} = \frac{I'_{i,j}}{q\gamma} = \exp(\rho_{i,j}) \frac{\sinh\left[\frac{1}{2}\left(\delta\eta_i - \delta\eta_j + \frac{1}{2}F\rho'_{i,j}\right)\right]}{\cosh\left[\frac{1}{2}\left(\varepsilon_i - \delta\eta_i\right)\right]\cosh\left[\frac{1}{2}\left(\varepsilon_j - \delta\eta_j\right)\right]} \times \frac{\frac{1}{4}F\rho'_{i,j}}{\sinh\left(\frac{1}{4}F\rho'_{i,j}\right)} . \qquad (2)$$

To eliminate the effect of the field-enhanced availability of hopping sites we set $\varepsilon_i = \varepsilon_j = \varepsilon$, and to represent realistic conditions for impurity conduction we make $\varepsilon \gg 1$, so $\cosh[1/2(\varepsilon - \delta\eta_i)] = 1/2 \exp[1/2(\varepsilon - \delta\eta_i)]$ (valid for semiconductors with small compensations). Since the systems is nearly homogeneous on a scale of a macrobond, we use periodic boundary conditions. This amounts to setting $\delta\eta_1 = \delta\eta_{\tilde{n}+1}$, where ñ is the number of sites in a macrobond. Also, we constrain the number of carriers on a macrobond to remain constant as the field changes. Performing the expansions $\delta\eta_i = A_i F + B_i F^2 + G_i F^3 + ...$, and $I = K_1 F + K_2 F^2 + K_3 F^3 + ...$ (where $I \equiv I'_{i,j}/q\gamma$ $= I'_{i,i+1}/q\gamma$), one arrives, to third order in F and after considerable algebra, at

$$K_1 = e^{-\varepsilon} \left[\sum_{i=1}^{\tilde{n}} \rho'_i\right]\left[\sum_{i=1}^{\tilde{n}} \exp(\rho_i)\right]^{-1} , \qquad (3)$$

$$K_2 = e^{-\varepsilon} \frac{1}{2} \left[\sum_{i=1}^{\tilde{n}} \rho'_i (A_i + A_{i+1}) \right] \left[\sum_{i=1}^{\tilde{n}} \exp(\rho_i) \right]^{-1}, \text{ and} \quad (4)$$

$$K_3 = e^{-\varepsilon} \left[\sum_{i=1}^{\tilde{n}} \exp(\rho_i) \right]^{-1} \times \left[\frac{1}{2} \sum_{i=1}^{\tilde{n}} \rho'_i (B_i + B_{i+1}) \right.$$

$$\left. + \frac{1}{4} \sum_{i=1}^{\tilde{n}} \rho'_i (A_i^2 + A_{i+1}^2) + \frac{1}{24} \sum_{i=1}^{\tilde{n}} \rho'^{2}_i (A_i - A_{i+1}) \right], \text{ where} \quad (5)$$

$$A_i = \frac{1}{(2\tilde{n})} \left\{ \left[\sum_{l=1}^{\tilde{n}} \rho'_l \right] \left[\sum_{l=1}^{\tilde{n}} \exp(\rho_l) \right]^{-1} \sum_{k=1}^{\tilde{n}} (\tilde{n}-k) \exp(\rho_{i+k-1}) - \sum_{k=1}^{\tilde{n}} (\tilde{n}-k) \rho'_{i+k-1} \right\}, \text{ and}$$

$$B_i = \frac{1}{\tilde{n}} \left\{ \sum_{k=1}^{\tilde{n}} (\tilde{n}-k) \exp(\varepsilon + \rho_{i+k-1}) K_2 / 2 - \rho'_{i+k-1} (A_{i+k-1} + A_{i+k}) / 4 \right.$$

$$\left. - (A_{i+k-1}^2 - A_{i+k}^2) / 2 - \sum_{k=1}^{\tilde{n}} A_{i+k}^2 \right\}.$$

It is of interest to point out that the expressions above contain "sequencing terms", which make the current dependent not only on the specific values of the ρ_i in the macrobond, but also on their specific arrangement. In particular, to second order, the current depends sensitively on how the values of ρ_i are arranged, i.e., when the sequence of values of ρ_i is somehow ordered, the magnitude of K_2 is large, and its sign depends on whether ρ_i increase or decrease in the direction of the field. However, for a random distribution, extensive ordering is improbable, in which case the magnitude of K_2 must be small. For this reason it is very important to expand the current to at least third order.

3. Computer Simulations.

The procedure chosen in the present work was to generate by computer a large number of macrobonds. The current-field characteristics of the individual macrobonds was then computed as described in section 2. From these, the macrobond conductances g for every F, and the distribution p (g; F), can be found. An averaging procedure can then be performed at every F, using the EMA, to arrive at the conductivity of the bulk σ_B (F).

The EMA performed here is based on that developed by Kirkpatrick[21], but, since the distribution p (g; F) changes with F, we perform the EMA at every value of F, i.e., for every F we evaluate (numerically) the integral

$$\int \frac{(g_m - g) p(g) dg}{\left[g + \left(\frac{z}{2} - 1\right) g_m\right]} = 0, \qquad (6)$$

where g_m is an "effective" or "average" macrobond conductance, such that σ_B (F) remains unchanged when every g is replaced by g_m. Here, a coordination number z=6 was used. This equation determines g_m, from which σ_B (F) can be found: σ_B (F) = g_m / L

The first problem is to find the number ñ of sites in a macrobond and how ñ is related to ρ_M, the maximum ρ_i allowed in the macrobonds. It should be noted that ñ can in principle be a function of the field.

One way to find ñ is by writing

$$\Lambda(\rho_M) = \frac{\zeta}{\rho_M - \rho_c}, \qquad (7)$$

and setting

$$\Lambda = (\tilde{n} - 1)\overline{\rho} + \rho_M, \qquad (8)$$

In these equations, Λ is the "twisted length" of the macrobonds, $\overline{\rho}$ is the average of ρ_i on a macrobond and n is the concentration of impurities. In equation (14), ρ_M has been separated from the other ρ_i because there is always at least one ρ_M in every macrobond.

In the present work, a Poisson distribution was used for $r_{i,j}$ (and therefore for ρ_i also), i.e., the locations of the sites are assumed to be uncorrelated.

In order to arrive at values of ζ which lead to reasonable values of ρ_M and $\tilde{n}$, equation (14) is first solved by setting $\rho_M = \rho_c + \delta$, $\delta = 2\nu$ (ν is the critical exponent of the correlation length), and $\tilde{n}_\delta \equiv \tilde{n}(\rho_M = \rho_c + \delta) = 15, 20, 25, 30$. The result can then be used in equation (13). This gives a value of ζ which can then be used to find ρ_M for any $\tilde{n}$. This procedure has the problem that $\tilde{n}_\delta$ is still left as some kind of "fitting parameter". It should be noted that this approach to find ρ_M is not equivalent to maximizing the current density with respect to ρ_M, so that when this procedure is used it does not necessarily follow that the $\tilde{n}$ for which the current density is maximum must be equal to $\tilde{n}_\delta$.

Once ρ_M has been determined, macrobonds with $\tilde{n}$ sites (for some $\tilde{n}$ to be fixed as described below) can be constructed in a random fashion. The z-direction is defined as that of the electric field. The azimuthal angles ϕ_i for the pairs of sites are chosen with the distribution $p(\phi) = 1/2\pi$, $0 \leq \phi \leq 2\pi$. The angles θ_i are chosen with the probability $p(\theta) = 1/2 \sin\theta$, $0 \leq \theta \leq \pi$. The ρ_i are chosen with probability $p(\rho) = \alpha_2 \rho^2 \exp(-\alpha \rho^3)$, $\rho_{min} \leq \rho \leq \rho_M$, where $\alpha = \pi n a^3/6$, and $\alpha_2 = 3\alpha \ [\exp(-\alpha \rho_{min}^3) - \exp(-\alpha \rho_M^3)]$. In these equations, ρ_{min} is the minimum allowed value of ρ in the macrobonds. In the present work, ρ_{min} is set equal to 4 so that carriers are well localized around the sites. The angles θ_i and ϕ_i were chosen in this manner except when they would lead to a shorting out of the path $i-1 \to i \to i+1$ by the path $i-1 \to i+1$, in which case θ_i and ϕ_i were recalculated with the above distributions until this problem disappeared.

Once a macrobond is so constructed, its macrobond current is determined by K_1, K_2, and K_3. The macrobond conductance $g(F)$ is then easily obtained and binned. This is done for a large number of macrobonds. From the binned conductances one can arrive at $p(g; F)$, from which $g_m(F)$ can be found by using the EMA. $\sigma_B(F)$ can then be easily found.

The process described so far is repeated, holding F constant, for all the $\tilde{n}$ in some interval centered about $\tilde{n}_\delta$. The largest $\sigma_B(F)$ thus obtained is the one kept, and the quantity reported in section 4. The process is repeated for F in the range in which the approximations can be justified.

We should note that $<\theta(I,F)>$, the average value of θ on the macrobonds was calculated. We found that the characteristics of the highly conducting macrobonds were the same at low and moderate fields. We concluded that θ_i did not have to be restricted to smaller values as the field was increased.

One important aspect of the computer simulations is the determination of the range of fields in which the present approach is valid. A number of approximations were made to find the macrobond currents, and it is difficult, from the approximations, to come to a definite criterion for the maximum value of F for which eqs. (3), (4), and (5) are valid, and for which three terms are sufficient to represent I(F). The criterion used in this work is as follows. For every macrobond, I(F) and I(F − dF) can be calculated (dF > 0). If, at field F, the current decreases as a function of F, the macrobond is disregarded. When the number of macrobonds thus rejected is of the order of 1/10 of the size of the statistical sample being considered the simulation is stopped. The approach taken here is not valid for fields larger than that. The reasoning behind this criterion is that the macrobond currents decrease when a depletion of carriers from the current-carrying part of the percolation cluster becomes important. Since this number of carriers is assumed (in the present treatment) to remain constant as the field is changed, a decrease in the macrobond currents with the field must be a result of a failure of our approximations and assumptions to reflect the physical situation.

4. Results and Conclusions.

Of the four nonohmic effects listed in section 1, namely, (i) the field-induced enhancement of available hopping sites, (ii) the field-induced redistribution of occupation probabilities, (iii) the backward flow due to the twistedness of the current paths, and (iv) the field-induced change in the carrier concentration on the current paths, we suppressed the first and last effects by an appropriate choice of a model. We chose this model to evaluate the relative importance of the second and the third effect in the moderate field regime and also because it is used often to represent impurity conduction. Our results can thus be compared with experiments on such systems.

The main results obtained are presented in figs. 1 and 2. The results plotted are within a range of F for which the expansion for the current to third order in F is valid. Fig. 1 shows the dependence of the conductivity on the field when only effect (ii) exists. This is because the twistedness of the paths has been eliminated in the computations leading to the results of fig. 1 by setting all $\theta_i = 0$. Fig. 2 shows the results for twisted paths, so they include both effects (ii) and (iii). It is seen that in contrast to fig. 1, the conductivity decreases as a function of the field, implying that in the regime of moderate

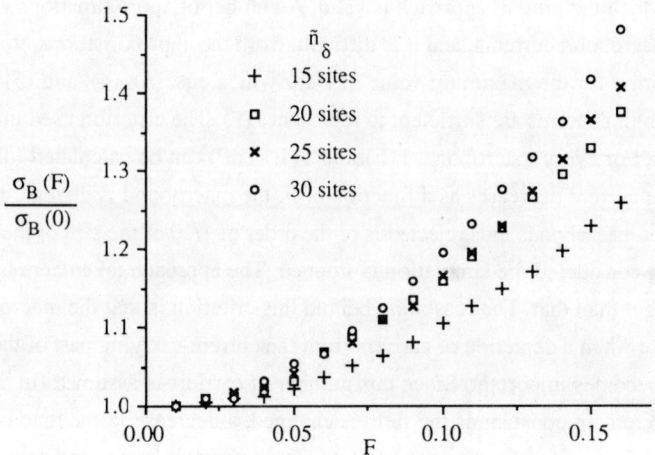

Figure 1. Field dependence of the bulk conductivity when the macrobonds are straight. The different plots are for different $\tilde{n}_\delta$.

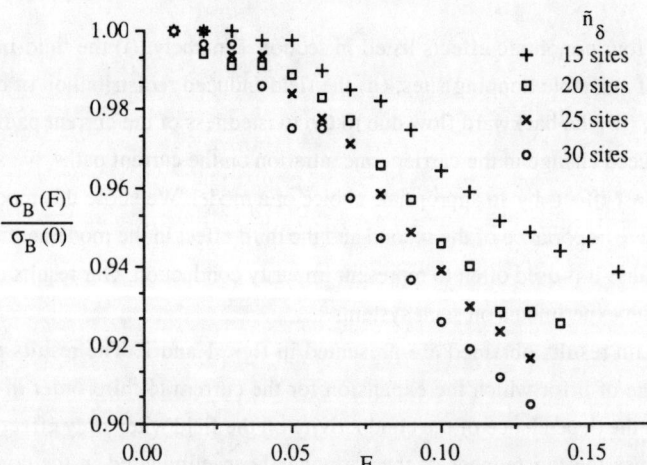

Figure 2. Field dependence of the bulk conductivity when the macrobonds are twisted. The different plots are for different $\tilde{n}_\delta$.

electric fields, the effect of backward flow dominates over the effect of redistribution of occupation probability.

The result of a sublinear conductivity is in agreement with the investigations of Böttger and Bryksin[5,6], the computational work of Levin et. al.[12], and the experimental results of Zabrodskii and Shlimak[17]. We think that our results are better founded for the moderate field regime than those of ref. 5. The reason is that, although a mean-field approximation is used in both works, we investigate in detail what occurs on the scale of a macrobond, whereas Böttger and Bryksin's analysis requires an investigation on the scale of single links. That scale is much smaller than that used in this paper. In this work, we constructed a procedure which treats the transport of carriers as a percolation problem on the scale where fluctuations are exponentially large, and as an effective medium problem on a scale where fluctuations are relatively small. We feel that this is the best method devised so far near the onset of nonlinear effects. Moreover, Böttger and Bryksin[5] assumed a binary distribution for the transition rates from one site to its nearest neighbor. This is to be contrasted with the more realistic distributions used here for ρ_i, θ_i, and ϕ_i. Also, we fell that our work provides a more detailed picture of the processes that occur at the onset of nonlinear behavior. The approach of Böttger and Bryksin has of course the distinct advantage that it applies to a much larger range of fields.

One of us (M. P.) would like to acknowledge helpful discussions on the subject of the paper with Prof. B. I. Shklovskii and with several other members of his group, and their lavish hospitality at the A. F. Yoffe Institute, and thank the USA-USSR Academies of Science exchange program for making the visit there possible.

REFERENCES

1) R. M. Hill. Phil. Mag. **26**, 1307 (1971).
2) N. Apsley and H. P. Hughes. Phil. Mag **31**, 1327 (1975).
3) M. Pollak and I. Riess. J. Phys. C: Solid State Phys. **9**, 2339 (1976).
4) B. I. Shklovskii. Sov. Phys Semicond. **10**, 855 (1976).
5) H. Böttger and V. V. Bryksin. phys. stat. sol. (b) **96**, 219 (1979).
6) H. Böttger and V. V. Bryksin. Phil. Mag. B **42**, 297 (1980).
7) H. Böttger and D. Wegener. phys. stat. sol. (b) **121**, 413 (1984).
8) H. Böttger and D. Wegener. Phil. Mag. B **50**, 409 (1984).
9) H. Böttger, P. Szyler, and D. Wegener. phys. stat. sol. (b) **128**, K179 (1985).
10) H. Böttger, P. Szyler, and D. Wegener. phys. stat. sol. (b) **133**, K143 (1986).

11) J. Talamantes, M. Pollak, and R. Baron. J. Non-Cryst. Solids **97** & **98**, 555 (1987).
12) E. I. Levin, V. L. Nguen, and B. I. Shklovskii. Sov. Phys. Semicond. **16**, 523 (1982).
13) E. I. Levin, and B. I. Shklovskii. Sov. Phys. Semicond. **18**, 534 (1984).
14) B. I. Shklovskii. Sov. Phys. Semicond. **13**, 53 (1979).
15) Nguen Van Lien and B. I. Shklovskii. Solid State Communications **38**, 99 (1981).
16) D. Redfield. Adv. Phys. **24**, 463 (1975).
17) A. G. Zabrodskii, and I. S. Shlimak. Sov. Phys. Semicond. **11**, 430 (1977).
18) P. J. Elliot, A. D. Yoffe, and E. A. Davis. Hopping Conduction in Amorphous Semiconductors, in AIP Conf. Proc. 20, ed. H. C. Wolfe (AIP, New York, 1974) pp. 311-318.
19) R. T. Phillips and A. D. Yoffe. J. Non-Cryst. Solids **88**, 167 (1986).
20) G. Timp, A. B. Fowler, A. Hartstein, and P. N. Butcher, Phys. Rev. B **34**, 8771 (1986).
21) S. Kirkpratrick, Rev. Mod. Phys. **45**, 574 (1973).

FREQUENCY DEPENDENT CONDUCTIVITY OF DISORDERED INSULATORS.

A. Hunt[#] and M. Pollak[*]
#Dept. of Physics, Univ. of California, Irvine CA 92717
*Dept. of Physics, Univ. of California, Riverside CA 92521

ABSTRACT.

The frequency dependent conductivity of Anderson localized disordered systems is investigated, with focus on the low frequency behavior. Contrary to other theories, it is found that in the conductivity is approximately proportional to the frequency, rather than to its second power, in the low frequency limit, and is thus nonanalytic at zero frequency. The results are compared, and found to be in good agreement, with recent detailed experiments by Long.

1. INTRODUCTION.

The frequency dependent phonon assisted hopping conductivity has been the subject of many theoretical studies. Currently, the conductivity is rather well understood in the high frequency (pair approximation) limit, and at dc. The low but finite frequency regime is not understood nearly as well. Both theory and experiment are more difficult here, the latter because the frequency dependent part $\sigma(\omega)$ at low frequency is difficult to separate from the dc conductivity, σ_{dc}. Recently, Long[1] reported on careful experiments in which he was able to do so.

Existing theories for low frequencies can be divided into effective medium type theories[2], and cluster expansion theories[3]. All of them result in $Re\sigma(\omega) \propto \omega^2$ (Re denotes the real part, ω is the frequency). Since the real part of the conductivity must be a symmetric function of frequency, an analytic behavior at zero frequency requires an even integer power. Effective medium theories, like mean field theories of critical phenomena, must be analytic, and the quadratic behaviour from such theories is a consequence of this. Cluster expansion theories, in analogy with scaling theories of critical phenomena, do not generally give such a result. In the context of the rate equation, commonly used for hopping transport, the quadratic behaviour implies the existence of a maximum finite relaxation time. This has been associated with the critical percolation resistance of the random network. We show below that there exists no maximum relaxation time for hopping in a disordered system, which results in a non-analytic behavior at zero frequency.

To understand the lack of a maximal relaxation time, consider the process of electron rearrangement following the sudden application of a field. In a system of two sites, the time to reach a new equilibrium occupation is proportional to the inverse effective hopping rate between the sites,

$$u_{ij} = f_i(1-f_j)w_{ij}, \qquad (1)$$

where i and j label the sites, and f and w are the equilibrium occupation rates and transition rates. Detailed balancing requires $u_{ij}=u_{ji}$. It is tempting to extend this consideration to a random system of sites by arguing that, because of exponentially large fluctuations among the u's, the time for rearrangement of the electrons in a cluster corresponds to the largest u_{ij}^{-1}, say u_m^{-1}, required to span the cluster. If this would be the case, then the equilibration time for very large clusters would be the inverse critical percolation value u_c of u. This argument would be correct if equilibration were achieved by a local rearrangement of charge between neighboring sites. But this does not happen in hopping. Equilibration of a cluster requires electrons to traverse the entire cluster. For a very large cluster, this means traversing a large number (say N) of links with u equal or close to u_m. The time of relaxation thus increases as N grows, and becomes infinite as the cluster size becomes infinite. Parenthetically, we note that magnetic polarization of a cluster in disordered magnetic systems happens by local relaxation, so there a finite maximal relaxation time could be expected.

A more formal way to see the absence of a maximum relaxation time is in terms of the random impedance network[3] (explained in more detail in the next section). In this representation of transport, each pair of sites is connected by a resistance, and each site has a capacitance connected to the applied potential. The circuit for a pair of sites is charachterized by a pole on the negative imaginary frequency axis at $\omega=-i/\tau$ $\tau=RC/2$. The closest pole to the origin corresponds to the longest relaxation time. For all the pairs on the percolation path the closest would then be at $\omega=-iu_m$. However, the pair circuits for adjacent pairs are coupled by the capacitance on the common site, and this shifts the poles (still keeping them on the negative imaginary frequency axis). Some of the shifting is towards the origin, and coupling over a long range (i.e. large clusters) shifts some poles very close to the origin. The situation is somewhat analogous to the localization problem. In that case, the poles for the decoupled system are (on the real frequency axis) at ω = E/h, where E is a site energy. Coupling between sites, provided by elastic hopping, shifts the poles. For reasons of space, we do not elaborate further on this analogy.

2. QUANTITATIVE ANALYSIS.

The calculation of the conductivity will be based on currents within frequency dependent clusters of sites. The frequency dependent conductivity $\sigma(\omega)$ can be calculated from

$$\sigma(\omega) = j(\omega)/F = (dp/dt)/F = i\omega p/F \qquad (2)$$

where p is the polarization. The current density can be constructed[4] from cluster currents, defined by

$$I_{clust} = \Sigma e(df_i/dt)X_i \qquad (3)$$

where the summation is over all sites on the cluster, and $X_i = r_i \cdot F/|F|$ The current density $j(\omega)$, and $\sigma(\omega)$ then are[4]

$$j(\omega) = \Sigma I_{clust}/V, \quad \sigma(\omega) = j(\omega)/F, \qquad (4)$$

V being the volume, and the summation is over all clusters. Pair clusters (i.e. clusters of two sites) have a single relaxation time τ in which case the cluster current can be written as

$$j(\omega) = (\alpha F/\tau)(\omega\tau)^2/[1+(\omega\tau)^2], \qquad (5)$$

where α is the polarizability of the pair. As was alluded to before, larger clusters have more than a single mode τ, but we shall see that for the clusters of interest here, one mode is dominant.

It is well known[3] that the ohmic transport equation for disordered insulators maps on a random network, with a resistance $R_{ij} = kT/e^2 u_{ij}$ between any two sites, and $C_i = e^2 f_i(1-f_i)/kT$, a capacitance connecting any site to its appropriate applied potential[4]; kT is the thermal energy and e the electronic charge. $1/R_{ij}$ measures the rate at which charge can be transferred between i and j per unit difference in electrochemical potential, and C_i measures the change of charge on site i induced by a unit change in the electrochemical potential. $C_i C_j/(C_i+C_j)$ times R_{ij} thus is the relaxation time for polarization of the pair i,j. For a wide range of conditions[4],

$$\begin{aligned}R_{ij} &= (kT/e^2)\exp(2r_{ij}/a)\exp(E_{ij}/kT) \\ C_i &= (e^2/kT)\exp(-E_i/kT)\end{aligned} \qquad (6)$$

where E_i is the energy of site i measured from the chemical potential E_F, a is the localization radius, r_{ij} is the site separation, and $E_{ij}(E_i,E_j)$ is a pair energy. It is usual to group the C's and R's by factors of, say, 2 or e. Because of the extremely wide spreads in values shown by eq.(6), the values within a group are considered all equal, while those in other groups are considered much larger or much smaller. Thus, in any particular situation, i.e. at any ω, the network includes resistances of one value, $R=R_\omega$; the smaller resistances are replaced by $R=0$ shorts, and the larger resistances can be deleted. Clusters are thus formed with $R=R_\omega$ and with shorts. (Clusters which can be spanned entirely via shorts contribute only to Imj, and need not be considered here.)

For simplicity, we treat here only the so called r-percolation case, where sites are positioned at random, but site energies vary by less than kT. Then C is uniform over all sites, and the clusters are formed just by specifying the maximal R in a cluster, say R_m. When R_m is much less then the avarage nearest neighbor resistance <R>, most clusters consist of pairs of sites, joint by R R_m. The pair currents are given by eq.(5) with

$$\alpha = X^2 C/2, \qquad (5a)$$

X being the projection of the pair length along the field direction. The pairs with $R=R_m$ dominate the real part of the conductivity at the frequency $2/R_m C$. As the frequency is lowered below $2/<R>C$, the typical clusters include more than two sites, and the analysis becomes more complicated. Here the focus will be on very low frequencies, where the conductivity involves very large clusters. These also make their major contribution at some particular frequency (but not at $2/R_m C$). To evaluate the low-frequency conductivity from eq. (4), the statistics of large clusters is required, and their I_{clust} need to be evaluated from eq.(3). The work of Stauffer[5] provides such statistics. Since large clusters have a very complicated structure, a simplified model is introduced below to facilitate the calculation of I_{clust}. In doing so we take care to preserve the essential physical features of the large clusters, in particular their proper critical behaviour. The simplifications made are

1. The shorts which replace resistances with $R<R_m$ connect the resistance side of (say) n_C C's. We also interconnect their other sides (so they are all in parallel) and place them on an appropriately avaraged applied potential.

2. It is assumed that all n_C are equal to the most probable value of n_C, calculable from the probability of $R<R_m$ on the percolation cluster.

3. The cluster is replaced by the longest chain of R's and C's, spanning the length of the cluster in the direction

of the applied field. This approximation seems rather radical, but can be justified[6] quite well. The model clusters thus are chains of N (say) resistances R_m, and n_C capacitances C, connected at one end between any two successive R_m's, and at the other end to the appropriate applied potential.

The frequency dependent current of such chains is known[7]. It's real part is

$$ReI_{clust}(N,\omega) = F(2/\pi^2)(N\lambda^2/R_m) \cdot \sum_{\gamma\ odd=1}^{N} \gamma^{-2}(\omega R_m n_C C)^2 / \{4[1-\cos(\gamma/N)]^2 + (\omega R_m n_C C)^2\} \quad (7)$$

λ is the projection along F of the distance between R_m's. Like n_C, λ is calculable from the probability for $R<R_m$ on the percolation cluster. The result for very large clusters is, approximately, $\lambda = 1/(n^{4/9} a^{1/3})$, $n_C = (\lambda/a)^\beta$ where n is the site concentration, a the localization radius, and β the fractal dimension of the percolation cluster. The first mode, $\gamma=1$, turns out to account for almost all of ReI_{clust}; the combined contribution from all $\gamma \neq 1$ is at most 20%, so they can be neglected. Then

$$ReI_{clust}(N,\omega) = $$
$$= (2/\pi^2)(N\lambda^2/R_m)(\omega R_m n_C CN/\pi^2)^2 / [1+(\omega R_m n_C CN/\pi^2)^2] \quad (8)$$

It is interesting to note that eq. (8), the dominant mode of eq. (7), is equivalent to eqs. (5), (5a), with $C \to C' = N'n_C C$, $R \to R' = N'R$, $X \to X' = N'\lambda$, $\tau \to \tau' = R'C'$, where $N' = N/\pi$. Thus, the dominant mode of a large cluster is equivalent to a single pair of length $\lambda N/\pi$, with $n_C N/\pi$ capacitances C lumped at the two sites, and $R_m N/\pi$ lumped resistances R_m.

$Re\sigma\ \omega)$ is now obtained from eq. (4), using cluster statistics[6]. The density n_s of clusters with s sites is

$$n_s = 1.6\ s^{-2.2}\ \exp(-z^2), \quad z = |p_c-p| s^{0.45} \quad (9)$$

The number of sites N on the chain is related to s by $s = N^\beta$, while (p_c-p) is related to the maximal resistance R_m on the chains via their length r, $(p_c-p) \sim (2\pi/9) n(r_c^3-r^3)$, where the subscript c stands for the critical percolation value, and, from eq(6), $r/a \sim \ln R$. Insertion into eq.(9) gives the distribution $n_N(N,R)$. Eq. (4) then can be written $\sigma(\omega) = \sum_N n_N \mathrm{ReI}_{\mathrm{clust}}(N,R_m,\omega)$, with eq.(8) for $\mathrm{ReI}_{\mathrm{clust}}$.

It is usual in ac conductivity that some particular cluster dominates $\mathrm{Re}\sigma(\omega)$ at any ω. This is also assumed here, as it simplfies the calculation. We first maximize I_{clust} for each N with respect to R_m. The maximum is

$$\mathrm{ReI}_{\mathrm{clust}} = N\lambda^2/R' \qquad (10)$$

and occurs at

$$R_m = R' = \pi^2/\omega C \lambda^\beta N^2. \qquad (11)$$

(λ varies but slowly with R_m, so we can assume at very low ω that λ has always its critical percolation value.) Now

$$\mathrm{Re}\sigma(\omega) = n_N N\lambda^2/R' = \omega \sum_N n_N N^3 \lambda^{\beta+2} C/\pi^2, \qquad (12)$$

which, with N, is maximal at the largest N. We notice the sharp cutoff at $z=N^{.45\beta}\ln(R_c/R)=1$ in eq.(9). It determines the maximum value N_m of N for any R_m and thus for R'. (This value clearly corresponds to the correlation length $\chi(R_m) = \lambda N_m$.) In eq.(12), $n_N N$ is the number of R' in a unit volume. Clearly it must be independent on whether the unit volume is chosen larger or smaller than a cluster. In the latter case, the number of R' per unit volume is $1/\chi^2 \lambda$, so

$$\mathrm{Re}\sigma(\omega) = \omega N_m^2 \lambda^{\beta+1} C/\chi^2 \pi^2 = \omega C \lambda^{\beta-1}/\pi^2. \qquad (13)$$

3. CONCLUSIONS.

The left hand side of eq.(13) physically represents the difference between the conductivity measured at ω and the dc conductivity. In contrast with previous theories, it is proportional to the first, not the second power of ω at small ω. It is comparable directly to Long's[1] measurements. At the lowest frequencies he can measure, the experiment is much closer to our result. It must be mentioned that the experimental system is better represented by variable range hopping (VRH) percolation than by r-percolation. However, we do not anticipate a different frequency dependence for VRH, and our yet unpublished results confirm this.

The above derivation was based on the assumption that certain charachteristic cluster dominate the conductivity at any frequency. Our preliminary studies, where we added the contribution from all possible (not necessarily separate) finite clusters confirm $\text{Re}\sigma(\omega) \alpha \omega$ at low enough frequencies. A more difficult problem is the implicit assumption that $j(\omega)$ and j_{dc} flow in separate clusters, which is clearly not the case. The optimal percolation cluster, which carries j_{dc}, can shunt some of the current through the finite ac clusters. This may raise the power s of in $\text{Re}\sigma(\omega) \alpha \omega^s$ from 1 to somewhat larger values at very low ω. Simple arguments suggest that $s \leq 1.5$, but we believe that s is much closer to 1, due to the capacitors on the optimal percolation cluster. This question needs to be addressed more adequately in the future.

REFERENCES.

1. A.R. Long, Hopping Transport in Solids, eds. M. Pollak and B.I. Shklovskii, North Holland, in the press.; A.R. Long, J. McMillan, N. Balkan, and S. Summerfield, Phil. Mag. B **58**, 153 (1988)

2. S. Summerfield and P. N. Butcher, J. Phys. C **16**, 295 (1983)

3. H. Böttger, V.V. Bryksin, and G.Y. Yashin, J. Phys. C**12**, 2797 and 3951 (1979); H. Böttger and V.V. Bryksin, Hopping Conduction in Solids, Akademie Verlag 1985.

4. M. Pollak, Metal Non-Metal transition in Disordered Systems, eds. D. Tunstall and L. Friedman, SUSSP 1978; Non-Crystalline Semiconductors, ed. M. Pollak, CRC Press 1987.

5. A. Miller and E. Abrahams, Phys. Rev. **120**, 745 (1960)

6. D. Stauffer, Physics Reports **54**, 1 (1979)

7. M. Pollak and H. A. Pohl, J. Chem. Phys. **93**, 2980 (1975)

HOPPING CONDUCTION AND LOCALIZATION IN

HIGH ELECTRIC FIELDS

Harald Böttger and Dieter Wegener

Technische Universität "Otto von Guericke" Magdeburg,
Sektion Physik, DDR-3010 Magdeburg, G.D.R.

ABSTRACT

In the current theory of hopping conduction in disordered systems in the presence of a high electric field, an evaluation of the hopping current requires solving a set of rate equations and performing a configuration average. The concept of using rate equations is based upon the assumption of Anderson-localized electron states. Recent localization studies show that an electric field can strongly affect Anderson localization. Possibly, this fact is important to the theory of non-ohmic hopping transport. In this paper, elements and results of the conventional theory of hopping in a high electric field are reviewed, effects of an electric field on localization are discussed, and first steps toward a generalized approach to non-ohmic transport in disordered systems are presented.

1. INTRODUCTION

Non-ohmic conduction can arise from different effects: (i) bulk effects (Poole-Frenkel effect, tunnelling, hopping), (ii) effects depending on the conditions at the electrodes (space charge limited currents), and (iii) heating effects (see, e.g., ref. 1). In this paper, we restrict ourselves to non-ohmic conduction due to hopping.

In recent years, the theory of high-field hopping transport has been characterized by a considerable progress (see, e.g., refs. 2,3). For sufficiently strong electrical fields E, the theory predicts an exponential increase of the hopping current j with increasing field.[1,4-13] For not too strong fields, sufficiently high temperatures, and strong electron-phonon coupling, the theory predicts that dj/dE decreases with increasing field up to some critical field, above which the current begins to increase exponentially with E.[3,10,11,13-16] Thus, unlike the band mobility, the hopping mobility in general strongly depends on the strength of the electric field. An investigation of this dependence may therefore supply relevant information on the transport mechanism. In recent years, progress has been achieved in the theory of non-ohmic hopping transport by using the percolation approach,[8,9,11,12,14,17] the effective-medium approximation,[3,10,18,19] and numerical computations.[13,14,20] A theoretical investigation of non-ohmic hopping transport requires solving a set of rate equations with field dependent hopping probabilities and performing a configuration average. The concept of using rate equations is based upon the assumption of Anderson-localized electronic states. Recent localization studies[21-28] show that an electric field can strongly affect Anderson localization. Possibly, this fact is also important to the theory of non-ohmic hopping transport. For the case of one-dimensional disordered systems in a finite electric field, it was shown that for smooth electron-impurity scattering potentials, all states are delocalized[23]; whereas for discontinuous potentials in weak fields, there is power-law localization, and in strong fields, there is a mobility edge past which the states are extended.[21,22-24,26] For higher dimensions, it was predicted that Anderson localization is not possible in finite electric fields.[26] These results were obtained by investigating the electron wave function (a one-particle property)[22-25,27,28] or the density-density correlation function in a time independent electric field in the absence of electron-phonon interaction. However, a study of hopping transport requires consideration of two-particle properties, well-defined conditions for switching-on the electric field, and account of electron-phonon interaction. Therefore, the significance of the above localization studies for non-ohmic hopping transport is not quite obvious. In certain systems, an electric field can also enhance localization. It is well known (see e.g., ref. 29) that in a one-dimensional periodic single-orbital tight-binding model an electric field applied along the direction of the chain lifts the degeneracy between the local orbitals, so that all eigenstates of the chain become localized. The states form a series of localized states centered about each of the sites of the chain. The existence of these "Wannier-Stark" states is a problem of great current interest,[30-37] which is also of importance to the problem of Anderson localization in the presence of an electric field and the problem of non-ohmic hopping transport.

This paper is organized as follows. In Section 2 we outline our work devoted to non-ohmic hopping transport (For papers on this subject by other authors, the reader is referred, e.g., to the monograph, ref. 2). The effect of an electric field on localization is discussed in Section 3. First steps towards a generalized approach to non-ohmic transport in disordered systems are presented in Section 4.

2. THEORY OF NON-OHMIC HOPPING TRANSPORT

2.1 Basic Formulae

In the theory of hopping transport, the electron subsystem is in general described in terms of the tight-binding approximation. Thereby, for the sake of simplicity, each atom is assumed to possess a single energy level only. On using a Fröhlich-type Hamiltonian for the electron-phonon interaction, after applying the small-polaron canonical transformation, for a hopping system, the Hamiltonian becomes[2]

$$H = \sum_m V_m a_m^+ a_m + \sum_{mm'} j_{mm'} a_m^+ a_m \hat{\phi}_{mm'} + H_{ph}. \qquad (1)$$

Here $j_{mm'}$ is the resonance integral between sites m and m', a_m^+ (a_m) is the creation (annihilation) operator of a charge carrier at site m, $V_m = \epsilon_m - eER$, e is the charge of an electron, R_m is the position vector of site m, $\epsilon_m = \tilde{\epsilon}_m - E_{pm}$ is the polaron shift,

$$E_{pm} = \frac{1}{2N} \sum_q \hbar\omega_q |\gamma_m(q)|^2 \qquad (2)$$

ω_q is the phonon frequency of wave-vector q, $\gamma_m(q)$ is the dimensionless electron-phonon coupling constant, $\hat{\phi}$ is the multiphonon operator defined by

$$\hat{\phi}_{mm'} = \exp\{\frac{1}{\sqrt{2N}} \sum_q [\gamma_{m'}(q)\exp(-iqR_{m'}) -$$
$$- \gamma_m(q)\exp(-iqR_m)]b_q - h.c.\}, \qquad (3)$$

and H_{ph} represents the hamiltonian of the phonon subsystem,

$$H_{ph} = \sum_q \hbar\omega_q (b_q^+ b_q + \frac{1}{2}), \tag{4}$$

where b_q^+ (b_q) are phonon creation (annihilation) operators. Here, effects of disorder on the phonon subsystem and effects of electron-electron interaction are disregarded. In order to derive an expression for the current, we must define the conditions for turning on the electric field. We assume the electric field to be suddenly switched on at time $t=0$, so that at $t<0$ the system is in thermodynamic equilibrium, while at $t>0$ the system is described by the Hamiltonian (1). At $t>0$, at first relaxational currents occur, but after a short time interval δt, the current becomes stationary. In what follows, we are interested in the stationary regime only. With the aid of the familiar relation between the current density j and the time derivative of the dipole moment, the current can be expressed as

$$j(t) = \frac{e}{\Omega} \sum_m R_m \, d\rho_m/dt, \tag{5}$$

where $\rho_m = \rho_{m;m}$ is the diagonal element of the one-particle density matrix $\rho_{m;m'}$ (5) and Ω is the volume of the system. The n-particle equal-time density matrix is defined as

$$\rho_{m_1 m_2, \ldots, m_n; m_1' m_2' \ldots m_n'}(t) = \frac{\text{Tr}(\exp[-\beta(H-\mu\hat{N})]U^+(t)a_{m_n'}^+ a_{m_n} \ldots a_{m_1}^+ a_{m_1} U(t)])}{\text{Tr}(\exp[-\beta(H-\mu \hat{N})])}, \tag{6}$$

where H is given by H for $E=0$, $\beta = 1/kT$, μ is the chemical potential, $\hat{N} = \sum_m a_m^+ a_m$ is the operator of the total particle number, and $U(t)$ is the usual time evolution operator

$$U(t) = T_t \exp\left(-\frac{i}{\hbar} \int_0^t dt' \, H(t')\right), \tag{7}$$

where T_t is the time-ordering operator.

To examine $\rho_m(t)$, we have used a perturbation expression which is obtained by expanding in (6) the operators $U(t)$, $U^+(t)$ and $\exp(-\beta H)$ in powers with respect to the second term in the Hamiltonian (1) (see ref. 2). Thereby, the electron-phonon interaction is not required to be weak. Each term of the perturbation series for $\rho_m(t)$ can be represented by diagrams on a contour in the complex t-plane consisting of two horizontal branches, separated from each other by a cut on the real axis from 0 to t, and a vertical branch on the imaginary axis from $-i\hbar\beta$ to 0.

The basic feature of this perturbation expansion is the fact that it contains growing powers of the parameter W/ω, where W is the intersite transition probability and ω is the frequency of the external electrical field. Accordingly, at least for small ω, an infinite number of diagrams must be summed up. This procedure leads to the following equations[2]:

$$\frac{d\rho_{m';m}(t)}{dt} = \frac{i}{\hbar} V_{mm'} \rho_{m';m}(t) + \sum_{\{m_i\}} \int_0^t dt' \times$$

$$\times (\rho_{m_1;m_2}(t') \omega^{(1)m_1m'}_{m_2m}(t',t) + \rho_{m_1m_2;m_3m_4}(t') \omega^{(2)m_1m_2m'}_{m_3m_4m}(t',t) + \ldots \quad (8)$$

where $V_{m'm} = V_m$, $-V_m$ and, in the language of the diagram technique, $\omega^{(n)}$ are irreducible blocks of nth order. An important property of the blocks ω is the fact that they remain finite in the limit $\omega \to 0$.

The analytical expressions for the first and second order blocks, respectively, are given by[2]

$$\omega^{(1)m_1m}_{m_2m'}(t',t) = \frac{i}{\hbar}[\tilde{J}_{m_2m'}\delta_{mm_1} - \tilde{J}_{mm_1}\delta_{m'm_2}]\delta(t'-t) \quad (9)$$

$$\omega^{(2)m_1m_2m}_{m_3m_4m'}(t',t) = \frac{1}{\hbar^2} \times \quad (10)$$

$$\times (\delta_{m_3m'}\tilde{J}_{mm_1}\tilde{J}_{m_2m'} e^{\frac{i}{\hbar}\int_0^{t'-t} d\tau V_{m'm_1}(\tau+t)} F^{m_3m_1}_{m_4m'}(t-t') +$$

$$+ \delta_{m_2 m} \tilde{J}_{m_3 m_1} \tilde{J}_{m_4 m'} e^{\frac{i}{\hbar} \int_0^{t'-t} d\tau\, V_{m_4 m}(\tau+t)} F_{m_4 m'}^{m_3 m_1}(t-t') -$$

$$- \delta_{m_1 m} \tilde{J}_{m_3 m'} \tilde{J}_{m_4 m_2} e^{\frac{i}{\hbar} \int_0^{t'-t} d\tau\, V_{m_3 m}(\tau+t)} F_{m_4 m_2}^{m_3 m'}(t-t') -$$

$$- \delta_{m_4 m'} \tilde{J}_{m_3 m_1} \tilde{J}_{mm_2} e^{\frac{i}{\hbar} \int_0^{t'-t} d\tau\, V_{m',m_2}(\tau+t)} F_{mm_2}^{m_3 m_1}(t-t')$$

where

$$\tilde{J}_{m'm} = m'm \exp[-s_T(m',m)], \tag{11}$$

$$S_T(m',m) = \frac{1}{2N} \sum_q |\gamma(q)|^2 (1 - \cos q(R_{m'} - R_m)) \times$$

$$\times \coth(\hbar \omega_q \beta|2) \tag{12}$$

$$F_{m_3 m_4}^{m_1 m_2}(t) = P_{m_3 m_4}^{m_1 m_2}(t) - 1, \tag{13}$$

$$P_{m_3 m_4}^{m_1 m_2}(t) = \exp\left\{ \sum_q \frac{|\gamma(q)|^2}{N \sinh(\hbar\omega_q \beta|2)} a_{m_3 m_4}^{m_1 m_2} \cos \omega_q(t + i\hbar\beta|2) \right\} \tag{14}$$

$$a_{m_3 m_4}^{m_1 m_2} = \frac{1}{2} \{\cos qR_{m_1 m_4} + \cos qR_{m_2 m_3} + \cos qR_{m_1 m_3} + \cos qR_{m_2 m_3}\} \tag{15}$$

and $R_{m'm} = R_{m'} - R_m$.

To proceed further, in (8) we restrict ourselves to the contribution of irreducible blocks up to the second order. Furthermore, we make use of the Hartree-Fock decoupling:

$$\rho_{m_1 m_3; m_2 m_4}^{m_1} \approx \delta_{m_2 m_3} \rho_{m_1; m_4} - \rho_{m_3; m_2} \rho_{m_1; m_4} + \rho_{m_1 m_2} \rho_{m_3 m_4}. \tag{16}$$

Inserting (16) into (8), we get a coupled set of equations for the diagonal elements $\rho_m(t)$ and nondiagonal elements $\rho_{m;m'}(t)$ ($m \neq m'$) of the density matrix. To the lowest order in the resonance integral, it holds that (for time independent fields)

$$\rho_{m;m'} \approx \tilde{J}_{mm'} (\rho_m - \rho_{m'})/i\, V_{m'm}. \tag{17}$$

correspondingly, for most pairs of sites, it holds that $\rho_{m;m'} \sim J(\rho_m - \rho_{m'})/\Gamma$, where Γ is the width of the spread of the site energies. For strong localization (the usual prerequisite for hopping transport), it holds that $J/\Gamma \ll 1$, so that $\rho_{m;m'} \ll \rho_m, \rho_{m'}$.

2.2 Rate Equation and Expression for the Current in Arbitrary Electrical Fields

Since we are interested in the stationary regime, in (8) we take the limit as $t/\delta t \to \infty$. To do this, in this equation we replace the variable of integration by $t' = t - t_1$. Thus, because $\omega(t_1-t,t)$ goes to zero in the time $t_1 \approx \delta t$, in the limit as $t/\delta t \to \infty$, in (8) the upper limit of the integration over t may be set equal to ∞. For $\omega t \ll 1$ (ω frequency of the external field), in (8) $\rho(t - t_1)$ may be taken out of the integral over t_1 after setting t_1 equal to zero in this quantity. In this manner we obtain the following Markovian rate equation (cf.^{2})

$$\frac{d\rho_m}{dt} = \sum_{m'} [\rho_{m'}(1-\rho_m) W_{m'm}(t) - \rho_m(1-\rho_{m'}) W_{mm'}(t)], \tag{18}$$

where

$$W_{m'm} = \frac{|\tilde{J}_{m'm}|^2}{\hbar^2} \exp\left\{\frac{\beta}{2} V_{m'm}(t) \int_{-\infty}^{\infty} dt_1 \cos(t_1 V_{m_1m}(t)/\hbar) \times \right.$$

$$\left. \times \left\{ \exp\left[\sum_q \frac{|\gamma(q)|^2 (1 - \cos q\, R_{m'm})}{N \sinh(\hbar\omega_q \beta/2)} \cos \omega_q t_1 \right] - 1 \right\} \right.. \tag{19}$$

Since $t_1 \omega \sim \delta t\, \omega \ll 1$, here we have used the approximation

$$\int_0^{t_1} dt'\, E(t'-t) \approx t_1 E(t).$$

With the aid of (19), it is easy to show that the hopping probability $W_{mm'}$ has the symmetry property

$$W_{m'm}/W_{mm'} = \exp(\beta V_{m'm}), \tag{20}$$

which reflects the principle of detailed balancing for hops between sites m and m' in an external electrical field. Making use of (5) and (18), for arbitrary electrical fields, the density of the hopping current can be expressed as

$$j = \frac{1}{2\Omega} \sum_{mm'} (R_m - R_{m'}) \, i(m'm). \tag{21}$$

Here $i(m',m)$ is the current running between sites m and m', given by

$$i(m'm) = e \, \tilde{W}_{m'm} [\rho_m (1 - \rho_{m'}) \exp(\beta V_{m'm}/2) - \rho_{m'}(1 - \rho_m) \exp(\beta V_{mm'}/2)], \tag{22}$$

where it holds that

$$\rho \frac{d\rho_m}{dt} = \sum_{m'} i(m'm). \tag{23}$$

Here

$$\tilde{W}_{m'm} = \tilde{W}_{mm'} = W_{m'm} \exp(-\beta V_{m'm}/2) \tag{24}$$

is the symmetrized two-site hopping probability. Introducing the change of the chemical potential, $\delta\mu_m$, by

$$\rho_m = 1/[e^{\beta(\epsilon_m - \delta\mu_m - \mu)} + 1] \tag{25}$$

then the expression (22) may be written as[2]

$$i(m'm) = e\tilde{W}_{m'm} \times \tag{26}$$

$$\times \frac{\sinh[\frac{\beta}{2}(-\delta\mu_m + \delta\mu_{m'} + eu_{m'} - eu_m)]}{2\cosh[\frac{\beta}{2}(\epsilon_m - \delta\mu_m - \mu)]\cosh[\frac{\beta}{2}(\epsilon_{m'} - \delta\mu_{m'} - \mu)]}$$

where $u_m = -ER$.

As in the case of weak fields, equations (22) and (23) for $d\rho_m/dt=0$, which holds for dc currents) may be thought of as Kirchhoff's first and second law, respectively. The expression for j (21) represents the equation for the external circuit of the corresponding network. For weak fields, it is sufficient to expand sinh(...)(in (26)) in powers of the argument up to the linear term and to put $\delta\mu_m = u_m = 0$ elsewhere. This yields the random resistor network of Miller and Abrahams, where $u_m + \delta\mu_m/e = \mathcal{U}_m$ is the electrical potential at the mth site. However, for fields of arbitrary strength, the expression (25) may in general not be interpreted as the current running through a nonlinear resistor, because it does not only depend on the difference between the potential $\mathcal{U}_m$ and $\mathcal{U}_{m'}$, but it depends separately on these quantities. Therefore, for strong fields, the hopping problem may not be reduced to a network of nonlinear resistors.

The latter fact was illustrated in ref. 11 by studying a one-dimensional R-hopping system on the basis of (22) for the case of Boltzmann statistic ($\rho_m \ll 1$). It turns out that

$$j \sim \left[< \tilde{W}^{-1}(R) \exp(-\tfrac{1}{2} \beta |e| ER) > \right]^{-1}, \qquad (27)$$

where the symbol <...> indicates the configuration averaging procedure and $W(R) \sim \exp(-2 \alpha R)$ (α reciprocal Bohr radius). Accordingly, for a random distribution of sites in the chain under consideration, the current vanishes, because it holds that

$$<\tilde{W}^{-1}(R)\exp(-\tfrac{1}{2}\beta|e|ER)> \to \infty$$

$$\text{for} \qquad 2\alpha > \frac{1}{\ell} + \frac{|e|E}{2kT}, \qquad (28)$$

where $\ell=<R>$ is the average spacing between nearest neighbors. On the other hand, for such a distribution, on the basis of (26), by neglecting $\delta\mu$, the nonlinear resistor model yields[38] a non-vanishing current $j \sim (|e|\ell E/kT)^{2\alpha\ell}$ ($\alpha\ell \gg 1$), due to the breakdown of hard hops.

The most important property of $W_{m'm}$ is $W_{m'm} \sim |J_{m'm}|^2 \cdot \exp(-2\alpha |R_{m'm}|)$, i.e., for $|\gamma(q)|^2 \gg 1$, $W_{m'm}$ only weakly depends on E of eER is less the height of the barrier, which is of the order of $E_a \sim E_p/2$ (2). Hence, according to (19) and (20), for strong interaction with phonons and small spread of the energy levels, we have

$$\tilde{W}_{m'm} = W_0 \exp(-2\alpha |R_{m'm}|), \qquad (29)$$

where W_0 only weakly depends on E, R and ϵ_m.

For weak interaction with phonons, i.e., for $|\gamma(q)|^2 \ll 1$, in (19) we may expand the exponential in powers of the coupling constant. In the lowest nonvanishing order, we find that

$$\tilde{W}_{m'm} = \frac{1}{\sinh(|V_{m'm}|\beta/2)} \frac{|J_{m'm}|^2}{\hbar^2} \frac{\pi}{N} \sum_q |\gamma(q)|^2 \times$$

$$\times [1 - \cos q R_{m'm}] \delta(\omega_q - |V_{m'm}|/\hbar). \qquad (30)$$

Hence, for weak interaction with phonons and small spread of the energy levels, we get

$$\tilde{W}_{m'm} = W_0' [\sinh(|V_{m'm}|\beta/2)]^{-1} \exp(-2\alpha |R_{m'm}|), \qquad (31)$$

where W_0' only weakly depends on E, R_m and ϵ_m.

For weak interaction with acoustic phonons (low temperatures), we may assume that

$$|\gamma(q)|^2 \sim \frac{1}{q^2} \frac{1}{\omega_q} \theta (\omega_m - \omega_q),$$

$$\omega_q = \begin{cases} q \omega_0 / q_0 & \text{for } 0 < q < q_0 \\ 0 & \text{otherwise,} \end{cases} \qquad (32)$$

where θ denotes the Heaviside step function, ω_D is the Debye frequency, q_D is the Debye wave-number, and ω_m is the maximum phonon frequency with which a charge carrier can interact. Inserting (32) into (30), and using $(1-\cos qR) \approx (qR)^2$, then we get $W_0' \sim 1/|V_{mm'}|$. Furthermore, since[40] $\omega_m \sim 2\alpha a \omega_D \pi$ (a lattice constant) a large radius impurity state can only interact with a very small fraction of the Debye energy. Therefore owing to the δ-function in (30), with the application of a large electric field, hops nearly parallel to the direction of the field are precluded since their energy difference is too great. This may lead to negative differential conductivity at sufficiently large electric fields.[40]

For very low temperatures and weak electron-phonon interaction, from (26) and (30) we find that

$$i(m'm) = \frac{|e|}{2} W_0' \exp\{-\frac{\beta}{2}|\epsilon_{m'} - \delta\mu_{m'} - \mu|\} \times$$

$$\times \exp\{\frac{\beta}{2}|\epsilon_m - \delta\mu_m - \mu|\} \times$$

$$\times \exp\{\frac{\beta}{2}|\epsilon_{m'} + eu_{m'} - \epsilon_m + eu_m| -$$

$$- |e|\left[\frac{\beta}{2}(\frac{\delta\mu_{m'}}{e} + u_{m'}' - \frac{\delta\mu_m}{e} - u_m)\right] - 2\alpha|R_{mm'}|\}. \quad (33)$$

For $\delta\mu_{m'}/e + u_{m'}' > \delta\mu_m/e + u_m$, the expression (33) represents the current running from site m' to site m (the elementary hopping event occurs from site m to m'), since $e = -|e|$). Setting $\delta\mu_m = \delta\mu_m' = 0$ and assuming $\epsilon_{m'} - \epsilon_m \gg u_{m'}, -u_m \gg 0$, then from (33) we get

$$i(m'm) = \frac{|e|}{2} W_0' \times$$

$$\times \exp\{-\beta[(\epsilon_{m'} - \epsilon_m) + |e|ER_{m'm}] - 2\alpha|R_{m'm}|\}, \quad (34)$$

which represents the expression used for i(m'm) in some papers devoted to high field hopping transport. According to (34), for a hop opposite to the direction of the external electric field, the energy barrier $(\epsilon_{m'} - \epsilon_m) > 0$ is lowered by the field by an amount $|e|ER_{m'm} > 0$.

Furthermore, for $(u_m - u_{m'}) \gg (\epsilon_{m'} - \epsilon_m) > 0$ and $\delta\mu_m - \delta\mu_{m'} = 0$, from (33) we get for a hop oppose to t field

$$i(m'm) = \frac{|e|}{2} W_0' \exp(-2\alpha|R_{mm'}|), \quad (35)$$

i.e., we get no exponential field-dependence. From (26) and (30), we see that for weak electron-phonon coupling at sufficiently strong fields the differential conductivity becomes independent of the field, while according to (26) and (29) for strong electron-phonon coupling, at sufficiently strong fields the current increases exponentially with increasing field, provided that $|e|ER<E_1$.

In the following part of this Section, we restrict ourselves to Boltzmann statistics ($\rho_m \ll 1$). In this case, from (22) and (23) we get

$$\frac{d\rho_m}{dt} = \sum_{m'} \widetilde{W}_{m'm} [\rho_{m'} \exp(\beta V_{m'm}/2) - \rho_m \exp(\beta V_{mm'}/2)]. \tag{36}$$

Introducing the quantity $P_{mm'}(t)$ (conditional probability function) that obeys the differential equation

$$\frac{dP_{mm'}(t)}{dt} = \sum_{m'} \widetilde{W}_{m_1 m} [P_{m_1 m'} \exp(\beta V_{m_1 m}/2) - P_{mm'} \exp(\beta V_{mm_1}/2)], \tag{37}$$

together with the initial condition

$$P_{mm'}(t=0) = \delta_{mm'}, \tag{38}$$

then we see that

$$\rho_m(t) = \sum_{m'} P_{mm'}(t) \rho_{m'}(t=0) \tag{39}$$

satisfies (36). The quantity $P_{mm'}(t)$ has the meaning of the probability that an electron is at time t at site m if it was placed at time t=0 at site m'. According to our assumption that the system is in thermodynamic equilibrium at times t<0 and that the field is suddenly switched on at t=0, it holds that

$$\rho_m(t=0) = \exp[-\beta(\epsilon_m - \mu)]. \tag{40}$$

As before, the chemical potential μ, is given by the condition

$$\frac{1}{N} \sum_m \rho_m = n_e, \qquad (41)$$

where n_e is the mean occupation probability of a site. Since, in the approximation studied in this paper, a Markovian rate equation governs hopping transport, the stationary regime does not depend on the initial state and we may put $\rho_m(t=0) = n_e$ in (39) for $t \to \infty$, i.e.,

$$\lim_{t \to \infty} \rho_m(t) = n_e \lim_{t \to \infty} P_{mm'}(t). \qquad (42)$$

Thus, by means of the well-known relation

$$\lim_{t \to \infty} f(t) = \lim_{s \to 0} s \int_0^\infty dt\, e^{-st} f(t) \qquad (43)$$

and with the aid of (5), (42), and $\sum_m P_{mm'}(t) = 1$, for $t \to \infty$ we get

$$\lim_{t \to \infty} j(t) = \lim_{s \to 0} \frac{s^2 e}{\Omega} \sum_m R_m \int_0^\infty dt\, e^{-st} [\rho_m(t) - \rho_m(0)]$$

$$= \lim_{s \to 0} \frac{s^2 e\, n_e}{\Omega} \sum_{mm'} (R_m - R_{m'}) P_{mm'}(s) \qquad (44)$$

where $P_{mm'}(s) = \int_0^\infty dt\, \exp(-st)\, P_{mm'}(t)$ denotes the Laplace transform of $P_{mm'}(t)$. Let us define

$$P(r,r',t) = \frac{1}{N} \langle \sum_{mm'} P_{mm'}(t)\, \delta(r-R_m)\delta(r'-R_{m'}) \rangle \qquad (45)$$

and

$$P(k,s) = \frac{1}{\Omega} \int_0^\infty dt\, e^{-st} \int dr\, dr'\, P(r,r',t) e^{-ik(r-r')}, \qquad (46)$$

where r and r' are continuous coordinates and $N = N/\Omega$. Then, from (44) to (46), we find that

$$j = |e| n_e\, N\, s^2 \frac{\partial}{\partial (ik)} P(k,s)\Big|_{k \to 0,\ s \to 0} \qquad (47)$$

2.3 Percolation Approach and Numerical Computations

On studying high-field hopping transport, in the spirit of the percolation theory, we assume[11] that the current runs along certain optimal one-dimensional paths from electrode to electrode. In the regime where Ohm's law holds, this assumption means that the current take paths with minimal resistance. In the general case, the current paths are determined by the requirement of maximum current flux. Assuming that such paths do exist and that the current along each path is conserved and equal to i, then we have j~i.

Let us consider such a percolation path and number the sites along it in increasing manner in the direction of the current flow, say, from the left to the right electrode (E is directed along the x-axis). Note that it does not at all hold that $(R_{m+1})_x > (R_m)_x$, however, it holds that $(R_{m+n})_x > (R_m)_x$ for $n \to \infty$. Restricting ourselves to the Boltzmann statistics ($\rho_m \ll 1$), then from (22) we obtain

$$i = -e\tilde{w}_{m+1,m} \times$$

$$\times (\rho_{m+1} \exp(\tfrac{1}{2}\beta V_{m+1,m}) - \rho_m \exp(-\tfrac{1}{2}\beta V_{m+1,m}). \tag{48}$$

On solving this equation with respect to ρ_{m+1} and iterating the resulting equation, then, for the case of R-hopping, we find that

$$i = en_e \{ \frac{1}{N} \sum_m \sum_{n=0}^{\infty} (\tilde{w}_{m,m-1})^{-1} \exp(\tfrac{1}{2}\beta e\, u_{m-1,m}) \times$$

$$\times \exp(-\beta e\, u_{m+n,m}) \}^{-1}, \tag{49}$$

where

$$n_e = \frac{1}{N'} \sum_m \rho_m \tag{50}$$

is the concentration of electrons in the percolation path and N' is the total number of sites along the path. For a current path, the expression in the denominator of (49) becomes minimum. Correspondingly, it holds that

$$i \sim \left[W_{m,m-1} \exp(-\tfrac{1}{2} \beta e\, u_{m-1,m}) \exp(\beta e\, u_{m+n,m}) \right]_{max} \quad (51)$$

where the right-hand site is an element of the optimal percolation path. In (51), the subscript "max" indicates that the optimal current path is determined by the requirement that the right-hand side of (51) should be as large as possible. Furthermore, also the third factor, which is governed by the largest backbends λ in the path, must be as large as possible.

For the percolation problem in question, the characteristic geometrical percolation figures are obtained by setting the exponents of the first two terms equal to a constant, which is chosen such that a closed path of overlapping figures occurs, whereby the maximum backbend is as small as possible. In the case of strong interaction with phonons (see (29)) the characteristic figure is an ellipsoid

$$2\alpha R - \frac{|e|ER}{2kT} \cos \zeta = \rho, \quad (52)$$

where ζ is the angle between $(R_m - R_{m'})$ and E and $R = |R_m - R_{m'}|$, while for weak interaction with phonons (see (30)) the characteristic figure is composed of a hemisphere (in the direction opposite to E) and a semi-ellipsoid described by (52) (in the direction along to E), that is

$$2\alpha R - \frac{|e|ER}{2kT} \cos \zeta \; \theta(\cos \zeta) = \rho, \quad (53)$$

where θ is the Heaviside step function.

The parameter ρ is determined by means of the following equation[11]

$$\rho_0 = \frac{|e|E\, \lambda(\rho_0)}{kT} \quad (54)$$

where $\lambda(\rho_0)$ denotes the maximum backbend for $\rho = \rho_0$.
According to (51), we have

$$j \sim \exp(-\rho_0) \quad (55)$$

For a given field E, the first percolation path opens at $\rho = \rho_c$. With increasing ρ, the backbend λ decreases, so that at $\rho = \rho_m$ the quantity λ turns to zero, i.e., for $\rho > \rho_m$, the percolation is directed. To solve (54) $\lambda(\rho)$ was assumed to by asymptotically given by

$$\lambda(\rho) = 2\delta \, R_c [(\rho-\rho_c)/\rho_c]^{-\nu} [(\rho_m-\rho)/\rho_c]^{\mu}, \tag{56}$$

where $R_c = (\rho_c/2\alpha)_{E \to 0}$, and δ, μ, ν are positive numbers. The field dependence of ρ_m and ρ_c was determined numerically.[14] The results obtained may be well fitted by the formula

$$\rho_i(E) = 2\alpha R_i (1 - |e|E/4\alpha kT)^\tau, \quad (i = c, m) \tag{57}$$

where $\tau = 0.65 \approx 2/3$ and $R_m = (\rho_m/2\alpha)_{E \to 0}$. With the aid of (54)-(57), the current becomes[14]

$$j \sim \exp(-2\alpha R_c) \times$$

$$\times \exp\left\{ -2\alpha\, R_c \left[(\delta \frac{|e|E}{\alpha kT})^{1/\nu} \gamma^{\mu/\nu} - \frac{|e|E}{6\alpha kT} \right] \right\}. \tag{58}$$

On the other hand, inserting (57) into $j \sim \exp(-\rho_m)$, we get

$$j \sim \exp[-2\alpha\, R_c(1 + \gamma)] \exp\left[\frac{|e|ER_c}{3kT}(1 + \gamma) \right], \tag{59}$$

where $\gamma = (R_m/R_c) - 1 \approx 0.07$.

The expression (59) may be thought of as describing the current at sufficiently high fields, where the current is governed by directed percolation, while the expression should apply to low electrical fields.

According to a numerical inspection of $\lambda(\rho)$ it holds that[14] $\delta=0.26$, $\nu=1.1$, and $\mu=1.0$ (with error of about 50%). Using these values of the parameters δ, ν, and μ, then in the braces of the exponent of (58) the second term dominates. Accordingly, for strong interaction with phonons, even at low fields above the Ohmic regime, the differential conductivity increases with the field, contrary to prediction of the effective-medium theory[10] (see also Section 2.4.) and numerical computations[13]. The latter computations indicate that at low fields the number n_e (50) of the charge carries in the percolation path rather strongly depends on the strength of the external field. Accordingly, the expression for the current (58) must yet be multiplicated by a factor $n_e(E)$, which improves the qualitative agreement with results of the effective-medium approximation and the numerical computations.

Computer studies for R-ϵ-hopping and strong electron-phonon interaction[15,16] show that at moderate temperatures the differential conductivity exhibits a (relative) minimum separated from the Ohmic regime by a relative maximum. On lowering the temperature, the height of the maximum increases, while the depth of the minimum decreases and completely vanishes at sufficiently low temperatures.

2.4 Effective-Medium Theory

The basic idea of the effective-medium approximation consists in replacing the disordered system in question by some effective ordered medium, the parameters of which are subsequently chosen such that it describes the properties of the actual medium as well as possible.

For R-hopping and strong electron-phonon coupling, a decrease of the differential conductivity with increasing field just above the Ohmic regime was first predicted by using an effective-medium theory.[10] In this theory, nearest-neighbor hopping is studied on a cubic lattice in which the hopping probability is a random quantity. In the absence of a spread of energy levels, according to (22), the current between sites m+g and m (g vector connecting nearest neighbors) is given by (for $\rho_m \ll 1$)

$$i_{m+g,m} = \widetilde{W}_{m+g,m} \times$$
$$\times (\rho_{m+g} \exp(-\tfrac{1}{2}\beta|e|Eg) - \rho_m \exp(\tfrac{1}{2}\beta|e|Eg)), \qquad (60)$$

where

$$\sum_g i_{m+g,m} = 0. \qquad (61)$$

According to the effective-medium concept, all hopping probabilities $W_{m+g,m}$ are placed by a uniform one denoted by w. Then all ρ_m are equal to each other: $\rho_m = f_0 = \exp(-\beta|\mu|)$, and for E parallel to the x-axis, the current density (21) becomes

$$j_x = \frac{2|e|}{|g|} \, w \, e^{-\beta|\mu|} \, \sinh\left(\frac{1}{2} \beta \, |e|E|g|\right). \tag{62}$$

w is determined by the condition that the average value of the local change of the occupation probability, $(\rho_m - f_0)$, vanishes if any w is replaced by the random quantity $W_{m+g,m}$ (see ref. 10).

It turns out that in very strong fields, the effective dimension of the hopping system approaches to unity. Evidently, at high fields the current cannot choose the optimal path by by-passing hard hops, but it is forced to run from electrode to electrode along a straight line. Recently, we have elaborated a mean-field theory of non-Ohmic R-ϵ- hopping transport.[3,19] This theory is based on the diagrammatic technique of Gochanour et al.,[41] which we have generalized such that energetic disorder can be taken into account, in addition to spatial disorder. Starting point of this theory is the equation governing the Laplace transform of the conditional probability function. From (37) we get

$$-\delta_{mm'} + s\, P_{mm'}(s) = \sum_{m_1} [W_{mm_1} P_{m_1 m'}(s) - W_{m_1 m} P_{mm'}(s)]. \tag{63}$$

The function P(k,s) needed for evaluating the current (cf.(47)) is related to $P_{mm'}(s)$ as

$$P(k,s) = \frac{1}{N} \left\langle \sum_{mm'} P_{mm'}(s) \exp[-ik(R_m - R_{m'})] \right\rangle. \tag{64}$$

Here, the symbol $\langle ... \rangle$ indicates averaging over the site position R_m as well as the site energies ϵ_m. The site position are assumed to be distributed at random. The site energies are assumed to be statistically independent quantities whose distribution is characterized by a function $N(\epsilon)$. We decompose the function P(k,s) as follows:

$$P(k,s) = \int d\epsilon \, P_d^\epsilon(s) N(\epsilon) + \iint d\epsilon d\epsilon' \, P_{nd}^{\epsilon\epsilon'}(k,s) \, N(\epsilon) N(\epsilon'). \tag{65}$$

Here

$$P_d^\epsilon(s) = \langle \frac{1}{N} \sum_m P_{mm}(s) \rangle'$$

(66)

$$P_{nd}^{\epsilon\epsilon'}(k,s) = \langle \frac{1}{N} \sum_{m \neq m'} P_{mm'}(s) \exp[-ik(R_m - R_{m'})] \rangle',$$

where the dash at the symbol of averaging indicates that in the first line of (66), the averaging does not include the energy of sites m and m', but these energies are kept fixed at $\epsilon_m = \epsilon$ and $\epsilon_{m'} = \epsilon'$.

To evaluate P_d^ϵ and $P_{nd}^{\epsilon\epsilon'}$, we iterate (63) and average the resulting series term by term. $P_d^\epsilon(s)$ and $P^{\epsilon\epsilon'}(k,s)$ can be written as a diagrammatic series. A self energy is defined as the sum of all irreducible diagrams, such that

$$P_{nd}^{\epsilon\epsilon'}(k,s) = P_d^\epsilon(s)\, \Sigma^{\epsilon\epsilon'}(k,s)\, P_d^{\epsilon'}(s) +$$

$$+ \int d\epsilon_1\, P_{nd}^{\epsilon\epsilon_1}(k,s)\, N(\epsilon_1)\, \Sigma^{\epsilon\epsilon'}(k,s)\, P_d^{\epsilon'}(s).$$

(67)

The function $P_d^\epsilon(s)$ is the Laplace transform of the probability of finding an electron at some (finite) time at some site, if it was placed at this site at time t=0. A complete description of the motion of this electron is given by the function $P(k,s)$ defined by

$$P^\epsilon(k,s) = P_d^\epsilon(s) + \int d\epsilon'\, N(\epsilon')\, P_{nd}^{\epsilon\epsilon'}(k,s).$$

(68)

With the aid of (67), from (68) we get

$$P^\epsilon(k,s) = \left[(P_d^\epsilon)^{-1} - \int d\epsilon'\, N(\epsilon')\, \Sigma^{\epsilon'\epsilon}(k,s) \right]^{-1}$$

(69)

It holds that

$$P^\epsilon(k=0,s) = \frac{1}{s},$$

(70)

which results from the fact that an electron starting at time t=0 at a site with energy will be placed at time t>0 either at this site or at another site with arbitrary energy ϵ'.

With the aid of (70), P_d^ϵ can be eliminated from (69). Thus we obtain

$$P^\epsilon(k,s) = \{s + \int d\epsilon' \; N(\epsilon') \; [\Sigma^{\epsilon\epsilon'}(0,s) - \Sigma^{\epsilon'\epsilon}(k,s)]\}^{-1}. \quad (71)$$

Making use of (71), (68) and (65), we find that

$$P(k,s) = \int d\epsilon \; \frac{N(\epsilon)}{s + \int d\epsilon' \; N(\epsilon') \; [\Sigma^{\epsilon'\epsilon}(0,s) - \Sigma^{\epsilon'\epsilon}(k,s)]}, \quad (72)$$

and the current (47) becomes

$$j = |e|n_e \; N \; \frac{\partial}{\partial(ik)} \; \Sigma \; (k,s)\Big|_{k\to 0, \; s\to 0}, \quad (73)$$

where

$$\Sigma(k,s) = \int d\epsilon \; d\epsilon' \; N(\epsilon) \; N(\epsilon') \; \Sigma^{\epsilon\epsilon'}(k,s). \quad (74)$$

To get an approximate expression for $\Sigma^{\epsilon\epsilon'}(k,s)$, the diagrammatic series for $\Sigma^{\epsilon\epsilon'}(k,s)$ must be partially summed up. Doing so, we seek an analytical expression for $\Sigma^{\epsilon\epsilon'}(k,s)$ that contains the conditional probability function (Green function) of an appropriately defined self-consistent medium. Since a crystal is the only system whose Green function can be evaluated analytically in a closed form, we choose a crystal as the effective reference system (for details see refs. 3,19, similarly as in our theory of phonon-like excitations in structurally disordered systems.[42]

For sufficiently high temperatures an strong electron-phonon interaction, our theory predicts a minimum in the differential conductivity, which disappears when the temperatures is lowered, in accordance with our numerical results.[15,16]

Concluding, Boltzmann statistics may be expected to apply to non-ohmic hopping only at not too high electric fields, because high fields cause charge redistribution, so that for most sites the site-occupation probability equals either zero or unity[13], which requires the use of Fermi statistics. Sufficiently high electric fields can affect the localization range of electric states, which possibly requires account of the non-diagonal elements of the density matrix (see also Section 4).

3. LOCALIZATION IN AN ELECTRIC FIELD

3.1 Wannier-Stark States

For the case of a single orbital periodic one-dimensional tight-binding model subjected to a uniform dc electric field there is a now a universal agreement that the energy spectrum forms a uniformly spaced ladder - the Wannier-Stark ladder.[29-37, 43-47]

Wannier's[43] proof of this ladder is as follows. The Schrödinger equation for a periodic one-dimensional system in the presence of an electric field E is given by

$$H \psi(x) = \epsilon \psi(x) \tag{75}$$

with

$$H = -\frac{\hbar^2}{2m}\frac{d^2}{dx^2} + V(x) + e\,Ex, \tag{76}$$

where ϵ is the energy eigenvalue and $V(x)=V(x-ma)$ (m integer, a lattice constant). Let $x \to x-ma$, then (75) becomes

$$H \psi(x-ma) = \epsilon_m \psi(x-ma) \tag{77}$$

where

$$\epsilon_m = \epsilon + me\,Ea \tag{78}$$

Accordingly, if $\psi(x)$ is an eigenfunction of H with eigenvalue ϵ, then $\psi(x-ma)$ is also an eigenfunction of H with eigenvalue ϵ_m. The spectrum (78) constitutes the Wannier-Stark ladder. Within the tight-binding theory, the eigenstate amplitude corresponding to the energy (78) is asymptotically given by (31)

$$|f_n| \sim (2J/|eEa|)^{|n-m|}/|n-m|! \;. \tag{79}$$

Hence, in an electric field, the electron is "factorially localized" about site m (J is the nearest-neighbor resonance integral). It is confined to a finite region of a size $2J/|eE|$. This confinement is produced by the band edges: The electric field causes the electron to move in k space at a rate proportional to the magnitude of the field. When it gets to the end of the Brillouin zone, it undergoes Bragg reflection, reenters on the other side of the zone, and its thus able to complete closed orbits in k space. The motion of the electron in a dc field is therefore periodic in time (Bloch oscillations; the period of oscillation is $\tau_0 = 2\pi \hbar/|eEa|$) if the energy band is periodic in k. The Bohr-Sommerfeld quantization applied to these closed orbits produced the Wannier-Stark states.

One objection to the simple theory of Wannier-Stark ladder is that the possibility of Zener tunneling from one band to another is neglected. It was shown[44] that the spectrum of the total Hamiltonian of an electron in an electric field is continuous with extended states and that any approximation with a finite number of bands leads to a pure point spectrum formed by interpenetrating Stark ladders with localized states (cf. ref. 34). Nevertheless, even if all bands are included, the density of states may still show some structure with resonance caused by the modulation that the periodic potential imposes on the electric field, i.e., the actual problem admits metastable states related these "Wannier-Stark resonances". Estimates for a few cases show that over a wide range of electric fields the interband transition rates are negligible, so that the persistence of the field induced localization may be expected.[46] For the case of sufficiently singular potentials, such as the δ-potentials in the Kronig-Penney model, it is believed that one can have actual bound states in the presence of an electric field.[30,33,45] The effect of disorder on Wannier-Stark states is another problem of current interest.[30,31,33] It has been shown[31] that the type of localization (79) is a generic property of any single orbital tight-binding model, irrespective of the form of the site energy ϵ_m. In the presence of disorder, the eigenvalue spectrum remains discrete, but is no longer uniformly spaced.[31,33] for strong disorder, due to the configuration average that must be performed, the Wannier-Stark resonances essentially disappear.[33] Furthermore, disorder is expected to affect the tunneling between the bands, which may qualitatively modify the properties of the Wannier-Stark resonances.[33]

As noted above, the Wannier-Stark states are localized over a length $2J/|eE|$ (cf.(79)). For sufficiently high electric field, they are localized in regions with a diameter less than the scattering length. In this case, scattering promotes transport between Wannier-Stark states (see, e.g., refs. 2,35,47). In the language of the density-matrix formalism described in Section 2.1, for sufficiently high electric fields, the tunneling contribution to the current, governed by the non-diagonal elements of the density matrix, describes hopping between the levels of the Wannier-Stark ladder, which leads to a negative differential conductivity at large electric fields (see refs. 2,35 and references therein).

Superlattice structures may provide the best opportunity to observe Wannier-Stark quantization[36,37] because the energy bands of superlattices tend to be very narrow due to their larger than usual lattice spacing, which may result in strong Wannier-Stark localization in fields of moderate strength (crystal: $a \sim 1 \text{Å}$, $2J \sim 1 eV$, $E \sim 10^6 V/cm$; superlattice: $a \sim 100 \text{Å}$, $2J \sim 0.01$ to $0.1 eV$, $E \sim 10^4 V/cm$; if a Wannier-Stark state is localized with a region with the diameter of the lattice constant a). According to[48,49], an ac electric field can produce localization whenever the magnitude and frequency of the field are in certain ratios to each other.

For recent theoretical approaches to Wannier-Stark localization, see refs. 47,50.

3.2 Anderson Localization and Delocalization in an Electric Field

In zero electric field it is believed that in a disordered system in dimensions d<3 all electronic states are localized, while in d>2 localized and extended states may occur, which are separated by a mobility edge. In particular, in one-dimensional disordered systems, all states are exponentially localized, the spectrum of the infinite system is pure point, i.e., discrete.

As already noted in Section 1., a finite electric field has a qualitatively dramatic effect on electronic states. In a one-dimensional system, all states are localized if the lattice potential is smooth, whereas for discontinuous potentials (Kronig-Penney model) in weak fields there is a mobility edge past which the states are extended.[21,22-24,26] In higher dimensions, finite electric fields cause delocalization.[26]

According to ref. 25, in a one-dimensional system of length L the effect of an electric field on localization is governed by the parameter $X=|eEL|/\epsilon$, which is the ratio between the electrostatic energy by an electron traversing the top the ramp potential produced by the electric field. For x<1, the kinetic energy gained by the electron from the field is small compared to its incoming energy. Therefore, the random potential is dominant. The results for the different potentials considered[25] (smooth, rectangular and δ-potentials) are qualitatively the same. The states are exponentially localized, depend on E, and lead to a linear correction in E to the resistance.

For x>1, the kinetic energy gained by the electron from the electric field is large compared to its incoming energy. Hence, the lattice potential becomes a small perturbation and the type of potential becomes important. In the smooth and rectangular potential cases for large L, the electronic states are extended even for small E. Unlike this, in the δ-potential case, for large L the electronic states are power-law localized, owing to the fact that the height of the δ-function potential is essentially infinite. No matter how much energy the electron gains from the field, it cannot overcome the potential profile. The finite-height potentials yield a nonlinear transmission coefficient as a function of E, while the power-law localized states in the singular potential give zero conductance in the thermodynamic limit.[25] Similar results have been obtained in[27,36].

With regard to transport in electron-phonon systems in the presence of a high electric field it is interesting to note that at finite temperatures, the kinetic energy gained by the electron from the field is eEL, with L replaced by the inelastic mean free path. After the electron travels a distance L it loses its energy to the phonon system, after that it gains again energy to the field, and so on.[25]

3.3 Influence of an Electric Field on Weak Localization

The study of the effect of a dc electric field on weak localization has been quite controversial (see, e.g., refs. 56, 57). On the other hand, an ac electric field, as any other potential that varies in time, destroys

the perfect phase coherence in systems with time reversal invariance, if
the potential energy changes appreciably in the time that an electron
takes to return to its starting point. This gives a characteristic cut-
off time and reduces weak localization (cf. refs. 56,58).

4. A NEW APPROACH TO HIGH-FIELD HOPPING

4.1 Preliminary Remarks

As discussed in Section 3., in a disordered system a finite electric
field can produce delocalization, while in an ordered or weakly disordered
system a high electric field can produce localization (Wannier-Stark
localization). With regard to a reformulation of the theory of non-ohmic
hopping transport, Wannier-Stark localization may be probably disregarded
if we restrict ourselves to electric fields of moderate strength and take
into account the fact that disorder and electron-phonon interaction may
suppress Wannier-Stark localization. However, field-induced delocalization is possibly important to non-ohmic transport in disordered systems.
The parameter $X=|eEL|/\epsilon$ (cf. Section 3.2.) may be expected to govern the
delocalization process (L is an appropriate inelastic scattering length).
Electrons in power-law localized states should move via hopping (cf. ref.
51) with hopping probabilities determined by the localized states whose
localization length increases with increasing field. The time-dependence
of the delocalization process is expected to be important to the transport
mechanism. Possibly, an electron leaves its state via hopping before the
corresponding state becomes delocalized due to the external electric
field.

Within the density-matrix formalism described in Section 2., field-
induced delocalization requires account of non-diagonal elements of the
density matrix. For a weak electric field and strong electron-phonon
coupling, the contribution of the non-diagonal elements to the current has
been studied in ref. 52. Here we do not follow this method, but we
orient ourselves by the paper[53] in which (for a weak electric field) the
self-consistent current-relaxation approach[54] is used for studying

phonon-induced delocalization. Unlike the density matrix formalism, the latter approach is able to describe the Anderson transition, which decided our choice.

4.2 Basic Formulae

We consider a tight-binding system described by (1), assume energetic disorder, and switch a dc electric field E adiabatically on at $t \to -\infty$. Then, with the aid of the von Neumann equation for the statistical operator, assuming the electric field to be directed along the x-axis. The x-component of the current can be written as (cf. ref. 55)

$$j = - \frac{\hbar \beta e^2}{2\Omega} E \sum_{mm'} (R_m - R_{m'})_x^2 \psi_{mm'}(z), \quad z = i\hbar\epsilon, \quad \epsilon \to 0, \tag{80}$$

where

$$\psi_{mm'}(z) = <n_m | \frac{i}{z-L} | \dot{n}_{m'}>. \tag{81}$$

Here $L=[H,...]$ is the Liouville superoperator, $\dot{n}_m = a_m^+ a_m$, $\dot{n}_m = (i/\hbar) L n_m$, and

$$<A|B> = \frac{1}{\beta} \int_0^\beta d\lambda <\text{Tr } (A^+ B(i\hbar\lambda)\rho_o)>_{\text{conf}}, \tag{82}$$

with

$$B(i\hbar\lambda) = e^{-H\lambda} B e^{H\lambda}, \tag{83}$$

and

$$\rho_o = e^{-\beta(H-\mu\hat{N})} / \text{Tr}(e^{-\beta(H-\mu\hat{N})}), \tag{84}$$

where H is given by H for E=0, and $<...>_{\text{conf}}$ denotes the configuration average. The expression (83) can be interpreted as a scalar product in the space of variables A and B. The quantity $\psi(z)$ in (81) is the "current-current" correlation function. We make the decomposition (cf. Ref. 53)

$$\dot{n}_m = \dot{n}_m^c + \dot{n}_m^h, \tag{85}$$

with

$$\dot{n}_m^c = \frac{i}{\hbar} \sum_{m'} \tilde{J}_{m'm} a_{m'}^+ a_m + h.c., \qquad (86)$$

$$\dot{n}_m^h = \frac{i}{\hbar} \sum_{m'} \tilde{J}_{m'm} \{\phi_{m'm} e^{S_T(m',m)} - 1\} a_m^+ a_m + h.c., \qquad (87)$$

where c stands for coherent and h for hopping. Then, up to the second order with respect to the resonance integral J, it holds that

$$\psi_{mm'}(z) \simeq \psi_{mm'}^c(z) + \psi_{mm'}^h(z), \qquad (88)$$

where

$$\psi_{mm'}(z) = \langle \dot{n}_m^j | \frac{i}{z-L} | \dot{n}_{m'}^j \rangle, \quad (j = c, h). \qquad (89)$$

We assume the electron-phonon interaction to be only weakly affected by the electric field and adopt the relation[53] between ψ^c and ψ^h valid for a weak electric field. To evaluate ψ^c, the Mori-Zwanzig reduction formalism is used, in combination with the mode-coupling approximation.[53,54] In this way a self-consistent scheme for computing ψ^c can be obtained. For a weak electric field, the formalism proposed here reduces to the approach[53] for describing phonon-induced delocalization, which in turn, in the absence of electron-phonon interaction, reduces to the self-consistent current-relaxation approach[54] for describing Anderson localization.

5. CONCLUDING REMARKS

In recent years, the interest in transport in high electric fields has been much stimulated by the research in the field of submicron systems and subpicosecond phenomena. As the dimensions of a system become smaller than the distance between inelastic scattering events, which destroys phase coherence, quantum-mechanical interference phenomena govern transport properties. This can lead to interesting non-ohmic effects such as effects due to hopping[59,60] or zener tunneling.[61,62] The theory of non-linear response is still an active field of research.[47,50,55,63-65]

REFERENCES

1. Mott, N.F., Phil. Mag. 24, 911 (1971).
2. Böttger, H. and Bryksin, V.V., "Hopping Conduction in Solids", Akademie-Verlag Berlin; VCH-Verlag, Weinheim, 1985.
3. Böttger, H. and Wegener, D., J. Non.-Cryst. Sol 97&98, 547 (1987).
4. Shklovskii, B.I., Fiz. Tekh. Poluprov. 6, 2335 (1972).
5. Hill, R.M., Phil. Mag. 24, 1307 (1971).
6. Apslay, N. and Hughes, H., Phil. Mag. 31, 1325 (1975).
7. Böttger, H. and Bryksin, V.V., phys. stat. sol. (b) 68, 285 (1975).
8. Shklovskii, B.I., Fiz. Tekh. Poluprov. 10, 1440 (1976).
9. Pollak, M. and Riess, I., J. Phys. C9, 2339 (1976).
10. Böttger, H. and Bryksin, V.V., phys. stat. sol (b) 96 216 (1976).
11. Böttger, H. and Bryksin, V.V., Phil. Mag. B42, 297 (1980).
12. van der Meer, M., Schuchardt. R., and Keiper, R., phys. stat. sol. (b) 110, 571 (1982).
13. Böttger, H. and Wegener, D., Phil. Mag. B50, 409 (1984).
14. Böttger, H. and Wegener, D., phys. stat. sol. (b) 121, 413.
15. Böttger, H., Szyler, P., and Wegener, D., phys. stat. sol (b) 128, K179 (1985).
16. Böttger, H., Szyler, P., and Wegener, D., phys. stat. sol (b) 133, K143 (1986).
17. Nguen van Lien and Shkovshii, B.I., Sol. State Commun. 38, 99 (1981).
18. Bryksin, V.V. and Yashin, G. Yu., Fiz. tverd. Tela 23, 3063 (1983).
19. Wegener, D. and Böttger, H., Phil. Mag. B57, 609 (1988).
20. Levin, E.I. and Shklovskii, B.I., Fiz. Tekh. Poluprov. 18, 856 (1984).
21. Prigodin, V.N., Zh. eksp. teor. Fiz. 79, 368 (1980).
22. Soukoulis, C.M., Jose, J.V., Economou, E.N., and Ping Sheng, Phys. Rev. Lett. 50, 764 (1983).
23. Bentosela, F., Carmona, R., Cuclos, P., Simon, B. Souillard, B., and Weber, R., Commun. Math. Phys. 88, 387 (1983).
24. Delyon, F., Simon, B.,d Souillard, B., Phys. Rev. Lett. 52, 2187 (1984).
25. Cota, E., Jose, J.V., and Azbel, M. Ya, Phys. Rev. B32, 6157.
26. Kirkpatrick, T.R., Phys. Rev. B33, 780 (1986).
27. Castello, D., Caro, A., and Lopez, A., Phys. Rev. B36, 3002 (1987).
28. Mato, G. and Caro, A., J. Phys. Cond. Mat. 1, 901 (1989).
29. Roy, C.L. and Mahapatra, P.K., Phys. Rev. B25, 1024 (1982).
30. Bentosela, F., Grecchi, V., and Zirony, F., Phys. Rev. B31, 6909 (1985).
31. Luban, M., and Luscombe, J.H., Phys. Rev. B34, 3674 (1985).
32. Jose, J.V., Monsivais, G., and Flore, J., Phys. Rev. B31, 6906 (1985).
33. Cota, E., Jose, J.V., and Monsivais, G., Phys. Rev. B35, 8229 (1987).
34. Emin, D. and Hart, C.F., Phys. Rev. B36, 7353 (1987).
35. Emin, D. and Hart, C.F., Phys. Rev. B36, 2530 (1987).
36. Leo, J. and Movagar, B., Phys. Rev. B38, 8061 (1988).
37. Voisin, P., Bleuse, J., Bouche, C., Gaillard, S., Alibert, C., and Regreny, A., Phys. Rev. Lett. 61, 1639 (1988).
38. Böttger, H. and Bryksin, V.V., phys. stat. sol (b) 68, 285 (1975).
39. Emin, D., Phys. Rev. Lett. 32, 303 (1974).

40. Emin, D. and Hart, C.F., Phys. Rev. B32, 6503 (1985).
41. Gochanour, C.R., Andersen, H.C., and Fayer, M.D., J. Chem. Phys. 70, 4254 (1979).
42. Wegener, D. and Böttger, H., phys. stat. sol (b) 138, 83 (1986).
43. Wannier, G.H., Phys. Rev. 117, 432 (1960).
44. Avron, J.E., Zak, J., Grossmann, A., and Gunther, L. J. Math. Phys. 18, 918 (1977).
45. Bentosela, F., Caliceti, E., Grecchi, V., Maioli, M., and Sachetti, A., J. Phys. A21, 3321 (1988).
46. Krieger, J.B. and Iafrate, G.J., Phys. Rev. B33, 5494 (1986).
47. Davis, J.H. and Wilkins, J.W., Phys. Rev. B38, 1667 (1988).
48. Dunlap, D.H. and Kenkre, V.M., Phys. Rev. B34, 3625 (1986).
49. Dunlap, D.H. and Kenkre, V.M., Phys. Rev. A127, 438 (1988).
50. Krieger, J.B., and Iafrate, G.J., Phys. Rev. B35, 9644 (1987).
51. Kaveh, M., Phil. Mag. B52, 521 (1985).
52. Bryksin, V.V., Fiz. tverd. Tela 28, 1731, 2981 (1986).
53. Müller, H. and Thomas, P., J. Phys. C 17, 5337 (1984).
54. Götze, W., Solid State Commun. 27, 1393 (1978); J. PHys. C12, 1279 (1979); Phil. Mag. B43, 219 (1981).
55. Van Vliet, C.M., J. Stat Phys. 53, 49 (1988).
56. Lee, P.A. and Ramakrishnan, T.V., Rev. Mod. Phys., 57, 287 (1985).
57. Hu, G.Y. and O'Connel, R.F., Physica A153, 114 (1988).
58. Chakravarty, S. and Schmid, a., Phys. Rep. 140, 193 (1986).
59. Webb, R.A., Hartstein, A., Wainer, J.J. and Fowler, A.B., Phys. Rev. Lett. 54, 1577 (1985).
60. Timps, G., Fowler, A.B., Hartstein, A. and Butcher, P.N., Phys. Rev. B34, 8771 (1986).
61. Gefen, Y. and Thouless, D.J., Phys. Rev. Lett. 59, 1752 (1987).
62. Blatter, G. and Browne, D.A., Phys. Rev. B37, 3856 (1988).
63. Argyres, P.N., Phys. Rev. B39, 2982 (1989)

Chapter 5

ELECTRON-PHONON INTERACTIONS AND POLARONS

LARGE BIPOLARONS: FORMATION, MOTION AND SUPERCONDUCTIVITY

David Emin
Sandia National Laboratories
Albuquerque, New Mexico 87185-5800

ABSTRACT

Ferroelectric materials, characterized by very large ratios of the static to optical dielectric constants, are unique in fulfilling a requirement for the formation of a multidimensional large bipolaron. Unlike the very compact "small" bipolarons that are found in various transition-metal oxides, "large" bipolarons are not so massive that they readily localize. Thus, a large bipolaron is a mobile charged boson. As such, large bipolarons may manifest superconductivity analogous to the superfluidity of mobile neutral bosons. Distinctive electronic and magnetic properties of large bipolarons in the normal-state and superconducting states are described and compared to observations in high-temperature superconductors.

INTRODUCTION

Superfluidity and superconductivity are both associated with resistanceless flow at sufficiently low temperatures. The similarity of these two phenomena led (circa 1950-1960) to consideration of a common quantum-mechanical origin of these phenomena. In particular, superfluidity results from the Bose-Einstein condensation of mobile neutral bosons and superconductivity can result from the Bose-Einstein condensation of mobile charged bosons.[1] This form of superconductivity is now termed "bipolaronic superconductivity," since the charged bosons are presumed to be bipolarons.

A bipolaron forms when two carriers are bound within a common potential well produced by shifts of the equilibrium positions of a solid's atoms from their carrier-free locations. The term bipolaron refers to the unit comprising the "self-trapped" pair of carriers and the atomic displacement pattern associated with their self-trapping within a common potential well. Bipolaron formation can occur because the atomic displacements about a pair of carriers are greater than those about a single carrier. As a result, the potential well induced by atomic displacements about a pair of carriers is deeper than that about a single carrier. Pairing of the carriers is energetically favorable if the lowering of carriers' energy resulting from their occupying a common potential well exceeds their Coulomb repulsion energy. Hence, bipolaron formation is associated with a "strong" electron-lattice interaction.

The pairing of two carriers to form a bipolaron is analogous to the pairing of electrons within an atomic orbital or the pairing of electrons to form the covalent bond of a hydrogen molecule. In particular, in all these instances pairing occurs because the lowering of the electronic energy arising from the electrons being closer to (positive) nuclear charge exceeds the increased Coulomb repulsion energy between the two electrons due to their sharing a common orbital. Thus, the pairing of electrons added to Ti^{4+} cations in Ti_4O_7 to form a small bipolaron (a Ti^{3+}-Ti^{3+} pair) is analogous to the formation of a covalent bond. Similarly, holes can be envisioned to pair among anions (e.g., O^{2-}).

It has only been through collateral electronic-transport and magnetic measurements that it has been demonstrated that charge carriers manifesting hopping transport in a variety of systems are localized singlet pairs, small bipolarons, rather than the commonly presumed spin-1/2 carriers, small polarons.[2-6] Some of these systems are the transition-metal oxides: Ti_4O_7, V_2O_5, WO_3 and $BaTiO_3$.[3-6] Furthermore, superconductivity has been observed in solids that have a familial relationship to those in which small bipolarons are reported.[7-10] For example, suitably doped $SrTiO_3$ is a superconductor[7] while the charge carriers in $BaTiO_3$ are reported to be immobile (small) bipolarons.[6] Furthermore, the estimated carrier density in the heavily "doped" $SrTiO_3$ superconductor, $\approx 10^{20}$ cm^{-3}, is far below that of conventional superconductors. These observations raise the possibility of bipolaronic superconductivity.[11]

The search for bipolaronic superconductivity is naturally directed toward materials that have been associated with an especially strong electron-lattice interaction. Ionic solids, transition-metal oxides and perovskite ferroelectrics are generally associated with polarons and bipolarons. As such, these materials are candidates for bipolaronic superconductivity. Indeed, even at their discovery, the CuO_2-based superconductors were considered possible bipolaronic superconductors.[11]

Bipolaronic superconductivity requires two exceptional circumstances. First, at least some charge carriers must form bipolarons. Second, these bipolarons must be mobile (move coherently). However, beyond quasi-one-dimensional electronic systems, clear evidence of bipolarons has been limited to cases in which the polaronic carrier is "small." Polarons and bipolarons are termed "small" when their electronic states are severely localized: e.g., confined to a single site or single bond. Small polarons and small bipolarons are generally so massive that they localize rather than move coherently.[12] Because a small polaron is associated with a localized spin while a singlet small bipolaron is not, magnetic measurements may distinguish between small polarons and small bipolarons. However, since small bipolarons are immobile, they are not suitable carriers for bipolaronic superconductivity. By contrast, less compact "large" polarons have a much smaller effective mass and are mobile. In analogy with large polarons, large bipolarons are expected to be mobile. These observations lead one to ask if and when large bipolarons can form in multi-dimensional electronic systems.

In this paper, results of studies of the formation, motion and superconductivity of large bipolarons are summarized.[13-15] A significant finding is that the formation of a large two- or three-dimensional bipolaron in an insulator requires an unusually large ratio of the static dielectric constant, ϵ_0, to the high-frequency dielectric constant, ϵ_∞: $\epsilon_0 \gg 2\epsilon_\infty$. Such a circumstance corresponds to having an exceptionally strong long-range component of the electron-lattice interaction. This situation occurs in ferroelectric-type ionic solids. Furthermore, the formation of an energetically stable large bipolaron requires the electronic bandwidth to lie within a limited domain. If it is too small, small bipolarons rather than large bipolarons will form. If the electronic bandwidth is too large, the large bipolaron will be unstable with respect to dissociating into two large polarons. The regime of stability of a large bipolaron with respect to dissociation increases as the ratio $\epsilon_0/2\epsilon_\infty$ increases and as the electronic anisotropy increases. Thus, these studies indicate that the formation of two- and three-dimensional large bipolarons is limited to materials with $\epsilon_0/\epsilon_\infty \gg 1$, "incipient ferroelectrics." Furthermore, electronically anisotropic ferroelectrics, such as the CuO_2-based materials, are prime candidates for the formation of large bipolarons.

Since the formation of large bipolarons is possible, it is meaningful to consider their properties. In the normal state, a large bipolaron is characterized by its large effective mass and by its weak scattering by phonons (long scattering time). In addition, a self-trapped carrier is characterized by photon absorption above a low-energy threshold (a few tenths of an electron volt). This absorption arises from exciting the bound electronic carriers of the bipolaron.

Interacting large bipolarons give rise to type II superconductivity. However, this superconductivity only occurs within a limited carrier concentration range. For too low a carrier density, large bipolarons do not condense into a superconducting state. For too high a carrier density, the atomic displacement patterns of the large bipolarons interfere causing the bipolarons to be destabilized with respect to forming free (nonpolaronic) carriers.

The superconducting condensate is characterized by an energy gap in its excitation spectrum and by the occurrence of the Meissner effect. The formation of a Bose condensate suppresses density fluctuations of the bipolarons. This change alters the solid's electronic states. As such, the positron annihilation rate and Van Vleck paramagnetism associated with a large bipolaron will be altered as the condensate forms. This change of the magnetism should produce a monotonic reduction of the NMR frequency and relaxation rate as temperature is lowered below T_c.

The superconducting transition temperature, T_c, is the Bose condensation temperature. Approximating T_c as the condensation temperature for a noninteracting gas of large bipolarons results in dependences of T_c on the size of the large bipolaron and on the masses of atoms that are displaced in forming the large bipolaron. These dependences are used to discuss how T_c varies between different lattice structures and how T_c is affected by isotopic substitutions.

This discussion of the properties of large bipolarons is only qualitative. Nonetheless, at this level the normal-state and superconducting properties of a large-bipolaronic superconductor appear consistent with observations of the high-temperature superconductors.

FORMATION OF A LARGE BIPOLARON

A bipolaron is a boson quasiparticle composed of two carriers that are self-trapped within a common potential well. Self-trapping occurs when carriers are bound within a potential well produced by displacing the equilibrium positions of atoms in condensed matter from their carrier-free locations to locations that are consistent with the presence of the charge carrier. Self-trapping is a nonlinear phenomenon in which the potential well experienced by the carriers depends upon the carriers' wavefunction. The nonlinearity of the self-trapping in solids admits two distinct types of self-trapped states. These two classes of solutions correspond to "small" and "large" polarons or bipolarons.[12,16]

The adjective "small" refers to the polaron's or bipolaron's electronic state being localized to the smallest unit compatible with the solid. In particular, the electronic state of a "small" polaron or bipolaron may be confined to a single atom, single bond or single molecule depending on the physical situation. Because the motion of a small polaron or bipolaron requires the tunneling of atoms through distances that are much larger than the atoms' zero-point displacements, these quasiparticles generally have effective masses that are huge multiples (e.g., $> 10^4$) of the bare electronic mass.[17] The combination of the compact electronic state and the large effective mass facilitate the localization of small polarons and bipolarons. Thus, small polarons and bipolarons are generally viewed as localized. Their transport is then described in terms of phonon-assisted hopping motion.[12]

By contrast, a large polaron or bipolaron is associated with a self-trapped state that extends over multiple sites. By itself, expanding the electronic wavefunction rapidly reduces the magnitude of the carrier-induced atomic strains. Then the motion of a large polaron only requires the tunneling of atoms through distances that are much smaller than the atoms' zero-point displacements. This effect leads to a drastic reduction of the mass enhancement of a large polaron or bipolaron from that of a small polaron or bipolaron. For example, with atomic displacements only in the immediate vicinity of the carrier, the mass enhancement for a small polaron at zero temperature may be written as $\exp(\delta^2/2z^2)$, where δ is the magnitude of the carrier-induced atomic displacement of a bond and z is the amplitude of the bond's zero-point atomic displacement.[17] Extending the electronic wavefunction equally over N bonds reduces the carrier-induced displacement of each bond by the factor N ($\delta \to \delta/N$) while increasing the number of stretched bonds from one to N. Taken together, these effects reduce the mass enhancement from $\exp(\delta^2/2z^2)$ to $\exp(\delta^2/2Nz^2)$. Thus, a large polaron with relatively strong electron-lattice coupling (e.g., electrons in the alkali halides) has only a modest mass enhancement, (e.g., $\lesssim 10$).

With its large size and moderate mass enhancement, a large polaron moves coherently rather than localizing in the manner of a small polaron. Nonetheless, the mass of a large polaron is significantly greater than the bare mass of a free carrier. Furthermore, as will be discussed subsequently, the large size and large mass of a large polaron renders its scattering by phonons relatively ineffective. Thus, a large polaron has a scattering time that is longer than that of a simple electronic carrier.[18] A large bipolaron is expected to be qualitatively similar to a large polaron albeit with even a larger mass and a longer scattering time.

The propensity of a small bipolaron to localize is incompatible with small bipolarons being the charge carriers for bipolaronic superconductivity. However, the coherent type transport that is presumed to characterize large bipolarons is consistent with such carriers manifesting bipolaronic superconductivity.

Therefore, the conditions under which a large bipolaron can form must be investigated. The formation of a large or small polaron is dependent upon the range of the electron-lattice interaction and on the dimensionality of the electronic system.[12,16,19] A similar situation prevails for the formation of a large bipolaron.[14]

In a covalent or molecular solid, the energy of a carrier only depends upon the positions of the atoms that have direct contact with the electronic wavefunction. With such a short-range electron-lattice interaction, it has been shown, within the adiabatic approximation, that a carrier either 1) does not self-trap or 2) self-traps to form a small polaron.[12,16,19] Similarly, with a purely short-range electron-lattice interaction, the only possible type of bipolaron is a small bipolaron.[14]

However, in an ionic solid a carrier's energy also depends upon the positions of the solid's ions far from the carrier via the long-range Coulomb interaction. This long-range component of the electron-lattice interaction is termed the Frohlich interaction.[20] The strength of the Frohlich interaction depends on $\beta = (\epsilon_0 - \epsilon_\infty)/\epsilon_0 \epsilon_\infty$, where ϵ_0 and ϵ_∞ are the static and optical dielectric constants, respectively. This parameter is a measure of the extent to which atoms of a solid are displaced in response to the application of a dc electric field. For covalent materials [$\epsilon_0 \approx \epsilon_\infty \gg 1$], the electronic susceptibility is dominated by the electronic polarization: $\beta \ll 1/\epsilon_\infty$. For common ionic solids, such as alkali halides [$\epsilon_0 \approx 2\epsilon_\infty$], β is very much larger: $\beta \approx 1/2\epsilon_\infty$ with $\epsilon_\infty \approx 2\text{-}4$. For ionic solids that are "ferroelectrics" [$\epsilon_0 \gg \epsilon_\infty$], β is still larger: $\beta \approx 1/\epsilon_\infty$ with $\epsilon_\infty \approx 2\text{-}4$.

It has been shown that with the Frohlich interaction as the sole component of the electron-lattice interaction, polarons are generally large polarons.[12,16] Furthermore, within the Frohlich model, a finite-radius bound state with two carriers within a common potential well can form if $\epsilon_0 > 2\epsilon_\infty$. However, within the Frohlich model, a large bipolaron is unstable with respect to forming two separated large polarons.

To understand the origin of these results, the energy of a Frohlich polaron associated with a self-trapped carrier of (dimensionless) radius R, $E_{pol}(R)$, is written as the sum of two terms:[14]

$$E_{pol}(R) = t/R^2 - v_c(\epsilon_0 - \epsilon_\infty)/2\epsilon_0\epsilon_\infty R, \tag{1}$$

where t is the electronic bandwidth parameter of a carrier and v_c is a constant associated with the Coulomb interaction. The first contribution to $E_{pol}(R)$ is the energy gain associated with the spreading of the self-trapped carrier. The second contribution to $E_{pol}(R)$ is the lowering of the polaron's potential energy that is produced by altering the atomic equilibrium positions. The proportionality of this term to $1/R$ characterizes the Coulomb-like self-trapping potential well produced by the Frohlich interaction. An explicit expression for R may be found by noting that a self-trapped carrier assumes the radius that minimizes the system's energy: $\partial E_{pol}(R)/\partial R = 0$. This polaron state is termed large because its minimum always occurs at finite R (at $R > 0$). The energy of this large polaron is $-[v_c(\epsilon_0 - \epsilon_\infty)/\epsilon_0\epsilon_\infty]^2/16t$. This situation differs from that of the small polaron within the continuum model for which the polaron minimum always occurs at $R = 0$.

The energy of a singlet bipolaron includes the energy of two equivalent carriers within a common potential well as well as the net Coulombic repulsion energy of two carriers:[14]

$$\begin{aligned} E_{bi}(R) &= 2t/R^2 - v_c(\epsilon_0 - \epsilon_\infty)/\epsilon_0\epsilon_\infty R + v_c/\epsilon_0 R \\ &= 2t/R^2 - v_c(\epsilon_0 - 2\epsilon_\infty)/\epsilon_0\epsilon_\infty R. \end{aligned} \tag{2}$$

The energy of this bipolaron is $-[v_c(\epsilon_0 - 2\epsilon_\infty)/\epsilon_0\epsilon_\infty]^2/8t$. This energy always exceeds the energy of two widely separated large polarons. Thus, this scaling argument indicates that the Frohlich model by itself does not permit the formation of an energetically stable large bipolaron.

However, in a real ionic solid the electron-lattice interaction contains a short-range component in addition to the Frohlich component. In this situation, a large bipolaron can be stable. To illustrate this circumstance, consider the energy functions for a polaron and a bipolaron in this expanded model:[14]

$$E_{pol}(R) = t/R^2 - v_{s-\ell}/2R^2 - v_s/2R^d - v_c(\epsilon_0 - \epsilon_\infty)/2\epsilon_0\epsilon_\infty R, \qquad (3)$$

and

$$E_{bi}(R) = 2t/R^2 - 2v_{s-\ell}/R^2 - 2v_s/R^d - v_c(\epsilon_0 - 2\epsilon_\infty)/\epsilon_0\epsilon_\infty R, \qquad (4)$$

where v_s and $v_{s-\ell}$ are constants and d is the dimensionality of the electronic system. The constant v_s is related to the short-range component of the electron-lattice interaction and $v_{s-\ell}$ ($\propto \sqrt{\beta}$) arises from a cross term associated with the coexistence of long- and short-range components of the electron-lattice interaction.

A large bipolaron only exists when there is a minimum of $E_{bi}(R)$ at finite R. In particular, a large bipolaron can only exist if the coefficient of the 1/R term in $E_{bi}(R)$ is negative: i.e., $\epsilon_0 > 2\epsilon_\infty$. In addition, for a two-dimensional electronic system, d = 2, a large bipolaron minimum can only exist if the coefficient of the $1/R^2$ term in $E_{bi}(R)$ is positive: $t/(v_{s-\ell} + v_s) > 1$. For a three-dimensional electronic system the situation is more complex, but the condition is similar.[14] In other words, the formation of a large bipolaron requires a sufficiently wide electronic band and $\epsilon_0 > 2\epsilon_\infty$.

Detailed analyses of these energy functions show that the energetic stability of a large bipolaron with respect to dissociation into two large polarons requires[14]

$$1 + (\epsilon_0 - 2\epsilon_\infty)^2/2\epsilon_\infty(2\epsilon_0 - 3\epsilon_\infty) > t/(v_{s-\ell} + v_s) \text{ for } d = 2 \quad (5)$$

and

$$1 + (\epsilon_0 - 2\epsilon_\infty)^2/2\epsilon_\infty(2\epsilon_0 - 3\epsilon_\infty) > t/v_{s-\ell} \text{ for } d = 3. \quad (6)$$

In other words, the stability of a large bipolaron requires that the electronic bandwidth, t, be less than a maximum value. The value of this maximum increases as the ratio of $\epsilon_0/2\epsilon_\infty$ increases. Furthermore, the value of this maximum is greater for a two-dimensional electronic system than for a three-dimensional electronic system. Thus, <u>electronically anisotropic ferroelectric systems are prime candidates for the formation of large bipolarons</u>.

The physics of the formation of a large bipolaron can be summarized simply. Within the Frohlich model, a large polaron is formed as the long-range portion of the electron-lattice interaction produces a r^{-1}-type potential well for a solitary carrier. With the addition of a second carrier to form a singlet bipolaron two competing effects occur. The carriers' potential energy is increased by their Coulomb repulsion. In addition, the atomic displacements are increased so as to deepen the r^{-1}-type potential well. This latter effect just reduces the effective Coulomb repulsion between the carriers. Namely, as a result of the long-range electron-lattice interaction, the dielectric constant that appears in the carriers' Coulombic repulsion is reduced by the static dielectric constant rather than the optical dielectric constant.[14] Nonetheless, in our treatment of the Frohlich model, the residual repulsive interaction of the carriers always destablizes a large bipolaron relative to two separated large polarons.[14] However, the inclusion of the short-range portion of the electron-lattice interaction contributes an additional lowering of the bipolaron's energy that can overwhelm the residual Coulomb interaction and stabilize the large bipolaron. As shown by Eq. (4), the large bipolaron may be viewed as an expanded helium atom: two carriers sharing a hydrogenic state.

So far only the formation of a singlet large bipolaron has been considered. This approach is justified by the observation that a triplet bipolaron is energetically less favorable than a singlet because of the electronic promotion energy associated with forming a triplet state.[15] Similarly, the electronic promotion energy associated with binding three or more carriers within a single polaronic state renders such entities (tripolarons, quadrapolarons, etc.) unstable with respect to forming large bipolarons and polarons.[15] Thus, within the generalized Frohlich model, only large polarons and bipolarons are envisioned.

The discussion to this point has described the conditions for two carriers in an insulator forming a stable large bipolaron. With a sufficiently high density of large bipolarons, the effects of their mutual interactions must also be considered. In particular, with a rising density of large bipolarons, the Coulombic forces of different bipolarons compete to displace the same atoms. This effect tends to reduce the effectiveness of the long range portion of the electron-lattice interaction in self-trapping. As a result, with an increasing density of large bipolarons, the energy of a large bipolaron rises. Ultimately, at a sufficiently large density, where a bipolaron's diameter exceeds the interbipolaron separation, large bipolarons cannot form. Since the radius of a large polaron exceeds that of an energetically stable large bipolaron, large polarons will also be destabilized at this density. Thus, at large carrier densities, bosonlike bipolarons will transform directly to nonpolaronic fermions.

In novel superconductors, such as perovskite-based materials, the carrier density is much lower than that of metals. For example, if each Sr "dopant" to La_2CuO_4 produces a hole, $La_{1.84}Sr_{.16}CuO_4$ corresponds to a holelike bipolaron being centered at only 4% [= 16%/(2×2)] of the oxygen sites in the CuO_2 planes. In addition, the ratio of the static to optical dielectric contants in the insulating parents of these superconductors is especially large, $\epsilon_0/\epsilon_\infty \gg 2$. The static dielectric constant also rises with "doping" of these materials.[21,22] Distinctively, the increase of ϵ_0 from ≈ 20 to ≈ 300

in passing from $YBa_2Cu_3O_6$ to $YBa_2Cu_3O_7$ is ascribed primarily to additional displaceable bound charge centers rather than to the Drude contribution of the charge carriers.[22] Thus, the low carrier densities and large static dielectric constants of the novel superconductors are consistent with the carriers being bipolarons.

NORMAL STATE PROPERTIES

The two charge carriers of the large bipolaron are bound together within a potential well produced by displacements of the solid's atoms from their carrier-free equilibrium positions. Therefore, the motion of a bipolaron requires motion of the solid's atoms.

If the binding energy of the bipolaron and the electronic bandwidth parameter in a particular direction are sufficiently large (> a phonon energy), bipolaron motion in that direction is adiabatic.[14,18] Within the adiabatic approach, the electronic charge carriers are treated as instantaneously adjusting to the motion of the solid's atoms. The effective mass for adiabatic motion of the bipolaron is then the atomic mass associated with the motion of the electronic carriers.[14,18] For example, in the simplest case, a one-dimensional solid, the bipolaron's effective mass tensor reduces to a scalar:[18]

$$m = \Sigma_i M_i (\Delta d_i/\Delta x)^2, \qquad (7)$$

where M_i is the mass of the i-th atom and Δd_i is the shift of the equilibrium position of the i-th atom when the centroid of the bipolaron moves the distance Δx.

The bipolaron's effective mass is dominated by atomic displacements close to the bipolaron. These displacements are primarily determined by the short-range component of the electron-lattice interaction.[14] Neglecting the contribution to the effective mass arising from the long-range component of the electron-lattice interaction, the effective mass is inversely proportional to the volume of the bipolaron.[14] This result may be understood in the one-dimensional example by noting that

$\Delta d_i/\Delta x$ is inversely proportional to the number of displaced atoms of the bipolaron, N_{bp}, while the summation has comparable contributions from N_{bp} atoms. As a result, $m \propto 1/N_{bp}$. Thus, the mass of the bipolaron is inversely proportional to its volume and proportional to the mass of the atoms involved in the polaronic displacements.

The morphology of a large bipolaron refects the electronic anisotropy of the solid. For example, for a quasi-two-dimensional solid with the electronic transfer energies largest for transfer within a two-dimensional sheet, the bipolaron will have a disklike morphology. However, the mass of the bipolaron need not reflect this electronic anisotropy. For instance, if the electronic transfer energies are sufficiently large in all three directions (> the Debye energy), the adiabatic approach is justified for motion in all directions. Then, for an elastically isotropic medium, the effective mass (arising from the short-range component of the electron-lattice interaction) only depends on the bipolaron's volume.[14] That is, with elastic isotropy, the adiabatic mass of a bipolaron is isotropic. Thus, electronic anisotropy need not, by itself, produce an anisotropic effective mass (with only the short-range component of the electron-lattice interaction). Of course, the mass becomes anisotropic in the extreme limit that an electronic transfer energy is so small as to invalidate the adiabatic approach. Then the mass associated with motion in a direction with an especially small electronic transfer energy is especially large.

With a significant density of large bipolarons, interference between the atomic displacement patterns of different bipolarons will become significant. As a result, the magnitude of the atomic displacements that on average characterizes a large bipolaron will decrease as the density of bipolarons increases. This suppression of the atomic displacements about a self-trapped carrier will reduce the average effective mass of a large bipolaron. Ultimately, at a sufficiently large carrier density, the competition between different carriers to displace a given atom is so severe that self-trapping is precluded.

The motion of a large bipolaron parallels that of a large polaron.[18] In particular, a large bipolaron is a heavy massed quasiparticle that moves through a solid with a velocity below the sound velocity. As with other bound electrons (such as shallow donors), a large bipolaron primarily interacts with phonons whose wavelengths are at least double the bipolaron's diameter. Such long wavelength phonons have momenta that are very small compared with the average (thermal) momentum of a large bipolaron. As a result, a large bipolaron is only weakly scattered by its interaction with such a phonon. Thus, a bipolaron is characterized by scattering times that are much longer than those characterizing free-carrier scattering. In fact, the scattering time is proportional to the mass of the bipolaron. It is this feature that accounts for large polarons having moderate mobilities despite their large effective masses.[18] In addition, the scattering rate of a very massive bipolaron by "light" phonons is primarily dependent on the density of such phonons. The density of these long wavelength phonons is proportional to the temperature provided the temperature exceeds that corresponding to the energy of the long wavelength phonons. Thus, at such temperatures, the scattering rate is proportional to the temperature.[18] In this regime, the bipolaron's mobility in the j-th direction is, apart from a numerical constant, just $(2|e|)L_j s_j/k_B T$, where L_j is the bipolaron's length and s_j is sound velocity in the j-th direction. Thus, with a constant density of carriers, the coherent motion of a large bipolaron is characterized by a resistivity that is proportional to temperature.

As described above, unlike the adiabatic mass, the adiabatic mobility of an anisotropic bipolaron is anisotropic. Scattering is generally enhanced in the direction in which the spatial extent of the large bipolaron is restricted. Thus, the scattering of a disklike bipolaron will be strongest in the direction perpendicular to the disk. If the scattering in the direction of small spatial extent is strong enough, small-polaron-like hopping motion[12,17] may even result.

Self-trapped carriers are generally characterized by the presence of a broad optical absorption band. This absorption corresponds to excitation of an electronic carrier from (or within) its self-trapping potential well. Designating the binding energy of a large polaron as E_p [= $-E_{pol}(R_{min})$, where R_{min} is the value of R that minimizes $E_{pol}(R)$], the Franck-Condon principle ensures that optically freeing a carrier from its self-trapping well requires a photon energy well in excess of E_p. In particular, the energy of a large polaron is the sum of its confinement energy, E_p, the electronic potential energy, $-4E_p$, and the strain energy associated with the atomic deformation, $2E_p$. In addressing the formation of a large polaron, the electronic potential energy and the strain energy were combined.[14] However, optical excitation only involves the electronic contributions to the polaronic binding energy. In particular, the threshold energy required for photoionization is the negative of the sum of the electronic contributions to the total energy, $-(E_p - 4E_p) = 3E_p$. Thus, the photon energy required to liberate a carrier from a large polaron greatly exceeds the (thermal) binding energy of the carrier, E_p. The optical absorption threshold for the large bipolaron we are considering is presumed to be at least comparable to that of a large polaron, $\approx 3E_p$. This energy greatly exceeds the (thermal) binding energy of a large bipolaron relative to two large polarons, $\ll E_p$.[15]

Summarizing, large bipolarons have large effective masses, long scattering times and a carrier-related sub-gap absorption. With the scattering being dominated by low-energy phonons, the scattering rate for itinerant motion is proportional to the temperature.

Reflectivity studies in the normal state of CuO_2-based superconductors indicate very large values of the effective mass per unit charge ($|e|$) and very long scattering times that are inversely proportional to temperature.[22] The resistivity of the CuO_2-based superconductors is also generally observed to be proportional to temperature. Although motion perpendicular to the CuO_2 layers often manifests a hopping-type conductivity, very clean samples of $YBa_2Cu_3O_7$ are reported to have a resistivity that is also proportional to temperature in this direction.[23] A sub-gap absorption has been reported for photo-induced carriers in the insulating parents of the superconductors[24,25] and in the normal state of the superconductors themselves.[22,26] These results are consistent with expectations for bipolaronic carriers.

SUPERCONDUCTING PROPERTIES

Consideration of superconductivity necessitates studying the condensate of a system of interacting particles. At its simplest, mobile bipolarons in a solid are modeled as a collection of mobile bosons of charge q ($|q|$ = 2e) that interact via their Coulomb repulsion. For overall charge neutrality, the bosons are presumed to be in a medium with a uniform charge density of opposite sign and equal net magnitude to that of the bosons. Each charged boson is assigned a mass equal to a bipolaron's mass, m. The Coulomb interaction between bosons is reduced by the static dielectric constant of the carrier-free medium, ϵ_0.

The ground state of an analogous system has been investigated previously.[27] Two distinct situations emerge. At low carrier densities, $n^{1/3} < (2mq^2/\epsilon_0 \hbar^2)$, the Coulomb interactions dominate the kinetic energy causing the bosons to condense into a Wigner crystal. In the complementary regime, $n^{1/3} > (2mq^2/\epsilon_0 \hbar^2)$, a Bose liquid is formed. The energy of an excitation of momentum p of this Bose fluid, calculated to lowest order in the Coulomb interactions, is[27]

$$E(p) = [(\hbar\omega_p)^2 + (p^2/2m)^2]^{1/2}, \qquad (8)$$

where $\omega_p = 4\pi q^2 n_0/\epsilon_0 m$ and n_0 is the density of bosons occupying the ground state. The excitation spectrum is separated from the ground state by the energy gap $\hbar\omega_p$. Due to the presence of this gap in the excitation spectrum, the Landau condition for the condensate possessing resistanceless flow is satisfied for this Bose fluid. That is, the minimum value of $E(p)/p$ is nonzero.[28] Thus, this system of interacting bosons will manifest resistanceless flow if the carrier density is sufficiently large so that the ground state forms a Bose fluid.

Expressing the excitation spectrum as

$$E(p) = [(2m\hbar\omega_p)^2 + p^4]^{1/2}/2m, \qquad (9)$$

it is noted that excitations only appear single-particle-like for wavelengths smaller than the coherence length $\xi = (\hbar/2m\omega_p)^{1/2}$. Noting that the London penetration depth is $\lambda_L = c/\omega_p$, where c is the speed of light, the ratio of the coherence length to the penetration depth is $\xi/\lambda_L = (\hbar\omega_p/2mc^2)^{1/2}$. The low carrier densities ($\simeq 10^{21}$ cm^{-3}), large effective masses ($> 10\ m_e$, where m_e is a free electron's mass) and large static dielectric constants ($\epsilon_0 > 15$) all contribute to make $\xi/\lambda_L \ll 1$. In other words, bipolaronic superconductors are expected to be type II.

The superconductivity of the charged Bose system is similar in many respects to conventional superconductivity. In particular, the superconductivity of the charged Bose system is described in terms of a two-fluid picture. The resistanceless superconducting fluid is associated with occupancy of the ground state. The normal fluid is associated with occupancy of excited states. As the temperature is raised from absolute zero, occupancy of the ground state of the boson system is decreased. The density of the superconducting fluid then falls as the temperature is raised. Furthermore, since the energy gap is proportional to the square root of the superfluid density, increasing the temperature also shrinks the energy gap.[15,29] Condensed charged bosons also display the Meissner effect.[1,15,29]

A distinctive feature of bipolaronic superconductivity is the limited range of carrier concentrations over which it occurs. For too low a boson density the boson system will form a Wigner solid rather than a superconducting fluid.[27] For too high a carrier concentration the large bipolarons that constitute the charged bosons required for superconductivity will not form.[13,14]

With the bipolaron density being within the regime in which bipolaronic superconductivity can occur, the superconducting transition temperature is the Bose condensation temperature, T_c. For a noninteracting gas of bipolarons of isotropic mass, $k_B T_c = 3.3\, \hbar^2 n^{2/3}/m$, where k_B is the Boltzmann constant.[28] If n is taken to be 10^{21} cm^{-3} and m = 20 m_e, where m_e is the free-electron mass, then T_c is 120 K. Furthermore, existing studies indicate that repulsive interactions between bosons further raise the Bose condensation temperature.[30] Thus, it is plausible that bipolaronic superconductivity can account for a relatively high superconducting transition temperature. Furthermore, the low carrier densities, large effective masses and large dielectric constants of the superconductors, combine to reduce the energy gap, $\hbar\omega_p$, to a modest value. For example, with $n_0 = 10^{21}$ cm^{-3}, m = 20 m_e and $\epsilon_0 = 20$, one has $\hbar\omega_p \approx 0.1$ eV. Thus, the energy gap may be expected to be a modest multiple of $k_B T_c$.

The properties of the solid affect T_c through the bipolaron's effective mass. In the preceding discussion of the bipolaron's effective mass it was presumed that the short-range component of the electron-lattice interaction provides the dominant contribution to the bipolaron's effective mass. Then the bipolaron's effective mass is proportional to the atomic mass that characterizes atomic displacements in the immediate vicinity of the bipolaron. In this model, the bipolaron's effective mass is also inversely proportional to the volume of the bipolaron. The implications of these results for the dependences of T_c on isotopic substitutions and on crystal structure are now addressed.

First consider a deformational-potential type short-range electron-lattice interaction of the electronic carriers with long wavelength acoustic vibrations. Here, the bipolaron's energy is altered by a general expansion or contraction of the lattice where it is in direct contact with the self-trapped electronic carriers. In this case the atomic mass that is proportional to the bipolaron's effective mass is that associated with acoustic phonons. That is, the bipolaron's effective mass is then proportional to the net mass of the atoms of a unit cell. In this case, the fractional change in T_c with an isotopic substitution for one species of atom will be the negative of the resulting fractional change in the net mass of the atoms of a unit cell: $\Delta T_c/T_c = - \Delta M/M$. Then T_c for a multielement solid will manifest only a slight shift with an isotopic change of one of the atomic species.[13,14]

A different situation prevails if the short-range electron-lattice interaction of the self-trapped electronic carriers is primarily with the relatively light atoms of a polyatomic solid. Then the atomic displacements that dominate the bipolaron's effective mass are associated with optical phonons. The atomic mass that is proportional to the bipolaron's effective mass is then the effective mass of the long wavelength optical vibration frequency. This atomic mass corresponds to that of an appropriate light atom. As such, this situation produces a stronger dependence of the bipolaron's effective mass on the mass of these light atoms than that characterizing a short-range interaction with acoustic phonons. That is, this example yields a larger shift of T_c with an isotopic change of these light atoms than does the example of the preceding paragaph.[13,14]

The effective mass of a large bipolaron is inversely proportional to its volume. This feature can produce a dependence of the bipolaron's effective mass, and hence of T_c, on the structure of an electronically anisotropic solid. For example, consider an electronically quasi-two-dimensional system of alternating layers in which the self-trapped carrier resides within one type of layer. The morphology of the electronic state of such a large bipolaron will be disklike provided that the thickness of the layer containing the bipolaron is less than the in-plane diameter of the large bipolaron. That is, here the thickness of this disklike bipolaron is determined by the thickness of the layer containing the bipolaron. Since the effective mass of the bipolaron varies inversely with its volume, the effective mass of this bipolaron is inversely proportional to its layer thickness. Because T_c is inversely proportional to the bipolaron's effective mass, T_c is proportional to the thickness of the layer containing the bipolaron.

The proportionality of T_c to the layer thickness will cease when the layer thickness exceeds the diameter of the large bipolaron. Then the large bipolaron will be three-dimensional and the layer thickness will no longer affect the volume of the bipolaron. This feature can account for the increase of T_c with the number of contiguous CuO_2 sheets in the CuO_2-based superconductors.[13,14] In particular, the thickness of the layer containing the large bipolaron is then taken to be the net thickness of contiguous CuO_2 sheets in a CuO_2-based superconductor.

With Bose condensation, local variations of the bipolarons' density are suppressed. This effect will generally affect probes of the electronic charge distribution in the solid. For example, the annihilation of injected positrons probes the solid's electronic charge density. The positron annihilation rate is proportional to the product of the positron and electron densities integrated over the solid. The electronic charge distribution is dependent on the configuration of (slowly moving) bipolarons. Thus, the injected positron responds to the instantaneous distribution of bipolarons. This positron annihilation rate is minimized when the electronic charge distribution of the solid is most uniform.[15] Thus, if the positrons are annihilated in regions of the solid occupied by bipolarons, the positron annihilation rate should fall as density fluctuations of the bipolarons are suppressed with their condensation. These features are in general agreement with measurements on CuO_2-based superconductors.[31]

Despite the pairing of the carrier spins, a large singlet bipolaron in which the orbital angular momentum is quenched may not be magnetically inert. In particular, the two carriers of a large bipolaron occupy a multiatom molecular orbital associated with the binding of the pair of carriers within the potential well produced by atomic displacements. Thus, a system of singlet large bipolarons is analogous to a molecular solid in which the ground state of each molecule has neither orbital nor spin angular momentum. As such, the application of a magnetic field can induce a magnetic moment for a bipolaron. If this paramagnetic effect exceeds the bipolaron's diamagnetism, the bipolarons will display Van Vleck paramagnetism.[32] The magnitude of the Van Vleck paramagetic term depends inversely on the electronic energetic separation between the ground state and excited states of a self-trapped carrier. Since the energetic separation expected of a large bipolaron's electronic ground state from its excited states (of the order of 0.1 eV) is much smaller than the energy gaps that typify molecules, the bipolaron's paramagnetic contribution may be unusually large.

The electronic structure of the self-trapped carriers is sensitive to atomic positions generally and to interactions between bipolarons in particular. As bipolarons condense, the allowed configurations of bipolarons is restricted. This effect will generally alter the Van Vleck paramagnetism. For example, the Van Vleck paramagetism may be dominated by extreme configurations of bipolarons (e.g. bunchings) with small electronic energy separations. With the suppression of these configurations upon condensation, the paramagetism of the bipolaronic carriers will fall as the temperature is lowered below T_c. Such an effect will reduce NMR frequencies. However, since a bipolaron-induced paramagnetism will persist even at absolute zero, some "Knight-like" shift of the NMR frequencies will always persist.

In this picture the magnetic-field-induced orbital angular momentum of the self-trapped carriers experienced by nuclear spins is determined by the instantaneous atomic configuration. That is, nuclear spins interact through the self-trapped electrons with the atomic vibrations. Thus, relaxation of a nuclear-spin is accomplished with the emission of a very low energy phonon. Since the energy of this phonon is much less than the thermal energy, the (stimulated) emission rate is just proportional to the temperature. As the temperature is lowered below T_c that portion of the bipolaronic system assigned to the bipolaronic condensate will cease to contribute the nuclear relaxation. In particular, the condensate can not participate in the nuclear relaxation since it cannot accept a phonon with an energy below that of the energy gap. Thus, the product of the nuclear relaxation time, T_1, and the temperature, T, will monotonically fall below its normal-state value and approach zero as the temperature is lowered below T_c. These features agree qualitatively with observations on high-temperature superconductors.[33-36]

The difficulty of addressing the cooperative properties of interacting particles is great. The problem is compounded immensely when the interacting particles are themselves quasiparticles such as bipolarons. Thus, this discussion of the superconducting properties of a bipolaronic liquid has been far from rigorous or quantitative. Here, the situation has first been simplified to one of charged bosons interacting via a screened Coulomb interaction. In this model, below the Bose condensation temperature, a sufficiently large density of bosons will manifest type II superconductivity. Furthermore, this model yields the Meissner effect and an energy gap that collapses as the temperature is raised from zero to T_c. Presuming the condensation temperature to be approximatable by that of a Bose gas and assigning the mass of a large bipolaron to a boson, the isotope effect and a novel dependence of T_c on the lattice structure have been addressed. Finally, qualitative arguments have been put forth to address the responses of two probes of the electronic state of the system (positron annihilation and NMR). With the uncertainties that accompany this discussion, one can only take the predicted features as being plausible manifestations of interacting large bipolarons. Thus, the consistency of many of these findings with experiments on high-temperature superconductors is only viewed as indicating that bipolaronic superconductivity is a real possibility in these systems.

SUMMARY

In analogy with the situation of large polarons, large bipolarons are distinguished from severely localized (small) polarons and bipolarons by their ability to move coherently. As a result, the formation of a superconducting condensate of bipolarons is only possible with large bipolarons. Large bipolarons can only be formed in ionic solids in which some atoms are exceptionally displaceable. Thus, a large bipolaron can only form in an ionic insulator that has an exceptionally large ratio of the static to high-frequency dielectic constants. Furthermore, the condition for the stable binding of two large polarons to form a large bipolaron is much less severe in electronically quasi-two-dimensional systems than for electronically isotropic solids. Thus, quasi-two-dimensional "ferroelectrics" are prime candidates for the formation of large bipolarons.

As a result of their being self-trapped through their interactions with a solid's atoms, large polarons and bipolarons have effective masses that are considerably larger than typical electronic masses. Because of their heavy masses and their large spatial extents, large polarons and bipolarons are not effectively scattered by phonons. Thus, these quasiparticles have exceptionally long scattering times. Furthermore, with the large polaron or bipolaron primarily scattered by acoustic phonons, its motion is like a heavy classical particle moving through a viscous medium. The resistance it experiences is simply proportional to the number of light particles (low energy acoustic phonons) it encounters. Thus, except at very low temperatures, the resistivity of a constant density of large polarons or bipolarons is simply proportional to the temperature. The electronic species that form polarons or bipolarons are bound within a potential well produced by atomic displacements. As such, photon absorption can excite these electronic species to higher lying self-trapped levels or liberate the self-trapped carriers from their potential wells. Thus, with their relatively shallow potential wells, large polarons or bipolarons are characterized by the presence of absorption above a modest threshold ($\simeq$ 0.1 eV).

Considering the bipolaron to be a charged boson leads to the possibility of bipolaronic superconductivity. In particular, above a critical density, the condensate of a gas of interacting charged bosons manifests 1) resistanceless current flow, 2) the Meissner effect, and 3) an energy gap that collapses as the temperature is raised from absolute zero to the Bose condensation temperature, T_c. This superconductivity is type II since the coherence length of the condensate is always less than the London penetration depth.

Bipolaronic superconductivitiy is confined to a limited range of carrier densities. If the carrier density is too low, the Bose condensate will not be the Bose fluid required for superconductivity. If the carrier density is too large, bipolaron formation can not occur. This limitation arises because each self-trapped carrier competes to displace the atoms surrounding it.

Qualitative considerations suggest that the condensation of large bipolarons is distinguishable from the opening up of a BCS gap in a metal. In particular, in the normal state the slowly moving bipolarons undergo local density fluctuations that produce centers for the annihilation of incident positrons. These fluctuations are suppressed in the Bose condensate. As a result, the positron lifetime increases as the temperature is reduced below T_c. Such behavior is reported for the CuO_2-based superconductors.[31] By contrast, the formation of overlapping Cooper pairs of rapidly moving electrons in a metal does not alter the electronic charge density presented to an incident positron. Thus, consistent with experiment, the positron lifetime for conventional superconductors is unaffected upon passage into the superconducting state.[31]

The condensation of singlet bipolarons should be associated with different magnetic properties than are associated with the BCS condensation. It is speculated that the small excitation energies of the self-trapped carriers of large singlet bipolarons will cause them to display an exceptionally large Van Vleck paramagnetic susceptibility. Different configurations of bipolarons and the concomitant differences of the self-trapping wells will produce inequivalent contributions to the Van Vleck susceptibility. The restriction of the distribution of self-trapping wells that occurs with Bose condensation should reduce the Van Vleck susceptibility and the NMR frequencies as the temperature is lowered below T_c. However, unlike the situation with simple spin-pairing, this "Knight-like" shift will not vanish as the temperature approaches absolute zero. Furthermore, this means of coupling nuclear spins to atomic configurations also provides a mechanism for spin-lattice relaxation. In particular, nuclear-spin relaxation can occur with the emission of a low energy phonon. However, the presence of an energy gap precludes the condensate acting as a reservoir for relaxation. Thus, the relaxation rate should fall increasingly rapidly as T is lowered below T_c. This behavior contrasts with nonmonotonic Hebel-Slichter behavior characterizing BCS superconductors.[37] However, these predictions for the superconductivity of large bipolarons are qualitatively consistent with observations on the CuO_2-based superconductors.[33-36]

The condensation temperature for a system of large bipolarons is approximated as the Bose condensation temperature for noninteracting bosons having the bipolaron's mass. Then, rather high values of the superconducting transition temperature, > 100 K, are readily obtained. Furthermore, the fractional change of this transition temperature with isotopic substitutions is the negative of the fractional change of the atomic mass that characterizes a bipolaron's short-range atomic displacements. In particular, if the atoms that are displaced in the immediate vicinity of a self-trapped carrier are primarily heavy (e.g., Cu, La), a small isotope effect results. Conversely, if the displaced atoms that are in contact with the carrier are relatively light (e.g., O), there is a relatively strong isotope dependence of T_c. Finally, it is noted that T_c is proportional to the volume of a self-trapped carrier-pair. This result affords an explanation of the increase of T_c in CuO_2-based superconductors with the number of contiguous CuO_2 layers. In particular, if carriers are constrained to reside within CuO_2 layers, the thickness of the self-trapped carrier is determined by the number of contiguous CuO_2 layers if the intrinsic diameter of the self-trapped carrier-pair exceeds the net thickness of the contiguous CuO_2 layers. Thus, the volume of a self-trapped carrier-pair, and hence T_c, increases with the number of contiguous CuO_2 sheets.

The mobile bipolarons required for bipolaronic superconductivity must be large bipolarons. Large bipolarons can only form in ferroelectric-type materials. These are solids characterized by an exceptionally large ratio of the static to optical dielectric constants. Solids with bipolaronic carriers possess distinctive normal-state and superconducting properties. The CuO_2-based superconductors appear both to satisfy the requirements for the formation of large bipolarons and to manifest normal-state and superconducting features expected of large bipolarons.

ACKNOWLEDGEMENT

This work was performed at Sandia National Laboratories with the support of the U. S. Department of Energy under Contract No. DE-AC-04-76DP00789. This review also appears in the proceedings of the First USSR-USA Meeting on Ferroelectrics published in Ferroelectrics.

REFERENCES

1. M. R. Schafroth, Phys. Rev., 100, 463 (1955).
2. L. J. Azevedo, E. L. Venturini, D. Emin and C. Wood, Phys. Rev. B, 32, 7970 (1985).
3. C. Schlenker, S. Ahmed, R. Buder and M. Gourmala, J. Phys. C, 12, 2503 (1979).
4. B. K. Chakraverty, M. J. Sienko and J. Bennerot, Phys. Rev. B, 17, 3503 (1979).
5. O. F. Schirmer and E. Salje, J. Phys. C, 13, L1067 (1980)
6. B. Ya. Moizhes and S. G. Suprun, Sov. Phys. Solid State, 26, 544 (1984).
7. G. Binnig, A. Baratoff, H. E. Honig, J. G. Bednorz, Phys. Rev. Lett., 45, 1352 (1980).
8. M. R. Harrison, P. P. Edwards and J. B. Goodenough, Phil. Mag. B, 52, 679 (1985).
9. A. W. Sleight, J. L. Gillson and P. E. Bierstedt, Solid State Commun. 17, 27 (1975).
10. R. J. Cava, B. Batlogg, J. G. Krajewski, R. Farrow, L. W. Rupp, Jr., A. E. White, K. Short, W. F. Peck, and T. Komentani, Nature, 332, 814 (1988).
11. J. G. Bednorz and K. A. Muller, Z. Phys. B, 64, 189 (1986).
12. D. Emin, Physics Today, 35 No. 6, 34 (1982).
13. D. Emin, Phys. Rev. Lett., 62 1544 (1989).
14. D. Emin and M. S. Hillery, Phys. Rev. B, 39, 6575 (1989).
15. D. Emin and M. S. Hillery, to be published.
16. D. Emin and T. Holstein, Phys. Rev. Lett., 36, 323 (1976).
17. T. Holstein, Ann. Phys. (N.Y.), 8, 343 (1959).
18. H.-B. Schuttler and T. Holstein, Phys. Rev. Lett., 51, 2337 (1983); Ann. Phys. (N.Y.), 166, 93 (1986).
19. D. Emin, Adv. Phys., 22, 57 (1973).
20. H. Frohlich, Adv. Phys., 3, 325 (1954).
21. S. K. Kurtz, A. Bhalla and L. E. Cross, Ferroelectrics (in press)

22. K. Kamaras, S. L. Herr, C. D. Porter, N. Tache, D. B. Tanner, S. Etemad, T. Venkatesan, E. Chase, A. Inam, X. D. Wu, M. S. Hegde and B. Dutta, (preprint).
23. A. Zettl, Proc. Int. Conf. on Materials and Mechanisms of Superconductivity: High-Temperature Superconductors, Physica C (in press).
24. Y. H. Kim, C. M. Foster, A. J. Heeger, S. Cox and G. Stucky, Phys. Rev. B, 38, 6478 (1988)
25. C. Taliani, R. Sambona, G. Ruani, F. C. Matacotta and K. I. Pokhodnya, Solid State Commun., 66, 487 (1988).
26. G. A. Thomas, J. Orenstein, D. H. Raphine, M. Capissi, A. J. Millis, R. N. Bhatt, L. F. Schneemeyer and J. V. Waszczak, Phys. Rev. B, 61, 1313 (1988).
27. L. L. Foldy, Phys. Rev., 124, 649 (1961).
28. L. D. Landau and E. M. Lifshitz, Statistical Physics (Pergamon, London, 1958) p. 169, pp. 202-206.
29. A. L. Fetter and J. D. Walecka, Quantum Theory of Many-Particle Systems (McGraw Hill. New York, 1971) pp. 500-501.
30. D. J. Thouless, The Quantum Mechanics of Many Particle Systems (Academic Press, New York, 1961) pp. 154-160.
31. Y. C. Jean, J. Kyle, H. Nakanishi, P. E. A. Turchi, R. H. Howell, A. L. Wachs, M. J. Fluss, R. L. Meng, H. P. Hor, J. Z. Huang and C. W. Chu, Phys. Rev. Lett., 60 1069 (1988).
32. C. Kittel, Introduction to Solid State Physics (John Wiley and Sons, New York, 1956) pp. 210-211, 578-588.
33. R. E. Walstedt, W. W. Warren, Jr., R. F. Bell, G. F. Brennert, G. P. Espinosa, R. J. Cava, L. F. Schneemeyer and J. V. Waszczak, Phys. Rev. B, 38, 9299 (1988).
34. T. Imai, T. Shimizu, H. Yasuoka, Y. Ueda and K. Kosuge, J. Phys. Soc. Jpn., 57, 2280 (1988).
35. C. H. Pennington, D. J. Durand, C. P. Slichter, J. P. Rice, E. D. Bukowski and D. M. Ginsberg, Phys. Rev. B, 39, 2902 (1989).
36. M. Takigawa, P. C. Hammel, R. H. Heffner and Z. Fisk, Phys. Rev. B, 39, 7371 (1989).
37. L. C. Hebel and C. P. Slichter, Phys. Rev., 113, 1504 (1961).

SMALL-POLARON HOPPING IN UO_{2+x} AND U_4O_{9-y} SINGLE CRYSTALS

P. Nagels
Rijksuniversitair Centrum, University of Antwerp
B-2020 Antwerpen, Belgium

ABSTRACT

Electrical conductivity, thermopower and Hall effect measurements have been made on UO_{2+x} and U_4O_{9-y} single crystals as a function of temperature. In UO_{2+x} the Hall coefficient showed a double sign reversal from n- to p-type at around 380 K and again to n-type at 1000 K. A large Hall mobility of $\simeq 2.5$ cm^{2}V^{-1}s^{-1} was measured above 1000 K. In U_4O_{9-x} the Hall coefficient remained negative between 200 and 700 K. The results are interpreted on the basis of small-polaron hopping.

1. INTRODUCTION

When stoichiometric UO_2 is slightly oxidized the excess oxygen enters interstitial positions in the cubic fluorite structure. The additional oxygen ions may be distributed at random forming the UO_{2+x} phase or in an ordered way forming a cubic superlattice with a unit cell four times that of UO_2. The incorporation of oxygen ions in interstitial positions in the UO_{2+x} phase results in a conversion of two adjacent U^{4+} ions to U^{5+} ions. In an ionic crystal such as UO_2 the hole may become trapped as a consequence of the displacement of adjacent ions. The carrier plus induced distortion then migrates through the crystal by phonon-assisted hopping. In the superstructure U_4O_9 conduction also arises when the composition deviates from stoichiometry. The removal of oxygen creates electrons bound to U^{4+} ions. Conduction can again occur by hopping of a small-polaron electron.

2. RESULTS

The electrical conductivity σ, measured as a function of temperature on a UO_{2+x} crystal[1] with an O/U ratio of about 2.004 is shown in Fig. 1. After determining σ at room temperature, the temperature was rapidly raised to 975 K and then slowly lowered to room temperature. Below 590 K there is a change in slope, which is related to the crossing from a UO_{2+x} single phase to a $UO_2 + U_4O_{9-y}$ two phase system. Below 450 K the slope of the σ curve remains constant and is identical to that of the single phase material. The phenomenon observed during cooling is associated with the previous treatment of the crystal. When alloying the electrical contacts, the crystal was rapidly quenched from about 1500°C to room temperature. As a consequence, the UO_{2+x} phase, which according to the $UO_{2+x}-U_4O_9$ phase diagram exists at higher temperatures, is frozen in. The room temperature conductivity before and after heating has decreased by a factor of 3.5. This is in good agreement with the difference in the concentration of electrically active oxygen interstitials in the UO_2 matrix (approximately four). Indeed, electron microscopy observations have shown that at room temperature the maximum oxygen solubility in UO_2 amounts to about 0.001, yielding an O/U ratio = 2.001. At higher concentrations, U_4O_9 precipitates under the form of small islands. The same crystal reduced in hydrogen at 800°C (O/U < 2.001) did not show a break in slope.

The Hall coefficient R_H, measured on the sample before and after reduction, is shown in Fig. 2. In both cases the Hall coefficient showed a sign reversal from n-type to p-type around 380 K. A break in the R_H curve of the $UO_{2.004}$ crystal occurs at the same temperature where the σ curve changes in slope (590 K). The thermopower measured between 200 and 800 K remained positive over the entire temperature range. The Hall mobilities, represented in Fig. 3 in the p-type region, are low and increase with temperature. The sign reversal of the Hall coefficient was observed on other samples and seemed to be a general feature of slightly oxidized UO_2. Figure 4 shows the measured $R_H \times \sigma$ data obtained on a single phase $UO_{2.001}$ crystal during a

Small-Polaron Hopping in UO_{2+x} and U_4O_{9-y} Single Crystals 379

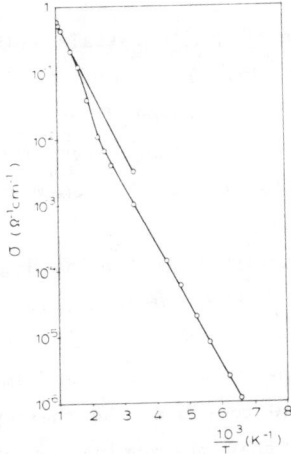

FIG. 1. Conductivity of a $UO_{2.004}$ single crystal as a function of reciprocal temperature.

FIG. 2. Hall coefficient R_H vs $1/T$ for a $UO_{2.004}$ single crystal (●,○), also measured after reduction (△)

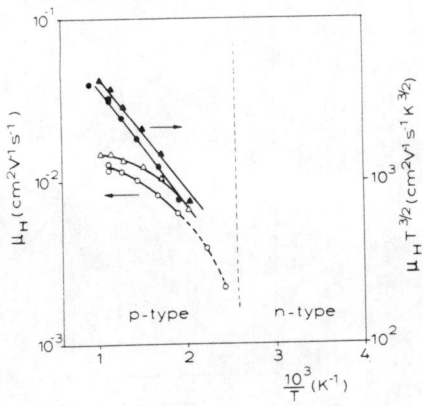

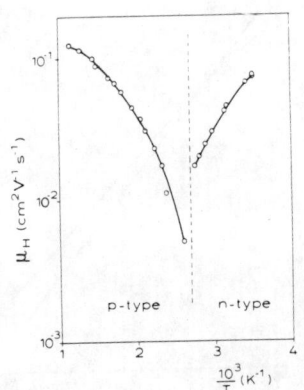

FIG. 3. Hall mobility μ_H and $\mu_H T^{3/2}$ vs $1/T$ for a $UO_{2.004}$ single crystal (○,●), also measured after reduction (△,▲).

FIG. 4. Hall mobility μ_H vs reciprocal temperature for a $UO_{2.001}$ single crystal.

heating cycle from room temperature to 870 K. Cooling yielded similar results.

Above 900 K, the conductivity curves of UO_{2+x} crystals exhibited a much steeper slope (not represented in Fig. 1) corresponding to an activation energy of roughly 1.5 eV. This increase is held to be caused by the onset of intrinsic conduction. We found a transition to a n-type Hall effect with a large Hall mobility of $\cong 2.5$ $cm^2V^{-1}s^{-1}$ at 1140 K.

Figure 5 shows the electrical conductivity plotted as log σ vs reciprocal temperature for a U_4O_{9-y} crystal with chemical composition $UO_{2.18}$. It can be seen that at higher temperature the curve deviates from the linear behaviour. The sign of the Hall coefficient remained negative in the temperature range from 240 to 650 K. The thermopower also showed a negative sign, indicating that the dominating charge carriers are electrons. The Hall mobility, plotted as log μ_H vs reciprocal temperature in Fig. 6, manifests a thermally activated type

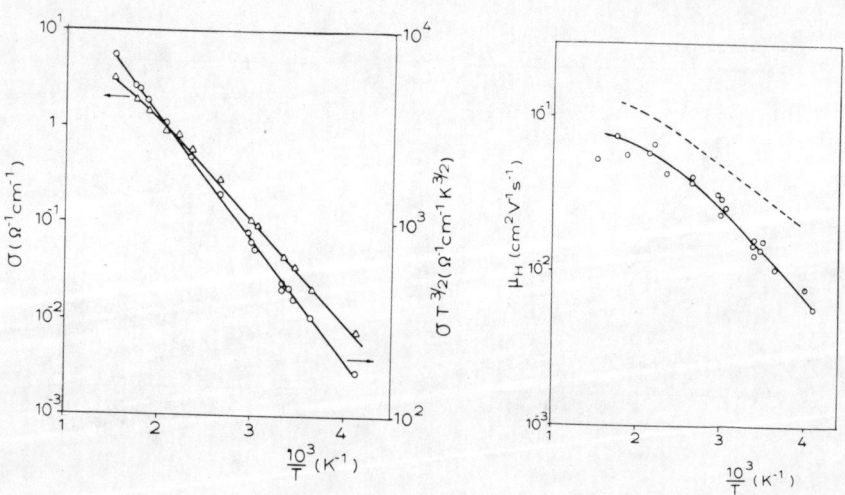

FIG. 5. Conductivity and $\sigma T^{3/2}$ vs reciprocal temperature for a U_4O_{9-y} single crystal

FIG. 6. Hall mobility μ_H vs 1/T for a U_4O_{9-y} single crystal; the dashed line is calculated.

temperature dependence below approximately 350 K and then increases with a continuously decreasing slope.

4. DISCUSSION

The previously described results will be interpreted on the basis of small-polaron hopping. The drift mobility (or conductivity mobility), μ_D, is small and thermally activated below some temperature limits. This theoretical expectation will serve as the principal criterion for showing the existence of small polarons in UO_{2+x} and U_4O_{9-y}. A rough estimate of μ_D can be obtained from the conductivity $\sigma = pe\mu_D$ and the excess oxygen concentration. The number of holes $p = 1.95 \times 10^{19}$ cm^{-3} for the $UO_{2.004}$ crystal, under the assumptions that all charge carriers are free and hence, equal to x. Thus we find that $\mu_D = 1.2 \times 10^{-3}$ cm^{2}V^{-1}s^{-1} at 300 K. This value is much lower than the limiting value ($\cong 0.1$ cm^{2}V^{-1}s^{-1}) usually taken as the minimum for conventional broad-band conduction. Information on the carrier concentration and its variation with temperature can be gained from the thermopower. We observed a very slow decrease with increasing temperature, the associated activation energy being equal to 0.06 eV. The thermopower can be expressed by $S = k/e \ln N/p$, where N is the number of available hopping sites, in UO_{2+x} equal to the concentration of the U^{4+} ions for a small x value (2.44 $\times$ 10^{22} cm^{-3}). At 300 K, we measured $S = 760$ μVK^{-1}, which yields $p = 3.6 \times 10^{18}$ cm^{-3}. This is a factor of five lower than the maximum hole concentration calculated from the chemical composition. The result of a nearly constant charge carrier concentration indicates that the temperature dependence of the drift mobility is close to that of the conductivity, i.e. has a thermally activated nature.

The transition probability for small-polaron hopping was theoretically studied in two cases, which are called the adiabatic and non-adiabatic regimes. Expressions were derived for the drift and Hall mobilities in terms of two characteristic parameters: J, the electron charge transfer integral and W_H, the activation energy for hopping. The magnitude of J yields a criterion for the applicability of the non-adiabatic small-polaron theory:

$$J < (\frac{2W_H kT}{\pi})^{1/4} (\frac{\hbar\omega_o}{\pi})^{1/2} \qquad (1)$$

For UO_2, the Debye temperature θ = 816 K and, hence, $\hbar\omega_o$ = 0.07 eV. Inserting this value in expression (1) together with W_H = 0.22 eV, yields J < 0.04 eV. Although we have no conclusive arguments, we shall interpret our results in the framework of the non-adiabatic theory. For this regime, the drift mobility and the four site Hall effect have been calculated. The adiabatic theory of the Hall mobility of a small polaron in a square lattice has not been worked out.

Holstein[2] derived the following expression for the drift mobility in the non-adiabatic case:

$$\mu_D = \frac{ea^2}{\hbar} [\frac{J^2}{kT} (\frac{\pi}{4kTW_H})^{1/2}] \exp(-W_H/kT) \qquad (2)$$

where a is the intersite distance.

The Hall mobility for the four-site non-adiabatic case has been calculated by Emin[3]:

$$\mu_H = \frac{4ea^2}{\hbar} (\frac{2J}{\sqrt{3}}) (\frac{\pi}{4W_H kT})^{1/2} \exp(-W_H/3kT) \times H \qquad (3)$$

where H is a dimensionless quantity which varies with temperature. The magnitude and temperature dependence of H has been discussed by Emin in two limiting cases: low and high temperature. In the high temperature limit ($W_H/kT \leq 6$) the Hall mobility can be approximated by:

$$\mu_H = \frac{4ea^2}{\hbar} \frac{2J^2}{3\sqrt{3}} (\frac{1}{kT})^{3/2} (\frac{\pi}{4W_H})^{1/2} C(T) \exp(-W_H/2kT) \qquad (4)$$

with C(T) a slowly varying function of temperature.

In the low temperature limit ($W_H/kT \gg 6$), H slightly increases with temperature.

We shall examine whether the results obtained on UO_{2+x} and U_4O_{9-y} are in agreement with the theoretical predictions.

UO_{2+x}

The most salient feature of the results is the sign reversal of the Hall coefficient around 380 K. The sign of the Hall effect for hopping motion depends upon the relative orientations of the sites and the local wave functions between which hopping occurs[4]. If the number of orbitals is even, i.e. a four site configuration, then the sign of the Hall coefficient is always normal. Because of the cubic structure, one expects a positive sign in UO_{2+x} and a negative sign in U_4O_{9-y}, as indicated by the thermopower results. The reason for the appearance of a negative sign below 380 K is not clear to us. A speculative interpretation might be that a three-site configuration becomes predominant below this temperature, which might be a consequence of small local rearrangements of ions in the lattice. Structural work by Willis[5] has shown that, as a result of electrostatic forces from two neighbouring uranium ions the extra oxygen atoms occupy interstitial positions which are displaced from the centre of the large interstices in the fluorite structure. The neighbouring uranium ions are also slightly displaced. It is known that at around 100°C the oxygen interstitial becomes mobile in the lattice and this might give some relaxation on the position of the neighbouring uranium ions. In U_4O_9 a phase transition was observed at about 350 K.

The drift and Hall mobilities differ in their temperature dependence. According to eq. (4), the Hall mobility activation energy is roughly half that of the drift mobility. We have made a comparison of the activation energies by plotting $\log \mu T^{3/2}$ (see Fig. 3) and $\log \sigma T^{3/2}$ reciprocal temperature. We found 0.15 eV for μ_H and 0.28 eV for σ. The temperature dependence of the thermopower yielded a small activation of 0.06 eV. The difference between the activation energies of the conductivity and the thermopower gives the hopping energy, W_H = 0.22 eV. Thus, it seems that Hall mobility activation is higher than half that of the drift mobility and, hence, not in agreement with the theoretical prediction. It must be mentioned, however, that the behaviour of the thermopower for hopping conduction is more complicated than that given by the expression $S = k/e \ln \frac{N}{p}$. As indicated by Emin[6], the thermopower is the sum of two terms, the first being associated with the activation energy for carrier creation and the second proportional to the average vibrational energy transported with the carrier. The second term might also depend on temperature and, therefore, influence the overall temperature dependence of the thermopower.

U_4O_{9-y}

As in the previous discussion, the conductivity data were also plotted as $\log \sigma T^{3/2}$ vs reciprocal temperature (Fig. 5). The activation energy deduced from the slope was 0.25 eV. The Hall mobility was measured in a lower temperature range than for UO_{2+x} and, therefore, equation (4) cannot be applied. Using the expression of H given in Emin's work for the low temperature approximation, the Hall mobility was calculated with W_H = 0.25 eV, J = 0.04 eV and a = 3.85 Å. The theoretical temperature dependence is represented by the dashed line in Fig. 6. The experimental Hall mobility curve manifests a rather good resemblance with the theoretical line. Its slope at low temperature (0.10 eV) is again somewhat higher than that of the calculated line (0.08 eV).

Above 1000 K, we observed a large n-type Hall effect (2.5 $cm^2V^{-1}s^{-1}$ at 1140 K). A similar transition from a p-type to a n-type Hall effect with a large electron Hall mobility of the order of 10 $cm^2V^{-1}s^{-1}$ was observed by De Wit and Crevecoeur[7] in Li-doped MnO. Furthermore, Gvishi et al.[8] reported that the Hall mobility above 1000 K seemed to be similar to the drift mobility of the electrons.

REFERENCES

1. W Van Lierde, R. Strumane, E. Smets and S. Amelinckx, J. Nucl. Mat. 5, 250 (1962).
2. T. Holstein, Ann. Phys. (N.Y.) 8, 343 (1954).
3. D. Emin, Ann. Phys. (N.Y.) 64, 393 (1971).
4. D. Emin, Phil. Mag. 35, 1188 (1977).
5. B. Willis, J. Phys. 25, 431 (1964).
6. D. Emin, Phys. Rev. Lett. 35, 882 (1975).
7. H. De Wit and C. Crevecoeur, Phys. Lett. 25A, 393 (1967).
8. M. Gvishi, N. Tallan and D. Tannhauser, Solid State Commun. 6, 135 (1968).

POLARONIC CONDUCTION IN OXIDE GLASSES CONTAINING V_2O_5

P. Nagels
Rijksuniversitair Centrum, University of Antwerp
B-2020 Antwerpen, Belgium

ABSTRACT

The electrical conductivity σ and thermopower S were measured as a function of temperature on vanadate glasses having the composition 90 % V_2O_5 - 10 % P_2O_5 and 60 % V_2O_5 - 40 % TeO_2. The thermopower was temperature independent, indicating a constant number of charge carriers. The σ and S data suggested that the mobility of the charge carriers, believed to be small polarons, is much higher in the V_2O_5-TeO_2 glass than in V_2O_5-P_2O_5. Attempts to measure the Hall effect were unsuccessful, yielding an upper limit of the Hall mobility of the order of 1×10^{-2} $cm^2 V^{-1} s^{-1}$ in V_2O_5-TeO_2 (300-500 K). This is still many orders of magnitude higher than the conductivity mobility evaluated from the conductivity and thermopower data.

1. INTRODUCTION

It is generally agreed that conduction in transition metal oxide glasses takes place by the movement of small polarons between metal ions of different valency state. The most salient feature of small-polaron hopping is a very low and thermally activated conductivity mobility, the so-called hopping mobility W_H. According to small-polaron theory, the Hall mobility is also thermally activated with an activation energy which is 1/3 that of the conductivity mobility.

Hall effect measurements on a V_2O_5-P_2O_5 glass have been reported by Vomvas and Roilos[1]. Although a thermally activated Hall mobility was observed, the authors concluded that their results were not in agreement with small-polaron theory. The aim of our study was to compare the transport properties of two vanadate glasses: V_2O_5-P_2O_5 and V_2O_5-TeO_2. The reason is the following: conductivity and thermopower data suggest that the mobility of charge carriers, believed to be small polarons, is much higher in the V_2O_5-TeO_2 system that in V_2O_5-P_2O_5 (or Fe_2O_3-P_2O_5).

2. RESULTS

All measurements were performed on two vanadate glasses of nominal composition 90 % V_2O_5 - 10 % P_2O_5 and 60 % V_2O_5 - 40 % TeO_2, prepared by pouring the appropriate melt into a thick-walled stainless steel cylinder having an inner diameter of 8 mm. The dc electrical conductivity of the 90 % V_2O_5 - 10 % P_2O_5 glass, measured between 150 and 425 K, is shown as a function of reciprocal temperature in Fig. 1. The slope of the log σ vs 1/T plot tends to a value of 0.33 eV at high temperature. The continuous curvature towards lower temperature over a broad range has been observed by many authors[1-3]. Figure 2 represents the temperature dependence of the dc electrical conductivity of the 60 % V_2O_5 - 40 % TeO_2 glass measured in the range 300-500 K. The slope of this curve yields an activation energy of 0.34 eV.

The thermopower of the two glasses is shown as a function of reciprocal temperature in Figs. 3 and 4. It can be seen that in both cases the thermopower is negative and temperature independent.

3. DISCUSSION

In stoichiometric V_2O_5, the vanadium metal is fivefold positively charged (V^{5+}). In substoichiometric V_2O_5 the removal of each oxygen ion (V_2O_{5-y}) converts two V^{5+} ions into V^{4+} ions. In these polar materials, the extra electrons located on the V^{4+} ions will polarize their surroundings creating a self-trapped state. There exists good evidence that the charge carriers form dielectric small polarons. The

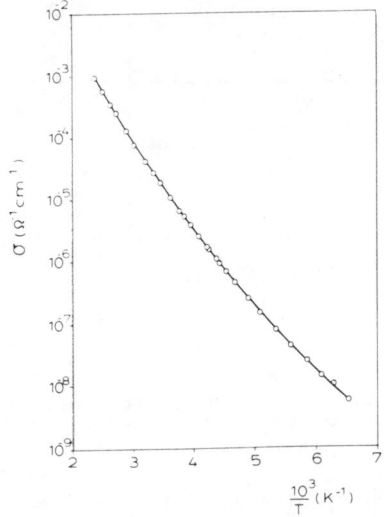
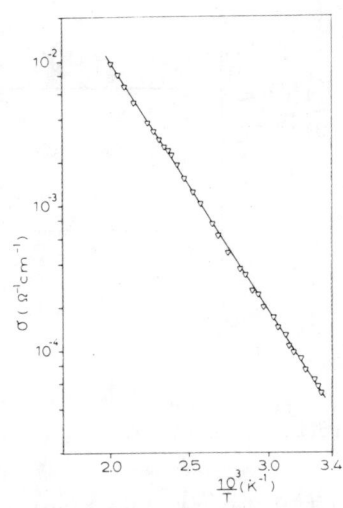

FIG. 1. dc conductivity of a 90 % V_2O_5 - 10 % P_2O_5 glass plotted as $\log \sigma$ vs $1/T$.

FIG. 2. dc conductivity of a 60 % V_2O_5 - 40 % TeO_2 glass plotted as $\log \sigma$ vs $1/T$.

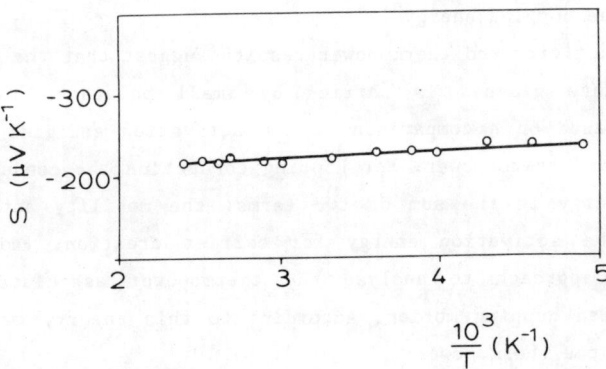

FIG. 3. Thermopower of a 90 % V_2O_5 - 10 % P_2O_5 glass plotted vs reciprocal temperature.

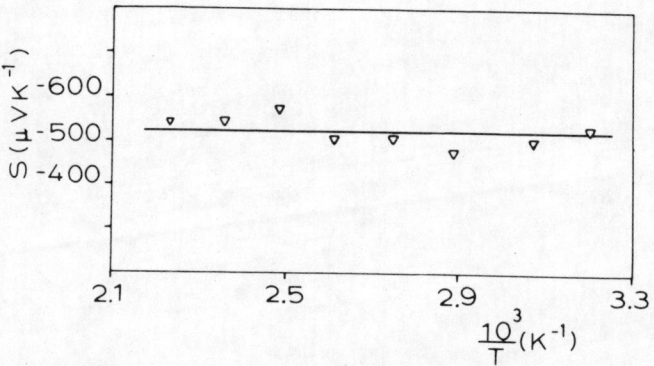

FIG. 4. Thermopower of a 60 % V_2O_5 - 40 % TeO_2 glass plotted vs reciprocal temperature.

charge transfer may take place by hopping of a small polaron electron from a V^{4+} ion to an adjacent V^{5+} ion. Such a process is characterized by a hopping energy W_H. Due to different environments of the ions with two valencies, one can expect that the polaron hopping energy will be increased by a disorder energy. In the vanadate glasses there exists some evidence that the disorder energy is small and can be neglected compared to the hopping energy[4].

Our conductivity and thermopower results suggest that the current in the vanadate glasses is carried by small polarons. The main argument is based on a comparison of the activation energies of the conductivity and thermopower. For hopping conduction the conductivity activation energy is the sum of two terms: the mobility activation energy and the activation energy for carrier creation. Emin[5] has described an approach to analyse the thermopower associated with polaron-assisted hopping motion. According to this theory, two terms contribute to the thermopower:

$$S = \frac{k}{e} \left(\frac{E}{kT} + \frac{E_T}{kT} \right) \qquad (1)$$

The first is the standard term in the expression of the thermopower and describes the activation energy for carrier creation. The second is proportional to the average vibrational energy transported with the carrier as it hops from a site to an adjacent one. This term vanishes for hopping between sites that are equivalent in energy or in electron-lattice coupling strength. Wood and Emin[6] reported that the behaviour of the thermopower in boron carbide, considered to be a small-polaron semiconductor - (they noticed an increase with increasing temperature) - can be explained with the help of Eq. (1). They showed that for a broad distribution of inequivalent hopping sites the second term yields a contribution to the thermopower which increases linearly with temperature. This contribution will dominate the overall temperature dependence of the thermopower in the case of a small activation energy for carrier creation.

The temperature independence of the thermopower indicates that (1) the vibrational energy transferred in a hop is small, which means that the energy of the sites is nearly identical and (2) that the density of small-polaron electrons is constant with temperature. In this case the thermopower can be expressed as:

$$S = \frac{k}{e} \ln \frac{N}{n} \qquad (2)$$

where N is the number of available hopping sites, given by $[V^{5+}]-[V^{4+}]$ and n the number of electrons equal to the concentration of V^{4+} ions. To make an estimate of N, the number of V^{5+} ions present in crystalline V_2O_5 (2.2×10^{22} cm^{-3}) was multiplied with the molar fraction of V_2O_5 in the two vanadate glasses. Some characteristics obtained from the conductivity and thermopower data are shown in Table I.

In the $V_2O_5-P_2O_5$ glass, the $[V^{4+}]/[V^{5+}]$ ratio calculated from the thermopower was equal to 0.05, which is in agreement with chemical determinations reported by other authors. By potentiometric titration Sayer et al.[2] found $[V^{4+}]/[V^{5+}] = 0.038$ in a 80 % V_2O_5 - 20 % P_2O_5 glass prepared in oxygen atmosphere.

From Table I it can be seen that, although the number of electrons is quite different, the conductivity of the two glasses is the same at room temperature. Using these data, the drift mobility (or conductivity mobility) can be calculated from $\sigma = ne\mu_D$. This yields

TABLE I. Characteristics of two vanadate glasses deduced from the experimental conductivity and thermopower data

	90 % V_2O_5 - 10 % P_2O_5	60 % V_2O_5 - 40 % TeO_2
$\sigma(\Omega^{-1} cm^{-1})$ at 300 K	5×10^{-5}	5×10^{-5}
ΔE_σ (eV)	0.33	0.34
$S(\mu V\ K^{-1})$	-260	-480
$N(cm^{-3}) \cong [V^{5+}]$	2×10^{22}	1.4×10^{21}
$n(cm^{-3})$ from Eq. (2)	1×10^{21}	5×10^{19}
$[V^{4+}]/[V^{5+}]$	5×10^{-2}	4×10^{-2}

$\mu_D = 3 \times 10^{-7}\ cm^2 V^{-1} s^{-1}$ for the V_2O_5 - P_2O_5 glass and $6 \times 10^{-6}\ cm^2 V^{-1} s^{-1}$ for the V_2O_5-TeO_2 glass. As a result, we see that the conductivity mobility is much higher (twenty times) in the V_2O_5 glass with TeO_2 as glassformer.

A number of theoretical papers have been concerned with the hopping motion of small polarons. These calculations have been carried out within two approximations, called the non-adiabatic and adiabatic regimes. As indicated by Holstein[7], the non-adiabatic theory may be applied when the electron transfer energy J characterizing the hopping is small. The extremely low conductivity mobilities found in the vanadate glasses suggest that the localization time of the charge carrier on a given site is long and that the carrier cannot the follow the vibrational motion of the lattice (the non-adiabatic regime).

In the non-adiabatic regime, the drift mobility in a hexagonal lattice is given by[7]:

$$\mu_D = \frac{3ea^2}{2\hbar} \frac{J^2}{kT} \left(\frac{\pi}{4W_H kT}\right)^{1/2} \exp(-W_H/kT) \qquad (3)$$

where a is the intersite hopping distance. Thus, when the drift mobility is plotted as $\log \mu_D T^{3/2}$ vs 1/T the hopping energy W_H can be deduced from the slope of this line. In our case the temperature dependence of the conductivity is that of the mobility. Equation (3) allows us to calculate the value of the transfer energy J. For the V_2O_5-TeO_2 glass, which exhibited the highest mobility, μ_D was found equal to 5×10^{-6} cm^{2}V^{-1}s^{-1} at room temperature. The V-V distance is of the order of 4 Å. A log $\sigma T^{3/2}$ vs $1/T$ plot yielded $W_H = 0.37$ eV. Incorporating these parameters in Eq. (3) gives $J = 0.08$ eV.

The Hall mobility for hopping in a three site configuration has been calculated by Friedman and Holstein[8] for the non-adiabatic regime:

$$\mu_H = \frac{ea^2}{2\hbar} J \left(\frac{\pi}{4W_H kT}\right)^{1/2} \exp(-1/3\ W_H/kT) \qquad (4)$$

The activation energy of μ_H is only one-third of that associated with the drift mobility. For a triangular lattice it follows from Eqs. (3) and (4) that the ratio μ_H/μ_D is given by:

$$\frac{\mu_H}{\mu_D} = \frac{kT}{3J} \exp\left(\frac{2}{3}\frac{W_H}{kT}\right) \qquad (5)$$

Equation 5 can be used to estimate the order of magnitude of μ_H. At room temperature we found $\mu_H = 1 \times 10^{-2}$ cm^{2}V^{-1}s^{-1} and $\mu_H/\mu_D = 2000$.

Although a high sensitivity was reached, all our attempts to measure the Hall effect in a magnetic field of 25.000 Gauss reversed every 30 sec were unsuccessful. This yields an upper limit to the Hall mobility roughly equal to 1×10^{-3} cm^{2}V^{-1}s^{-1} for the V_2O_5 - P_2O_5 system and 1×10^{-2} cm^{2}V^{-1}s^{-1} for the V_2O_5 - TeO_2 system in the temperature range from 300 to 500 K. Measurements of dc conductivity, thermopower and Hall effect on a 80 % V_2O_5 - 20 % P_2O_5 have been reported by Vomvas and Roilos[1]. Their conductivity curve showed a continuous curvature in the temperature range 225-500 K. The Hall mobility measured on three samples was roughly equal to 1×10^{-4}, 3×10^{-4} and 7×10^{-4} cm^{2}V^{-1}s^{-1} at room temperature and increased exponentially with temperature. The Hall mobility activation energy was in the range from 0.15 to 0.21 eV with a mean value of 0.19 eV. This was one-half that of the conductivity activation energy (0.44 eV). Therefore, the authors concluded that their results were not in agreement with small-polaron theory. To our opinion, it does not seem convincing to reject the small-polaron model on the basis of the arguments presented by the authors. The conductivity activation energy of 0.44 eV was calculated as a mean slope between 300 and 500 K from a log σ vs 1/T plot. In this temperature range the conductivity curve is continuously bent. According to the theoretical predictions, small-polaron hopping generates a thermally activated mobility with a temperature-dependent prefactor. Assuming non-adiabatic small-polaron hopping to be applicable, μ_D varies as $\mu_D \propto T^{-3/2} \exp(-W_H/kT)$. To our estimate the slope of the $\sigma T^{3/2}$ curve should tend to approximately 0.52 eV at the highest temperature. Moreover, due to the difficulty in measuring the Hall effect, the accuracy of the Hall mobility data is low (no error bars are given in the figure). Consequently, the activation energies deduced from the Hall mobility curves can vary over broad limits. It seems to us that, within the ranges of accuracy, the typical criterion of small-polaron motion, which predicts a smaller activation energy of the Hall mobility than of the drift mobility, is quite well fulfilled.

REFERENCES

1. A. Vomvas and M. Roilos, Phil.Mag. 49, 143 (1984).
2. M. Sayer, A. Mansingh, J. Reyes and G. Rosenblatt, J. Appl. Phys. 42, 2857 (1971).
3. G. Linsley, A. Owen and F. Hayatee, J. Non-Cryst. Solids, 4, 87 (1970).
4. T. Allersma and J. McKenzie, J. Chem. Phys. 47, 1406 (1967).
5. D. Emin, Phys. Rev. Lett. 35, 882 (1975).
6. C. Wood and D. Emin, Phys. Rev. B 29, 4582 (1984).
7. T. Holstein, Ann. Phys. (N.Y.) 8, 343 (1954).
8. L. Friedman and T. Holstein, Ann. Phys. (N.Y.) 21, 494 (1963).

DYNAMICS OF THE FORMATION AND MIGRATION OF THE SELF-TRAPPED HOLE IN SILVER CHLORIDE

Lawrence Rowan and Lawrence Slifkin
Department of Physics and Astronomy
University of North Carolina
Chapel Hill, NC 27599-3255

ABSTRACT

This paper reviews results of the past several years on the self-trapped hole (STH) in AgCl, using electron paramagnetic resonance to monitor the concentration. It was shown that the transition from the free Bloch state to the localized self-trapped state involves surmounting an energy barrier of 1.7 meV; ie, a reasonable fraction of the energy of the pertinent acoustical phonons. At temperatures above 35K, the STH migrates by thermally-activated hopping, with an energy of 62 meV, implying a thermal trap depth of about 0.12 eV. The estimated pre-exponential factor is very small, consistent with a highly localized state. With a high density of hole traps, the decay kinetics is first-order, but for a low trap density, an accurate fitting requires a diffusion-limited kinetics formulation. For temperatures below 30K, the migration rate begins slowly to rise again, indicating a tunneling process which can be fit to a T^{-n} behavior with n dependent on the history of the crystal.

THE SELF-TRAPPED HOLE

In silver chloride (but, interestingly, not in silver bromide) the ground state of the hole at the top of the valence band is not the free Bloch polaron, but rather is a localized self-trapped hole (STH), centered on a silver ion and stabilized by an axial Jahn-Teller distortion[1,2]. The fact that the positively charged hole resides on a cation, instead of on one or two anions (as in the alkali halides), is a reflection of the strong mixing of silver d-states with chloride p-states, which also causes the top of the valence band to occur at the L-point instead of at the zone center.

Because a potential well in three dimensions has no bound state unless it is sufficiently deep, the lattice must first be locally distorted before a free hole can "fall" in. Thus, if the coupling of the hole to the appropriate lattice modes is not very strong, it has been argued that there must be an activation energy barrier in the self-trapping transition[3-5]. This energy comes mainly from components of short-wavelength transverse acoustical lattice modes, which in AgCl are in the range of 4 to 8 meV[6]. One therefore expects the height of any barrier to be of the order of a few meV. If such an activation energy is indeed necessary, then one result would be that the trapping rate for free holes would <u>decrease</u> as the temperature is decreased, in contrast to the usual behavior of photocarriers.

The presence and magnitude of the proposed energy barrier was established[7,8] by determination of the temperature dependence of the fraction of photoholes in AgCl that reach a set of fixed traps before becoming self-trapped in the perfect regions of the crystal. The concentration of STH thus formed was monitored by means of electron paramagnetic resonance. In such an experiment, it is essential that electron-hole recombination be prevented. This was accomplished in two different ways: (a) the crystal was doped with both Cu^+, which served as hole traps, and Cu^{2+}, which served as a source of holes upon low-temperature irradiation with sub-bandgap blue light[9]; (b) a different crystal was doped with Pd^{2+}, which introduced both the free dopant ion (an electron trap) and also impurity-vacancy-complexes (which can also trap holes at low temperatures[10]); in this case, band-to-band excitation was produced by near UV illumination, and the photoelectrons were scavenged by the free Pd^{2+} ions.

In each crystal, it was found that more STH were produced by irradiation at higher temperatures (say, 30K) than at lower temperatures. It also appeared that the capture of holes at many of the traps was not markedly dependent on temperature; this was especially well-established for the copper-doped crystals. Thus, one could draw the qualitative conclusion that the transition from free hole to STH requires some thermal activation. Analysis of the data by means of an Arrhenius-type representation then showed that for temperatures above 10K this activation energy is 1.7 meV; this is a substantial fraction of the energies of the acoustical phonons that are believed to be involved. Below about 10K, however, the Arrhenius plot was found to tail off, indicating the onset of tunneling transitions. Actually, if one estimates[4] the tunneling rate to be about 10^{12} sec^{-1} × exp [-E_{act}/(hυ/4)], then for a phonon energy of 5 meV, the tunneling and thermally activated rates would be about equal at 10K, in agreement with observation.

MIGRATION OF THE SELF-TRAPPED HOLE

It was not possible to measure directly the hopping rate of the STH, but useful information could nevertheless be obtained by studies of the decay of the electron paramagnetic resonance (EPR) signal[8]. The observation that during decay the EPR signals from trapped electron centers decreased, while those due to other trapped hole centers increased, showed that this decay was really the result of migration to traps of the STH and not due to recombination with thermally-freed electrons. In particular, small amounts of Fe^{2+} impurity were converted into the trigonal Fe^{3+} center[11], which required also the aquisition of an additional cation vacancy. Since the silver ion vacancy is not mobile at these temperatures (50K and below), this leads to the interesting conclusion that the capture of a STH by a Fe^{2+}-vacancy complex must also cause ejection of a neighboring substitutional Ag^+ into the interstitial sites.

It was found that for a sufficiently large concentration of hole traps (several tens of ppm), the decay of the EPR signal from the STH followed first-order kinetics reasonably well, with the possible exception of the very early time period. The temperature-dependence of the decay rate showed two quite distinct regions: above 35K, the time constant decreased with increasing temperature according to an Arrhenius relation, while below 30K, at which it exhibited a maximum, the time constant was almost athermal, and actually decreased slowly with decreasing temperature. The two regions were sharply delineated.

The Arrhenius behavior in the high-temperature region is that expected for phonon-assisted hopping. The time constant had an activation energy of 61 - 63 meV and a pre-exponential factor slightly less than 10^{-5} sec. Now, for temperatures in the range of 35 to 50K, a major share of the acoustical phonons are almost classically excited. Then to a good approximation[12-14], the activation energy for hopping is about half the trap depth, yielding a value for the depth of about 0.12 eV (much larger, of course, than the energy barrier against the self-trapping transition). This trap depth for AgCl may be compared to the activation energy in AgBr of about 0.1 eV for the high-temperature mobility of the photohole -- a process interpreted by Sumi and Toyozawa[15] as being limited by thermally-activated transitions between the free hole and the metastable (in AgBr) self-trapped state.

What can we learn from the value of the pre-exponential factor in the phonon-assisted range? For random migration, one expects the time constant to have the form $1/4\pi r$ ND, where we take the capture radius to be 1.5×10^{-8} cm for uncharged traps, and we estimate the trap density to be about $10^{18}/cm^3$. This gives a value for the pre-exponential factor of the diffusivity, $D_o \approx 7 \times 10^{-7}$ cm^2/sec, which must also equal $(1/6) \lambda^2 z \upsilon_o P$, where the jump distance λ for the hopping hole is about 3×10^{-8} cm, the coordination number z is 12, the attempt frequency υ_o is taken to be about 3×10^{12}/sec, and P is the probability that the hole can execute the hop within one vibrational period if the lattice locally has the necessary thermal excitation. We get $P \approx 10^{-4}$, and from Emin[12] we can then estimate the electron transfer integral to be about 1% of the appropriate phonon energy, or 0.06 meV, giving a bandwidth for the STH states of almost 2 meV. Clearly, the STH in AgCl is quite localized.

In later experiments, in which the density of hole traps (mainly Cu^+) was only 1 - 2 ppm, it was found[16] that, as expected, the decay kinetics did not at all resemble first-order. Second-order kinetics, in which the initial ratio of concentrations of STH and traps was allowed to vary as a fitting parameter, gave a reasonable fit, but with systematic deviations that were outside experimental error. The diffusion-limited kinetics formulation of Waite[17], however, gave excellent fits for the entire series of twelve decay runs over the temperature range 20 - 50K, with no systematic deviations, and with reasonable values for the fitting parameters appearing in Waite's equation. In this sort of treatment, it is recognized that close pairs will combine first, thus changing the shape of the joint probability function. In contrast, ordinary chemical kinetics assumes that only the amplitude, and not the shape, of this function changes during the reaction; this "works" in the chemical case because the success probability in a collision is much less than unity. The significance of the present work is that for defect reactions in solids, the success probability is often near to unity, and one expects that ordinary chemical kinetics should be valid only in the long-time limit. These measurements thereby provide one of the very few quantitative demonstrations of the appropriateness of the diffusion-limited formulation.

Let us now consider the temperature range below 30K, which must involve a tunneling migration within the narrow band of small polaron states. It is not at all obvious that this should be observable in a real crystal, with its distribution of lattice strains[18]. One possibility, however, is that here we are seeing the migration only of those STH which are far from any edge dislocation. It is known that except at high temperatures, edge dislocations in AgCl are negatively charged[18], and one expects that any holes localized nearby would quickly be drawn into the dislocation and would disappear from view of the EPR spectrometer. For screw dislocations, however, there is no charge, and one expects to see the effects of their strain fields.

Four sets of decay measurements[20] over the temperature range 4.5K to 40K were made on two crystals, one doped with copper and the other with palladium. The kinetics was more complicated than a simple first- or second-order decay, and the hopping rate for the STH in a given specimen was taken to be proportional to the initial slope of the concentration vs. time plot. All four series of runs showed a minimum hopping rate near 30K, with a slow rise as the temperature was lowered toward 10K; below 10K, however, the results were greatly perturbed by what appears to be the effect of a set of very shallow traps, apparently about a few meV deep.

For each series of measurements, the data over the range 10 to 30K are well fit by a T^{-n} dependence of the hopping rate, where $\underline{n}$ ranged from near zero for one series to 1.9 ± 0.2 for another. Presumably, this variation reflects the uncontrolled effects of strains, perhaps from uncharged screw dislocations, which act to inhibit the low-temperature rise in the tunneling rate. On these grounds, one proposes that the series of runs yielding a $T^{-1.9}$ dependence of hopping rate must represent our "best" crystal.

The overall temperature dependence of the hopping rate of the STH -- a high-temperature phonon-assisted region and a low-temperature tunneling region, with a minimum mobility between the two regions -- is reminiscent of the hopping of the positive muon in high-purity copper and aluminum[21]. In these cases also, the tunneling rate follows a T^{-n} law except at the lowest temperatures, where it levels off. The values of the exponent $\underline{n}$ for metals seem to be approximately 0.7 or less, and they have been interpreted in terms of a theory of Kondo[22], in which the tunneling rate is determined by electronic excitation across the Fermi surface. Such a mechanism clearly cannot be operating in ionic crystals; moreover, our largest value of the exponent $\underline{n}$ is considerably greater than the maximum of 1.0 allowed by the Kondo theory.

Recently, the hopping rate of muons has been measured in an ionic crystal, KCl, and a qualitatively similar temperature-dependence was observed[23]. In this case the exponent $\underline{n}$ is even higher, 3.3, presumably reflecting both the absence of conduction electron drag and also a low level of lattice strains. A detailed theory, however, for quantum diffusion of the STH and the muon in ionic crystals is not yet available.

ACKNOWLEDGEMENTS

This work was supported by the National Science Foundation under Grant DMR-8722476. The single crystals used in the EPR experiments were prepared from boules kindly grown by Mr. Charles B. Childs of this department.

REFERENCES

1. M. Höhne and M. Stasiw, Phys. Stat. Solidi **33**, 405 (1969).

2. A review of the STH in AgCl is given in L. Slifkin, in "The Physics of Latent Image Formation", edited by A. Baldereschi et al (World Scientific), 35 (1984).

3. Y. Toyozawa, Semicond. and Insul. **5**, 175 (1983).

4. N. Mott and A. Stoneham, J. Phys. **C10**, 3391 (1977).

5. E. Rashba, Izv. Akad. Nauk SSR, Ser. fiz **40**, 1973 (1976).

6. P. Vijayaraghavan, R. Nicklow, H. Smith and M. Wilkinson, Phys. Rev. **B1**, 4819 (1970).

7. E. Laredo, L. Rowan and L. Slifkin, Phys. Rev. Lett. **47**, 384 (1981).

8. E. Laredo, W. B. Paul, L. Rowan and L. Slifkin, Phys. Rev. **B27**, 2470 (1983).

9. D. Burnham and F. Moser, Phys. Rev. **136**, A744 (1964).

10. R. Eachus and R. Graves, J. Chem. Phys. **65**, 5445 (1976).

11. K. Hay, D. Ingram and A. Tomlinson, J. Phys. **C1**, 1205 (1968).

12. D. Emin, Adv. Phys. **22**, 57 (1973); **24**, 305 (1975).

13. N. Mott, Mater. Res. Bull. **13**, 1389 (1978).

14. E. Gorham-Bergeron, and D. Emin, Phys. Rev. **B15**, 3667 (1977).

15. A. Sumi and Y. Toyozawa, J. Phys. Soc. Jpn. **35**, 137 (1973).

16. W. B. Paul, S. Goldenberg, L. Rowan and L. Slifkin, Cryst. Latt. Defects **15**, 197 (1987); W. B. Paul, Ph.D. Thesis, University of NC (1987).

17. T. Waite, Phys. Rev. **107**, 463 (1957).

18. D. Emin, J. Solid State Chem. **12**, 246 (1975).

19. R. Whitworth, Adv. Phys. **24**, 203 (1975).

20. S. Goldenberg, Ph.D. Thesis, University of NC (1988); S. Goldenberg, L. Rowan and L. Slifkin, to be published.

21. O. Hartmann, et al, Phys. Rev. **B37**, 4425 (1988); R. Kadono et al, Phys. Rev. **B39**, 23 (1989); S. Cox, J. Phys. - **C20** 3187 (1987) and references therein.

22. J. Kondo, Physica **125B**, 279 (1984); **126B**, 377 (1984).

23. R. Kiefl, R. Kadono, J. Brewer, G. Luke, H. Yen, M. Celio and E. Ansaldo, Phys. Rev. Lett. **62**, 792 (1989).

Chapter 6

RELAXATION, DRIFT, AND DIFFUSION

ENERGY AND PHASE RELAXATION IN DISORDERED SEMICONDUCTORS

R. Fischer, E.O. Göbel, G. Noll, P. Thomas and A. Weller
Fachbereich Physik
and
Zentrum für Materialwissenschaften
Philipps Universität Marburg
Renthof 5, D–3550 Marburg, FRG

ABSTRACT

The relaxation of optically generated excitations in disordered semiconductors differs from that observed in crystalline semiconductors because of an extremely wide spectrum of relaxation rates. Experimental methods to study relaxation in disordered materials include time resolved luminescence, pump and probe measurements and transient four–wave mixing experiments. Typical experimental results obtained on mixed crystals and amorphous semiconductors are discussed. Theoretical approaches towards an understanding of the underlying microscopic processes are reviewed. In particular, it is shown that phase relaxation in disordered semiconductors is strongly influenced by Anderson localization.

1. INTRODUCTION

Information about quasiparticle interactions in semiconductors can be obtained from a study of the relaxation of the optically excited sample towards equilibrium. Since the availability of very short laser pulses, the underlying microscopic processes can be studied in the time domain. Typical experimental methods include time resolved luminescence, pump and probe experiments, photon echo and transient four–wave mixing experiments.

Very short time scales of the order of ps down to fs are typical for intraband relaxation processes in crystalline semiconductors. A relaxation of the optically excited nonequilibrium distribution towards a thermalized distribution, subsequent cooling of this thermal distribution and, finally, recombination is observed. In addition, relaxation of the optically induced orientation can be detected in semiconductors where optical excitation leads to an unisotropic population of states in k–space [1,2]. The relevant interaction processes are schematically depicted for the case of a band–to–band excitation in a perfect crystalline semiconductor in Fig. 1.

An entirely different situation is encountered in disordered semiconductors. In particular, relaxation in amorphous semiconductors is characterized by an extremely wide spectrum of relaxation rates, the corresponding times ranging from fs up to days. A qualitative interpretation of this remarkable property can

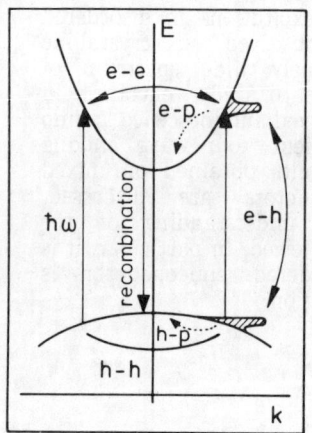 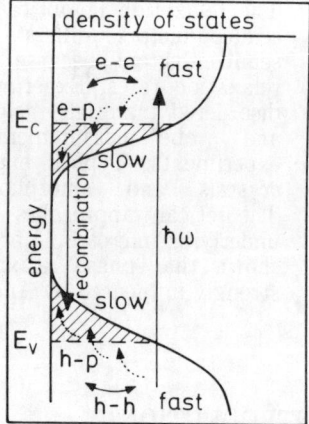

Fig. 1   Schematical representation of optical generation, recombination and interactions of carriers in crystalline semiconductors.

Fig. 2   Schematical representation of optical generation, recombination and interactions of carriers in dis–ordered semiconductors. Interaction processes within delocalized states are fast, those in localized states are slower. Note that in contrast to the crystalline case excitations are created in all states separated by the photon energy $\hbar\omega$ due to the lack of the k–selection rule.

be obtained on the basis of Fig. 2 showing a schematic density of states diagram of a disordered semiconductor. Localized states are formed at the band tails, separated from the delocalized states by mobility edges (E_c and E_v). Carriers initially located in, or scattered into, the localized region can relax further down in energy by performing a hopping process or by being reemitted above the mobility edge with subsequent trapping into a lower level (multiple trapping). At low temperatures hopping prevails. The corresponding hopping rate depends on the density of available states and therefore can be quite small. The observed very longest relaxation times, however, are certainly not of electronic origin alone. Stuctural relaxation, which acts back on the electronic properties, is known to exist in amorphous semiconductors [3].

The relaxation processes related to hopping and multiple trapping dominate on a time scale much longer than picoseconds. They have been investigated for a variety of disordered systems like amorphous semiconductors, mixed crystals and molecular crystals. The relevant hopping theory is well developed [4,5], and has been tested succesfully against computer simulations. This theory represents a sophisticated and complete approach to the general hopping relaxation problem, its application to real situations at arbitrary temperatures, however, is not easy to perform. We will give an example for the special situation where the temperature is low and only hops downward in energy need to be considered, which allows an appreciable simplification of the theoretical treatment.

The mobility edge looses its meaning as a strict boundary between localized and extended states at finite temperatures [6]. Electron–phonon coupling, which is also responsible for the hopping processes, induces in addition a delocalization if the inelastic scattering length is smaller that the localization length. Anderson localized states close to the mobility edge have a large localization length. In fact the localization length diverges at the mobility edge. States close to the mobility edge consequently cannot be classified as either extended or localized if the electron–phonon interaction is taken into account. Since the phonon induced delocalization mechanism is not an instantaneous process, it is of extraordinary interest to study the carrier–phonon interaction in this regime. Furthermore, it has been argued, that due to the fractal nature of the pure electronic eigenstates close to the mobility edge their interaction with phonons is dramatically increased [7]. The formation of small polarons has been predicted, which should influence both the dynamics as well as the equilibrium properties like the dc−conductivity. Although it appears that there is no need to invoke small polaron formation for the interpretation of equilibrium transport data [8], the question is not completely settled. Time resolved studies of the interactions taking place close to the mobility edge may give an answer to this important problem.

The interaction of optically excited carriers or excitons amongst each other or with the lattice is expected to contribute to a fast phase relaxation of the excitations. Therefore, dephasing experiments provide an alternative method to gain information about possible interactions in crystalline and amorphous solids. The notion of phase relaxation or dephasing is used in different meanings in the literature. A clear−cut definition is, therefore, needed. In the context of spin dynamics dephasing is easily defined in terms of the Bloch Equations. On the other hand, nonlinear optical experiments can be described in the framework of the Optical Bloch Equations. Following Yajima and Tahira [9] we shall use this approach to introduce dephasing and to identify it as the decay of the nonlinear polarization as a function of the time separation of two excitation laser pulses. Dephasing is due to carrier−carrier (or exciton−exciton) interaction and carrier−phonon interaction in crystalline semiconductors. We shall denote these processes as quasiparticle interactions in the following. Phase relaxation is at least as fast as energy relaxation. The dephasing rates are known to be extremely fast in crystals [10]. Dephasing can, however, also be studied in systems where independent two−level absorbers are excited optically. In fact the classical photon echo experiment by Kurnit and coworkers [11] on ruby belongs to this case. Nevertheless, quasiparticle interaction again causes dephasing. The situation may be different in disordered semiconductors. We will show that, in contrast to the linear optical response regime [12], the nonlinear response is strongly affected by Anderson localization [13]. The relevant theory [14] and the interpretation of the most recent experimental data in terms of this disorder induced dephasing is still in a preliminary state. We think, however, that the traditional style of this meeting allows for a somewhat speculative interpretation of our data.

In chapter 2 we shall first present the theoretical background of the experiments, which are discussed in chapter 3. It will become clear that the study of nonlinear transient optical processes in disordered semiconductors is an exciting and promising field, which may yield in the near future new contributions to so far unsolved problems concerning interactions in disordered solids.

2. THEORETICAL CONSIDERATIONS

We first discuss energy relaxation having in mind time resolved luminescence, pump and probe and transient four–wave mixing experiments. We then introduce the notion of dephasing within the framework of the Optical Bloch Equations applied to the special case of inhomogeneously distributed uncoupled two–level absorbers. We also briefly discuss the case of momentum dephasing which may be confused with optical dephasing. The more general case of coupled two–level absorbers, i.e. the most simple tight binding model for a disordered semiconductor, is then treated and the third order response to the optical excitation field is calculated. We begin by briefly discussing various kinds of disorder which have to be considered in disordered semiconductors.

2.1 Disorder

Disorder is usually envisaged as a fluctuation of local potentials (diagonal disorder). One distinguishes short–range correlated and long–range correlated potentials. Both kinds of disorder are present in amorphous semiconductors [8]: the short–range correlated potential fluctuates on a length scale of typically 5 Å and gives rise to the tails of both the conduction band and the valence band [15], to Anderson localization, to the mobility edges and to the relaxation of the k–selection rule. The long–range potential certainly exists in doped amorphous semiconductors and is then of electrostatic nature [16]. It is characterized by a correlation length of the order of 100 Å, or more, and can therefore be taken as the origin of a local spatial modulation of the band structure. In mixed semiconductor crystals the situation is different. Consider, e.g., CdS_xSe_{1-x}, where the conduction band is formed from states related to Cd and is, therefore, unaffected by the fluctuating composition of Se and S atoms, which is assumed to be random. Holes in the valence band, on the other hand, will be affected by the compositional disorder, in particular if they are close to the valence band edge. The k–selektion rule is well preserved for transitions exceeding the optical gap. Excitonic transitions in this system require a special consideration [17,18]. The exciton has a Bohr radius a_b being larger than the correlation length of the compositional disorder. Two cases can be envisaged: The exciton averages the fluctuating potentials over regions of linear dimensions of the order of a_b. The effective disorder potential then has a correlation length of the same order. This situation can be assumed for excitons of higher energy. The exciton can then be localized in this potential fluctuation as a whole. On the other hand, an energetically deep lying exciton can be bound to a short range fluctuation, where the hole becomes trapped. It is then plausible that the strength of the interaction between excitons themselves and between excitons and phonons depends on the energy [18]. Further theoretical work, however, is certainly needed to put these considerations on a more solid basis.

2.2 Energy Relaxation Involving Localized States

Once a carrier is initially excited into a localized state with energy E or has been trapped there it may relax further into states with even lower energy. Two paths are possible: the carrier can be released from the trap across the mobility edge with subsequent retrapping into a state with lower energy or it can find its way towards a lower lying trap by performing a sequence of hops. Both possibilities will contribute at sufficiently high temperatures. In fact, a clear distinction between these two processes is questionable because at high temperatures a well defined mobility edge does not exist and the mobility within "localized states" is comparable to that of delocalized states close to the mobility edge. Nevertheless, the concept of a reasonably well defined boundary $\tilde{E}_c$ between highly conducting states and those with a small mobility may still be applicable, although it will in general not coincide with the mobility edge defined for $T = 0$.

We may then still use the multiple trapping expression for the energy relaxation of the carrier packet within the localized states. The maximum of the carrier distribution $E_d(t)$ shifts with time according to [19]

$$E_d(t) = \tilde{E}_C + kT \ln \nu_0 t \tag{1}$$

The photoluminescence signal as a function of time and photon energy then can be calculated if the momentary distribution is evaluated on the basis of eq. (1).

The situation becomes conceptually simpler at low temperatures because on the one hand the boundary between weakly conducting and highly conducting states becomes identical with the mobility edge at $T = 0$ (note, however, that emission of phonons is always possible which may blur the mobility edge.) and on the other hand a release across the mobility edge will be slower than a hop to a state with energy close to or lower than the initial energy. This is the regime of energy relaxation due to hopping. The theory has been elaborated in ref. 4 and 5. The equilibrium hop rates are usually taken to be

$$\Gamma_{ij} = \nu_0 \, e^{-2\alpha R_{ij}} \, e^{-\frac{(\Delta \epsilon_{ij} + |\Delta \epsilon_{ij}|)}{2 \, kT}} \tag{2}$$

where the attempt to escape frequency ν_0 is of the order of $10^{12}/s$. There is an exponential dependence of the rate on the spatial separation R_{ij} of the localized states involved. The decay length of these states is denoted by α^{-1}. The dependence on the energy difference $\Delta \epsilon_{ij}$ is described by a Boltzmann factor for hops upwards in energy and is unity otherwise. It should be noted that under high excitation conditions the phonon temperature T may be higher than the bath temperature.

In order to extract microscopic parameters like decay length, slope parameters of the band tails, etc., from the experiments, the hopping theory has to be evaluated for a realistic model system. This can be done most simply for low temperatures where upward hops can be completely neglected. The luminescence intensity at time t in general can be expressed in terms of a conditionally averaged Green's function $\langle G_{ij}(t) \rangle_{ij}$ [4,5]. Let us consider as an example excitons localized in a tail with tailing parameter ϵ_0. The luminescence signal at energy ϵ_i is then proportional to

$$L(\epsilon_i) \sim N(\epsilon_i) \int \langle G_{ij}(t) \rangle_{ij} N(\epsilon_j) \, d^3R_{ij} \, d\epsilon_j \qquad (3)$$

If the experimental time window is short enough, all higher order contributions to the Born series for the Green's function except the first two terms can be neglected (i.e. sequences with more than one hop are neglected). The total Green's function can then be expressed by the diagonal term only, resulting in [20]

$$L(\epsilon_i) \sim N(\epsilon_i) \, e^{-t(\tau_r^{-1} + gn_i)} (1 - tg(1 - n_i)) \qquad (4)$$

with

$$n_i = \int_{-\infty}^{\epsilon_i} N(E) \, dE \qquad (5)$$

$$n = \int_{-\infty}^{0} N(E) \, dE \qquad (6)$$

$$g = n \int d^3R \, \nu_0 \, e^{-2\alpha R} \qquad (7)$$

where a phenomenological recombination rate τ_r^{-1} has also been introduced.

These expressions are applicable if the excitation (charge carriers or excitons) definitely stays within the localized regime for experimentally relevant times. Initial relaxation from delocalized states into the localized states can be neglected provided the time resolution of the experiment is long compared to the time scale of this initial process. On the other hand, this initial relaxation of carriers across the mobility edge can now be resolved experimentally (see chapter 3).

2.3 Energy Relaxation Involving Extended States

Extended state relaxation in crystalline as well as in amorphous semiconductors is generally studied by femtosecond excite and probe experiments [21,22,23,24,25] with the transmission and/or the reflectivity as the probe. The evaluation of the data, however, very often requires a thin film analysis taking into account interference effects in optically thin samples. The presence of the carriers within the extended states generated by the "excite–pulse" leads to an increased absorption (photoinduced absorption) in amorphous semiconductors, opposite to the case of crystalline materials, where a photoinduced bleaching is found. This difference can be attributed to the lack of the k–selection rule for optical transitions in amorphous semiconductors, which gives rise to a strong intraband free carrier absorption overcompensating the bleaching of the interband transitions [21]. The relaxation and thermalization of carriers within the extended states is usually described by a simple Drude model with a dielectric function according to

$$\epsilon = \epsilon_\infty - \frac{\omega_p^2}{\omega\,(\omega - i/\tau)} \tag{8}$$

where τ is a momentum relaxation time, which has been determined experimentally to be of the order of $0.5 \cdot 10^{-15}$ to $3 \cdot 10^{-15}$ s in a strongly disordered material [22,23] and ω_p is the plasma frequency. Energy relaxation of carriers within the extended states is provided by optical phonon interaction with an energy relaxation rate of about 0.5 eV/ps [21]. The relevant mechanisms, however, have not yet been investigated in great detail (for a brief discussion see ref. 21, page 104). Finally, the carriers will be trapped into localized states, which should result in a substitution of the Drude contribution to the dielectric function by a respective localized carrier contribution, most simply described e.g. by an ensemble of Lorentz–oszillators.

2.4 Optical Dephasing

2.4.1 The Optical Bloch Equations.

In this section we follow the discussion by Yajima and Tahira [9]. Let us consider an ensemble of mutually independent two-level absorbers with inhomogeneously distributed energy separations ω_i. The distribution function $g(\omega_i)$ has a width $\delta\omega$. The equation of motion for the density matrix of one absorber with diagonal element ρ_d (actually the difference between the upper and lower level diagonal elements) and nondiagonal element ρ_{nd} can be written as

$$\frac{\partial \rho_d}{\partial t} = -\frac{2i}{\hbar}(H'\rho_{nd} - \rho_{nd}{H'}^*) - (\rho_d - \rho_{do})/T_1 \qquad (9a)$$

and

$$\frac{\partial \rho_{nd}}{\partial t} = -\frac{i}{\hbar}{H'}^*\rho_d - i\omega_i\rho_{nd} - \rho_{nd}/T_2 \qquad (9b)$$

where H' is the interaction Hamiltonian containing two light pulses at t and $t+\tau$, respectively (a δ-function pulse shape is assumed for simplicity), with photon energy $\hbar\omega$ travelling in the directions $\mathbf{k_1}$ and $\mathbf{k_2}$, respectively. The relaxation of the population ρ_d towards the equilibrium ρ_{do} is described by the rate T_1^{-1}, while the dephasing rate is T_2^{-1}. A configuartional averaged polarization of the form

$$\langle P(t) \rangle \sim e^{i\mathbf{k_3}\cdot\mathbf{r}} e^{-t/T_2} g(t-2\tau) \qquad (10)$$

is obtained from eqs. (9a,b) in third order of the external electric field. For $\tau > 0$, the polarization appears only for $\mathbf{k_3} = 2\mathbf{k_2} - \mathbf{k_1}$. Here $g(t)$ is the Fourier transform of the distribution function $g(\omega_i)$. If we do not introduce the phenomenological dephasing mechanisms, $T_2^{-1} = 0$, we find a polarization which at time $t = 2\tau$ does not depend on the puls separation τ. This simple example provides a definition of optical dephasing valid for more general cases: Theoretically we calculate $\langle P(2\tau) \rangle$. If we find an exponential decay, the dephasing rate T_2^{-1} is given by $\exp(-2\tau/T_2)$. Experimentally, a signal proportional to the time-integrated square of $\langle P(t) \rangle$ is measured, which essentially also decays exponentially like $\exp(-4\tau/T_2)$.

The signal is not symmetric in time in a three beam experiment (stimulated photon echo) where a third pulse travelling in the direction corresponding to k_3 hits the sample after a time $t + T$ (A more detailed description of the experimental configurations is given in section 3.1.). For $\tau < 0$ a signal is observed only in the direction $k_4' = k_1 + k_3 - k_2$, whereas for $\tau > 0$ the signal is emitted into the direction corresponding to $k_4 = k_2 + k_3 - k_1$. In contrast, the signal is symmetric in time in the homogeneous case, with a decay according to $\exp(-2\tau/T_2)$. The stimulated photon echo experiment therefore is a sensitive probe for the amount of inhomogeneous broadening of an optical transition [26].

2.4.2 <u>Momentum dephasing</u>. Sometimes the notion of dephasing is used in a different context [27]. Consider a particle (wave packet) brought to a particular site in a system described by the usual Anderson Hamiltonian with diagonal disorder

$$H = \sum_i \epsilon_i n_i + J \sum_{ij} c_i^\dagger c_j, \qquad (11)$$

where ϵ_i are the site energies distributed according to a distribution function $g(\epsilon_i)$ and J is the nearest neighbour coupling. The width of $g(\epsilon)$ is W, so $\eta = W/J$ is the relevant parameter of this problem. We ask for the propagation of the particle in the course of time, i.e. we consider the time dependence of the mean square displacement $\langle R^2(t) \rangle$. Using the mode–coupling theory [28,29] of Anderson localization we find [27,30] that $\langle R^2(t) \rangle \sim t$ for long times provided the parameter η is smaller than a critical value, $\eta < \eta_{critical}$. Otherwise the long time limit of $\langle R^2(t) \rangle = \xi_{loc}^2$, where ξ_{loc} is the localization length. For short times, the particle moves like a quantum mechanical wave packet, i.e. $\langle R^2(t) \rangle \sim t^2$. The cross–over from the wave–like propagation to diffusive motion appears at a time τ_{deph}, which is the momentum dephasing time. Only for times shorter than τ_{deph} the k–vektors of the partial waves building up the wave packet are sufficiently well preserved, and the coherence of the partial waves still exists. For larger times the elastic scattering destroys the coherence of the partial waves completely and the propagation of the particle is no longer wave–like but becomes diffusive.

It is evident that the mechanism leading to momentum dephasing is in general not identical with that producing optical dephasing, as introduced in the last section. Consider, e.g., a semiconductor model [31] where both the valence band and the conduction band are described by Anderson Hamiltonians like that of eq. (11) (c.f. Fig. 3a). If we increase the disorder parameters beyond their

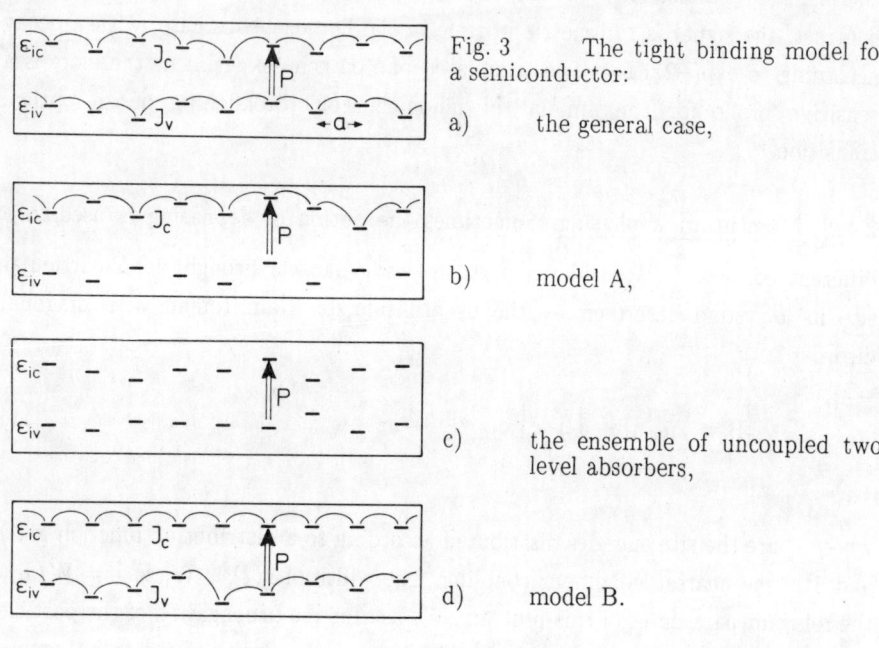

Fig. 3 The tight binding model for a semiconductor:

a) the general case,

b) model A,

c) the ensemble of uncoupled two level absorbers,

d) model B.

critical values we finally end up with an ensemble of completely decoupled two–level absorbers (Fig. 3c). This system then shows no optical dephasing, if we omit the phenomenological rate T_2^{-1}, see eq. (10). On the other hand, the momentum dephasing rate increases with increasing disorder parameter η. In the localized regime it in fact diverges and it makes no more sense to talk about momentum dephasing.

2.4.3 <u>Optical dephasing due to disorder.</u> We now turn to the question how the disorder parameter η influences optical dephasing in the sense of section 2.4.1. We consider a semiconductor model [31] built up from two Anderson Hamiltonians, describing the valence band (disorder parameter η_v) and the conduction band (disorder parameter η_c) respectively, see Fig. 3a. The energy distributions of the valence band and the conduction band site energies are assumed to be mutually uncorrelated. We do not consider any interaction between excitations and quasiparticles in order to emphasize the role of static disorder. We here also neglect the interband electron–hole coupling and treat the excitations as mutually independent electron–hole pairs (no excitons).

Our task now is the calculation of the averaged polarization $\langle P(2\tau)\rangle$ to third order in the external field. From time dependent perturbation theory we get for the relevant term

$$\langle P(2\tau)\rangle \sim \langle \mathrm{Tr}(P\, e^{-iH_v\tau}\, P\, e^{-iH_c\tau}\, P\, e^{iH_v\tau}\, P\, e^{iH_c\tau})\rangle \tag{12}$$

where

$$P = p \sum_i c_i^\dagger c_i + \mathrm{HC} \tag{13}$$

is assumed to be site–diagonal for simplicity and p is the intrasite dipol element. It turns out, that the model depicted in Fig. 3a can not easily be treated. However, a basic physical insight can be gained [14] by considering two somewhat simplified models called model A (Fig. 3b) and B (Fig. 3d). In model A, the coupling between the disordered states at lower energies is zero, consequently we have $\eta_v = \infty$. The upper levels have arbitrary disorder and coupling. Model A reduces to the limit of uncoupled two–level absorbers for $\eta_c = \infty$ (Fig. 3c). In model B the upper levels are all at the same energy and they are coupled, i.e. $\eta_c = 0$. The lower levels have arbitrary disorder and coupling. This model contains the perfect crystal for the special case $\eta_v = 0$. Both the uncoupled absorbers and the perfect ordered system do not show optical dephasing. In the latter case the system can be viewed as an ensemble of uncoupled two–level systems classified by k.

We find as a result that the decay of the nonlinear polarization for model A is essentially determined by the decay of the conduction band density–density correlation function. Applying the mode coupling theory to this function, we can calculate its time dependence and the well known results of Anderson localization are obtained. Results of our calculation are depicted in Fig. 4. The nonlinear polarization decreases with decreasing η_c. It decays to zero for $\eta_c < \eta_{c\ critical}$ ($\eta_{c\ critical} = 11.8$ for a box shaped $g(\epsilon_i)$ [30]). The decay of the envelope can be approximately determined and we find that

$$\langle P(2\tau) \rangle \sim e^{-\frac{6D\tau}{a^2}} \tag{14}$$

where D is the dc–diffusion constant and a the lattice constant. The decay rate, therefore, is given by $T_2^{-1} = 3D/a^2$. Note that this dephasing is observed even when the phenomenological rate (c.f. eq. (9b)) is absent. The above relation yields a value of $T_2 = 160$ fs if numbers for D and a, as generally accepted for, e.g., a–Si:H [32,8], are inserted. It should be noted, however, that due to the schematic

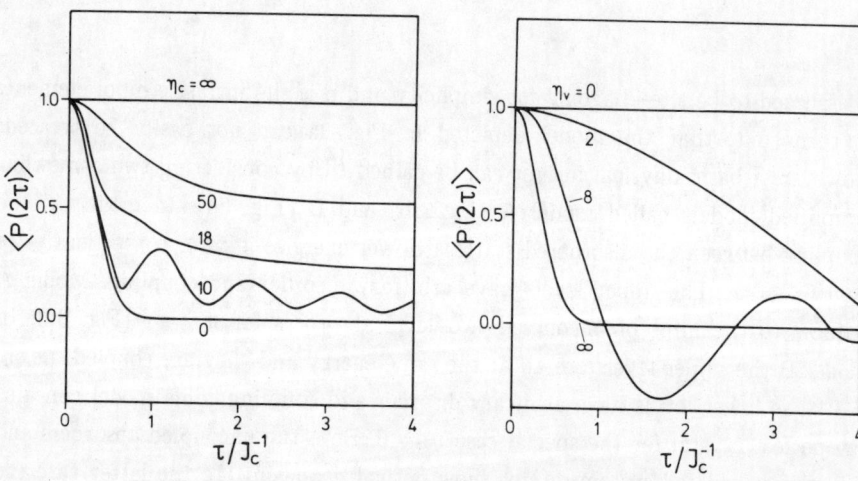

Fig. 4   Decay of the nonlinear polarization for model A, $\eta_v = \infty$. The disorder in the conduction band is varied. $\eta_c = \infty$ corresponds to the completely uncoupled case.

Fig. 5   Decay of the nonlinear polarization for model B, $\eta_c = 0$. The coupling in the valence band is varied. $\eta_v = 0$ corresponds to the perfectly ordered case.

nature of our model this result for T_2 should only be taken as a rough estimate. It applies if at least one type of carriers (electrons) is delocalized. On the other hand, for $\eta_c > \eta_{c\ critical}$, the polarization reaches a saturation value roughly proportional to ξ_{loc}^2 if the electrons are localized. No decay at all is found if $\eta_c = \infty$ in accordance with the model of uncoupled two-level absorbers (eq. 10). In this model, which starts from a completely decoupled ensemble of two-level absorbers, the origin of dephasing is the coupling between the individual absorbers. This case has been studied for the strong disorder case already by Root and Skinner [33].

The evaluation of model B shows that the decay of the nonlinear polarization is dominated by the current-current correlation function (for the holes) instead of the density-density correlation function. The results depicted in Fig. 5 are obtained using the appropriate methods based again on mode-coupling arguments [34]. In contrast to model A the dephasing rate now increases with increasing disorder (η_v). The current does not decay for vanishing disorder and the corresponding correlation function is constant, as is the nonlinear polarization. In this model, which starts from a perfect crystalline case, the origin of momentum dephasing and that of disorder induced optical dephasing are identical, namely elastic scattering [35]. More generally one can state that dephasing occurs whenever a selection rule is relaxed. In model A optical dephasing results from a relaxation of the "real-space" selection rule allowing for a transfer of the excitation between different centers due to coupling. In model B, on the other hand, the k-selection rule becomes relaxed due to scattering of carriers between states with different k.

3. EXPERIMENTAL RESULTS

3.1 Experimental Configurations

Experimental configurations frequently applied to study the ultrafast relaxation processes in semiconductors are sketched in Figs. 6 to 8. In a luminescence experiment (Fig. 6) a short light pulse with photon energy $\hbar\omega_{exc}$

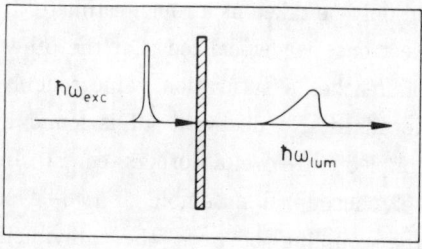
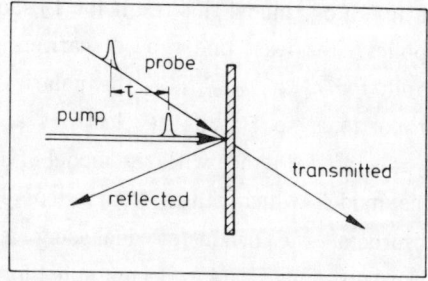

Fig. 6 Schematical representation of a luminescence experiment.

Fig. 7 Schematical representation of a pump and probe experiment.

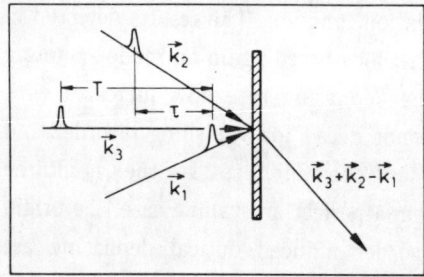
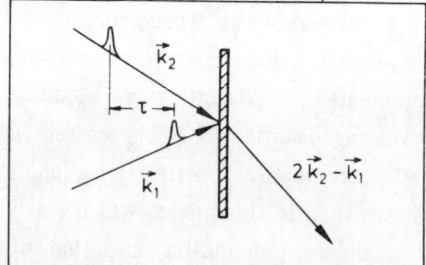

Fig. 8 Schematical representation of a transient grating experiment.

Fig. 9 Schematical representation of a photon echo experiment.

excites the sample and the resulting luminescence signal can be recorded time and energy ($\hbar\omega_{lum}$) resolved. The pump and probe experiment utilizes two excitation pulses with relative delay τ. The photoinduced changes of the transmitted and the reflected probe signals are monitored. In a transient grating configuration (Fig. 8), two pulses reach the sample with a time delay τ, while a third pulse at time T is diffracted by the grating produced by the first two pulses. For $\tau = 0$ this experiment is called incoherent transient grating while for finite τ (coherent transient grating) this technique allows to investigate the optical dephasing taking place during the delay time τ. Optical dephasing can also be studied by a photon echo experiment (Fig. 9) which is a special case ($\mathbf{k_3} = \mathbf{k_2}$, $T = \tau$) of the transient grating (or stimulated photon echo) configuration. For more details concerning transient grating experiments see ref. 36.

3.1 Trapping

The initial trapping of photoexcited carriers from extended into localized states has been studied by excite and probe techniques (see, e.g., 21–25, 37). Recently, it has been demonstrated that this trapping process can also be measured by incoherent transient grating experiments [38,39]. The interference pattern formed by the two impinging pulses is transformed into a corresponding pattern of optically excitet states (carriers, excitons, etc.), which contribute to the dielectric function. Thus the dielectric function of the laser excited region is periodically modulated giving rise to diffraction of the probe beam, hitting the sample with a variable delay T. The decrease of the diffracted probe signal with increasing delay time T reflects the decay of the photoinduced grating due to recombination and diffusion. In addition, however, spectral relaxation causing changes of the dielectric function can also be monitored. This is particularly true for relaxations of carriers from excited to localized states. One of the advantages is the coherent nature of this experiment allowing better signal to noise ratios. In addition, Fabry–Perot effects do not obscure the experimental results. Fig. 10 depicts as an example the results of a transient grating experiment on a a–Si:H film at three different temperatures. An initial fast decay of the transient grating signal to a quasi–stationary value is observed in all samples. The decay constant

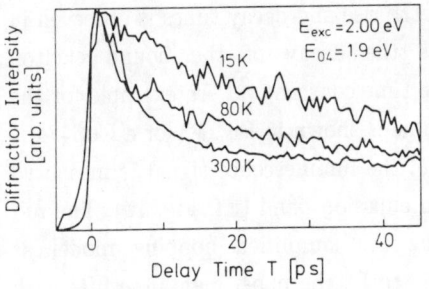

Fig. 10 Incoherent transient grating experiment in a–Si:H. Plotted is the first order diffracted light intensity vs. delay time. Data are shown for three different temperatures of 15K, 80K and 300K. The photon energy of the excitation and probe pulses was 2 eV, the energy of the optical gap of the a–Si:H film amounts to 1.9 eV.

of this initial decay depends slightly on temperature and density of localized states. The initial decay is attributed to the refractive index changes related to the trapping of the carriers [38,39]. The trapping time in a–Si:H amounts to 10 ps at room temperature in accordance with the values obtained by pump and probe experiments [23,25,40]. The slight increase of the trapping time with decreasing temperature can be explained by partial saturation of the shallow localized states due to decrease of the energy relaxation rate within the localized states. The trapping times from extended to localized states decrease with increasing defect density to about 1 ps or even shorter values [24,40].

3.3 Relaxation within Localized States

Relaxation within localized states has been studied in great detail on a timescale longer than 1 ns by time resolved luminescence [41,42] and time of flight experiments [43]. Excite and probe techniques [21] as well as picosecond luminescence have been employed to study the initial relaxation within localized states on a picosecond time scale. Disordered crystalline materials like heavily doped or alloy–semiconductors are of particular interest, because optical selection rules can still be applied [44,45,46,47]. The effect of exciton localization by disorder in the II–VI mixed crystal semiconductor CdS_xSe_{1-x} on the optical spectra is demonstrated in Fig. 11. The emission is dominated by the recombination of excitons bound to neutral donors (I_2–line) in the binary CdSe and for $x > 0.97$ and $x < 0.2$, while in the intermediate composition range, $0.2 < x < 0.97$ localized exciton recombination is observed [20,48]. The time behaviour of the luminescence in this composition range is significantly different from the bound exciton regime. A pronounced dispersion of the rise and decay times is observed in the "localized exciton" regime, whereas the decay of the bound exciton luminescence can be described by one single time constant [20]. An example for the time behaviour in the localized exciton regime is shown in Fig. 12 for a CdS_xSe_{1-x} crystal with $x = 0.42$. The time variation of the luminescence signal is measured at various photon energies within the main emission band (c.f. Fig. 11). The full lines represent calculated curves applying the simplified hopping model as described in section 2.2. This model is able to fit the experimental results with reasonable values for the respective microscopic parameters, which are the

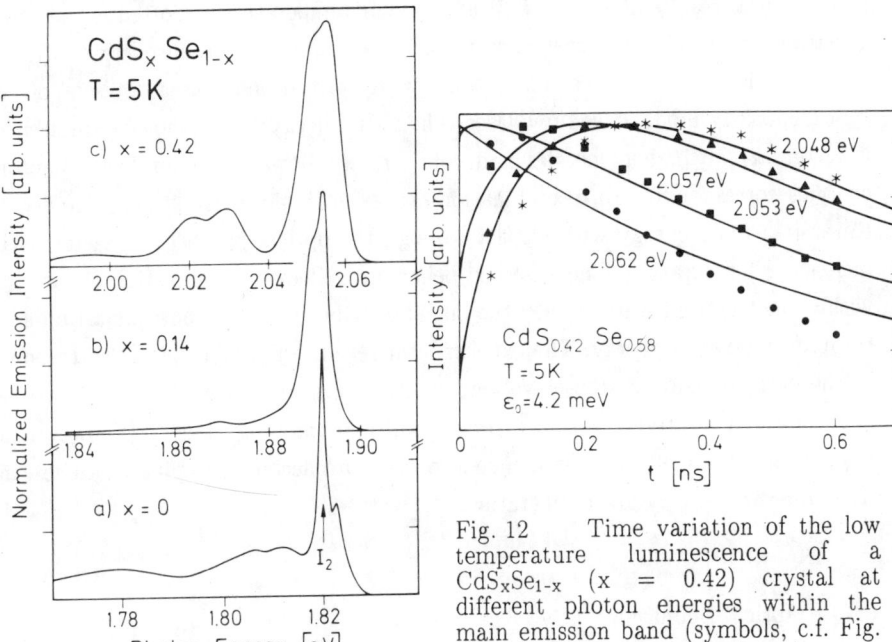

Fig. 11 Low temperature photo-luminescence spectra of CdS_xSe_{1-x} for different values of x. The emission line labelled I_2 corresponds to recombination of a neutral donor bound exciton, while for x = 0.42 the emission originates from excitons localized by disorder.

Fig. 12 Time variation of the low temperature luminescence of a CdS_xSe_{1-x} (x = 0.42) crystal at different photon energies within the main emission band (symbols, c.f. Fig. 11) and calculated curves (lines).

localization length, tail density of states parameter, and the attempt to escape frequency.

Picosecond luminescence studies of amorphous semiconductors have been recently performed in great detail by Fischer et al.[49,50]. Fig. 13 depicts experimental results of a–Si$_{1-x}$C$_x$:H at different temperatures. Plotted is the time variation of the photoluminescence intensity at different photon energies within the emission band after excitation with picosecond laser pulses of a synchroneously mode locked dye laser. The initial decay of the photoluminescence is faster at the higher energies and continuously slows down for lower photon energies corresponding to recombination deeper in the localized states. The decay times at a fixed energy with respect to, e.g., the optical gap become faster with increasing temperature. The observed behaviour reflects the relaxation of carriers within the localized state regime: the initial distribution after photoexcitation and trapping relaxes in energy. The experimental results shown in Fig. 13 correspond to the early relaxation steps involving shallow localized states. The data can be described quantitatively within a simple multiple trapping model [50], however, a more realistic model should of course also take into account hopping processes, in particular at the very low temperatures.

3.4 Optical Dephasing

Dephasing experiments are performed in the stimulated photon echo configuration which corresponds to the coherent transient grating experiments (see Fig. 8). The two pulses creating the interference grating are now delayed with respect each other ($\tau \neq 0$), while the " probe beam" is at a fixed time delay T, which has to be sufficiently large. For a more detailed description of this technique see, e.g., [26]. Recently, the dephasing of excitonic [2] and free carrier excitations [10] in GaAs has been determined by the stimulated photon echo technique. It has been shown that optical dephasing occurs on a time scale of a few femtoseconds if free carrier states high in the conduction and valence band, respectively, are optically coupled [10]. Resonantly excited free, i.e. delocalized, exciton states in III–V and II–VI semiconductors exhibit dephasing times of the order of a ps at low temperatures [2,51,52]. On the other hand, dephasing times of the order of ns are observed when isolated absorbers in a solid matrix are optically excited, like in the classical photon echo experiments on, e.g., ruby [11]. The spectrum from delocalized to strongly localized states can be covered in a single sample of a disordered semiconductors if different energies are considered.

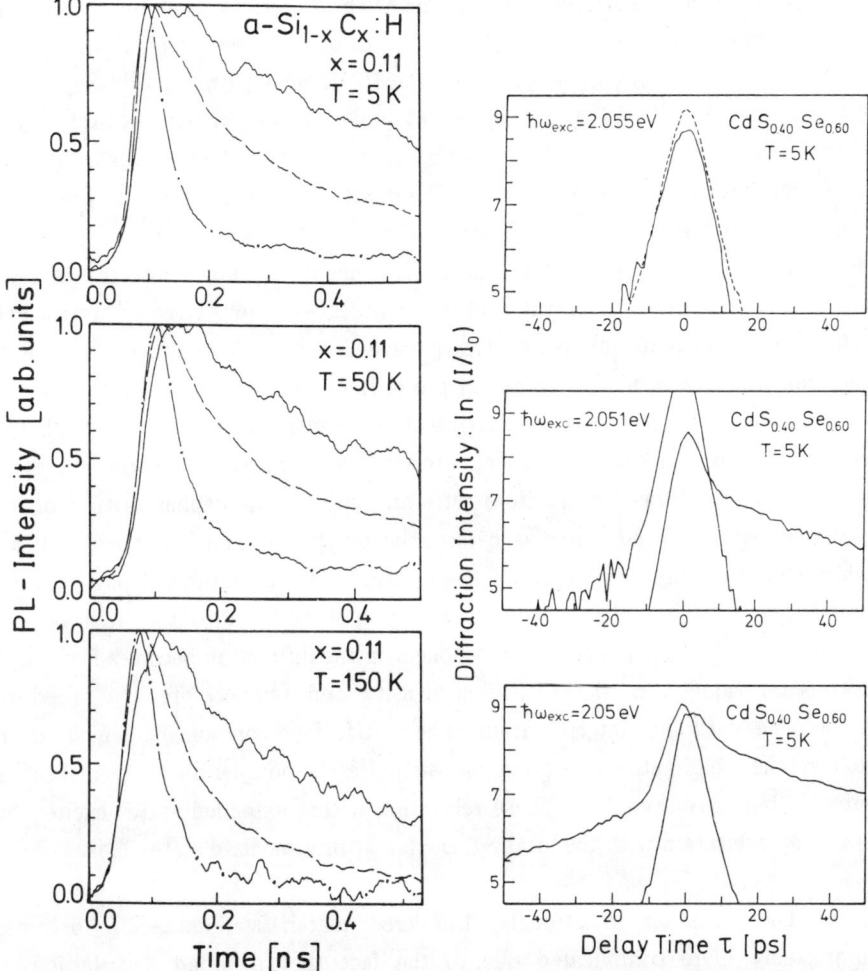

Fig. 13 Time resolved photo-luminescence of an a–Si$_{1-x}$C$_x$:H (x = 0.11) film at three different temperatures. The three traces at each temperature correspond to different photon energies within the high energy side of the cw–photoluminescence spectrum:
———— 1.55 eV,   – – – – 1.68 eV,
– · – · – 1.88 eV.

Fig. 14 Coherent transient grating experiment (stimulated photon echo) of a CdS$_{0.4}$Se$_{0.6}$ mixed crystal for different laser photon energies corresponding to transitions within the localized exciton band. The two traces shown at each photon energy correspond to the diffraction order +1 and −1 respectively.

We, therefore, have performed optical dephasing experiments in the stimulated photon echo configuration (see Fig. 8) in disordered semiconductors. As an example we have chosen and compared the II–VI mixed crystal CdS_xSe_{1-x} ($x \simeq 0.4$) and a–Si:H [53]. In Fig. 14 we present some typical low temperature results obtained on CdS_xSe_{1-x} with the excitation photon energy tuned through the band of the localized exciton absorption [54]. The figures show in a semilogarithmic plot the simultaneously detected diffraction intensity of order 1 and −1 versus the delay time of the two excitation pulses. The signals are not symmetric in time and are complementary for the diffraction order +1 and −1, apart from some differences in absolute intensities, clearly revealing the inhomogeneous nature of the transition involved. The decay of the diffraction signal basically follows the field amplitude autocorrelation function of the excitation pulses at high photon energies, whereas a slow component is clearly resolved as the photon energy is lowered and more localized exciton states are created. The dephasing time of the lowest energy localized states as determined by this technique is about 140 ps, which is appreciably larger than for free excitons. This increase in dephasing time of excitons is attributed to a reduced scattering of localized excitons with respect to that of free excitons. The slow component in the diffraction becomes faster and disappears rapidly as the excitation density and temperature are raised in accordance with our interpretation. The initial fast component, which again follows the field auto–correlation of the pulses is not definitely identified at present, but may arise from phase relaxation within extended states excited by two–step excitation with the localized exciton as intermediate state [53].

The situation in strongly disordered materials like a–Si:H becomes appreciably more complicated due to the fact that a broad distribution of electronic states is always excited because of the lack of **k**–selection rules. Figure 15 depicts the experimental result of a phase coherence experiment in a–Si:H obtained using excitation pulses with an intensity autocorrelation function width of 0.3 ps and a field autocorrelation width of 0.15 ps [53]. A time resolution for the determination of the phase coherence time of about 30 fs can be obtained under these conditions from the predicted temporal shift of the diffraction maximum of the +1 and −1 order [26]. The experimentally observed shifts are of the order of this temporal resolution and consequently at present we are only able

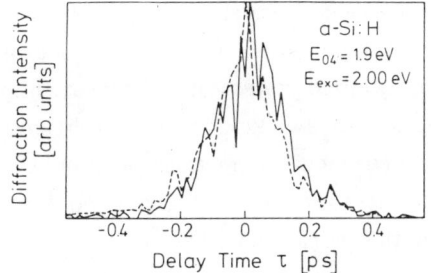

Fig. 15 Coherent transient grating experiment in a–Si:H at $T = 5$ K.

to give an upper limit for optical phase relaxation times of $T_2 \leq 30$ fs under the present excitation conditions. This value is lower than the value of about 150 fs determined from eq. (14) and thus it may be concluded that quasiparticle scattering still dominates compared to disorder. This point, however, needs further experimental, and theoretical, investigations.

4. SUMMARY

In this lecture we have discussed time resolved nonlinear optical experiments, which give information about relaxation processes on a very short time scale. In particular, energy and phase relaxation in disordered semiconductors are expected to show a rich variety of features because of the simultaneous presence of delocalized and localized states. We have given some experimental results and have presented the present state of their interpretation. The theoretical approaches towards a full understanding of the observed phenomena is still in a preliminary state. It has to take into account the full apparatus of Anderson localization theories, as we have shown by treating the most simple semiconductor model. We have obtained a rough estimate for the disorder induced dephasing time for a–Si:H. Experimental results yield lower values which may indicate that quasiparticle scattering still is the main phase destroying mechanism. Dephasing times of localized excitons in mixed semiconductor crystals are much longer than those of free excitons or even free carrier excitations, again demonstrating the fundamental effects of localization.

Trapping of carriers from delocalized into localized states can be studied by incoherent transient grating experiments. A trapping time of the order of 10ps is determined for a–Si:H. Relaxation within localized states can be conveniently investigated by time resolved luminescence. These experimental techniques alltogether are capable to provide information about interactions in disordered semiconductors which are otherwise difficult to obtain. At the same time, however, it has become clear that much theoretical work has to be done to unequivocally identify the microscopic processes that determine relaxation of excitations in disordered materials.

ACKNOWLEDGEMETS

We are indebted to our coworkers J. Feldmann, Ch. Lonsky and G. Peter for their contributions and helpful discussions. Fruitful cooperation with the group of Prof. Klingshirn, Kaiserslautern, is also gratefully acknowledged. This work is supported by the Deutsche Forschungsgemeinschaft.

REFERENCES

1. J.L. Oudar, A. Migus, D. Hulin, G. Grillon, J. Etchepare, A. Antonetti, Phys. Rev. Lett. **53**, 384 (1984)

2. J. Kuhl, A. Honold, L. Schultheis, C.W. Tu, in *Festkörperprobleme*, **29**, ed. by U. Rössler (Vieweg, Braunschweig), (1989), in press

3. J. Kakalios, R.A. Street, Phys. Rev. **B34**, 6014 (1986)

4. B. Ries, H. Bässler, M. Grünewald, B. Movaghar, Phys. Rev. **B37**, 5508 (1988)

5. M. Grünewald, B. Movaghar, J. Phys.: Condensed Matter **1**, 2521 (1989)

6. D.J. Thouless, Phys. Rev. Lett. **39**, 1167 (1977)

7. M.H. Cohen, E.N. Economou, C.M. Soukoulis, J. Non–Cryst. Solids **59&60**, 15 (1983); Phys. Rev. Lett. **51**, 1202 (1983); Phys. Rev. **B29**, 4496 and 4500 (1984)

8. H. Overhof, P. Thomas, *Electronic transport in hydrogenated amorphous semiconductors*, Springer Tracts in Modern Physics, Vol. 114, (Springer, Berlin, Heidelberg, New York, 1989)

9. T. Yajima, Y. Taira, J. Phys. Soc. Japan **47**, 1620 (1979)

10. P.C. Becker, H.L. Fragnito, C.H. Brito Cruz, R.L. Fork, J.E. Cunningham, J.E. Henry, C.V. Shank, Phys. Rev. Letters **61**, 1647 (1988)

11. N.A. Kurnit, I.D. Abella, S.R. Hartmann, Phys. Rev. Lett. **13**, 567 (1964); I.D. Abella, N.A. Kurnit, S.R. Hartmann, Phys. Rev. **141**, 391 (1966)

12. U. Dersch, M. Grünewald, H. Overhof, P. Thomas, J. Phys. **C20**, 121 (1987)

13. P.W. Anderson, Phys. Rev. **109**, 1492 (1958)

14. Ch. Lonsky, P. Thomas, A. Weller, to be published

15. C.M. Soukoulis, M.H. Cohen, E.N. Economou, Phys. Rev. Lett. **53**, 616 (1984); E.N. Economou, C.M. Soukoulis, M.H. Cohen, S. John, in *Disordered Semiconductor*, ed. by M.A. Kastner, G.A. Thomas, S.R. Ovshinsky (Plenum, New York), 681 (1987)

16. J. Tauc, Mat. Res. Bull. **5**, 721 (1970)

17. S.D. Baranovski, A.L. Efros, Fiz. Tekh. Poluprovdn. **12**, 2233 (1978) (Sov. Phys.–Semicond. **12**, 1328 (1978))

18. E. Cohen, M.D. Sturge, Phys. Rev. **B25**, 3828 (1982)

19. T. Tiedje, A. Rose, Solid State Commun. **37**, 49 (1980)

20. S. Shevel, R. Fischer, E.O. Göbel, G. Noll, P. Thomas, C. Klingshirn, Journ. Luminesc. **37**, 45 (1987)

21. J. Tauc, *Festkörperprobleme* **22**, ed. by P. Grosse (Vieweg, Braunschweig), 85 (1982); Z. Vardeny, J. Tauc, in: *Disordered Semiconductors*, ed. by M.A. Kastner, G.A.Thomas, (Plenum Press, New York, London), 339 (1987)

22. J. Heppner, J. Kuhl, Proc. 18th Int. Conf. Phys. Semicond., ed. by O. Engström (World Scientific, Singapore), 1033 (1987)

23. C. Tanguy, D. Hulin, A. Mourchid, P.M. Fauchet, S. Wagner, Appl. Phys. Lett. **53**, 680 (1988)

24. J. Kuhl, E.O. Göbel, Th. Pfeiffer, A. Jonietz, Appl. Phys. **A34**, 105 (1984)

25. P.M. Fauchet, D. Hulin, A. Migus, A. Antonetti, J. Kolodzey, S. Wagner, Phys. Rev. Lett. **57**, 2438 (1986)

26. A.M. Weiner, S. DeSilvestri, C.P. Ippen, J. Opt. Soc. **B2**, 654 (1985)

27. R.F. Loring, S. Mukamel, J. Chem. Phys. **85**, 1950 (1986)

28. W. Götze, Phil. Mag. **B 43**, 219 (1981)

29. W. Götze, P. Thomas, J. Non–Cryst. Solids **97&98**, 217 (1987)

30. P. Thomas, A. Weller, J. Non–Cryst. Solids **97&98**, 245 (1987); A. Weller, Diploma Thesis, (Marburg, 1987)

31. S. Abe, Y. Toyozawa, J. Phys. Soc. Jpn. **50**, 2185 (1981)

32. S. Komuro, Y. Aoyagi, Y. Segawa, S. Namba, A. Masuyama, H. Okamoto, Y. Hamakawa, Appl. Phys. Lett. **42**, 807 (1983)

33. L. Root, J.L. Skinner, J. Chem. Phys. **81**, 5310 (1984)

34. P. Prelovšek, Phys. Rev. **B23**, 1304 (1981)

35. J. Hegarty, M.D. Sturge, J. Opt. Soc. Am. **B2**, 1143 (1985)

36. H.J. Eichler, P. Günther, D.W. Pohl, *Laser Induced Dynamic Gratings*, Springer Series Optical Science **50**, (Springer, Berlin, Heidelberg, New York) (1986)

37. P.M. Fauchet, D. Hulin, J. Optical Soc. Am. **B6**, 1024 (1989)

38. G. Noll, E.O. Göbel, J. Non–Cryst. Solids **97&98**, 141 (1987)

39. G. Noll, E.O. Göbel, U. Siegner, in *Ultrafast Phenomena IV*,(Springer Series in Chemical Physics), in press

40. Th. Pfeiffer, J. Kuhl, E.O. Göbel, L. Palmetshofer, J. Appl. Phys. **62**, 1850 (1987)

41. R.A. Street, Adv. Phys. **30**, 593 (1981)

42. B. Wilson, in : *Disordered Semiconductors*, ed. by M.A. Kastner, G.A. Thomas, (Plenum Press, New York, London), 349 (1987)

43. P.G. LeComber, W.E. Spear, Phys. Rev. Lett. **25**, 509 (1970); for a review see: J.M. Marshall, Rep. Prog. Phys. **46**, 1235 (1983)

44. E. Cohen, M.D. Sturge, Phys. Rev. **B25**, 3828 (1982)

45. E.O. Göbel, W. Graudszus, Phys. Rev. Lett. **48**, 1277 (1982)

46. J.A. Kash, Phys. Rev. **B27**, 7069(1984)

47. J.A. Kash, A. Ron, E. Cohen, Phys. Rev. **B28**, 6147 (1983)

48. S. Permogorov, A. Reznitskii, S. Verbin, G.O. Müller, P. Flögel, M. Nikiforova, phys. stat. sol. (b)**113**, 589 (1982)

49. R. Fischer, E.O. Göbel, G. Noll, J. Non–Cryst. Solids **98&98**, 519 (1987)

50. R. Fischer, E.O. Göbel, Proc. Intern. Conf. on Amorphous and Liquid Semicond., (1989), in press

51. L. Schultheis, J. Kuhl, A. Honold, C.W. Tu, Phys. Rev. Lett. **57**, 1797 (1986)

52. J.M. Hvam, C. Dörnefeld, H. Schwab, phys. stat. sol. (b) **150**, 387 (1988); C. Dörnefeld, J.M. Hvam, IEEE Journ. Quant. Electr. **25**, 904 (1989)

53. G. Noll, PhD Thesis, Marburg (1989)

54. similar experiments performed in a two beam configuration have been reported by
    H.E. Swoboda, F.A. Majumder, C. Weber, R. Renner, C. Klingshirn, G. Noll, E.O. Göbel, S. Permogorov, A. Reznitskii, Proc. 19th Int. Conf. Phys. Semicond., ed. by W. Zawadski (Inst. of Phys. Polish academy of Sciences), 1323 (1988);
    H.–E. Swoboda, F.A. Majumder, R. Renner, C. Weber, M. Sence, Lu Jie, G. Noll, E.O. Göbel, J. Vaitkus, C. Klingshirn, phys. stat. sol. (b) **150**, 749 (1988);
    C. Dörnefeld, H. Schwab, C. Weber, R. Renner, C. Klingshirn, J.M. Hvam, G. Noll, E.O. Göbel, A. Reznitskii, S.A. Pendjur, O.N. Talensky, phys. stat. sol (b), in press;
    see also ref. 52

LOW TEMPERATURE TRANSPORT IN a-Si:H
by M.Kemp and M.Silver

Department of Physics and Astronomy, University of North Carolina,
Chapel Hill, North Carolina 27599-3255, USA.

Abstract

The low temperature drift mobility of a-Si:H is interpreted in terms of a hopping model. The model predicts a critical temperature above which the transport level is pinned at the conduction band edge giving an activated behavior of the drift mobility and below which the transport level falls in the gap leading to non-dispersive transport. It also predicts a rise of the drift mobility below the critical temperature when the transport level falls faster than the temperature decreases.

1. Introduction

Electron transport in a-Si:H as measured by time-of-flight is dispersive in the range 300-125°K. The drift mobility at the transit time is activated and has been shown[1] to be consistent with multiple trapping between a well defined trapping level and the conduction band. It is believed that this is so because of a sharp drop in the density of states (DOS) at an energy equal to the activation energy of the drift mobility. Recent measurements[2,3] in the range 125-10°K have revealed a sharp transition from dispersive to non-dispersive transport below 100°K accompanied by a sharp increase of the drift mobility. The data has been analyzed in terms of a two channel model[3], transport being through the bottom of the conduction band in the high temperature range and through a well defined hopping path 0.01 eV above the sharp drop of the DOS at low temperatures. Silver and Spear[4] have derived an estimate of the temperature for the onset of non-dispersive transport on the basis that a critical temperature below which no deep trapping occurs during a transit time can be defined for diffusion limited capture. However, that model does not address the origin of the mobility rise and predicts a voltage dependent critical temperature, not in agreement with the observations. Nevertheless, it provides a starting point for a more careful analysis of the data by stressing that the absence of trapping leads to non-dispersive transport.

In this paper, we show that both temperature regimes can be imbedded in a single model based on a hopping transport level[5,6]. We show that at high temperature, the transport level is pinned at the conduction band edge because of the strong overlap of these states with the localized states, giving rise to multiple trapping. As the temperature is lowered below a critical value, the transport level starts to fall in the band gap leading to less and less dispersive transport; if it drops faster than the temperature decreases, the drift mobility increases. Eventually, the transport level reaches the sharp drop in the density of states where it gets pinned yielding saturation of the drift mobility. We conclude by showing that the above behavior is extremely sensitive to the shape of the DOS.

2. The model

Hopping transport following the introduction of a group of carriers in an amorphous semiconductor proceeds in two stages[5]: 1) at early times, transport occurs by carriers hopping down in energy and 2) at times longer than the segregation time, by carriers hopping up to a transport level. For an exponential DOS, the segregation time increases exponentially with decreasing temperature[6] reaching low temperature values incompatible with the observed transit time. However, a DOS with a sharp drop will pin the transport level and freeze the segregation time at a value smaller than the transit time. For such a density of states, one needs only be concerned with the hop up regime. Our model calculations therefore proceed in two stages: 1) for a given DOS, we calculate the position of the transport level and 2) we relate the transport level to the drift mobility.

2.1 The transport level

A transport level can be defined as that level where most carrier transport takes place. In the case of delocalized transport, carriers trapped in the localized states are emitted to the bottom of the conduction band where they drift under the action of the electric field. In the case of band tail hopping, trapped carriers can hop to any state but at a rate that depends on both the distance between the initial and the final state and on their energy difference if hopping is up in energy. On the average, a carrier will hop more frequently to a state where the hopping rate is large and such a state will in fact be used more frequently as a level for transport. It is the purpose of a transport level calculation to look for states where this rate is maximum, thus to look for states where transport occurs predominantly.

The hopping rate ν_h from an initial state i to a final state f where the final state lies in energy above the initial one is

$$\nu_h(\varepsilon_i,\varepsilon_f)=\nu_o\exp(-2\gamma R_{if}-\beta(\varepsilon_f-\varepsilon_i)) \tag{1}$$

where ν_o is a typical phonon frequency ($10^{13}s^{-1}$), γ is the reciprocal of the localization length and R_{if} the distance between the states. A typical hopping rate is calculated following the introduction of an average lattice where independent of the initial state, the final state lies at a distance that is a function of its energy only

$$R_{if}=R_f(\varepsilon_f)=(g(\varepsilon_f)\delta)^{-\frac{1}{3}} \tag{2}$$

where $g(\varepsilon_f)$ is the density of states at the final level and δ is an energy spread around the level introduced after a proper discretization of the DOS; in our calculations, we use $\delta=0.01eV$. The reciprocal of the localization length γ is energy dependent and is given by [7]

$$\gamma=\gamma_o(\varepsilon_c-\varepsilon)^\lambda \tag{3}$$

reflecting the delocalization of the wave function at the band edge and the increased localization as one proceeds down in energy. In our calculations we use $\lambda=0.6$ and γ_o such that $\gamma^{-1}=5a$ at $\varepsilon_c-\varepsilon=0.8eV$.

A typical transport level calculation[5] would proceed by assuming a constant γ and by maximizing the hopping rate using $dv_h/d\varepsilon_f = 0$. This procedure is valid as long as one has a monotonically increasing DOS and a constant γ but is not if an energy dependence such as equation 3 is introduced. In fact, two cases arise

$$1 - \frac{dv_h}{d\varepsilon_f} = 0, \quad \frac{d^2 v_h}{d\varepsilon_f^2} < 0 \quad \text{and} \quad v_h(\varepsilon_f) < v_h(\varepsilon_c) \tag{4}$$

in which case because of the strong overlap of the states at the band edge with the localized states, the hopping rate is maximum at the band edge and ε_f is only a local maximum, and

$$1 - \frac{dv_h}{d\varepsilon_f} = 0, \quad \frac{d^2 v_h}{d\varepsilon_f^2} < 0 \quad \text{and} \quad v_h(\varepsilon_f) > v_h(\varepsilon_c) \tag{5}$$

for which ε_f is an absolute maximum. The first case is effective for all temperatures above a critical value T_{crit} given by

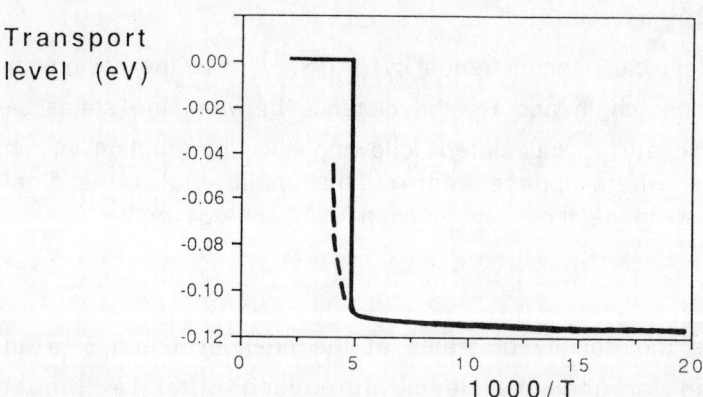

Figure 1: the transport level as a function of temperature for the density of states given by equation 10. The full line is the transport level and the dashed one, the local maximum as obtained from equation 7. Above T_{crit}, the transport level is pinned at the conduction band edge; below it, it discontinuously falls on the local maximum line.

$$T_{crit} = \frac{|\varepsilon|}{2\gamma R k_b}\bigg|_{\varepsilon=\varepsilon_{crit}} \tag{6}$$

$$\varepsilon_{crit} = 3(1-\lambda)\frac{g(\varepsilon_{crit})}{g'(\varepsilon_{crit})}$$

At all temperatures below T_{crit}, the transport level ε_t is obtained from the solution of

$$k_b T\,(2\gamma R)\left(\frac{g'}{3g} + \frac{\lambda}{|\varepsilon|}\right)\bigg|_{\varepsilon=\varepsilon_t} = 1 \tag{7}$$

A typical application of these equations is presented in figure 1. A striking feature of our model is the fact that the transport level falls abruptly in the gap at the critical temperature.

2.2 The drift mobility

The low field hopping mobility between states i and f is

$$\mu_h(\varepsilon_i,\varepsilon_f) = \frac{qR_{if}^2}{6k_b T}\,\nu_h(\varepsilon_i,\varepsilon_f) \tag{8}$$

The drift mobility is a weighted average of the carriers' mobility. In the hop up regime, most carriers lie below the transport level and hop up to it. If we now consider that enough trapping/detrapping events have occured to define a demarcation level ε_d above which the occupation function f is Boltzmann-like, and if we neglect the contribution to transport of all carriers above the transport level, then

$$\mu_d = \frac{\int_{\varepsilon_d}^{\varepsilon_t} d\varepsilon\, g(\varepsilon)\, f(\varepsilon)\, \mu_h(\varepsilon,\varepsilon_t)}{\int_{\varepsilon_d}^{\varepsilon_t} d\varepsilon\, g(\varepsilon)\, f(\varepsilon)} \tag{9}$$

Equations 6,7 and 9 form the backbone of our analysis of the drift mobility data.

3. Discussion

We now consider the application of the model to a density of states similar to the one advocated by Spear[2]

$$\frac{1}{g} = \frac{1}{g_1} + \frac{1}{g_2} \qquad (10)$$

$$g_1 = g_c \exp(\frac{\varepsilon}{k_b T_1})$$

$$g_2 = g_c \exp(\frac{\Delta}{k_b T_1}) \exp(\frac{\varepsilon - \Delta}{k_b T_2})$$

with $\Delta = -0.12$eV, $T_1 = 1100^\circ K$, $T_2 = 30^\circ K$ and $g_c = 2.5 \times 10^{21} cm^{-3} ev^{-1}$. The transport level for this DOS is shown in figure 1 and the drift mobility in figure 2. Above $200^\circ K$ the mobility is activated with an energy of 0.10eV. Below this critical temperature, the mobility rises up and reaches a maximum of 0.06 $cm^2 V^{-1} s^{-1}$ at $75^\circ K$. Below it, the mobility decreases with an activation energy of 5 meV. Therefore, the model reproduces the major features of the observed datas. We now proceed to the identification of the mechanisms responsible for this behavior and give estimates of the parameters that control the shape of the curve.

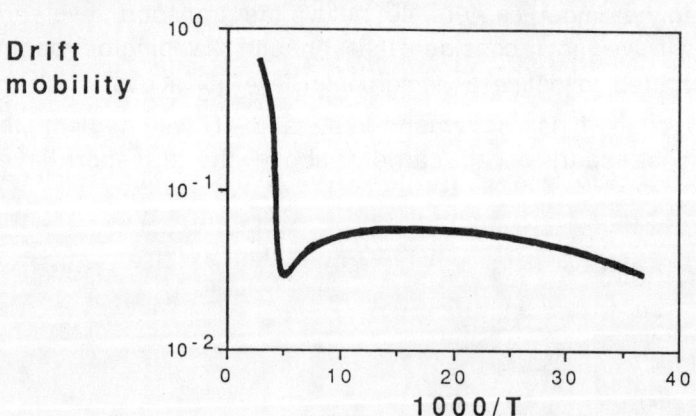

Figure 2: drift mobility as a function of temperature for the density of states given by equation 10.

Above the critical temperature, the transport level is pinned at the conduction band edge. Because of the sharp drop in the density of states at Δ, the trapped carrier population peaks there and <u>not</u> at the demarcation level. Therefore, the carrier exchange proceeds mainly between Δ and ε_c. and as was shown by Marshall et al.[1], the problem reduces to single level multiple trapping for which the drift mobility is decreasing with an of activation energy Δ. The dispersion of the packet of carriers also increases with decreasing temperature because of the larger spread of release times around Δ.

At the critical temperature, the transport level moves in the band gap giving smaller release times from the traps. To understand the mobility rise, consider the expression of the drift mobility (equation 9) with a constant density of states: integrating and expanding the exponential to second order yields

$$\mu_d = \left(\frac{v_o}{6k_bT} R^2(\varepsilon_t) \exp(-2\gamma R(\varepsilon_t))\right) \frac{1}{1+\frac{\beta(\varepsilon_t-\Delta)}{2}} \qquad (11)$$

The origin of the rise is contained in the second term of the expression: if $\varepsilon_t-\Delta$ decreases faster than β increases, the drift mobility increases. In other words, the rise is the result of the competition between the decrease of the transport level and the decrease of the temperature. If the former is faster, then the mobility rises; if the latter wins, the mobility decreases. Since the evolution of the transport level is solely determined by the shape of the density of states, a rise or a continual decrease of the drift mobility can be achived with different DOS. Our simulations have in fact showed that this behavior is an extremely sensitive function of the DOS as is illustrated in figure 3 which was obtained with $T_1=800°K$ and $T_2=30°K$.

Independent of the evolution of the drift mobility, carrier propagation undergoes a sharp transition from dispersive to non-dispersive transport at the critical temperature whose origin is as follows. When the transport level moves down in energy, the

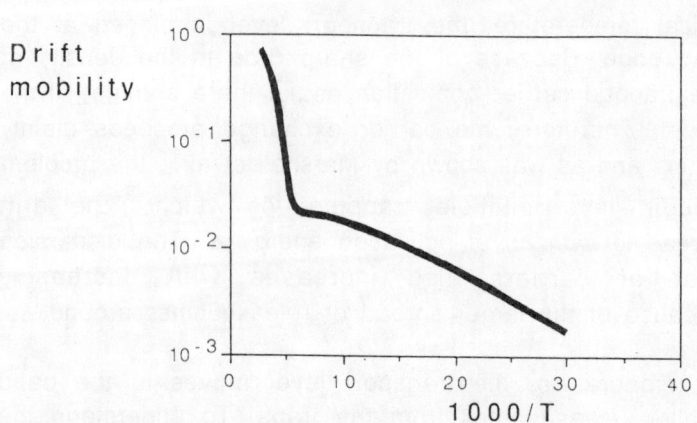

Figure 3: drift mobility for the density of states given by equation 10 with $T_1=800°K$ and $T_2=50°K$.

number of states available for trapping decreases. Because of the sharp drop in the density of states at Δ, trapping below it is essentially precluded. Therefore, carrier exchange proceeds essentially between Δ and ε_t which means that the spread in the release time distribution is small and that carriers propagate at essentially the same average velocity.

An estimate of the critical temperature is provided by using equation 6 for $g = g_c \exp(\beta_1(\varepsilon - \varepsilon_c))$ yielding

$$T_{crit} = \frac{1}{2\gamma_o R(\varepsilon_c) k_b} \left(\frac{3k_b T_1^*(1-\lambda)}{e}\right)^{1-\lambda} \qquad (12)$$

which for $T_1 = 1100°K$, that is in the case where a rise in the mobility is predicted, gives $200°K$. Note that this estimate is independent of the applied voltage in agreement with the experimental observations. An estimate of the temperature when the mobility reaches a maximum if there is a rise is more difficult to obtain because it depends crucially on the shape of the DOS around Δ via the position of the transport level. We however can say that the steeper the drop in the density of states at Δ, the steeper the mobility rise: this indeed follows because a steep drop will pin the transport level at a higher temperature.

4. Summary

We have shown that hopping can account for the observed low temperature drift mobility behavior if one properly takes into account the position of the dominant transport level and have provided an estimate of the critical temperature separating dispersive from non-dispersive transport. We predict that the mobility rise below the critical temperature is a highly sensitive function of the density of states.

Acknowledgements

This work was partially supported by SERI and by FCAR.

5. References

1. J.M. Marshall, R.A. Street and M.J. Thompson, Phil. Mag. B 54, 51 (1986).
2. C. Cloude, W.E. Spear, P.G. Le Comber and A.C. Hourd, Phil. Mag. B 54, L113 (1986).
3. W.E. Spear and C. Cloude, Phil. Mag. B 58, 467 (1988).
4. M.Silver and W. Spear, to be published.
5. D. Monroe, Phys. Rev. Lett. 54, 146 (1985).
6. M. Silver and H. Bassler, Phil. Mag. Lett. 56, 109 (1987).
7. R.A. Abram and S. Edwards, J. Phys. C 5, 1183 (1972).

HYDROGEN GLASS BEHAVIOR IN AMORPHOUS SEMICONDUCTORS

James Kakalios
School of Physics and Astronomy
University of Minnesota
Minneapolis, MN 55455

ABSTRACT

The recent proposal that the motion of the bonded hydrogen network underlies the glass-like behavior of the electronic properties of amorphous silicon is reviewed. Studies of the hydrogen diffusion coefficient lead to an identification of the microscopic mechanisms responsible for stretched exponential relaxation and the cooling rate dependence of the fictive temperature in amorphous silicon, and also provides insight into the Vogel-Fulcher temperature dependence which is often associated with the glass transition.

I. INTRODUCTION

Glasses are thermally arrested liquids which display a glass transition temperature T_G; a kinetically defined temperature which separates a high temperature regime where the glass is in thermal equilibrium from a low temperature regime in which the material is frozen into a non-equilibrium metastable state. Since T_G does not represent a thermodynamic phase transition, the glass transition is usually defined empirically as either the fictive temperature which depends on cooling rate; as the temperature for which the viscosity η becomes greater that 10^{13} poise; or as the temperature at which structural relaxations become comparable to experimental time scales.[1] These three definitions are equivalent, and result from a distribution of barriers which inhibit transitions between accessible states. A distribution of relaxation times arises from the disorder of the fluid state, and leads to a stretched exponential time dependence which describes the relaxation of the glass below T_G to its lowest energy equilibrium configuration. Despite much research effort, the nature of the microscopic mechanisms governing the glass transition remain poorly understood.

It has recently been demonstrated that the electronic properties of hydrogenated amorphous silicon (a-Si:H) are described by a metastable thermal equilibrium.[2,3] The time for the defect structure to reach equilibrium is thermally activated, becoming longer at lower temperatures. This temperature dependent equilibration rate results in electronic properties which resemble a glass, in that below an equilibration temperature T_E the defect structure depends on the cooling rate following a high temperature anneal, but above T_E the material's properties are independent of thermal history. The four-fold coordination of the amorphous silicon matrix gives it a very rigid structure, which makes it unlikely to exhibit glassy behavior. We have therefore proposed that the microscopic mechanisms which underlie the equilibration arise from the motion of bonded hydrogen. As the hydrogen diffuses through the amorphous silicon it can change the bonding configurations of dopants and silicon atoms. Measurements of the hydrogen diffusion coefficient D_H show that signfificant diffusion indeed occurs where the defect structure equilibrates.[4] The bonded hydrogen in a-Si:H can be considered a separate disordered network, distinct from the fixed amorphous silicon network, and it is the bonded hydrogen network which displays glassy behavior. a-Si:H can be viewed as a glassy analog of certain superionic conductors (such as AgI) for which one lattice melts while the other sublattice remains rigid. In this model the equilibration temperature is equivalent to the glass transition temperature.

The glass-like behavior of the hydrogen network in a-Si:H provides a novel system with which the properties and theories of glasses can be investigated. We will show that T_E satisfies all three definitions of a glass transition temperature mentioned above and moreover that measurements of D_H provide insight into the physical mechanisms which underlie T_G. We have found that D_H is not uniquely determined but rather displays a power-law time dependence.[4] This dispersive D_H can quantitatively account for the stretched exponential relaxation of the electronic properties of a-Si:H.[3] In addition, the time dependent D_H provides insight into the Vogel-Fulcher temperature dependence of the relaxation time, which is often associated with the glass transition.

II. COOLING RATE DEPENDENCE OF THE GLASS TRANSITION

Glass transition temperatures are typically obtained from plots of structural parameters, such as volume, against temperature, as illustrated in fig. 1. Above the melting temperature T_m the material is a liquid. Cooling below T_m leads to either crystallization, characterized by a sharp drop in volume at T_m, or supercooling below T_m. As the temperature of the supercooled liquid is further lowered, the shear viscosity increases until it becomes comparable to that of a solid. At this lower temperature, defined as the glass transition temperature, the material is said to form a glass. The plastic flow for a material with viscosity $\eta > 10^{13}$ poise will not be detectible on a standard observational time scale, so this η value is also associated with the glass transition. The structural rearrangements which keep the material in thermal equilibrium can follow a slow cooling rate to lower temperatures, so T_G will depend on the rate at which the liquid is cooled A specific point, termed the fictive temperature T_f is defined as the temperature at which the extrapolated glass and liquid lines cross; that is, at T_f the glass would be in metastable equilibrium if brought to that temperature instantly.[5]

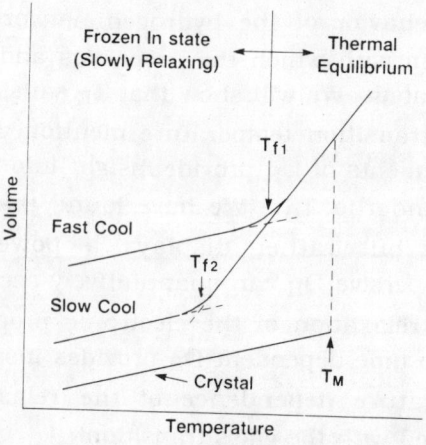

Fig. 1. Schematic diagram of volume against temperature for a glass. The fictive temperature where the thermal equilibrium and frozen-in curve intersects decreases as the cooling rate is lowered.

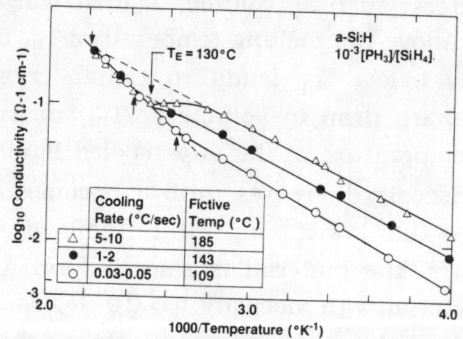

Fig. 2. Arhennius plot of the dark conductivity of a n-type a-Si:H sample, measured during warming for various cooling rates following a 200°C anneal. Above the equilibration temperature the conductivity becomes independent of thermal history. The fictive temperatures where the low and high temperature σ curves intersect are indicated by the arrows.

The temperature dependence of the dark conductivity of doped a-Si:H is quite similar to the above description, as shown in fig. 2. The co-planar dark conductivity σ for n-type a-Si:H (1000 p.p.m. gas phase doping with phosphine PH₃) is plotted against $1/T$ for various cooling rates following a high temperature anneal at 200°C The σ curves are measured after thermal quenching as the sample is warmed at a rate of 2-3°C/min. In fig. 2 the equilibration temperature T_E is defined as that temperature for which the equilibration time becomes comparable to the experimental time scale, and the conductivity becomes independent of thermal history as the sample returns to thermal equilibrium. All of the σ curves are reversible upon annealing above T_E, while the "frozen-in" σ is not affected upon warming to temperatures below T_E. Details of the conductivity measurements are given elsewhere.[2]

The fictive temperature where the low temperature "frozen-in" σ curve intersects the high temperature "thermal equilibrium" σ curve is indicated in fig. 2 for each cooling rate. The fictive temperature increases as the cooling rate following a high temperature anneal is increased. If we assume that the relaxation time for the glassy state is thermally activated at T_f then we can write

$$\tau = \tau_0 \exp[\Delta H/k_B T_f] \qquad (1)$$

where τ_0 is the preexponential factor, ΔH is the activation enthalpy, and k_B is Boltzman's constant. At the fictive temperature this relaxation time is comparable to the time interval $\Delta t = \Delta T/R$ where ΔT is an infinitesimal temperature change and R is the cooling rate.[5,6] We can therefore write $\tau(T_f) = \kappa \Delta t$, where κ is a proportionality constant. This relationship gives

$$d \ln R / dT_f^{-1} = -\Delta H/k_B = C^{-1} \qquad (2)$$

Hence the slope of a plot of lnR against $1/T_f$ should yield the activation enthalpy of the relaxation time controlling the glass transition.[5,6] The cooling rate dependence of the fictive temperature for the data of fig. 2 is shown in fig. 3. The data is consistent with a logarithmic dependence of R as observed for g-As_2Se_3 and g-As_2S_3 or with a small curvature as found for g-Se.[7] Experiments are in progress to decrease the uncertainty in the cooling rate, which is the major source of error in fig. 3. The slope C = 8.7 x 10^{-5} °K^{-1} from fig. 3 corresponds to an activation enthalpy ΔH = 0.99 eV per relaxing atom. The slopes found for g-Se, g-As_2Se_3 and g-As_2S_3 yield $\Delta H \sim$ 2.88 eV/atom.[1] In the hydrogen glass model for a-Si:H, only one chemical bond is broken as the hydrogen diffuses through the silicon network, while in the chalcogenide glasses the average coordination of the diffusing atoms is m = 2 to 2.4. The activation energy *per chemical bond* is then 1.0 - 1.4 eV for both a-Si:H and the chalcogenides. The lower ΔH for the glass transition in a-Si:H reflects the lower coordination of the hydrogen

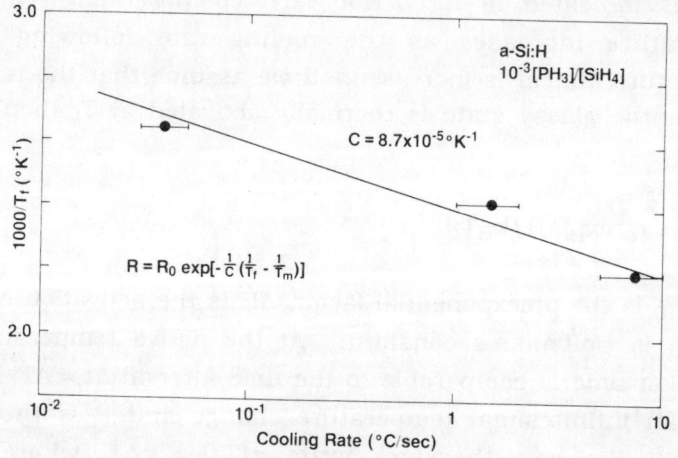

Fig. 3. Reciprocal fictive temperature as a function of cooling rate for the data in fig. 2. The slope c gives the relaxation time activation enthalpy from eq. (2).

glass network. We will show that this same activation enthalpy is found from studies of the stretched exponential relaxation of the electronic properties of a-Si:H

III. STRETCHED EXPONENTIAL RELAXATION

The distribution of barriers which inhibit transitions between accessible states also manifest themselves through the slow, non-exponential relaxation of the glass below T_G. A strikingly wide variety of disordered systems, ranging from oxide and polymeric glasses,[1] spin glasses,[8] charge density wave systems[9] and dipole glasses[10] exhibit a stretched exponential time dependence as they relax towards their equilibrium configuration. As illustrated in fig. 4, the normalized change in density of electrons residing in band tail states Δn (which is linearly proportional to σ for a-Si:H)[11] accurately follows a stretched exponential time dependence,[3] as given by

$$\Delta n = \Delta n_0 \exp[-(t/\tau)^\beta] \qquad (3)$$

where τ is the relaxation time and β is the dispersion parameter.

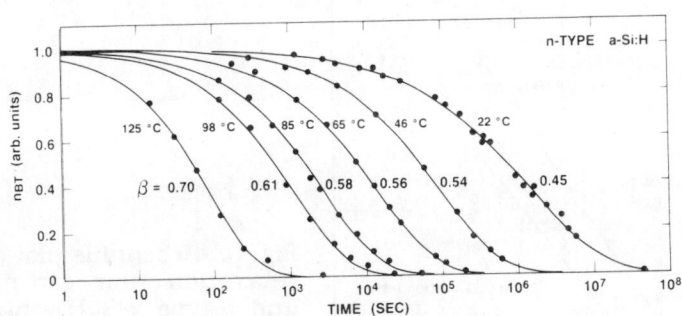

Fig. 4. Normalized decay of occupied band tail density for n-type a-Si:H for increasing temperature. The dispersion parameter β is found by fitting the data to a stretched exponential time dependence (solid lines).

The relaxation time, obtained by fitting the data of fig. 4 to eq. (4) for both n-type and p-type a-Si:H is thermally activated, as shown in fig. 5, with an activation energy of 0.95 eV for both curves and a preexponential factor of 10^{-10} sec for n-type and 10^{-11} sec for p-type a-Si:H.[2] This is identical to the activation enthalpy found earlier, indicating that the same process is responsible for the stretched exponential decay and the cooling rate dependence of the fictive temperature. As illustrated in fig. 6, the dispersion parameter β increases linearly with temperature, that is, $\beta = T/T_0$ with $T_0 = 600°K$.

Theoretical attempts to account for stretched exponential decay (also known as Kohlrausch[12] or Williams-Watts[13] decay) usually invoke a distribution of relaxation times.[14,15] The data in Fig. 4 can be fit by a thermally activated τ with an asymmetric distribution of activation energies peaked at 0.95 eV with a width of 0.1 eV.[3] However, this analysis immediately raises the question of

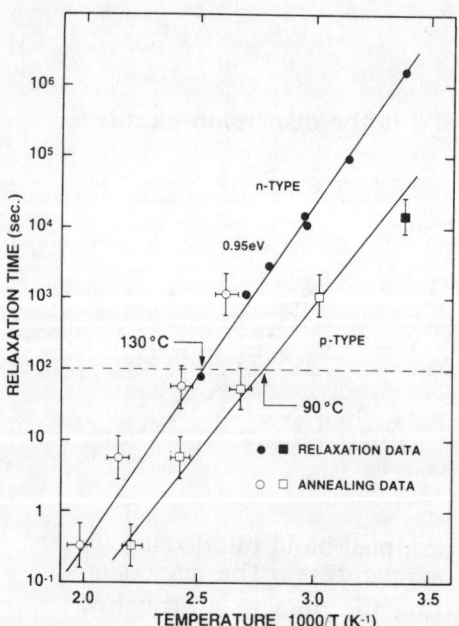

Fig. 5. Arhennius plot of the relaxation time for n-type and p-type a-Si:H, obtained from stretched exponential decay data as in fig. 4.

the physical origin of this distribution of relaxation times. An alternative approach involves a time dependent relaxation rate.[16-18] This second approach is appealing because it involves a continuous time random walk with a physically plausible distribution of trapping sites. For small departures from equilibrium the time dependence of Δn is given by

$$d\Delta n/dt = -\nu \Delta n \qquad (4)$$

where ν is the relaxation rate. In the hydrogen glass model for a-Si:H, ν is the rate of equilibration of the defect structure and is given by[19]

$$\nu = 4\pi N_H R_C D_H \qquad (5)$$

where N_H is the density of bonded hydrogen, R_C is the capture radius of a site which can trap a diffusing hydrogen atom, and D_H is the hydrogen diffusion coefficient. If ν decreases as a power law in time (i.e. $\nu \sim t^{-\alpha}$ with $\alpha < 1$) then the solution of eq. (5) will be the Kohlrausch expression with $\beta = 1-\alpha$. Any time dependence in ν will

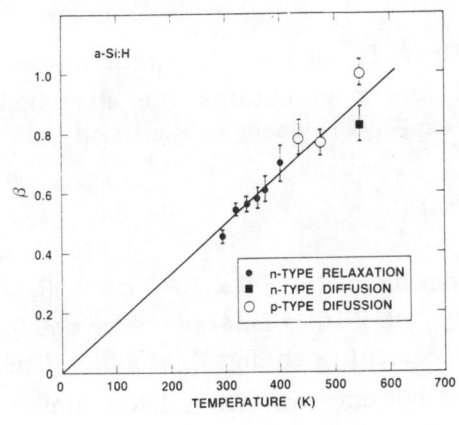

Fig. 6. Temperature dependence of β from fig. 4 and of $1-\alpha$ from measurements of the time dependent hydrogen diffusion coefficient.

be reflected in D_H. Studies of D_H, described in detail elsewhere, do indeed find that D_H has a power law time decay,[3,4] where

$$D_H = D_{oo}(\omega t)^{-\alpha} \tag{6}$$

with D_{oo} being the microscopic diffusion coefficient and ω the attempt-to-hop frequency. As shown in fig. 6, the temperature dependence of β from the Kohlrausch relaxation data and $1-\alpha$ from measurements of D_H (open circles and square data points in fig. 6) are identical, supporting the proposal that the motion of bonded hydrogen underlies the glass-like properties of a-Si:H.

The time dependent D_H can result from a distribution of Si-H bonding sites, which arise from variations in bonding configurations in the amorphous network. One possible distribution consistent with the data would be a density of Si-H trapping sites exponentially distributed in energy $\exp[-E/kT_0]$ where kT_0 is the characteristic width of the distribution.[3] This trap distribution would then lead to $D_H \sim t^{-\alpha}$ where $\alpha = 1-T/T_0$, following the analysis of dispersive transport.[20] The temperature dependence of β would then arise from the motion of hydrogen through an exponential distribution of bonding sites

The dependence of τ on D_H is found upon integration of eq. (4), using eqs. (5) and (6). One then obtains the stretched exponential function with $\tau = \tau_0 \exp[E_\tau/kT]$ where $\tau_0 = \omega^{-1}$ and

$$E_\tau = k_B T_0 \ln[\beta\omega/4\pi N_H R_C D_{oo}] \tag{7}$$

Taking $\omega = \tau_0^{-1} = 10^{10}$ sec^{-1} from fig. 5, $N_H = 5 \times 10^{21}$ cm^{-3}, $R_c \sim 3A$, $T_0 = 600°K$ from fig.6 and $D_{oo} = 5 \times 10^{-14}$ cm^2/sec,[19] we obtain $E_\tau = 0.95$ eV, in excellent agreement with figs.3 and 5. The hydrogen glass behavior of a-Si:H not only provides a determination of a microscopic mechanisms for stretched exponential relaxation, but allows an evaluation of the activation enthalpy governing the glass transition in terms of measured materials properties.

IV. VISCOSITY AT THE GLASS TRANSITION

As noted above, a frequently employed empirical definition of T_G is that it is that temperature for which η of the supercooled liquid exceeds 10^{13} poise.[1] In conventional glasses viscous flow is achieved via bond rearrangements which clearly involve the self-diffusion coefficient. Since only the hydrogen is mobile during equilibration, we cannot directly observe the viscosity of the hydrogen glass. However, η is related to D_H through the Stokes-Einstein relation

$$\eta D = k_B T / 6\pi R \tag{8}$$

where R is the effective radius of the self-diffusing species. Taking $T_E \sim 100^\circ C$ and $R \sim 3A$, eq. (9) gives $\eta D = 10^{-7}$ dyn. That is, at the glass transition temperature when $\eta = 10^{13}$ poise, $D_H = 10^{-20}$ cm^2/sec. If we associate T_G with the temperature at which the thermally activated D_H achieves this value, then we obtain $T_G = 70^\circ C$ in p-type, $110^\circ C$ in n-type and $160^\circ C$ in undoped a-Si:H. This is to be compared with $T_E = 90^\circ C$ for p-type, $130^\circ C$ for n-type and $210^\circ C$ for undoped samples, as determined from σ curves as in fig. 2. There is thus good agreement with T_G defined in this way and T_E, both in the absolute magnitude and in the trend with doping.[2]

Another definition of a glass transition temperature is that it is that temperature for which the experimental observational time becomes comparable to the material's relaxation time. If we identify this time scale as 100 sec, indicated by the dashed line in fig. 5, whereby faster times describes the "thermal equilibrium" state while slower times characterize the "frozen" state, then the glass transition so defined occurs at $130^\circ C$ for n-type and $90^\circ C$ for p-type a-Si:H, in excellent agreement with T_E obtained from σ data (for a heating rate of 2-3°C/min). The equilibration temperature therefore satisfies all three definitions of T_G described earlier.

As shown above, the Arhennius dependence of τ results from the time dependent D_H as given by eq. (6). However, D_H will still be thermally activated if the time t_L to diffuse a constant distance L is kept fixed as the temperature is varied. In that case, it is straightforward to show that[19]

$$L^2 = 4 \int_0^{t_L} D_H(t')dt' \qquad (9)$$

Inserting eq. (6) into eq. (9) and integrating yields

$$D(t_L) = L^2(1-\alpha)/4t_L \qquad (10)$$

and if $\omega t_L = \exp[E_D/k_B T]$ then D_H will be thermally activated, that is, $D_H = D_0 \exp[-E_D/kT]$ with

$$D_0 = \omega L^2(1-\alpha)/4 \qquad (11)$$

and

$$E_D = k_B T_0 \ln(\omega L^2 T/4T_0 D_{00}) \qquad (12)$$

For a typical diffusion distance L = 1000A, and using the same values for ω, T_0, D_{00} as above, we obtain $D_0 = 0.05$ cm^2/sec and $E_D = 1.45$ eV. This is in excellent agreement with the observed diffusion coefficient activation energy and prefactor, as shown in fig. 7. If we had kept the annealing time t_A constant as the temperature is varied, then the diffusion coefficient, given by eq. (6) with $t = t_A$, would exhibit upward curvature on an Arhennius plot, as illustrated in fig. 7. This curvature could be interpreted as evidence of a Vogel-Fulcher-Tamman temperature dependence,[21] i.e.

$D_H = D_0 \exp[-A/(T-T^*)]$ where T^* is a characteristic temperature less than T_G. Since measurements of η (α $1/D$) which are described by the Vogel-Fulcher expression are usually experimentally restricted to regions far from T^*, the singularity in η is not directly observed.[22] It is thus possible that the observed curvature of η on an Arhennius plot is simply a manifestation of a time dependent relaxation rate when η is measured for isochronal annealing.

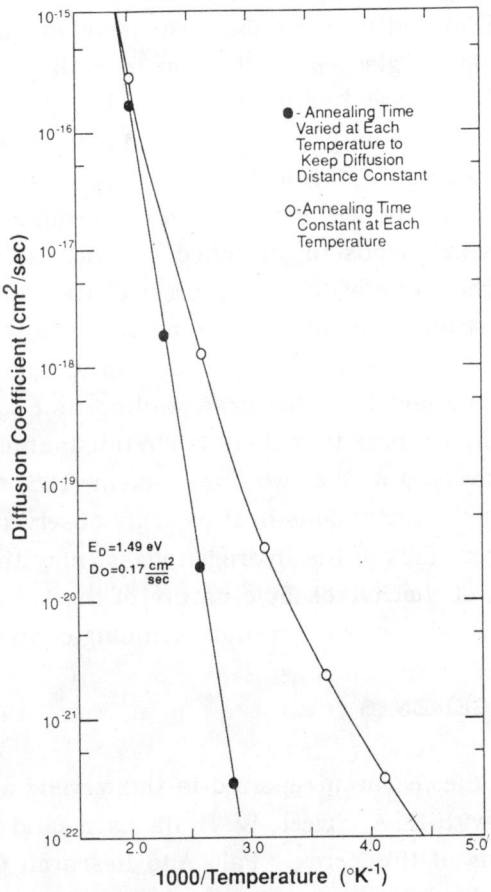

Fig. 7. Plot of the hydrogen diffusion coefficient against reciprocal temperature for isochronal annealing (open circles) and for a constant diffusion distance (closed circles).

V. CONCLUSIONS

We have argued that the bonded hydrogen in a-Si:H can be considered a separate network which behaves like a glass As the hydrogen diffuses it can change the bonding coordination of dopants and silicon atoms, which is reflected in the electronic properties of the a-Si:H films. A distribution of Si-H trapping sites leads to both Kohlrausch decay and the cooling rate dependence of properties which depend on hydrogen motion. The issue of frustration, crucial in any discussion of glasses, would arise from the density of weak Si-Si sites (which serve as hydrogen traps) being less than N_H. Rough estimates of the density of weak bonds capable of trapping a diffusing hydrogen is $\sim 10^{20}/cm^3$, a factor of 50 less than the hydrogen density. The situation is akin to a "trap amidst a swarm of walkers",[18] which leads to stretched exponential relaxation. The reverse situation, "one walker in a forest of traps" as in the case of time of flight studies, results in a power law relaxation.[18] The two cases cannot be transformed into each other by an exchange of labels due to the fact that the mean waiting time for the walkers is infinite. Finally we note that there is circumstantial evidence[23] that hydrogen is involved in the two level system (TLS) tunneling modes responsible for the anomalous heat capacity observed in a-Si:H below 1°K. Further studies of the hydrogen glass may therefore enable a determination of the microscopic nature of the two level systems in a-Si:H, and should improve our understanding of glasses in general.

ACKNOWLEDGEMENTS

Most of the research reported in this review was performed in collaboration with R. A. Street, W. B. Jackson, and C. C. Tsai when the author was at the Xerox - Palo Alto Research Center, and was supported by the Solar Energy Research Institute.

REFERENCES

1) See, for example, Amorphous Materials, edited by R.W. Douglas and B. Ellis (Wiley-Interscience, New York) (1972).

2) R.A. Street, J. Kakalios, C.C. Tsai and T.M. Hayes, Phys. Rev. B **35**, 1316 (1987)

3) J. Kakalios, R.A. Street and W.B. Jackson, Phys. Rev. Lett **59**, 1037 (1987).

4) R.A. Street, C.C. Tsai, J. Kakalios and W.B. Jackson, Philos. Mag. B **56**, 305 (1987).

5) H.N. Ritland, J. Amer. Ceram. Soc. **37**, 370 (1954).

6) C.T. Moynihan, A.J. Easteal, J. Wilder and J. Tucker, J. Phys. Chem. **78**, 2673 (1974).

7) S.O. Kasap and S. Yannacopoulos, J. Mater. Res. **4**, 893 (1989).

8) R.V. Chamberlin, G. Mozurkewich and R. Orbach, Phys. Rev. Lett. **52**, 867 (1984).

9) G. Kriza and G. Mahaly, Phys. Rev. Lett. **56**, 2529 (1986).

10) R.M. Ernst, L. Wu, S.R. Nagel and S. Susman, Phys. Rev. B **38**, 6246 (1988).

11) R.A. Street, J. Kakalios and M. Hack, Phys. Rev. B **38**, 5603 (1988).

12) R. Kohlrausch, Ann. Phys. (Leipzig) **12**, 393 (1847).

13. G. Williams and D.C. Watts, Trans. Faraday Soc. **66**, 80 (1970).

14. K.L. Ngai, Comm. Solid State Phys. **9**, 27 (1979) and **9**, 141 (1980).

15) M.H. Cohen and G.S. Grest, Phys. Rev. B **24**, 4091 (1981).

16) R.G. Palmer, D.L. Stein, E. Abrahams and P.W. Anderson, Phys. Rev. Lett. **53**, 958 (1984).

17) M.F. Shlesinger and E.W. Montroll, Proc. Natl. Acad. Sci. U.S.A. **81**, 1280 (1984).

18) M.F. Shlesinger, J. Stat. Phys. **36**, 639 (1984).

19) W.B. Jackson, J.M. Marshall and M.D. Moyer, Phys. Rev. B **39**, 1164 (1989).

20) H. Scher and E.W. Montroll, Phys. Rev. B **12**, 2455 (1975).

21) S.R. Nagel and P.K. Dixon, J. Chem. Phys. **90**, 3885 (1989).

22) J. Jackle, Philos. Mag. B **56**, 113 (1987).

23) M. Stutzmann, Phys. Rev. B **33**, 7379 (1986).

Chapter 7

POLYMERS AND BIOLOGICAL SYSTEMS

IONIC DIFFUSION IN POLYMER ELECTROLYTES: DYNAMIC DISORDER HOPPING MODELS

Stephen D. Druger and Mark A. Ratner,
Department of Chemistry and Materials Research Center,
Northwestern University, Evanston, IL 60208

ABSTRACT:

Hopping models can be used to describe ionic motion in so-called solid electrolytes, which are solid materials exhibiting ionic diffusivities of a magnitude more characteristic of liquids ($D > 10^{-7}$ cm^2/sec.). While in glassy and ceramic solid electrolytes the hopping occurs on a statically disordered lattice, in polymeric electrolytes, which are ordinarily studied above the glass transition temperature, the material is locally liquid-like. Under these conditions, the disorder in the polymer host has dynamic as well as static aspects; the dynamic disorder arises from large-scale orientational and translational motions of the polymer host chain.

To describe ionic motion in a dynamically disordered medium, we have studied generalized dynamic disorder hopping models in which each site-to-site jump changes its status between allowed and forbidden as a function of time. General results are given for the diffusivity in dynamic disorder hopping models, including formal relationships for the diffusion coefficient, parallels to continuous time random walk models, the distinction between ballistic and diffusive transport, and some consideration of the role of correlated renewal, or reinitialization, in the polymeric lattice host.

1. INTRODUCTION

While most of the contributions to this meeting are centered on the hopping process involving electrons, holes, or excitons, other particles also undergo diffusive hopping motion. There will be some discussion at this meeting of hopping by protons and ions. Additionally, atoms and molecules can also move by hopping, particularly in impurity systems or on surfaces. The term "hopping" is meant in this context to indicate that the motion can be characterized by two time scales, a residence time and a jumping time, and that the latter is considerably smaller than the former. The sites between which the particles hop may be very well defined, as they are in glassy or alloy materials, but may themselves evolve in time as they will in liquids (including polymeric liquids) above the glass transition temperature. Nevertheless the intuitive notion of hopping as involving relatively rapid motions between well-defined sites unifies the description of all the physical systems considered in this conference. An alternative model for defining, in terms of a generalized growth law, precisely what is meant by a hopping system, is inherent in much of the recent theoretical work on hopping systems, and is addressed in Sec. 5 below.

Since most participants at this meeting are probably not intimately familiar with the problems involved in discussing hopping transport in polymers, some experimental background is presented in sec. 2. Section 3 discusses a model based on dynamic hopping which has been widely applied to understanding diffusive and other behaviors in polymer electrolyte systems. Some results using this model for frequency-dependent conductivity are included in sec. 4. Section 5 presents a more general formulation of the dynamic disordered hopping problem, and a comparison with other standard models including continuous time random walk pictures. It also considers reexpression of hopping phenomena in terms of general growth laws, and contrasts between systems in which reinitialization increases diffusion rates (diffusive systems), and others in which re-initialization reduces diffusion rates (ballistic systems). The issue of correlations, either among the hoppers, or between the hoppers and the lattice, or within the motions of the lattice itself, is a vexing and important one; the introduction of the Coulomb gap has greatly changed the interpretation of standard hopping materials involving electrons, and similar behavior might be expected in certain polymer systems. Some preliminary discussion of lattice correlations is also included in Sec. 5, followed by general remarks in Sec. 6.

2. BACKGROUND: ION HOPPING IN POLYMER ELECTROLYTES

Solid ionic materials can generally be divided into those in which hopping occurs in systems of static disorder, and those in which hopping occurs in a dynamically disordered host.[1] The statically disordered systems are similar to those considered in standard hopping problems: ions move among well-defined sites, overcoming activation barriers. Typical systems of this kind include silver iodide, beta alumina, lead fluoride, and lithium aluminosilicate, either in crystalline or in glassy form. The polymer electrolytes differ from these rigid materials in that they are always studied at temperatures above the glass transition temperature, where ionic motion occurs in a host, the polymer itself, which is undergoing large-scale fluctuations and is, locally, liquid-like.[2] The polymer electrolytes come in several varieties, including complexes of a salt and a polymer host, and in particular include so-called polyelectrolytes, in which ionic charge of one sign moves while ionic charge of the other sign is bound to the backbone.

Electroactive polymers, including both polymer electrolytes and electron hopping polymers, can be visualized according to the cartoon in Fig. 1. The polymer host is the long connected chain, and the electroactive sites lie either pendant on the chain or encapsulated by the chain. In the electronically conductive redox polymers, the subject of the contributions to this conference by Facci and Murray, the sites are localization positions for electrons, which move along the chain or among the chains by activated hopping processes. In the polymer electrolyte materials with which we are concerned, the localization sites are simply relatively stable locations for the mobile ions (which can be cations or anions).

Polymer electrolytes are discussed very extensively in the literature,[2] and our discussion here will be highly abbreviated. Typically, a host polymer material should be disordered and have low

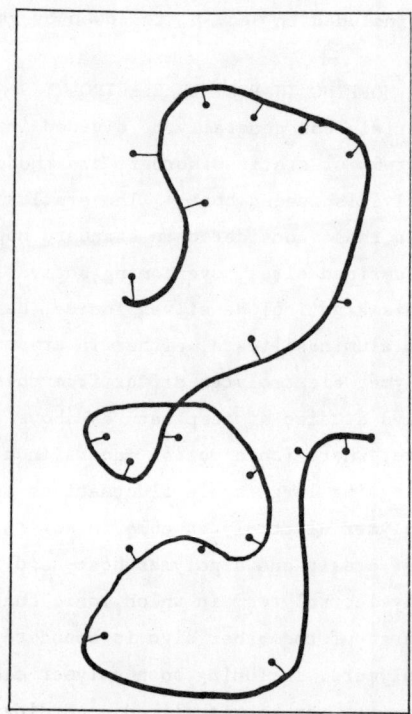

Figure 1: Cartoon indicating the general geometry of electroactive polymers. The long chain is the polymer backbone, and the round spots are the local electroactive sites, which can be solvation sites in the case of polymer electrolytes, or electron localization sites in redox polymers.

cohesive energy density, so that large-scale excursions are energetically relatively facile. The most standard materials are based on a polyether host such as polyethylene oxide, $(-CH_2CH_2O-)_n$, a host material in which a uni-univalent salt, such as LiI, NaSCN, AgCF$_3$SO$_3$ or some other salt, is dissolved in a particular stoichiometric ratio.[3] The electrical conductivity then arises from long hopping among favorable sites in the dynamically disordered lattice. A major role in determining ionic conductivity of these materials is played by the dynamics of the host itself; and ion-ion correlations are an important but secondary effect, determining both the number of carriers and a part of the potential in which those carriers move.

While high ionic conductivity is in fact observed only in the amorphous, elastomeric phase,[4] structural information has been obtained only in the crystalline phases of these polymers. Figure 2 shows the

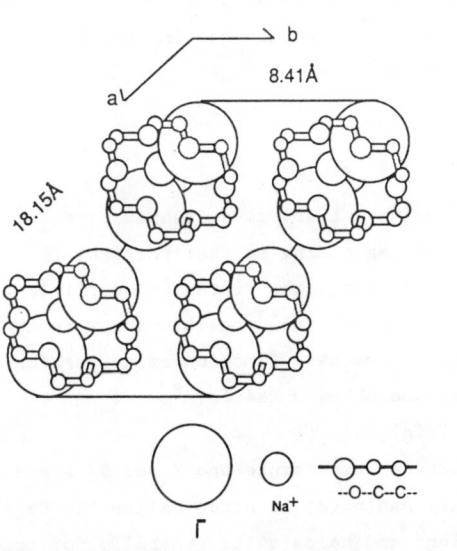

Figure 2: The crystal structure of crystalline PEO.NaI complex, with an oxygen/PEO/Na ratio of three. Note the surrounding of the Na by I⁻ and O in the chain. From Ref. 5.

crystal structure determined for a complex of PEO with NaI, as determined by Chatani.[5] In this structure the salt density is extremely high, but one expects the general motifs shown to characterize most polyether complex electrolytes. Note that both cation and anion are

surrounded by the chain, which moves through the system in such a way
that each cation is closely surrounded by several donor oxygen species
from the polymer chain and retains a residual coordination to the
anions. One striking thing about the structure is that there are no
obvious holes: for ion transport to occur, it is clear that void sites
will have to be created so that ions can in fact be facilitated in their
motion within the phase. This strongly suggests that free volume
concepts, in which the mobility is determined solely by the availability
of local free volume, will be useful for describing these materials.
Indeed, the free volume picture is still in some ways the most
satisfying one[6] for understanding ion hopping in these materials.

In the free volume picture,[7] carriers hop into sites of sufficient
free volume, and the relative availability of such sites is determined
by the form

$$D \sim \exp(-v^*/v_f), \qquad (1)$$

where v_f is the free volume at any given temperature and v^* is the
critical free volume corresponding to that particular ion hop. In the
absence of correlation effects, the diffusion coefficient D is
proportional, by the Nernst-Einstein relationship, to the conductivity
By expanding the free volume as a function of temperature around a given
critical temperature, one obtains the form[8]

$$D = D_0 \exp[-B/(T-T_0)], \qquad (2)$$

where T_0 is the reference temperature and B and D_0 are empirical
constants. The relationship (2) is often called the Vogel-Tamman-
Fulcher (VTF) equation[9] and holds quite generally for transport in many
disordered materials (indeed, it was first developed to discuss behavior
in glasses and ceramics as well as polymers). The VTF relation itself
was entirely empirical, but a rationalization for such behavior can be
found based on the free-volume picture.

If the system indeed follows VTF behavior, one expects the
diffusion coefficient to be curved if plotted in ordinary Arrhenius
coordinates, but straight if plotted in Arrhenius coordinates corrected

for T_0. If T_0 is related empirically to the glass transition temperature, then one obtains the plots shown in Figs. 3 and 4. In the graph in Arrhenius coordinates, as reproduced in Fig. 3, the conductivity in fact decreases with increasing salt concentration. That this is almost entirely due to increased host viscosity is indicated

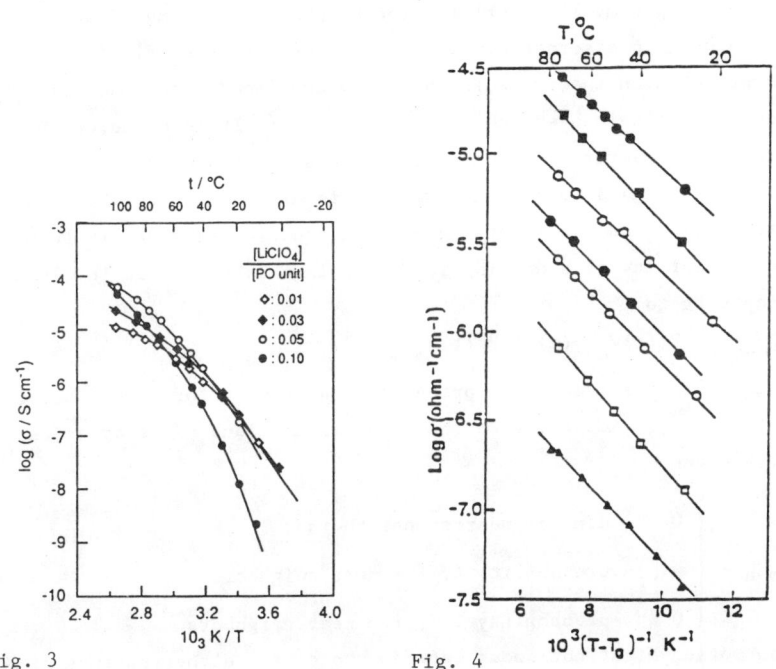

Fig. 3   Fig. 4

Figure 3: Dependence of conductivity on inverse temperature, for a salt complex electrolyte of lithium perchlorate in polyethylene oxide. From Ref. 6.

Figure 4: The same data as in Figure 3, but replotted to reflect the increase in conductivity with salt concentration, once corrected for the change in viscosity (or glass transition temperature) of the material.

both from Fig. 4, in which the effect of T_0 is removed, or by direct measurement of the glass transition temperature. Thus the transition temperature T_0, or equivalently the glass transition temperature T_g, characterizes the intrinsic viscosity of the host in the complex material: lower glass transition temperatures result in higher local mobility of the host, and in higher hopping rates for the diffusing ions.

3. DYNAMIC DISORDER HOPPING MODELS

A microscopic model describing ion hopping in such materials is offered by dynamic disorder hopping theory, or more specifically by dynamic percolation theory, which has been developed extensively in Evanston and Tel Aviv over the past decade.[10-17] It is based on the use of the ordinary master equation approach to percolation problems, generalized to take into account the dynamic disorder of the host medium found above T_g. The fundamental evolution equation is given in terms of P_n, the probability of a hopping ion occupying a site n. It involves in time according to

$$\dot{P}_n = \sum_{m \neq n} [P_m(t) W_{m \to n}(t) - P_n(t) W_{n \to m}(t)]. \tag{3}$$

Here $W_{m \to n}(t)$ is the transition probability at time t for ion to hop from site m to site n. According to the usual assumptions of percolation theory, we take

$$W_{m \to n} = \begin{cases} 0 & m,n \text{ not nearest neighbors} \\ w & \text{probability } f, \text{ for near neighbors.} \\ 0 & \text{probability } 1-f, \text{ for near neighbors} \end{cases} \tag{4}$$

That ion motion might not occur between two sites, either because of lack of availability of free volume at the acceptor site, or due to the interposition of a large barrier between the two, is reflected in the fact that the transition probability can be zero.

Equations (3) and (4) themselves define the ordinary percolation problem. Dynamic disorder hopping (DDH) models take the liquidlike motion of the host into account by generalizing the behavior of Eq. (3) to allow the probability $W_{n \to m}(t)$, to evolve in time. Several models for this evolution have in fact been taken: Harris et al. assumed[13] that bond availability, like the particles themselves, evolves from site to nearby site, so that the assignment of a particular hop as permitted or unpermitted (value for the transition probability of w or 0) moves from site to site, itself following something like a master equation. Druger, Nitzan, and Ratner have,[10-12] in contrast, defined a dynamic bond percolation (DBP) by assuming that the reassignment, or renewal, of bonds as available or unavailable changes globally throughout the lattice, with a characteristic renewal time τ_{ren}. Harrison and Zwanzig[18] assumed independent renewal, or reassignment, of each bond as a random process. The actual distribution of renewal processes has been extensively discussed:[15-17] the simplest assumption is that it follows either a delta function distribution (renewal occurring at fixed intervals) or that it follows a Poisson process (constant relative probability of renewal during the next time interval). Simple assumptions of the straightforward DBP model are then that the renewal time distribution is independent of the state of the moving ion, and that there are no temporal or spatial correlations either in bond reassignment or in the motion of the ions themselves. The renewal time τ_{ren} is determined by the dynamics of the polymer host, and corresponds to rearrangement of the polymer so as to change the relative probabilities of ion hopping.

Physically, the global renewal process, or dynamic disordered hopping model, assumed by Druger, Nitzan, and Ratner, is perhaps more appropriate for globally disordered materials such as polymer electrolytes, while the strongly correlated bond reassignment model of Harris et al. may be more appropriate for ion hopping processes in such linear polymers as polyphthalocyanines.[19]

The master equation with dynamic disorder hopping conditions has been analyzed extensively. Within an effective medium approximation, very comparable results are found for the Harrison-Zwanzig independent bond renewal process. Some of the important results include:

1. For times long compared to the mean renewal time, the system is always diffusive: that is, the mean-square displacement is always proportional to t for all integer dimensionality and all values of the completeness parameter f. No percolation threshold exists in the dynamic renewal process. This is very schematically indicated in Fig. 5, which shows the mean-square displacement as a function of time for the specific case of a delta-function distribution of renewal times.

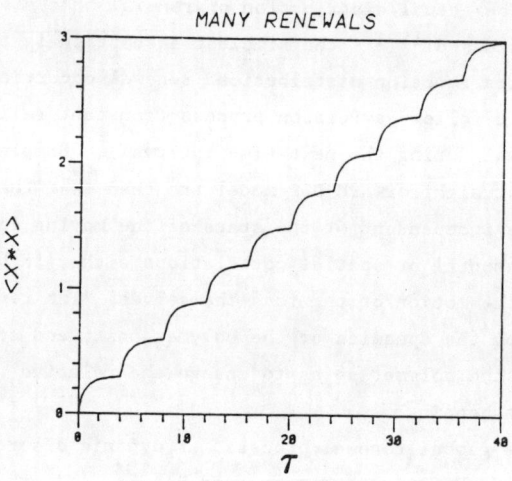

Figure 5: Calculated mean square displacement as a function of time, for one dimensional dynamic percolation model. The renewals are evenly spaced, in units of four. Note that the long time behavior is diffusive, although within each renewal cycle, the mean-square displacement becomes asymptotically flat.

This is a situation in which f is smaller than the static percolation threshold, so that within each renewal interval the mean-square displacement in fact becomes asymptotically flat; once renewal occurs, however, the local cluster in which the ion is located has changed its complexion, and motion can begin again. Thus within each renewal cycle, the behavior may not be diffusive, but over many renewal cycles diffusion is indeed linear in time.

2. The value of the diffusion coefficient is fixed by
$$D \sim \overline{\langle r^2 \rangle}/\overline{\tau}_{ren}, \tag{5}$$
where the numerator is the mean-square displacement within the average renewal cycle, and the denominator is the mean renewal time. This result holds again in all integer dimensionalities, and indicates that in general reduction in the mean renewal time increases the diffusion coefficient. Since the renewal time is directly determined by the fluidity of the polymer host, it then follows that the diffusion coefficient depends largely on how fluid the polymer host is, as reflected empirically in the WLF relationship.

3. A simple analytic continuation formula
$$D(\omega) = D_{static}(\omega - i/\overline{\tau}_{ren}) \tag{6}$$
exists between diffusion in the static lattice, described by D_{static}, and the diffusion in the dynamically disordered lattice. This analytic continuation rule simply says that the frequency dependence in the static lattice, if analytically continued to the complex frequency value, gives the diffusion coefficient in the dynamically-disordered lattice.

4. For the particular case of one dimension, the analytic form
$$D = \frac{a^2 f}{(1-f)^2 \overline{\tau}_{ren}} \tag{7}$$
holds, where a is the hopping distance between lattice sites. This form is determined by the inverse renewal time, but increases linearly with the probability of bond availability, for small bond availabilities. Figure 6 shows this behavior for one-dimensional systems, where any value of f less than unity lies below the static percolation threshold. Notice that while the mean-square displacement within a renewal interval

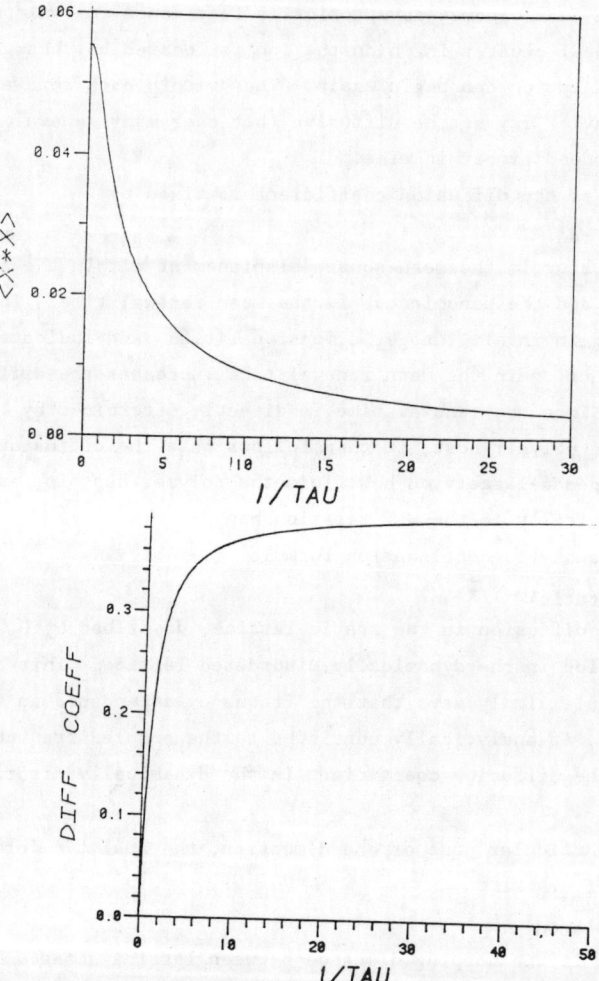

Figure 6: Mean-square displacement within a given renewal cycle, as a function of inverse renewal time. Figure 6A shows mean-square displacement itself, which increases with increasing renewal time. Figure 6B shows the diffusion coefficient, which exhibits threshold behavior for infinite renewal time, increases monotonically with decreasing renewal time, and becomes asymptotically flat at very short renewal times.

falls as the renewal time decreases, the actual diffusion coefficient increases as the renewal process becomes faster, becoming asymptotically flat for extremely fast renewals. Once again, this shows the characteristic result that the lattice dynamics, or the dynamics of the polymer host, as reflected in the renewal time, completely determines the ionic motion.

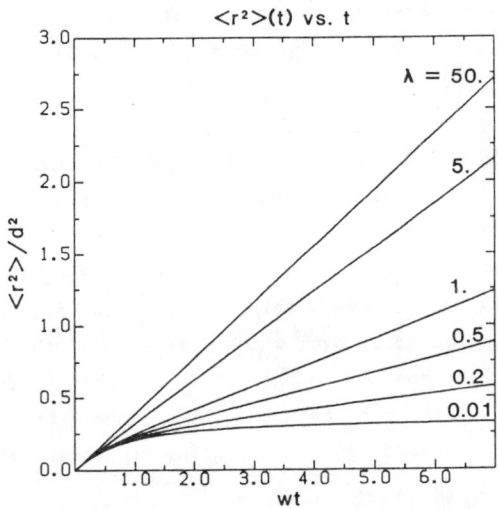

Figure 7: Calculated mean-square displacement, in units of lattice distance squared, for one-dimensional dynamic bond percolation problem. The lines show the limiting behavior of slow renewal (asymptotically flat) and fast renewal (diffusive at all times).

Figure 7, again for a one-dimensional system, shows the calculated mean-square displacement in units of the lattice distance squared, for different values of the inverse product of the hopping rate and the renewal time. For slow renewal, the mean-square displacement rapidly becomes asymptotic, as it reaches the end of the connected cluster and waits for the renewal process to permit further motion; for fast renewal, the mean-square displacement becomes linear with the time at all times, since renewal occurs before the hopping particle can explore the local connected cluster. In higher dimensionality, the behavior is qualitatively the same, again with the size of the connected region and the renewal time determining the diffusion. Scaling forms for the critical behavior near the static percolation threshold are available,[11] as are the relationships between the characteristic parameters of the dynamic disordered hopping model, renewal time τ_{ren}, bond probability f, and ion hopping rate w, and the comparable characteristic parameters in free volume theory, v^*, v_f, compressibility, and the masses and lengths.[7]

4. A TYPICAL RESULT: MICROWAVE CONDUCTIVITY OF POLYMER ELECTROLYTES

Conductivity at finite frequencies has been used to characterize the nature of the ionic motion in polymer electrolytes. These measurements are difficult to make, but by studying the response over a broad frequency range, one is able to deduce the frequencies, and indirectly the distances, over which diffusion of the dipoles in the host chain and the charges on the mobile ions can in fact diffuse. Important early work by Brodwin and his collaborators at Northwestern substantially increased our understanding of polymer electrolyte systems.[20]

In considering the overall conductive response of the material, one can separate the total conductivity into the sum of contributions from free charges (the ions), and from bound charges (dipolar species on the polymer host backbone).

$$\sigma_{total}(\omega) = \sigma_{ion}(\omega) + \sigma_{dipole}(\omega) \quad (8)$$
$$= \sigma_{free}(\omega) + \sigma_{bound}(\omega) \quad (9)$$

The free ions are described by the dynamic bond percolation picture, while the bound dipoles undergo finite-path-length hopping--that is, they hop on a percolating lattice without renewal. Formally, then, the percolation problem involves renewal for the ion, but does not involve renewal for the host (in which the dipoles are bound covalently to the very long polymer chain). At very low frequencies, one might expect to see reptation motion of the backbone chain itself, but this is unimportant for our considerations here.

A computation of the frequency-dependent diffusion coefficient is then extremely simple--one sums the contribution from the ions, which have a finite mean renewal time $\bar{\tau}_{ren}$, and the dipoles that have infinite renewal time. Precisely such calculations have in fact been carried out,[20] and the results, compared with experiment, are shown in Fig. 8.

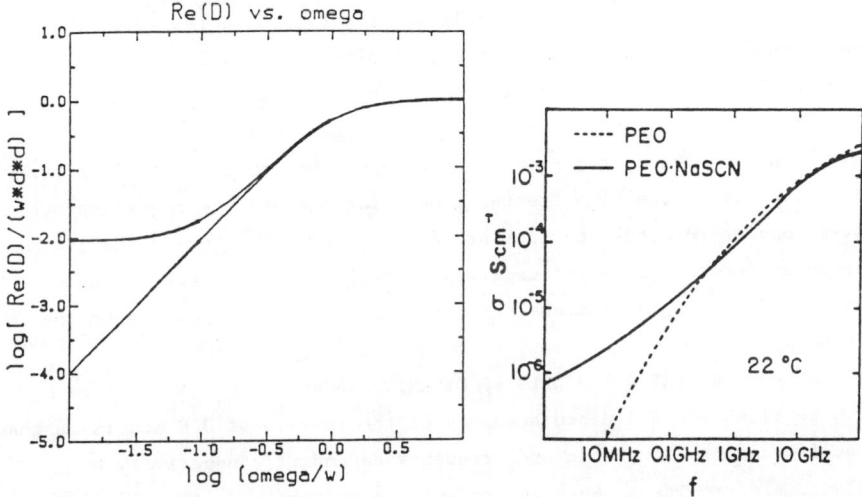

Figure 8: Frequency dependent conductivity of polymer salt complex material, and of neat polymer host. The experimental data is shown in Figure b, and a simulation involving two sorts of percolation, static for the host and dynamic for the guest, is shown in Figure a. From Ref. 20.

Figure 8a shows the calculated diffusion coefficient, in units of lattice constant squared times hopping frequency, as a function of the frequency in reduced units of the hopping frequency. For ω/w larger than 0.3, the static and dynamic responses are identical: at such high frequencies, the response is dominated by the stoichiometric majority species, the dipoles on the lattice. As the frequency is further reduced, the diffusion, or conductivity, of the bound static percolating system approaches its dc value of zero. On the other hand, the ionic contribution becomes flat at very low frequencies, being fixed by the mean-square displacement divided by the mean-square renewal time. Precisely such behavior is also seen in experiments on NaSCN polymer electrolytes, as shown in Fig. 8b. Here one sees essentially identical behavior of the neat host material and the polymer salt complex above one 1GHz, and a substantial disparity at lower frequencies, with the ionic conductivity evening out to some finite value while the host conductivity at dc goes essentially to zero.

Use of simple dynamic percolation concepts thus permits understanding of the frequency dependence of the conductivity in polymer salt complexes: the high frequency behavior mirrors the renewal process corresponding to actual motions of the lattice host, whereas the low frequency dependence, which is asymptotically flat, reflects behavior which is diffusive over much longer time scales, corresponding to the renewing problem of ionic motion.

5. GENERAL FORMULATION: MODELS OF DYNAMIC DISORDER

Sections 3 and 4 discussed some of the results of DDH models in the specific context of static and frequency-dependent conductivity of polymer electrolytes. Here we concentrate on some more formal results, some generalizations of hopping behavior in terms of growth laws, some effects of correlations in the renewal process, and some comparison among models.

A number of different models have been considered for transport in systems where microscopic structural host evolution occurs concurrently with site-to-site classical hopping. We have referred to these generically as "dynamic disorder hopping" (DDH) models.

Usually, such models start by assuming a random distribution of hopping rates between possible sites for the hopper, corresponding at this point to static disorder (Eqs. (2) and (3)). Next, random reassignment (or "renewal") of the hopping rates (and possibly also of allowed site locations) is assumed to occur at certain random instants, thus describing the random host reorganization, and corresponding to dynamic disorder. Average behavior over an ensemble of possible systems is sought in all cases, this procedure being especially warranted when the experimental data result from the superposed response of many independent carriers in microscopically different regions of the sample.

Specific DDH models differ in how the random assignment and reassignment of hopping rates occurs, with different assumptions applying best to different physical problems.

The important subclass of DDH models discussed in Secs. 3 and 4 starts with a periodic lattice of sites having hopping rates 0 and w statically assigned in some random way between nearest neighbors (with non-nearest-neighbor hops forbidden). Without hopping rate reassignment, the resulting model corresponds to ordinary bond percolation, where if the probability of a bond (i.e., an allowed pathway) between nearest neighbors is less than the percolation threshold value, there are only finite clusters of interconnected sites for the charge carrier and therefore no dc conduction across the system, while a bond fraction greater than the percolation threshold results in many finite clusters and a single infinite cluster, implying a nonzero dc conductivity. In this case also, dynamic disorder is introduced next by prescribing just how the hopping rates are randomly reassigned at random instants. A charge carrier trapped in the finite clusters might then, under the influence of dynamic disorder, move from one finite cluster to another, resulting in nonzero dc conductivity even below the percolation threshold. We referred to DDH based on percolation models of this sort as "dynamic bond percolation" (DBP).

Our own initial studies described above, aimed at applications to polymer electrolyte transport, assumed each reassignment event to randomize the hopping rates according to some fixed (and possibly continuous) distribution of hopping rates between sites but without correlation to any previous hopping rate assignment. The average time $\bar{\tau}_{ren}$ between reassignment (or "renewal") events corresponds in this case to a typical time of the polymer strand motion that interchanges allowed and forbidden pathways in the polymer host. The hopping time $\bar{\tau}_{hop}$, on the other hand, is the inverse of the success rate for carrier hops between nearby sites in favorable relative position for a hop. A third parameter, the observation time τ_{obs}, enters less directly; in a frequency-dependent measurement, it is simply the inverse frequency at which the response is observed. Evidently, the host reorganization will have little physical consequence for measurements in which $\bar{\tau}_{ren} \gg \tau_{obs}$.

An alternative model based on dynamic bond percolation in which the reassignment events are the independent switching on and off of hopping rates between individual sites, rather than global random reassignments of all hopping rates throughout the system, has been examined by Harrison and Zwanzig.[18]

Granek and Nitzan have dealt with spatial correlations of a specific kind in a different DBP model where each bond can flip at a characteristic rate α from one joining the neighboring sites (a,b) to one joining the neighboring sites (a,c), as well as having a characteristic rate $\bar{\tau}_{ren}^{-1}$ for its existence to be randomly reassigned; this then describes a system in which a given nearest neighbor pathway most probably appears only through the loss of a different nearest neighbor pathway. The spatial correlations thereby introduced were treated using an effective medium theory.

Hilfer and Orbach[21] study ion transport when randomly-located heavier ions block the motion of a lighter and more rapidly rearranging species and thereby provide a host background that defines which sites are available to the lighter ions at any given time. The heavier ions rearrange on some timescale $\bar{\tau}_{ren}$, while the lighter species moves from site to site on a faster timescale $\bar{\tau}_{hop}$. The model takes a DBP form as an approximation, and an effective medium theory is invoked in the end to obtain explicit results.

Hernandez-Garcia, Pesquera, Rodriguez, and San Miguel[22] have considered various models in which the hopping rates between nearest neighbor sites are all equal for any system in the ensemble, but this hopping rate can reassign randomly between the two allowed values w_1 and w_2. They have also studied one-dimensional models in which hopping rates $W(n,n+1)$ between neighboring sites vary independently except for either the constraint $W(n,n+1)=W(n,n-1)$ or $W(n,n+1)=W(n-1,n)$, and in addition have examined the role of anomalous diffusion ($<r^2>(t)$ not proportional to time) in the model that we have considered.

Finally, the Exxon group[23] has studied charge transport in microemulsions using a dynamic disorder model. Charges are assumed free to move from one water molecule to another within any given water cluster (at rate $\bar{\tau}_{hop}^{-1}$). Occasional reassignment (at rate $\bar{\tau}_{ren}^{-1}$) of which water molecules associate with one another to form clusters allows the charge to move from cluster to cluster to produce a non-zero dc conductivity.

B. Results implied by DDH models

Our own work[10-17] based on global reassignment events has now been extended[15,16] to include temporal correlation, in which one reassignment event affects when the next is likely to occur, to take into account the non-periodicity of the arrangement of sites, and consider general conditions similar to Eq. (6), relating transport in a dynamically disordered system to that for the statically disordered system. We consider each arrangement consisting of a carrier at a specified site surrounded by a (possibly randomly located) set of sites to constitute a configuration, and we assume that any reassignment event randomizes which system is in which configuration, with the sole restriction that the site of carrier just before reassignment must remain an allowed site immediately after reassignment. To allow for temporal correlation we assume a probability $\psi(t)$ dt for reassignment to occur in the interval (t, t + dt) *after a previous reassignment*. This requires introducing a second waiting time function $\phi(t)$ defined as the probability density for reassignment measured *from the arbitrary initial time of observation* (which can fall randomly between reassignments); $\psi(t)$ and $\phi(t)$ bear a simple relation to each other, as discussed elsewhere.[24] For any sequence of reassignments at specified instants $(t_1, t_2, t_3, \cdots)$, we can show that the growth of mean-square carrier displacement from its initial location (at zero applied field) behaves as shown in Fig. 9,

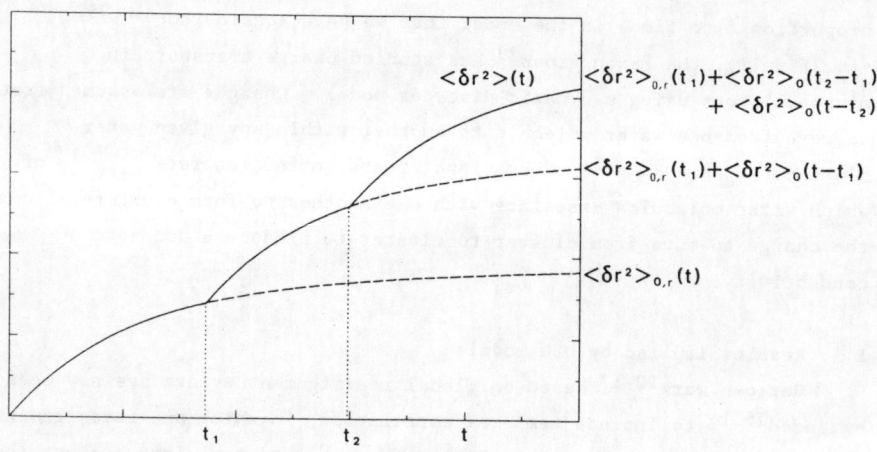

Figure 9: Behavior of the mean-square displacement as a function of time, showing general growth law dynamics. Note that within each renewal time cycle, there is an increase in the mean-square displacement depending on the length of that renewal time. This general growth law behavior characterizes dynamic disorder hopping models. From Ref. 16.

with each reassignment event adding on a segment of the $<r^2>_0(t)$ curve, which describes the carrier mean-square displacement averaged over an ensemble of systems that reorganize at the initial instant and then undergo no further hopping-rate reassignments. We refer to the behavior illustrated in Fig. 2 as the "growth law rule."

Now, because the growth-law rule might apply even more generally, at least as an approximation, than for the model based on random configurations, we take the growth law rule as our actual starting point for further results. The analysis yields a generalization of Eq. (6) giving the frequency-dependent diffusion coefficient $D(\omega)$ for the dynamically disordered system in terms of the growth of mean-square displacement in the statically disordered system

$$6\, D(\omega) = \frac{\tilde{\phi}(i\omega)}{\tilde{\Psi}(i\omega)}\, L\left(\Psi \frac{d}{dt} \langle r^2 \rangle_0\right) + i\omega\, L\left(\Phi \frac{d}{dt} \langle r^2 \rangle_{0,r}\right) \qquad (10)$$

$$= (\tilde{\phi}/\tilde{\Psi})\, L\{(\psi+i\omega\Psi) \langle r^2 \rangle_0\} + i\omega\, L\{(\phi+i\omega\Phi) \langle r^2 \rangle_{0,r}\} \qquad (11)$$

where

$$\Phi(t) = 1 - \int_0^t \phi(\tau)\, d\tau \;;\qquad \Psi(t) = 1 - \int_0^t \psi(\tau)\, d\tau \qquad (12)$$

are the probabilities for no renewal until t starting, respectively, from an arbitrary time at $t = 0$ or from a previous reassignment at $t = 0$. This result in turn gives the frequency dependent conductivity response to an applied oscillating field $\vec{E}(\omega)$ through the generalized Nernst-Einstein relation $\sigma(\omega) = [ne^2/k_B]\, D(\omega)$. In Eqs. (10)-(12), $L(F)$ and $\bar{F}$ denote the Laplace transform of a function $F(t)$ evaluated at the imaginary Laplace variable $s = i\omega$.

By considering the $\omega \to 0$ limit, we obtain the important result of Eq. (6) previously derived for the more restrictive dynamic bond percolation case, namely

$$\lim_{\omega \to 0} D(\omega) = \frac{\overline{\langle r^2 \rangle_0}}{6\,\overline{\tau}_{ren}} \tag{13}$$

where $\overline{\tau}_{ren}$ is the mean-time between reassignments and $\overline{\langle r^2 \rangle_0}$ is the mean-square displacement attained between reassignments. The existence of a finite limit for $D(\omega)$ as $\omega \to 0$ implies, by the usual Frobenius theorems, that $\langle r^2 \rangle(t) \propto t$ at large times, so that the carrier motion is diffusive. This result then applies also to the special case of transport in a dynamic bond percolation model, in agreement with the intuitive picture (below the percolation threshold) of host reorganization producing nonzero conductivity by allowing hopping to sites previously belonging to different clusters of interconnected sites.

Important results for the limit of zero temporal correlation now also follow more generally than in our previous DBP theory. When reassignments occur independently of each other with average rate λ, the waiting time density is of Poisson form $\psi(t) = \lambda \exp(-\lambda t)$, and we obtain (as in Eq. (6))

$$D(\omega) = D_{static}(\omega - i\lambda) \tag{14}$$

at angular frequency ω. Then the applicability of the growth-law rule and the assumption of zero temporal correlation between reassignment events is sufficient to imply the analytic continuation rule of Eq. (6). Moreover, when these generalized conditions apply, $6D(\omega)$ is given by

$$\lim_{\epsilon \to 0} (i\omega)^2 \int_0^\infty e^{-(\epsilon + i\omega)t} \langle r^2 \rangle(t)\, dt = (\lambda + i\omega)^2 \int_0^\infty e^{-(\lambda + i\omega)t} \langle r^2 \rangle_0(t)\, dt \tag{15}$$

The equality of these two expressions at all ω determines the mean-square displacement $\langle r^2 \rangle(t)$ for the dynamically disordered system in terms of $\langle r^2 \rangle_0(t)$ for the statically disordered system. The significance of these newer results is that the periodic lattice assumption (which is clearly inapplicable for polymer electrolytes) is now relaxed, and that results for statically disordered systems, which are more widely and more easily studied, can be converted into results for dynamically disordered systems under more general conditions than previously.

Application to the special DBP case illustrates the expected physical consequences of dynamic disorder. By obtaining an exact analytic solution for $<r^2>_0(t)$ in the one-dimensional static bond percolation problem based on a periodic lattice, and following the prescription just outlined in Eq. (15), we obtain the exact analytic solutions plotted in Fig. 7 for the dynamic bond-percolation $<r^2>(t)$ in one dimension without temporal correlation (calculated for average bond fraction f=0.5). With λ the reassignment rate and w the hopping rate, very large values of λ/w imply random reassignment before a typical hop, so that further increase in λ/w has little effect. Slow enough reassignment ($\lambda/w \simeq 0$) leads to quasi-saturation behavior of $<r^2>(\tau)$ [where $\tau = wt$]; the carriers are then all trapped in the finite clusters of interconnected sites, producing no long-range dc conductivity. At somewhat higher reassignment rates, $<r^2>(\tau)$ increases linearly with τ at long τ and has slope proportional to λ/w, corresponding to a renewal-controlled parameter regime that our physical model, as well as our efforts to fit experimental data for the polymer electrolytes, suggest applies for the polymer electrolytes. Descriptively, we may also regard the long-τ linear part of the plots in this range to describe renewal assisted diffusion, in the sense that carriers typically reach the end of the clusters and thereby suffer some impediment in motion, but, solely because of reassignment of hopping rates, can continue on and produce further increase in $<r^2>(\tau)$; the overall effect at long times is diffusion, but with a lower dc diffusion coefficient. The occurrence of renewal-assisted diffusion depends on the magnitude of the observation time τ_{obs} relative to other pertinent time scales; even very small non-zero reassignment rates produce observable renewal-assisted diffusive motion rather than apparent saturation when measured over a long enough τ_{obs}, and short enough τ_{obs} implies that the long-time diffusive regime resulting from the reassignment of the host will not be observed. The description given here is expected to carry over from DBP to generalized DDH models to a large extent.

Temporal correlation is expected on intuitive grounds in polymeric systems because the occurrence of a reassignment event should make an immediate second event less likely. The significance of temporal correlation between reassignment events can be estimated by assuming a form

$$\psi(t) = \rho \frac{(\rho t)^n}{n!} \exp(-\rho t) \qquad (16)$$

for the waiting time function $\psi(t)$ between reassignment events. This describes each event possibly delaying when the next occurs (but making it more likely at later times so that the average waiting time remains fixed at $\bar{\tau}_{ren} = \lambda^{-1}$). In the same spirit, a simplified form can be assumed for $<r^2>_0(t)$ appropriate to the kind of system considered; for static percolation below threshold

$$<r^2>_0 = A_0 [1 - \exp(-\alpha t)] \ . \qquad (17)$$

describes qualitatively the expected behavior of $<r^2>_0(t)$. Such an approach appears warranted because while $\psi(t)$, and certainly $<r^2>_0(t)$, cannot be specified precisely without considering a specific system (and for the highly disordered polymer electrolytes both functions are likely to depend on sample preparation and sample history), still the general behavior expected is known over a wide class of systems and should not depend seriously on whether the assumed $\psi(t)$ and $<r^2>_0(t)$ are merely rough approximations with the right general behavior.

Following this approach, it has been shown that parameter values giving long range ionic transport behavior consistent with experimental data for $\text{Re}[\sigma(\omega)]$ involve small λ/w values that (unexpectedly) involve fairly negligible effects from temporal correlation. The important implication then is that, for polymer electrolyte conduction, the analytic continuation rule should hold also as an approximation even with <u>temporal</u> correlation between reassignment events, the applicability of the growth law rule alone being sufficient. The growth law rule, which has been proven strictly only for global reassignment events, could in turn apply at least as an approximation even when some degree of <u>spatial</u> correlation between successive configurations of the hopping rate assignments persists during reassignment, thus considerably generalizing the theory.

Other dynamic disorder models not based on global reassignment have in fact been shown to yield the analytic continuation rule also, but only as an approximation. The DBP model of Harrison and Zwanzig,[18] for example, in which individual bonds fluctuate on and off independently, leads to this result in lowest order effective medium theory, as does the Exxon group's[23] model for transport in oil continuous microemulsions, and the model of Hilfer and Orbach[21] for transport of lighter ions within a background of heavier ions.

We may next consider a number of practical consequences suggested by the applicability of a dynamic disorder description. Kimball and Adams[25] have shown that for any hopping model, $\text{Re}[D(\omega)]$ (and correspondingly $\text{Re}[\sigma(\omega)]$) is monotone increasing as a function of ω, and we have shown that this implies an $<r^2>_0(t)$ vs. t curve that is concave down (i.e., has negative second derivative); thus the kinds of curve shown in Fig. 10a are characteristic in this respect. The characteristic shape of the growth law curve for hopping transport implies then that the carrier can very generally attain a greater displacement at a given time if reassignment occurs (Fig. 10b). More generally, it can be shown that if diffusive motion or saturation of $<r^2>_0(t)$ applies for the static system for times $\tau > \tau_0$, then $Re[D(\omega)]$ is always monotone increasing with λ for DDH at any $\omega < 2\pi/\tau_0$, including of course for the dc conductivity at $\omega = 0$. Stated simply, dynamic disorder fairly generally enhances transport by hopping.

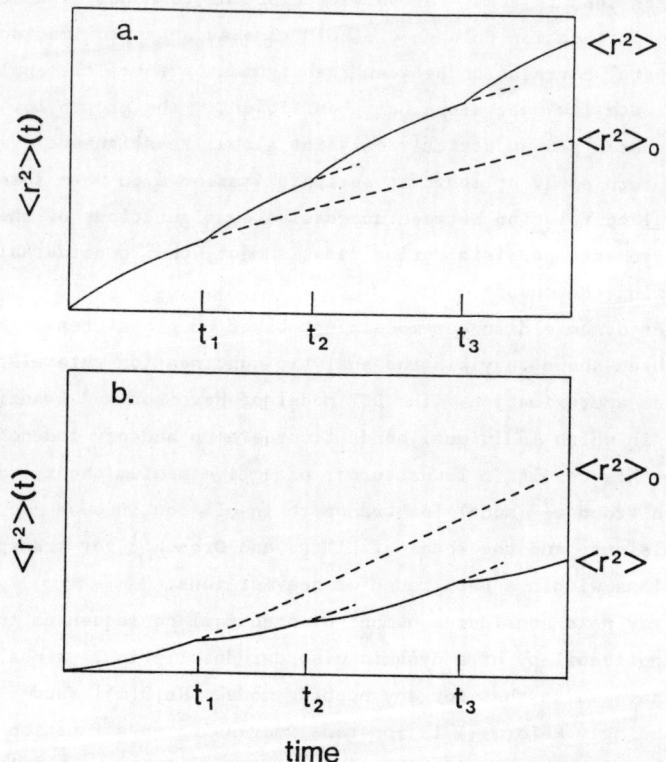

Figure 10: Specific cases of growth law behavior for diffusive transport (10a) and ballistic transport (10b). In Fig. 10a, the renewal process increases the local mean-square displacement, while in Fig. 10b it reduces it. In both cases, however, general growth law dynamics is obeyed. From Ref. 16.

But the picture of transport described in terms of the growth law of mean-square carrier displacement is in fact not limited to hopping. For ballistic motion of carriers impeded by collisions that randomize the average velocity, for example, the mean square displacement also obeys a growth law rule, except that in the simplest model $<r^2>_0(t)$ without the randomizing collisions is proportional to t^2; more generally in this case, $<r^2>_0(t)$ as a function of t is concave up (i.e., has a positive second derivative). It is seen from Fig. 10b that the renewal events at a sequence of specified times $(t_1, t_2, t_3, \cdots)$ impede transport, precisely because of the concave-up shape. Also, in contrast to hopping transport, we can show that $Re[D(\omega)]$ and therefore $Re[\sigma(\omega)]$ is monotone decreasing with increasing ω. These observations would appear of possible value in analyzing experimental data for mixed conductors, such as for the polyiodides, in which both ionic and electronic transport are believed to occur simultaneously.

A formal application of the theory arises by relating it to a continuous-time random walk (CTRW) on a lattice, a problem that has gained widespread interest especially through Scher and Lax's application of the CTRW theory to impurity hopping conduction in semiconductors such as Si and Ge. The starting point of the Scher/Lax theory[27] is site-to-site hopping of carriers among impurities randomly located on a lattice, and ultimately this problem is approximated by hopping of carriers on a periodic lattice with all sites available. We start with the latter hopping problem. A formal similarity to the growth law rule is immediately evident in that a sequence of hopping events causes an increase in $<r^2>_{CTRW}(t)$ for the carrier, except that the increase occurs by some quantity $<d^2>_0$ being added onto $<r^2>_{CTRW}(t)$ abruptly at the instant of each hop in a stepwise way, and not by a new curve segment $<r^2>_0$ being added on to produce a continuous curve as in the case of DDH models. We assume for generality that $<d^2>_0(\tau)$ is a function of time τ after the previous hop; specifically,

$$\langle d^2 \rangle_0(\tau) = \sum_{\vec{s}} \psi(\vec{s}, \tau) s^2 \bigg/ \sum_{\vec{s}} \psi(\vec{s}, \tau) \qquad (18)$$

is an average hop distance for a hop at time τ, with the denominator giving the total probability density for a hop of any distance at time τ, and with $\psi(\vec{s},\tau)$ the average waiting-time function for a hop with displacement $\vec{s}$. The similarity of the CTRW in a lattice without hopping-rate reassignment to the DDH problem allows the previous formal methods to be followed, yielding a somewhat (but not completely) analogous expression for the diffusion coefficient $D(\omega)$ and the conductivity $\sigma(\omega)$.

The question now arises whether it is possible to represent the solution to the DDH problem as a continuous-time random walk on a non-reorganizing lattice, in the sense that both give the same $D(\omega)$ (and $\sigma(\omega)$). Equating the two $D(\omega)$ expressions leads naturally to a mapping in which the two $\psi(t)$ waiting time functions are the same for the CTRW and the DDH, as are also the two $\phi(t)$ waiting time functions, with the requirement

$$\langle d^2 \rangle_0(t) = \frac{1}{\lambda} \frac{d}{dt} \langle r^2 \rangle_0^{(DDH)}(t) \qquad (19)$$

then implied. For a given DDH problem, it is usually impossible to satisfy these conditions with a simple constant distance $\langle d^2 \rangle_0(\tau) = a^2$, so that the hopping described in the mapping must be of variable range when the DDH problem is given and the corresponding CTRW is to be determined. Moreover, since the previous arguments show that the second derivative of $\langle r^2 \rangle_0$ is nonpositive for hopping transport, the mapping described has each DDH reassignment, on the average, being mapped into a hop (the waiting time functions being the same) with the range of the hops $\langle d^2 \rangle_0(\tau)$ decreasing on the average with time, so that longer range hops become less probable. The physical interpretation of the mapping must be approached cautiously, however, because it is based on a mathematical equivalence, not necessarily a physical one.

A particularly enlightening interpretation of the relation between transport in corresponding dynamically disordered and statically disordered systems arises by noting that, as a direct consequence of the Kubo linear response theory[26,17] applied to the conductivity $\sigma(\omega)$, $\sigma(\omega)$ for the dynamically disordered system is exactly equal to $\sigma(\omega-i\lambda)$ for the dynamically disordered system if and only if the corresponding velocity autocorrelation functions for the carrier in the two cases are related to each other (for zero applied field in both cases) by

$$<v_\xi(0) \ v_\eta(t)> = <v_\xi(0) \ v_\eta(t)>_0 \exp(-\lambda t) \ . \tag{20}$$

with (ξ,η) denoting cartesian components. Of course, some caution in needed in dealing with a velocity auto-correlation function for a hopping model, but it should be possible to circumvent this difficulty in practice by care in taking various limits. The application to DDH transport arises by considering the source of non-zero correlation in the global renewal case. In the ensemble of systems averaged over, whenever a reassignment occurs, complete randomization of the hopping rates reduces that contribution to the average velocity autocorrelation to zero. The nonzero part then arises solely because of memory built into the structural assignment of hopping rates (with $<v_\xi(0) \ v_\eta(t)>_0$ tending to be negative for small positive time), and it is damping of this memory by the $\exp(-\lambda t)$ factor in the ensemble average that gives the analytic continuation rule and thereby the relation between static and dynamic disorder for DDH transport. Here again, the damping can arise also as an approximation in models where the complete loss of structural memory of hopping assignment rates does not occur in a reassignment event.

In addition, the basic results just described extend beyond classical DDH hopping. A form of the essential result equating the condition $\sigma(\omega) = \sigma_0(\omega-i\lambda)$ to a damping relation between the corresponding autocorrelation functions holds also quantum mechanically, and the physical idea applies to carrier scattering in semiconductor band transport, where $\sigma_0(\omega)$ describes transport of a carrier limited by some particular set of scattering processes and $\sigma(\omega)$ includes an additional set of scattering process; if the effect of the additional processes is to damp the carrier velocity autocorrelation function by $\exp(-\lambda t)$, then the relation $\sigma(\omega) = \sigma_0(\omega - i\lambda)$ holds.

6. REMARKS

The use of percolation models is now very widespread for describing transport phenomena in disordered systems. While the assumptions of standard percolation theory are very stringent, nevertheless the general behavior, including percolation thresholds, scaling behavior, dimensionality effects and frequency dependences, seem to describe many phenomena quite well. In polymeric materials, especially those studied above their glass transition temperature, the static percolation arguments are untenable, since the host material itself is undergoing large-scale orientational and translational motions. A dynamic picture, in which the hopping process among sites is determined by probabilities that are themselves time dependent (reflecting the time dependence of the host) seems an appropriate picture for such electroactive polymer systems.

We have described some models of the dynamic disordered type, and some results obtained using them. The generalities, including diffusive behavior at long times, analytic continuation formulas from the static lattice, scaling behaviors, and effects of correlation in the lattice itself, are now fairly well understood. Direct relationships between dynamic disorder hopping models and the free volume picture have been described, and analysis of several experiments using the concepts of dynamic percolation have been advanced. The challenge remaining in the theoretical description is now similar to the Coulomb gap problem found in disordered semiconductors: ionic correlations among the carriers can substantially affect the conductivities, and their description, in these materials where the mean separation between Coulomb charges can be less than 10Å, will be complex.

While the dynamic disorder hopping models have been applied to polymer electrolytes and to microemulsions,[23] there are a number of other situations in which the physical idea of dynamic disorder is an attractive one. These include solutions, particularly viscous solutions, soft and orientationally disordering solids, and mixed phases involving both ceramic and polymeric components. Biological systems also show such behavior, for example the ionic motion through membrane channels should be describable in terms of motions of the membrane host occurring on time scales comparable to those of the diffusing system itself.[27]

<u>Acknowledgements</u>: We are grateful to the Northwestern Ionic Conductors Group, particularly D.F. Shriver and A. Nitzan, for close collaborations. We thank the NSF MRC Program, through the Northwestern Materials Research Center, (DMR 88 21571) for support of this research.

References

1. (a) P. Vashista, J.N. Mundy, G.K. Shenoy, Eds. *Fast-Ion Transport in Solids*, North-Holland: Amsterdam, (1979); (b) J.B. Bates, C.G. Farrington, Eds., *Fast-Ion Transport in Solids*, North-Holland: Amsterdam (1981); (c) M. Kleitz, B. Sapoval, D. Ravaine, Eds. *Solid State Ionics*, North-Holland: Amsterdam, (1983); (d) J. Boyce, L. DeJonghe, R.H. Huggins, Eds., *Solid-State Ionics*, North-Holland: Amsterdam, (1985).

2. M. A. Ratner and D. F. Shriver, *Chem. Revs.*, **80**, 109 (1988); J. R. MacCallum and C. A. Vincent, eds., *Polymer Electrolyte Reviews*, (Elsevier London, 1987); C. A. Vincent, *Prog. Sol. St. Chem.*, **17**, 145 (1987); *British Polym. J.* **20**, No. 13 (1988); J. S. Tonge and D. F. Shriver, in J. Lai, ed., *Polymers for Electronic Applications*, CRC Press, Boca Raton, in press.

3. P. V. Wright, *Brit. Polym. J.*, **7**, 319 (1975); D. E. Fenton, J. M. Parker and P. V. Wright, *Polymer*, **14**, 589 (1973); M. B. Armand, J. M. Chabagno and M. Duclot, Extended Abstracts, Second International Meeting on Solid Electrolytes, St. Andrews, 1978; in *Fast Ion Conduction in Solids*, ed. P. Vashishta, J. N. Mundy, G. K Shenoy (North-Holland: 1979).

4. C. Berthier, W. Gorecki, M. Minier, M. B. Armand, J. M. Chabagno and P. Rigaud, *Solid State Ionics*, **11**, 91 (1983); M. Minier, C. Berthier and W. Gorecki, *J. Physique*, **45**, 739 (1984).

5. Y. Chatani, S. Okamura and Y. Fujii, *ACS Div. Poly. Chem.*, *Poly. Preprints*, **30**, 404 (1989).

6. H. Cheradame, in *IUPAC Macromolecules*, ed., H. Benoit and P. Rempp (Pergamon, London, 1982); A. Killis, J. F. LeNest, A. Gandini, H. Cheradame, J. P. Cohen-Addad, *Solid State Ionics*, **14**, 231 (1984); M. Watanabe and N. Ogata in ref. 13, Chap. 3; *Brit. Polymer J.*, **20**, 181 (1988).

7. M.H. Cohen, D. Turnbull, *J. Chem. Phys.* **31**, 1164 (1959); M.H. Cohen, G.S. Grest, *Phys. Rev. B.* **B21**, 4113 (1980); S.D. Druger, A. Nitzan, M.A. Ratner, *Solid State Ionics* **9/10**, 1115 (1983).

8. M.A. Ratner, in J.R. MacCallum and C.A. Vincent, Eds., *Polymer Electrolyte Reviews I.*, Chap. 7.

9. H. Vogel, *Phys. Z.*, **22**, 645 (1922); G. Tammann and W. Hesse, *Z. Anorg. Allg. Chem.*, **156**, 245 (1926); G. S. Fulcher, *J. Am. Ceram. Soc.*, **8**, 339 (1925).

10. M. Ratner and A. Nitzan, *Disc. Far. Soc.*, in press.

TRANSPORT AND RELAXATION OF EXCITATIONS IN RANDOM ORGANIC SOLIDS:
MONTE CARLO SIMULATION AND EXPERIMENT

Heinz Bässler
Fachbereich Physikalische Chemie, Philipps-Universität
D-3550 Marburg, FRG

ABSTRACT

A survey is given on the energetic relaxation and diffusion of electronic excitations in random organic solids. After introducing the concept of time dependent stochastic random walks in systems characterized by static energetic disorder of the hopping sites, modelled by a Gaussian density of states, the results of Monte Carlo simulations and analytic effective medium theory are presented. Experimental results on time dependent diffusion of triplet excitations in a benzophenone glass, charge carrier migration, as well as recombination of geminately bound electron-hole pairs in polyvinylcarbazole support the model. Finally it will be shown that the concept can be extended to treat dynamic processes in supercooled melts near the glass transition as well.

1. INTRODUCTION

The lack of lattice periodicity in random solids, such as amorphous semiconductors, atomic and molecular glasses, or non-crystalline polymers, implies that the constituent atoms, molecules, or molecular sub-units, henceforth called sites, are located in statistically different environments. As a consequence, inter-site interaction energies are subject to a distribution. Determining the spectrum of eigen-energies of random solids involves diagonalizing a matrix containing the disorder-modulated self-energies of the sites as diagonal elements and the disorder-modulated exchange terms as off-diagonal elements, respectively. This is an extremely intricate problem, particularly if self-energies and exchange energies are of comparable magnitude as in amorphous inorganic semiconductors. Basically, it leads to a manifold of delocalized states above a certain demarcation energy, called the mobility edge and a distribution of localized states below.[1] Anticipating that the disorder potential can be modelled by a Gaussian profile John et al.[2] were able to derive the exponential energy dependence of the density of localized states (DOS) often seen in amorphous semiconductors.[3] A recent computer simulation, however, showed that the same result is recovered by invoking statistical distribution of coulombic centers as sources of the random potential.[4]

Contrary to what one might guess on a first glance basis, the situation is less complex in organic random solids. Due to only weak van der Waals coupling among the sites the exchange terms in the energy matrix are small, typically of order 10^{-2} eV only, and so must be the disordered-induced fluctuations.[5] The dominant disorder effect is, therefore, the modulation of the (static) lattice contribution to the site energies determined by the electronic polarization energy of a charge carrier or Frenkel exciton. These fluctuations are typically of order 0.1 eV for charge carriers and 0.03 eV for Frenkel excitons, i.e. about one order of magnitude larger than the corresponding transfer integrals. As a consequence, electronic states in random organic solids can to good approximation be considered as completely localized, their distribution reflecting the distribution of the diagonal elements of the energy matrix. (See fig.1 for a schematic illustration). Depending on a large number of configurational coordinates it is straightforward to describe it by an envelope function of Gaussian shape. A direct experimental probe is the absorption spectrum of organic glasses[6] or of chromophores encorporated in a glassy matrix.[7] Unfortunately, in organic solids - except for conjugated polymers with degenerate ground state such as trans-polyacetylene - transitions among valence and conduction states are extremely weak and completely masked by localized exciton transitions.[8] Therefore, no direct probe exists for mapping

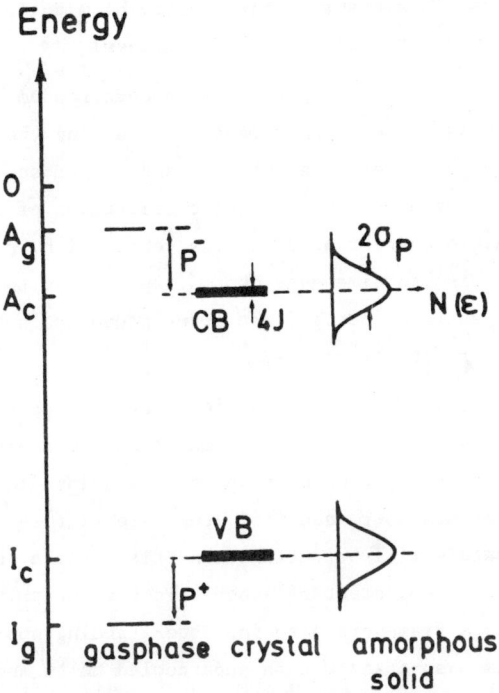

Fig.1 Energy level spectrum of ionic states in an organic molecule, the corresponding crystal and the amorphous phase, respectively. I_g, I_c, A_c and A_g are ionization potential and electron affinity in the gas-phase and the crystal, respectively, P^+ and P^- are the polarization energies of positive or negative charge carrier in the solid, J is the exchange integral and σ is the standard deviation of the probability density of the polarization energies.

the DOS for charge carriers. Since, however, the polarization excitons or charge carriers induce in their environment have a common origin it is legitimate to conclude on Gaussian shaped DOSs for charge transporting states as well.

Random organic solids, therefore, represent the case of intrinsic energetic disorder onto which extrinsic disorder due to specific chemical or physical defects may be superimposed. Since structural randomness is not connected with the breaking of covalent bonds they are devoid of dangling bonds present in amorphous inorganic semiconductors. This ensures that density of mid-gap states is low enough to rule out hopping at the Fermi level.

Their simple electronic structure renders random organic solids attractive as proto-type systems for studying the effect of diagonal energetic disorder on electronic and spectroscopic properties, its key feature being the energetic relaxation of an excitation, be it a charge carrier or an optical excitation, after generation. It gives rise to time dependent diffusion and to a distribution of the rates for kinetic first order phenomena not found in crystalline counterpart structures.

After describing the essential features of electronic transport as revealed by Monte Carlo (MC) simulation and analytic theory, experiments probing (i) motion of triplet excitations in an organic glass, (ii) charge transport and (iii) geminate pair recombination in polyvinylcarbazole will be outlined in this article. Finally it will be shown that the concept of stochastic transport in a random potential affords a framework also for understanding non-electronic phenomena such as mass-transport in supercooled melts and the occurrence of the glass transition.

2 MODEL CONSIDERATION
2.1 Transport Of Excitations

Transport within a manifold of localized sites is incoherent, i.e. phase memory is lost after every step. The dynamics can therefore be described by a generalized kinetic master equation

$$\frac{dn_i}{dt} = -\sum_j \nu_{ij} n_i(t) + \sum_j \nu_{ji} n_j(t) - n_i(t)/\tau_0 \qquad (1)$$

where n_i is the occupational density of site i characterized by position $\vec{R}_i$, energy ε_i and intrinsic decay time τ_0 if the moving particle is an optical excitation. The simplest ansatz for ν_{ij} compatible with the principle of detailed balance in a system where inter-site coupling is via exchange interaction is that of Miller and Abrahams[9]

$$\nu_{ij} = \begin{cases} \nu_0 \exp(-2\gamma|\Delta R_{ij}|)\exp-\frac{\varepsilon_j - \varepsilon_i}{kT} &, \quad \varepsilon_j > \varepsilon_i \\ \nu_0 \exp(-2\gamma|\Delta R_{ij}|) &, \quad \varepsilon_j < \varepsilon_i \end{cases} \quad (2)$$

γ is the inverse wavefunction localization radius. Eq.(2) ignores polaronic effects. If non-negligible, they could be accounted for by an additional Boltzmann factor containing the polaron binding energy. The asymmetry of the exchange rates ν_{ij} causes a major problem for any analytic treatment, in particular, if the DOS is of Gaussian shape. Recall that the continuous time random walk formalism (CTRW) introduced by Montroll, Scher and Lax[10,11] maps the true random walk onto a regular lattice assuming that all disorder can be accounted for by a single site waiting time distribution function without considering energetic relaxation.[12] The only approach available to date for considering the energetic relaxation implied by the asymmetry of the hopping rates is the effective medium approximation (EMA) developed by Movaghar and coworkers.[13-16] The higher order correlation effects appearing when summing over all paths a particle can take on going from site i to j lead in this approximation to an effective reduction of the site density after each jump. For details the reader is referred to the original work.

The alternative way to treat hopping within a random potential is MC-simulation. It can be considered as an idealized experiment carried out on samples of arbitrarily adjustable degree of complexity. It not only allows determining which level of sophistication is required for modelling a real world sample but is able to check the validity of mathematical approximations involved in analytic treatments that are based on identical physical premises, such as the form of the jump rates or the type of disorder. The MC simulations of the Marburg group were done on cubic lattices (lattice constant a=6Å) with periodic boundary conditions consisting of at least 50x50x70 sites. In case of charge transport maximum sample thickness considered was 4000 sites. To find out whether real world systems, like molecularly doped polymers used in electrophotography, can approximately by modelled by a system with exclusively diagonal disorder of the hopping sites, all the disorder was put into the distribution of site energies resembling a Gaussian DOS with standard deviation σ and centered at $\varepsilon_0 = 0$,

$$g(\varepsilon) = (2\pi\sigma^2)^{-1/2} \exp(-\frac{\varepsilon^2}{2\sigma^2}) \ . \tag{3}$$

To determine the effect of superimposed off-diagonal disorder, the parameter 2γa governing overlap among nearest neighbor sites, has been subjected to a distribution resembling a Gaussian density profile as well. An excitation was started at an arbitrary site and allowed to execute a random walk under either zero bias (exciton case) or under the influence of an external electrostatic potential. From each site a particle can in principle jump to acceptor sites contained within a cube containing $(7a)^3$ sites. Details of the simulation procedure have been published elsewhere.[17] Quantities the computer can keep track of are the energy of the site an excitation visits after a time t, its position and mean square displacement as a function of time as well as the number of new sites it has visited after a certain time. Averaging usually involves 50 lattice configurations and 20 excitations started independently on each.

In fig.2 the temporal evolution of the occupational density of states is portrayed for a disorder parameter $\hat{\sigma} = \sigma/kT = 2$. Time has been normalized to the dwell time $t_0 = [6\nu_0 \exp(-2\gamma a)]^{-1}$ of an excitation in a lattice composed of iso-energetic sites. While relaxing in energy space the width of the distribution becomes slightly asymmetric because the higher states relax more quickly. After a time t_{rel}, called the relaxation time, the mean energy settles at

$$<\varepsilon_\infty> = -\sigma^2/kT \tag{4}$$

as shown on fig.3 and the occupational DOS acquires the Gaussian

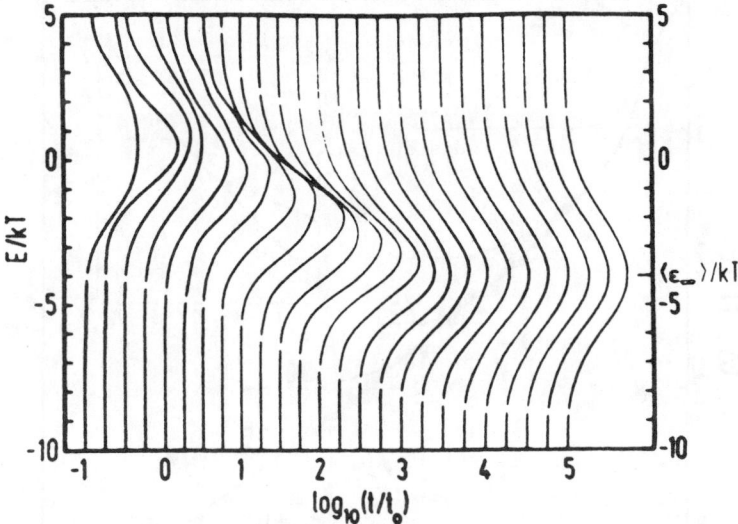

Fig.2 Temporal evolution of the distribution of carrier energies in a Gaussian DOS of width $\hat{\sigma} = \sigma/kT = 2$. All profiles are broken at the same carrier density illustrating the different relaxation patterns of the carriers in higher or lower energy states, respectively.

width σ as predicted by simple analytic theory. The temporal decay of the mean energy follows a logarithmic law, $<\varepsilon(t)> \propto \ln t/t_0$, to first order approximation only. It is gratifying to note that the analytic EMA result is virtually identical with the MC result.[16] This demonstrates that the EMA is indeed appropriate and sufficient for treating the E-hopping case quantitatively, its only shortcoming being the mathematic complexity which prevents casting the solution into a convenient analytic form.

Of key importance for the analysis of experimental results is the relaxation time of the system. Evaluating the data of fig.3 yields the empirical relation

$$t_{rel}/t_0 = 10 \exp[-(1.07\hat{\sigma})^2] \ . \qquad (5)$$

Remarkably, t_{rel} increases faster with disorder parameter $\hat{\sigma}$ than

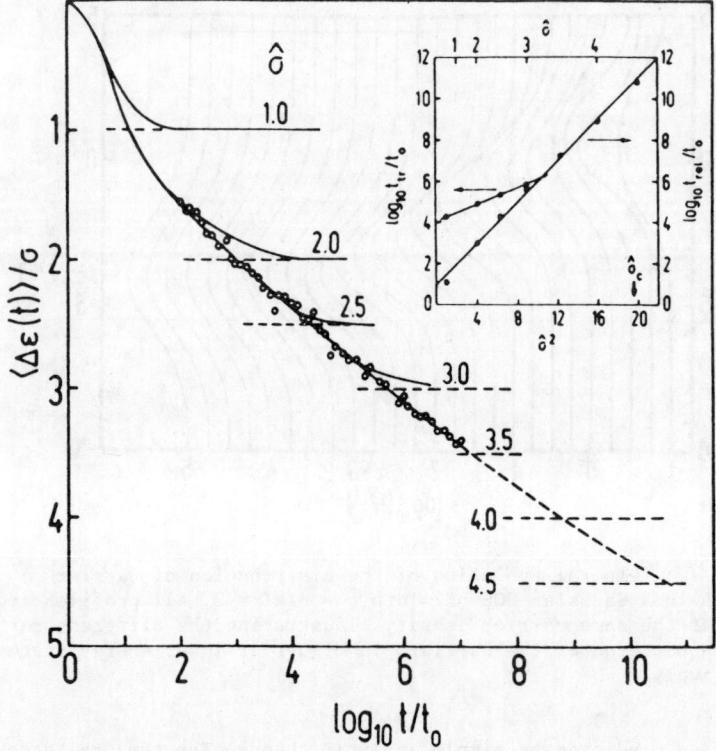

Fig.3 Simulation data for the energetic relaxation of a packet of charge carriers in course of hopping in a Gaussian DOS (width σ) as a function of normalized time. The inset shows the relaxation time for energetic equilibration as well as the transit time of a sheet of carriers across the computer sample at a field of 10^5 Vcm^{-1} as a function of $\hat{\sigma} = \sigma/kT$.

does the mobility of a packet of charge carriers in the dc-limit,

$$\mu_\infty(\hat{\sigma}) = \mu_0 \exp[-(0.69\hat{\sigma})^2] \quad . \qquad (6)$$

Here $\mu_0 = ea^2/6kTt_0$ is the charge carrier mobility in an energetically ordered counterpart hopping system. This has the important consequence that a time of flight signal produced by a packet of charge carriers drifting across a sample of length d under the action of an applied electric field E must become dispersive above a certain degree of disorder, i.e. below a certain temperature at otherwise constant system parameters, because eventually t_{rel} will exceed the carrier transit time. This criterion can be cast into an expression for the critical disorder parameter,

$$\hat{\sigma}_c = 1.22 \, (\ln \frac{6kTd}{ea^2E})^{1/2} \quad . \tag{7}$$

The transition from dispersive to non-dispersive transport, often seen in experiment[19,20], is illustrated in fig.4 for a simula-

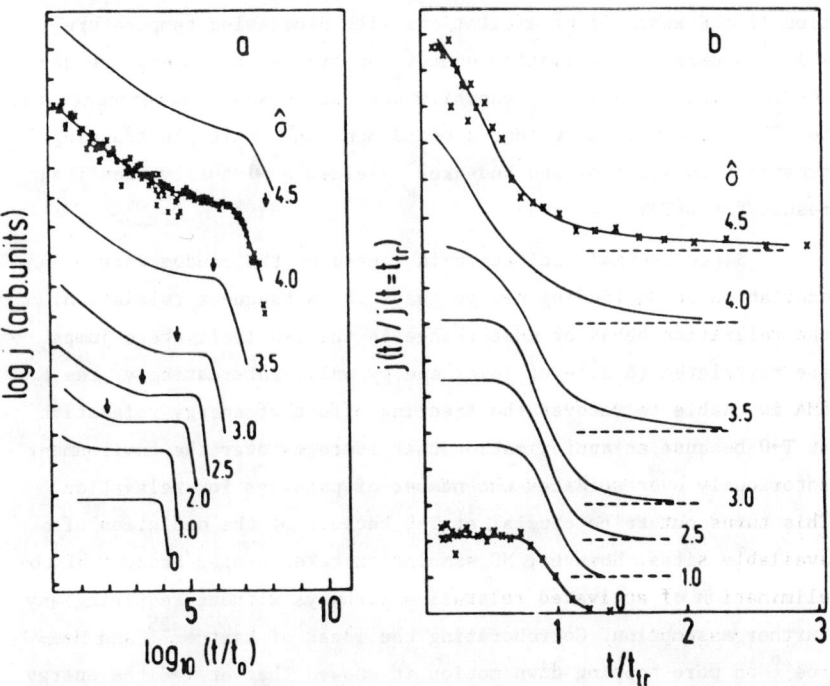

Fig.4 Simulated TOF signals, parametric in $\hat{\sigma}$, in log-log (section a) and lin-lin (section b) representation. In b the time has been normalized to the average carrier arrival time at the exit contact. Section a reveals increasing spreading of the tail of the TOF signal with increasing $\hat{\sigma}$ even if the dispersive to non-dispersive transition occurs at times $\ll t_{rel}$. This is a manifestitation of deviations from the Einstein relation, recently shown to be a characteristic feature of hopping in a Gaussian DOS (ref.38).

tion sample of length d = 300a.[21] It also demonstrates that µ and, hence, the diffusivity, do not follow a strict power law $D(t) \propto t^{-(1-\alpha)}$ as predicted by earlier dispersive transport theory.[11] To fit data collected over more than, say, two decades, by such a law formally required indroducing a time dependent dispersion parameter α.[22]

Of particular interest is the non-Arrhenius-type temperature dependence of the diffusivity in the non-dispersive transport regime predicted by eq.(6). It is the result of the energetic relaxation of the ensemble of excitations with decreasing temperature which renders the activation energy for transport temperature dependent. This effect has meanwhile been discussed in other contexts, too.[23] It is surprising that a one-dimensional multiple trapping treatment by Arkhipov and Rudenko[24] yielded a virtually identical result for D(T).

Since thermal excitation in course of the random walk of an excitation helps finding new pathways for subsequent relaxation, the relaxation behavior must change in the T→0 limit where jumps are restricted to site of lower energy only. Interestingly, the EMA is unable to recover the freezing effect of energy relaxation at T→0 because an approximation that averages over the environment notoriously overestimates the number of pathways for relaxation. This turns out to be crucial at T→0 because of the depletion of available sites. However, MC simulation takes proper account of the elimination of activated relaxation pathways without requiring any further assumption. Corroborating the ideas of Kastner[25] and Monroe[26] on pure hopping down motion it showed that at T→0 the energy relaxation function approaches

$$(\Delta\varepsilon/\sigma)^2 \simeq 3 \ln\ln(t\nu_0) \qquad (8)$$

in the long time limit. The ν_0 is defined via eq.(2). Subsequent work by Movaghar et al.[27] has laid out the basis for understanding this phenomena in terms of a rigorous analytic theory.

Stimulated by earlier experiments bearing out the surprising result that the diffusivity of triplet excitations can be much larger in a topologically disordered system, e.g. a melt[28,29] or a molecularly doped polymer matrix[30] as compared to the counterpart molecular crystal we investigated the role of off-diagonal disorder on excitation transport as well. It can arise from positional or orientational fluctuations of the interacting molecules in a random matrix and can be quantified by letting the intersite coupling parameter $2\gamma a$ vary statistically. For computational purposes the overlap parameter Γ_{ij} parameter characterizing exchange interaction among sites i and j was split into two terms specifying sites i and j, respectively, $\Gamma_{ij} = \Gamma_i + \Gamma_j$, and assuming that Γ_i and Γ_j vary randomly in an uncorrelated fashion following a Gaussian weight distribution $g(\Gamma)$ with standard deviation σ_Γ. This yields a standard deviation $\Sigma = \sigma_\Gamma \sqrt{2}$ for Γ_{ij}.

MC simulations were carried out for diagonal disorder parameter $\hat{\sigma} = 0$ [31] and $\hat{\sigma} = 3$, respectively. The major result is that inclusion on off-diagonal disorder does not increase the dispersion seen, for instance, in a charge carrier transit pulse, but increases the diffusivity and, concomitantly, the charge carrier mobility in the steady state limit drastically (see fig.5). This is because a symmetric variation of Γ_{ij} translates into an asymmetric distribution of overlap factors $\exp\Gamma_{ij}$. Statistical misorientations or distance fluctuations thus establish faster transport paths, reminiscent of percolation except that no threshold effect occurs. If the sample were a homogeneous medium, one could replace the distribution of overlap factors by the ensemble-averaged diffusion coefficient $<D> = D_\infty <\exp[-2g(\Gamma)]>$ yielding

$$<D> = D_\infty \exp(\Sigma^2/2) \quad . \tag{9}$$

Comparison with the MC data indicates that eq.(9) provides an order of magnitude estimate only for the enhancement factor because the large amplitude variations of Γ_{ij} contribute most to diffusion. They help establishing detour routes allowing an excitation to arrive

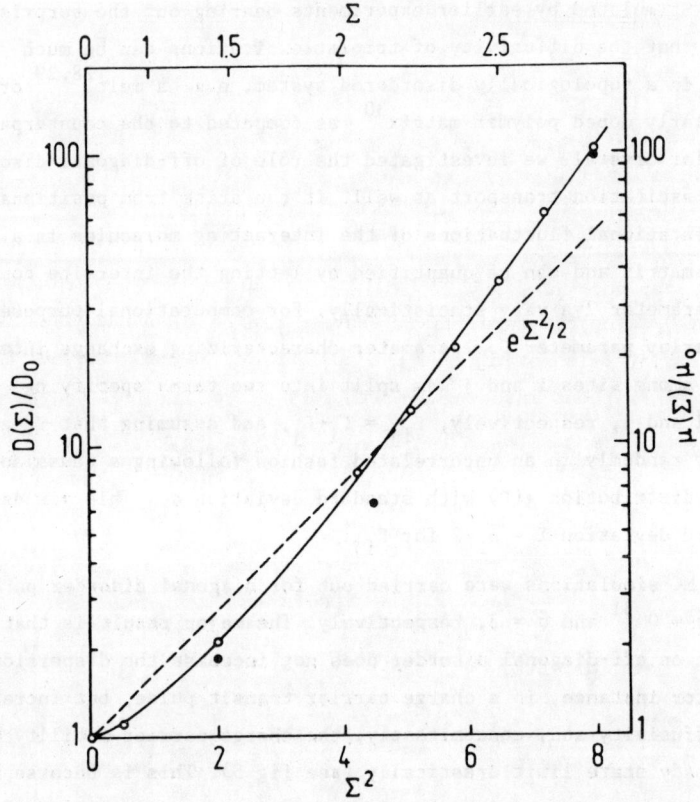

Fig.5 Left scale (open circles): Normalized diffusion coefficient of an excitation moving in an energetically degenerate hopping system where the wavefunction overlap parameter $2\gamma a$ is subject to Gaussian fluctuation with standard deviation Σ. Right scale (closed circles): Normalized mobility of a charge carrier moving in a hopping system with both diagonal disorder ($\hat{\sigma} = \sigma/kT = 3$) and off-diagonal disorder characterized by Σ.

at an acceptor site earlier than it would if executing nearest neighbor jumps with ensemble-averaged rate only. In any case do the simulation confirm that omission of off-diagonal disorder in the previous simulations did not effect the functional dependencies describing relaxation and diffusion of excitations in the early time domain because for realistic system parameters the dispersion generated by diagonal disorder dominates. Neither did it affect the temperature dependence of the diffusivity in the long time limit if restricted to small electric fields only. In the high field limit, off-diagonal disorder does turn out to influence functional dependences as will be shown elsewhere.

2.2 A Selection Of Experimental Results On Excitation And Charge Carrier Motion In Random Organic Solids

Probing energetic relaxation of excitations during their random walk in a random potential requires excitations amenable to spectroscopic techniques. Contrary to inorganic semiconductors, charge carriers in organic solids do not fulfil this requirement because (i) interband transition have vanishing oscillator strength and (ii) intraband transitions are absent as well because of the narrowness of the density of states involved. Experiments were therefore done on triplet excitations in a benzophenone glass. This system offers several advantages[22]: (i) Glasses are easily prepared by either quenching the melt or vapor deposition, (ii) exciting a vibrationally hot singlet state by an N_2-laser pulse results in a random population of the triplet manifold (T_1) via rapid intersystem crossing that decays to the ground state by phosphorescence emission, (iii) the hopping time of T_1 states in the crystal is of order 10 ns [32], their intrinsic lifetime is 5 ms affording a conveniently accessible temporal observation window. Monitoring the temporal shift of phosphorescence spectra measured by a gated optical multichannel analyser as a function the delay time between the laser pulse and opening of the gate window yields a direct image of the time evolution of the occupational density of T_1 states.[33] Results are shown in fig.6. Their evaluation yields the time dependence of the mean energy of the moving packet of T_1-excitations at 6 K and 80 K, respectively (fig.7). Not only is the

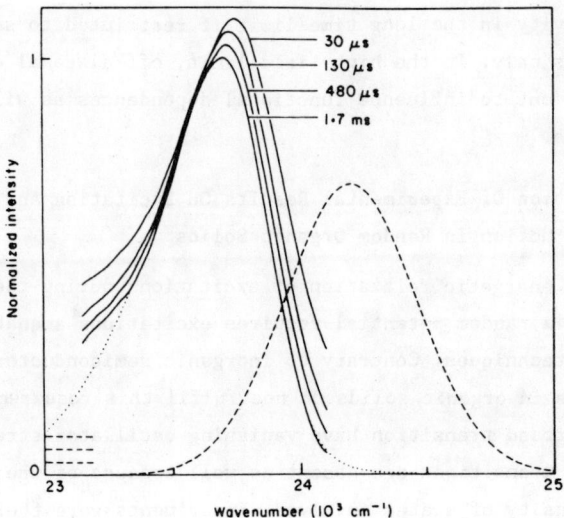

Fig.6 The 0-0 section of the phosphorescence spectrum emitted from glassy benzophenone at the indicated delay times between excitation and spectra recording. Dotted curve: profile after correction for the contribution from the first vibronic transition. Dashed curve: density-of-state profile of triplet states. All the spectra were recorded at 6 K.

expected slowing down of relaxation at low temperatures recovered, the functional dependences are also in full accord with the simulation results and can be fitted adopting the crystal value (10 ns) for the jump time of an excitation in the ordered counterpart structure and a width of the DOS $\sigma = 260$ cm^{-1} which is in good agreement with emission profiles. The mutual agreement between experiment, simulation, and analytic theory proves that the applied model of the random walk of an excitation within a Gaussian random potential in the limit of weak electron-phonon coupling is both appropriate and sufficient for treating transport of excitations that move via exchange interaction.

To monitor excitation transport spectroscopically one has to add defects to the sample acting as traps outside the DOS of intrinsic hopping states and record the decrease of the host luminescence as a function of time. This has been done by adding bromo-naphthalene

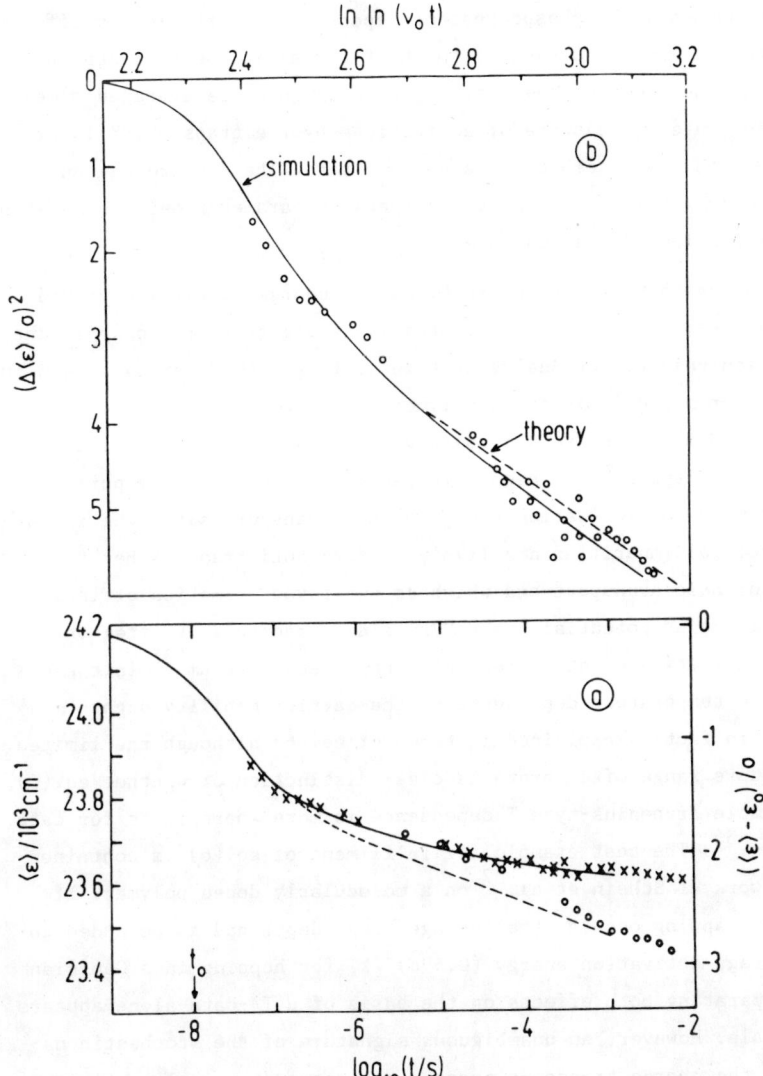

Fig.7　a) The crosses and open circles indicate how the center of the $T_1 \rightarrow S_0$ 0-0 phosphorescence band of a benzophenone glass shifts with delay time at 6 and 80 K, respectively. The full and dashed curves are simulation results for the parameter set $t_0 = 10^{-8}$ s and $\sigma = 260$ cm^{-1}.
b) Representation of the 6 K data on a $(\Delta\langle E\rangle/\sigma)^2$ against $\ln\ln\nu_0 t$ scale including predictions of simulations and analytic theory.[27]

to benzophenone. Its phosphorescence spectrum is sufficiently offset from the $T_1 \to S_0$ 0-0 band of the host to permit spectral separation. Analysing the decay of the latter yields an ensemble averaged time-dependent rate for capture of excitations by the traps which is proportional to the ensemble-averaged jump rate. The data reveal an approximate $\nu(t) \propto t^{-(1-\alpha)}$ law the dispersion parameter being consistent with excitation relaxation data.[34]

Since both triplet excitations and charge carriers move via exchange interaction it is conceptually straightforward to apply the simulation results for analysing time of flight (TOF) signals seen in polymers or molecularly doped polymers after generating a sheet of carriers. Usually one sees dispersive TOF signals which eventually become non-dispersive at high temperatures. In extensively purified systems with low ionization energy of the transport sites where residual chemical impurities are likely to form anti-traps rather than traps for hole transport and which do not favor formation of incipient dimers as potential traps "Gaussian" transport is often observed indicative of attainment of dynamic equilibrium.[35] In these cases the temperature dependence of the carrier mobility ought to be and can in fact be explained in terms of eq.(6) although the limited temperature range often prevents clear distinction of whether eq.(6) or a simple Arrhenius-type T-dependence is more appropriate for data analysis.[36] The best example for fulfilment of eq.(6) is contained in the work of Schein et al.[37] on a molecularly doped polymer. If shallow trapping occurs, the average trap depth had to be added to the average activation energy $(0.69\sigma)^2/kT$ for hopping in a Gaussian DOS. Separating both effects on the basis of $\mu(T)$-data alone appears impossible. However, an unambiguous signature of the stochastic nature of the charge transport process as opposed to simply activated transport governed by, for instance, a discrete trapping level or by polaron formation is the long tail TOF signals carry even if a well-developed plateau region in the TOF signal is suggestive of Gaussian transport statistics.[38]

Field dependences of charge carrier mobilities in random organic solids often feature a Poole-Frenkel behavior, $\ln\mu \sim E^{1/2}$, although it is unrealistic to assume presence of adventitious charged impurities in chemically very different systems.[33] This observation has been used as an argument against applying the stochastic transport concept outlined above. It predicts $\ln\mu \propto E$ in first order approximation, reflecting the linear decrease of the energy separations of sites along the field direction. However, recent MC simulations incorporating the effect of random inter-site overlap indicated that with increasing Σ (see above) a Poole-Frenkel like behavior can be recovered even including an inversion of sign of the coefficient as reported by Peled and Schein[39] without invoking existence of traps that are charged when empty. For further detail the reader is referred to future work.

3 GEMINATE PAIR RECOMBINATION

In molecular solids dielectric constants are typically around 3. Consequently, coulombic capture radii are of order 200 Å at 300 K as compared to ~50 Å in a-Si. In conjunction with the localized character of the optical transitions photo-stimulated charge carrier generation proceeds via the dissociation of a coulombically bound e...h pair with initial intra-pair distance ≈20...25 Å, originally generated via charge transfer transitions.[41] At sufficiently low light intensities adjacent geminate e...h pairs (GPs) do not overlap and both recombination and dissociation are kinetically first order processes. This is different from amorphous inorganic semiconductors where it is still open to conjecture whether recombination of e...h pairs is predominantly of geminate[42] or non-geminate character.[43] Investigating a kinetically "clean" system is, therefore, of interest for unravelling intricacies of more complex systems.

The 3D Onsager problem[44] for a structurally discrete system with built-in disorder has not been solved so far. The closest approach was that of Noolandi et al.[46] who incorporated the effect of disorder by including tunnelling transitions among localized sites in addition to diffusion prevailing at elevated temperatures. However, this approach neglects energetic relaxation, shown above to be the most essential feature of random organic systems. This shortcoming on the theory side prompted us to conduct a MC simulation study of GP recombination in a system with diagonal disorder.

The simulation system was cubic lattice consisting of $(101)^3$ sites with a fixed positive charge positioned in the center. It generates a coulombic potential superimposed onto the site energies chosen according to a Gaussian distribution. At a certain distance r_0 away from the positive charge a negative charge is created and its random walk is followed in energy and spatial coordinates. Fig.8a shows the survival probability of GPs as a function of time normalized to t_0 for various temperatures and $\sigma = 0.1$ eV. The sudden decrease at a time that becomes shorter with increasing temperature is due to pair dissociation. Counting the number of recombination events occurring per unit time at a given time after pair generation yields the recombination rate $R(t) = -k_{rec}(t)n(t)$ (fig.8b). It decreases according to a power law, $R(t) \propto t^{-s}$, with $s \approx 1.3$ for $t/t_0 \leq 10^4$ and $s \approx 1.05$ for longer times independent of both temperature, strength of inter-site coupling, and initial pair distance.[45] In the short time domain $R(t)$ is controlled by both the depletion of the pair reservoir and the decrease of the mean carrier hopping rate while at long times the latter effect prevails, the number of surviving pairs changing only slightly. The result is at variance with that of Noolandi et al.[46] predicting $R(t) \propto t^{-3/2}$. The $R(t) \propto t^{-1.05}$ dependence can be rationalized in terms of Movaghar et al's[27] theory of relaxation and diffusion of excitations in a random potential at T→0 (see section 2.1). It predicts a time dependent diffusivity $D(t) \sim [t \ln(t/t_0)]^{-1}$ which can be approximated by $D(t) \sim t^{-1.04}$ for long times. Obviously, the energetic relaxation of the mobile sibling carrier is the major origin for slowing down GP recombination, presence of the coulombic potential playing a minor role only. The disappointing message of the MC work is, however, that genuine GP recombination in presence of energetic disorder is functionally degenerate with distant pair recombination treated by Dunstan.[43]

Experimentally, GP recombination in random organic solids can

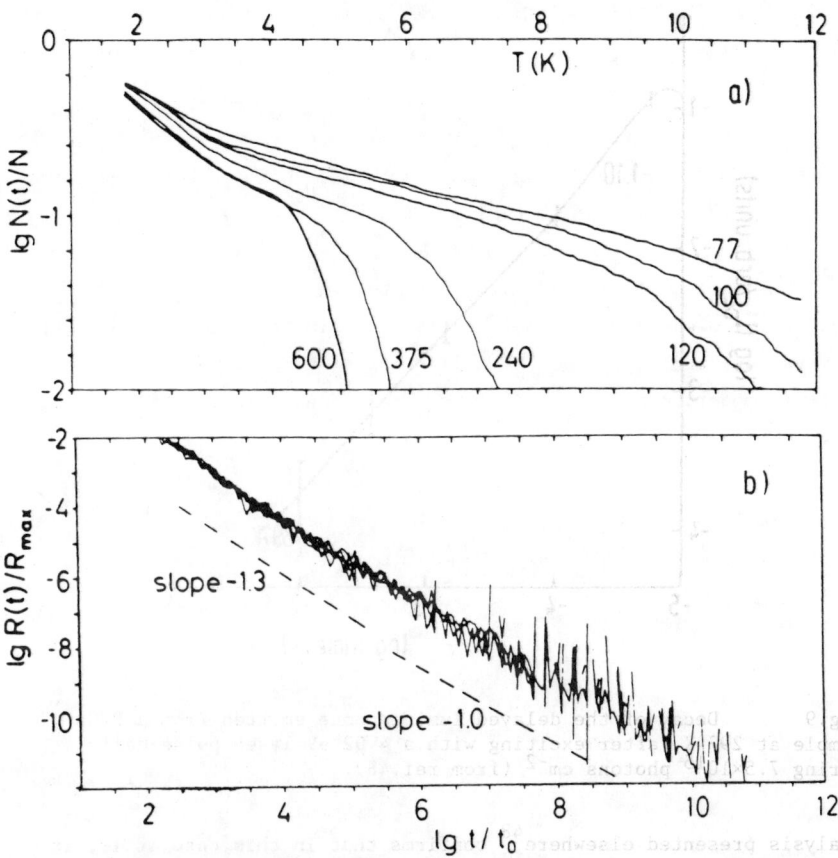

Fig.8   a) Fraction of geminate pairs surviving both recombination and dissociation as a function of time and temperature at fixed width of the DOS (σ = 0.1 eV). b) Computer printout for the recombination rate R(t) normalized to the maximum rate R_{max}, parametric in temperature (from ref.45).

be probed via delayed luminescence following the collaps of an e...h pair, provided that one manages to exclude conventional delayed luminescence caused by the fusion of two triplet excitations.[47,48] In polyvinylcarbazole (PVCA), a prototype pendent group polymer, GPs were excited by 4.02 eV photons of an excimer laser and decay of the delayed luminescence was monitored by phosphoroscopic detection. A typical decay pattern recorded at 298 K is presented in fig.9. More detailed

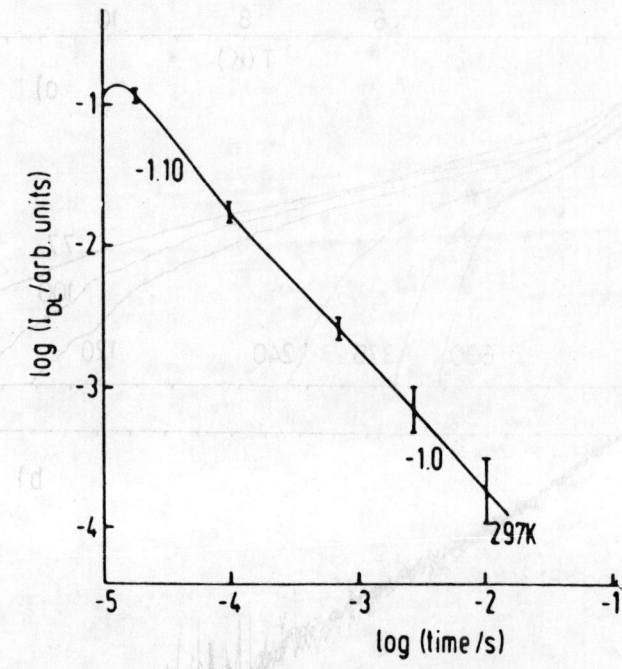

Fig.9 Decay of the delayed luminescence emitted from a PVCA sample at 297 K after exciting with a 4.02 eV laser pulse delivering 7.5×10^{15} photons cm^{-2} (from ref.48).

analysis presented elsewhere[48] confirms that in this case it is, in fact, GP recombination rather than triplet-triplet fusion that controlls delayed luminescence. The functional agreement between experiment and the prediction of MC simulation is obvious, although the absolute magnitude of the GP lifetime requires invoking shallow trapping by incipient dimers formed by the overlap of carbazole chromophores.[49]

4   TRANSPORT PROCESSES IN SUPERCOOLED MELTS

Before presenting arguments supporting the application of the concept of stochastic transport in a random potential to mass transport in supercooled melts, a few of their relevant properties shall briefly be recalled.[50] Upon supercooling a liquid under prevention of crystallization, its contraction is determined by the thermal expansion coefficient α_e of the liquid. At a certain temperature, T_g, called the glass transition temperature, dV/dT changes and below T_g the thermal expansion coefficient of the crystal is approximately attained. T_g depends on the cooling rate q. The larger q is, the higher is T_g. Hence T_g is not solely a material property as expected in case of a thermodynamical phase transition but reflects the history of the sample. In the supercooled phase relaxation processes occurring in response to an external perturbation are non-exponential in time, often following a Kohlrausch-Williams-Watts law $\phi(t) \propto \exp[-(t/t_0)^\beta]$ where ß is often of order 0.5.[51] Moreover, the temperature dependence of the mean relation rate $\nu_0 = t_0^{-1}$ is non-Arrhenius-like. Analysis in terms of the Vogel-Fulcher law,

$$\nu_0 \propto \exp - \frac{B}{T-T_0} \qquad (10)$$

or its equivalent, the Williams-Landel-Ferry equation turned out rather successful provided that the divergence temperature T_0 is assumed to be about 20% below T_g and close to the so-called Kauzmann temperature.[52] The latter is a fictitious temperature at which the entropy of the glass-structure matches that of the crystalline counterpart structure. Eq.(10) is unable to account for the occurrence of the glass transition at T_g without additional assumptions.

The model usually invoked to rationalize eq.(10) is the free volume concept developed by Cohen et al.[53,54] It is based on the idea that in the supercooled melt each elementary unit, be it a molecule or a molecular subunit, has a larger volume at its disposal that it had in a crystal. If it exceeds a certain critical value the excess volume can be redistributed among the "cells" of the liquid without cost of thermal energy. This permits activation-less exchange processes of the structural units, believed to be the elementary step for both viscous flow and relaxation processes. A Boltzmann-like statistical treatment for the distribution of the total free volume of the system on the individual cells in conjunction with the phenomenological assumption that the total free volume is the difference between that of the liquid and the extrapolated crystal volume, i.e. $\Delta v \propto (\alpha_e - \alpha_c)(T-T_0)$, immediately leads to eq.(10) without, however, explaining the kinetics of the process nor addressing the question of cooperativity of the molecular motions involved.

Microscopically a liquid can be characterized by a large number of possible intermolecular configurations each differing in energy. At moderately low viscosities rapid tumbling occurs involving a feedback between the motion of an individual molecule and that of its environment. Götze et al.[55,56] have developed a mode coupling theory for treating this kind of collective motion which avoids the shortcoming of the free volume concept and is able to explain bot the temperature dependence of the viscosity η as well as other glass properties remarkably well.[57] However, data fitting suggests a divergence of, say, η at a temperature $T_c > T_g$, at variance with experiment. It is the temperature range $T < T_c$ where we apply the random walk concept for treating mass transport.

The idea is as follows. If η is large enough collective motion is slowed down to an extent that the environment of an element of the liquid can be viewed as being static on the time scale of an elementary act. Therefore collectivity of the motion is lost yet elementary jump processes continue to occur, albeit on a molecular or weakly cooperative level. Since an element still sees a large variety of configurational states, its motion is reminiscent of that of a random walker migrating incoherently in a random potential except that now motion occurs in configuration space rather than in real space. Provided that the ratio of the magnitude of the potential fluctuation to kT, i.e., the disorder parameter, does not exceed, say, $\sigma/kT \approx 5$, the occupational density of configurational states can be regarded as being in dynamic equilibrium. Following eqs.(4) and (6) this leads to a temperature dependence of the elementary transition of the type

$$\nu(T) = \nu_0 \exp[-(T_0/T)^2] \quad . \tag{11}$$

T_0 is a characteristic temperature proportional to the width of the

DOS. Plotting viscosity data on a $\ln\mu \propto T^{-2}$ scale does, in fact, show that eq.(11), which, in contrast to the VF equation, is a one parameter equation as far as the slope is concerned, is able to explain η data within the range $10^3 \ldots 10^5$ Poise$<\eta<\eta(T_g) \approx 10^{13}$ Poise. The lower bound is determined by onset of collective effects tractable in terms of the mode coupling theory (see fig.10).

A signature of dynamic equilibrium of a glass forming element is the decrease of its mean energy relative to the center of the DOS, upon lowering the system temperature, given by $<\varepsilon_\infty> = \sigma^2/kT$. However, dynamic equilibrium can only be established if the experimental time scale of the experiment exceeds the relaxation time t_{rel} (see section 2.1). Upon subjecting a system to a cooling rate q it must fall out of equilibrium, i.e., become non-ergodic, once the cooling rate exceeds the rate at which the system relaxes towards equilibrium. Occurrence of glass transition below which the system preserves the overall structure it had just above T_g is therefore a necessary consequence of the stochastic transport model. The above condition is readily cast into an expression relating the glass transtion temperature T_g to both cooling rate and the width σ of the DOS

$$T_g = 1.07 \frac{\sigma}{k}[-\ln(qt_0 r_0)]^{-1/2} . \qquad (12)$$

Here t_0 is a typical relaxation time in the liquid, i.e., $t_0 \sim 10^{-12}$ s, and r_0 is a normalization factor of order unity. The fact that eq.(12) is able to fit experimental data for the $T_g(q)$ dependence of a variety of supercooled melts with only one system-specific parameter, namely σ which is obtained from experimental T_0 values (eq.(11)), represents another test for the success of the model (see fig.11).

Further support comes from the change of the temperature dependence of transport processes below T_g. The experiments of Ehlich and Sillescu[65] probing the diffusion of a photochromic label in a polymer matrix holographically demonstrated that the temperature dependence of the diffusion coefficient becomes Arrhenius-like below T_g; the magnitude of the activation energy being in excellent agreement with an estimate based on the $D(T)$-behavior above T_g (fig.12).

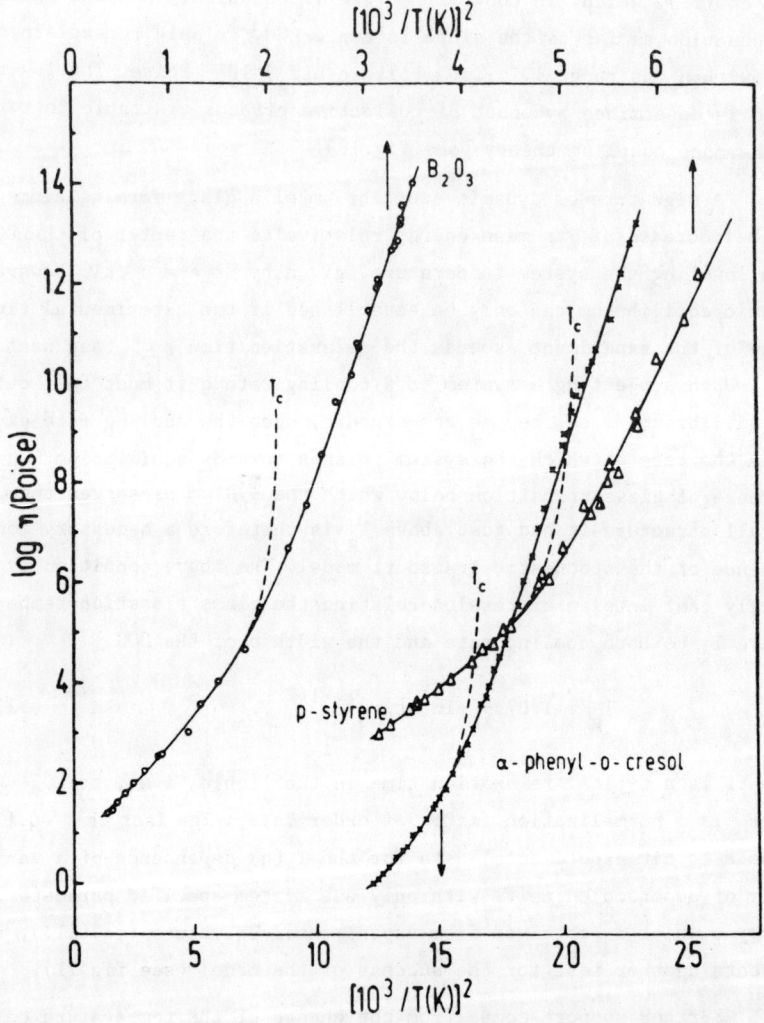

Fig.10 Viscosity data for B_2O_3 [58], p-styrene[59] and α-phenyl-o-cresol[60] in log η against T^{-2} representation. Dashed lines are extrapolations according to the mode coupling theory for a critical exponent γ≈3.3 (from ref.61).

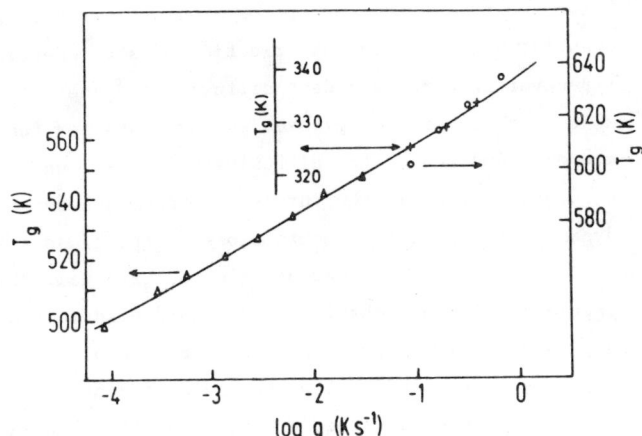

Fig.11 Glass transition temperature T_g as a function of cooling rate q. The full curve is theoretical (eq.12), data points are for borosilicate glass[62], arcolor[63] and P_2O_5.[64] The adjustable parameter is the width of the DOS derived from T_g measured at a fixed cooling rate.

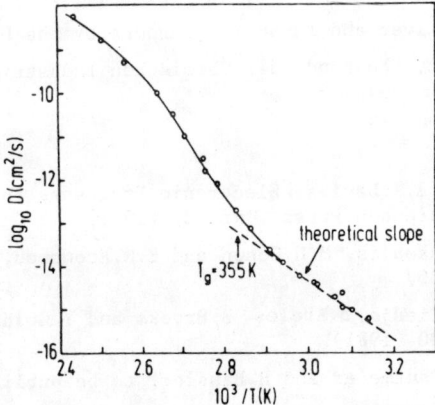

Fig.12 Diffusion coefficient for tetrahydrothiophenindigo in a polycarbonate matrix plotted on a lgD against T^{-1} scale. The slope of the dashed line has been calculated from the temperature dependence of D(T) at $T>T_g$ (from ref.61).

5 CONCLUDING REMARKS

It has been shown that the concept of stochastic random walks in a random potential with Gaussian distribution of site energies is sufficient for recovering time dependent diffusion and energetic relaxation of excitations in organic glasses and accounts for the anomalous temperature dependence of the diffusivity in the long time limit. Of particular interest is the success of the model for rationalizing the dynamics of supercooled melts close to the glass transition temperature. It confirms the conclusion of Götze et al. that the fundamental singularity in the behavior of a supercooled melt occurs at critical temperature $T_c > T_g$ where non-linear mode coupling vanishes. At $T_c > T > T_g$ dynamic is controlled by temperature dependent equilibration of the structural elements within a random potential while at $T < T_g$ the system falls out of ergodicity because relaxation times exceed experimental time scales.

ACKNOWLEDGEMENT

I gratefully acknowledge the contribution of L.Pautmeier, R.Richert and B.Ries to this work, the stimulating discussions with B.Movaghar and M.Silver and financial support by the Deutsche Forschungsgemeinschaft, the Fonds der Chemischen Industrie and NATO.

REFERENCES

1. N.F.Mott and E.A.Davis, Electronic Processes in Non-Crystalline Material (Clarendon Press, Oxford) (1971).
2. S.John, C.Soukoulis, M.H.Cohen and E.N.Economou, Phys.Rev.Lett. 57, 1777 (1986).
3. G.D.Cody, T.Tiedje, B.Abeles, B.Brooks and Y.Goldstein, Phys.Rev. Lett. 47, 1480 (1981).
4. M.Silver, H.Pautmeier and H.Bässler, to be published.
5. H.Bässler, phys.stat.sol.(b) 107, 9 (1981).
6. R.Jankowiak, K.D.Rockwitz and H.Bässler, J.Phys.Chem. 89, 4569 (1985).
7. A.Elschner and H.Bässler, Chem.Phys. 112, 285 (1987).
8. see, e.g., M.Pope and C.E.Swenberg, Electronic Processes in Organic Crystals (Clarendron Press, Oxford) (1982).

9. A.Miller and E.Abrahams, Phys.Rev. 120, 745 (1960).
10. H.Scher and M.Lax, Phys.Rev.B 10, 4491, 4502 (1973).
11. H.Scher and E.W.Montroll, Phys.Rev.B 12, 2455 (1975).
12. J.Klafter and R.Silbey, Phys.Rev.Lett. 44, 55 (1980).
13. B.Movaghar, J.Phys.C 13, 4915 (1980).
14. B.Movaghar and W.Schirmacher, J.Phys.C 14, 589 (1981).
15. M.Grünewald, B.Pohlmann, B.Movaghar and D.Würtz, Phil.Mag.B 49, 341 (1984).
16. B.Movaghar, M.Grünewald, B.Ries, H.Bässler and D.Würtz, Phys.Rev.B 33, 5545 (1986).
17. G.Schönherr, H.Bässler and M.Silver, Phil.Mag.B 44, 47 (1981).
18. B.Ries, H.Bässler, M.Grünewald and B.Movaghar, Phys.Rev.B 37, 5508 (1988).
19. G.Pfister and C.H.Griffith, Phys.Rev.Lett. 40, 659 (1978).
20. E.Müller-Horsche, D.Haarer and H.Scher, Phys.Rev.B 35, 1273-II (1987).
21. L.Pautmeier, R.Richert and H.Bässler, Phil.Mag.Lett. 59, 325 (1989).
22. R.Richert and H.Bässler, J.Chem.Phys. 84, 3567 (1986).
23. U.Larson, Phys.Lett. 105A, 307 (1984).
24. V.I.Arkhipov and A.I.Rudenko, Phil.Mag.B
25. M.Kastner, J.Non-Cryst.Solids 35/36, 807 (1980).
26. D.Monroe, Phys.Rev.Lett. 54, 146 (1985).
27. B.Movaghar, B.Ries and M.Grünewald, Phys.Rev.B 34, 5574 (1986).
28. H.Bässler, J.Chem.Phys. 49, 5198 (1968).
29. P.Holzman and R.C.Jarnagin, J.Chem.Phys. 51, 2251 (1969).
30. R.D.Burkhart and E.R.Lonson, Chem.Phys.Lett. 54, 85 (1983).
31. L.Pautmeier, B.Ries, R.Richert and H.Bässler, Chem.Phys.Lett. 143, 459 (1988).
32. T.F.Hunter, R.D.McAlpine and R.M.Hochstrasser, J.Chem.Phys. 50, 1140 (1969).
33. R.Richert, H.Bässler, B.Ries, B.Movaghar and M.Grünewald, Phil.Mag.Lett. 59, 95 (1989).
34. J.Lange, B.Ries and H.Bässler, Chem.Phys. 128, 47 (1988).
35. H.J.Yuh and M.Stolka, Phil.Mag.B 58, 539 (1988).
36. M.Abkowitz, M.Stolka, R.Weagley, K.McGrane and F.E.Knies, Adv. In Silicon Based Polymer Science, ACS book.

37. L.B.Schein, A.Rosenberg and S.L.Rice, J.Appl.Phys. 40, 4287 (1986).
38. L.Pautmeier, R.Richert and H.Bässler, Phys.Rev.Lett. 63, 547 (1989).
39. A.Peled and L.B.Schein, Chem.Phys.Lett. 153, 422 (1988).
40. P.M.Borsenberger, L.E.Contois and D.C.Hoesterey, J.Chem.Phys. 68, 637 (1978).
41. L.Sebastian, G.Weiser, G.Peter and H.Bässler, Chem.Phys. 95, 13 (1983).
42. D.Engemann and R.Fischer, Proc.12th Int.Conf.Physics of Semiconductors, Stuttgart (1974), p.1042.
43. D.J.Dunstan, Phil.Mag.B 46, 579 (1982).
44. L.Onsager, Phys.Rev. 54, 554 (1938).
45. B.Ries and H.Bässler, J.Molec.Electron. 3, 15 (1987).
46. J.Noolandi, K.M.Hong and R.A.Street, Solid State Comm. 34, 45 (1980).
47. F.Stolzenburg, B.Ries and H.Bässler, Ber.Bunsenges.Phys.Chem. 91, 853 (1987); J.Molec.Electron. 3, 149 (1987).
48. F.Stolzenburg and H.Bässler, Mol.Cryst.Liq.Cryst., in press.
49. S.Nespurek and V.Cimarova, 6th Int.Seminar on Polymer Physics: Relaxation in Polymers, Gomadingen, 1988.
50. for reviews see S.R.Elliot, Physics of Amorphous Solids (Longman, London) (1983); J.Wong and C.A.Angell, Glass Structure by Spectroscopy (Marcel Dekker, New York) (1976); J.Jäckle, Rep.Progr.Phys. 49, 171 (1986).
51. S.A.Brawer, J.Chem.Phys. 81, 954 (1984).
52. Y.H.Jeong, S.R.Nagel and S.Bhattacharya, Phys.Rev.A 34, 602 (1986).
53. M.H.Cohen and D.Turnbull, J.Chem.Phys. 31, 1164 (1959).
54. M.H.Cohen and G.S.Grest, Phys.Rev.B 20, 1077 (1979), ibid 24, 4091 (1981).
55. U.Bengtzelius, W.Götze and A.Sjölander, J.Phys.C 17, 5915 (1984).
56. W.Götze and L.Sjögren, J.Phys.C 21, 3407 (1988).
57. W.Götze and L.Sjögren, Proc.ILL Workshop on Dynamics of Disordered Materials, Grenoble 1988 (Springer Verlag) to be published.
58. P.B.Macedo and A.Napolitano, J.Chem.Phys. 49, 1887 (1968).
59. G.C.Berry and T.G.Fox, Adv.Pol.Sci. 5, 261 (1968).
60. W.T.Laughlin and D.R.Uhlmann, J.Phys.Chem. 76, 2317 (1972).
61. R.Richert and H.Bässler, J.Phys.C, submitted.

62. H.N.Ritland, J.Am.Ceram.Soc. 37, 370 (1954).
63. S.E.B.Petrie, J.Polym.Sci.,Polym.Phys.Ed. 10, 1255 (1972).
64. S.W.Martin and C.A.Angell, J.Phys.Chem. 90, 6736 (1986).
65. D.Ehlich and H.Sillescu, Macromolecules, in press.

PERCOLATION, HOPPING AND DISSIPATIVE QUANTUM TUNNELLING OF PROTONS IN HYDRATED PROTEIN POWDERS

G. Careri

Dipartimento di Fisica, Universita' di Roma "La Sapienza",
Piazzale Aldo Moro, 2 - 00185, Rome Italy

ABSTRACT

Previous work from this laboratory has shown that hydrated lysozyme powders exibit dielectric behaviour owing to proton conductivity, and that this behaviour can be described in the frame of percolation theory. Long range proton displacement appears only above the critical hydration for percolation h_c (g water/g dry weigth), when the 2-dimensional motion takes place on fluctuating clusters of hydrogen-bonded water molecules adsorbed on the protein surface. The emergence of biological function, (enzyme catalysis), was found to coincide with the critical hydration for percolation h_c.

More recentently we have measured the protonic conductivity in hydrated lysozyme powders, from room down to liquid N_2 temperature. In the high temperature limit we find a classical isotopic effect, and the conductivity follows the thermal hopping law. In the low temperature region the conductivity increases with temperature as $\exp T^4$, in agreement with prediction by the theory of dissipative quantum tunnelling.

1. INTRODUCTION

A general statistical-physical approach, called the percolation model, has been shown to be applicable to a wide range of processes where spatially random events and topological disorder are of intrinsic importance. A typical physical application of the percolation theory[1,2] is to the electrical conductivity of a network of conducting and non-conducting elements. One of the most appealing aspects of the percolation process is the presence of a sharp transition, where long-range connectivity among the elements of a system suddenly appear at a critical concentration of the carriers. My aim here is to show that this model can successfully describe the emergence of biological function in some very simple bio-materials. This is because biological systems are often disordered at microscopic scale; then long-range connectivity between subunits must be established according to statistical laws, and the presence of a threshold should be expected according to the percolation model.

In the typical example of a network of conducting and non conducting elements, percolation theory predicts the critical concentration P_c of the conducting elements for the onset of the percolative process, and the critical exponent t for the conductivity σ dependence on P above this threshold

$$\sigma = \sigma_c + k(P-P_c)^t \qquad (1)$$

In eq. (1) the kinetic coefficient k depends on the specific process in question, while P_c and t are universal quantities which are only dependent from the dimensionality D of the system. For a given system hydrated with H_2O or D_2O, if the charge carriers are protons we must expect

$$(k_{H_2O}/k_{D_2O}) = 2^{1/2} \qquad (2)$$

Thus percolation theory allow specific predictions to be made on the nature of the moving changes and on the dimensionality of the conduction process.

2. PROTONIC PERCOLATION

Lysozyme is a comparatively simple enzyme, and almost everything is known about its hydrated powders, thanks to I.R. spectroscopy, E.P.R. relaxation, heat capacity and other thermodynamic and dynamic properties, and especially the enzyme activity towards appropriate substrates[3]. We may summarize these data by saying that the hydration-stepwise process consists of three well-defined stages: 1) from 0 to about 60 H_2O molecules/macromolecule, dominated by the interaction of water with the charged groups of the protein; 2) from 60 to about 220 H_2O molecules/macromolecule, where some major changes in surface water arrangements take place; 3) from 220 to about 300 or more H_2O molecules/macromolecule, where the enzymatic activity starts and grows with increasing hydration, together with the condensation of water molecules onto weakly interacting unfilled patches of surface where the molecules are in rapid motion. It is important to note that no structural transitions in the adsorbed water or in the protein itself have been detected in the range from 60 to 220 H_2O molecules/macromolecule (or hydration h included between 0.07 and 0.25 g H_2O =/g dry weight).

Recent work by our group has shown that powders of lysozyme at low hydration display protonic conductivity[4] and that the conduction process follows the percolation model[5]. In this picture, the conductivity reflects motion of protons along threads of hydrogen-bonded water molecules adsorbed on the surface of the macromolecule, with long-range proton movement developing along with the extended network at the percolation threshold. In our more recent work we have been able to detect the critical exponents of this process[6].

In native Lysozyme powders the dielectric capcitance displayed a sharp increase at a water content threshold h=0.150 $\pm$ 0.016 g/g, followed by saturation at increased hydration[5]. From the capacitance data at different frequencies one can derive the d.c. conductivity σ, which displays a similar sharp increase. Since the hydration of one monolayer is h=0.38 $\pm$ 10% g/g, the experimental volume ratio for surface percolation is 0.40 10%, a value very close to the 0.45 $\pm$ 0.03 predicted by theory[1]. Notice that for three- dimensional networks, regardless of their structure, the conduction threshold predicted by theory is 0.16 $\pm$ 0.02, and this rules out connectivity through the protein interior, where water molecules are know to be very sparse. Moreover, the threshold h_c was found to be constant from pH 3 to ph 8, indicating that the local geography of water clusters about ionizable sites of the protein surface is not of primary importance. Thus only the number of water molecules acting as interconnected conductivity sites is relevant; and as a matter of fact, the same threshold is found for both H_2O - and D_2O- hydrated samples.

From equation (1) one can easily derive that above the threshold the conductivity σ must follow the power law

$$\sigma(h) - \sigma(h_c) = k(h - h_c)^t$$

where t depends on the dimensionality of the system. Result of this analysis are in very good agreement with the theoretical prediction for a 2D conduction process[6]. The previous analysis of the dielectric data reached independently the same conclusion about the dimensionality of the percolation from the close agreement between the measured value of h_c and the prediction from theory for a surface process[5]. The dielectric response at hydration levels near h_c reflects protonic conduction over pathlengths of the order of the diameter of a single macromolecule.

For lysozyme-saccharide complexes a higher value of the percolation threshold has been found, suggesting that the presence of a "foreign body", where the water bridges may not be favorable for proton transfer, must affect the long-range connectivity on the protein surface. This hydration level, $h_c=0.25$, is so close to the critical level for the onset of enzymatic activity in Lysozyme powders[3] that it suggests protonic percolation is involved in Lysozyme catalysis. The value of t found for the saccharide complex suggests that the protonic conduction remains a surface process. This observation is in agreement with the suggestion that preferred paths of proton movement pass through the active site; substrate would be expected to block these without changing the surface character of the percolation.

3. HOPPING AND DISSIPATIVE QUANTUM TUNNELLING

Here let me report some preliminary results[7] on the low temperature conductivity of hydrated lysozyme powders, to investigate the possible occurence of proton quantum of tunnelling in extended networks of hydrogen bonded water molecules. In this work our capacitor was cooled to liquid N_2 temperature by conventional cryogenics, and the dielectric data from 10 kHz to 1MHz have been recorded when rising the temperature at a rate of about 1 degree per minute. A typical run lasted about 6 hours and included about 300 conductivity v/s temperature data. Native lysozyme was at pH 7.

We first consider the Arrhenius plot in the high temperature range. The isotopic factor in the high temperature limit is found close to the classical value $2^{1/2}$, in agreement with previous work, as it can be seen by comparing the reduced conductivities of H_2O-versus D_2O-hydrated samples at nearly the same hydration levels. We conclude that the room temperature conductivity of this system obeys the transition state theory, and displays the dependence on the hydration level typical of most biomaterials. Actually a carefull study near h_c, has shown that for $h \to h_c$ the room temperature conductivity follow the parabolic scaling law with the appropriate critical exponent expected for a 2-dimensional system[6].

Next we consider the temperature region where tunnelling may prevail. A general theory of a quantum system which can tunnell out of a metastable state and which interacts with an environment at temperature T has been produced by Grabert, Weiss and Hanggi (GWH)[8], with the finding that for damping of arbitrary strenght the tunnelling decay rate always matches smoothly with the Arrhenius factor, and that heat enhances the tunnelling probability at T=0 by a factor exp [A(T)]. For undamped system A(T) is exponentially small, whereas for a dissipative system A(T) grows algebrically with temperature. Of particular interest here is the case of tunnelling centers in solids, where A(T) increases proportional to T^4 at low temperature. We have plotted the conductivity data versus T^4, and we have found that a remarkably simple description can be offered as follows. In this T^4 plot, the conductivity data ln $\sigma(h,T)$ can be fitted by straight lines originated near T = 0 K, and after a break adjust themselves on parallel lines. This break is slightly hydration dependent, and very close to the glass transition temperatures reported by other authors in proteins[9]. By comparing data of H_2O- and D_2O samples at nearly the same hydration level in a T^4 plot, the slope of H_2O- is found to be higher than that of D_2O-hydrated ones, as one would expect for a tunnelling phenomenon. The explicit mass dependence is not easy to derive from GWH theory in the case of tunnelling centers in solids.

In conclusion, we believe these preliminary results to be interesting from different viewpoints. Our data show that the GWH theory can be used as a guide to describe dissipative quantum proton tunnelling in extended H_2O clusters both below and above the percolative transition. Moreover our data show that proton tunnelling in hydrated proteins can be relevant up to temperatures not far from room temperature. Finally our tecnique can offer a simple and direct way to investigate phase transitions on the hydrated protein surface.

REFERENCES

1) R. Zallen, The Physics of Amorphous Solids (John Wiley and Sons: New York, 1983)

2) D. Stauffer, Introduction to Percolation Theory (Taylor and Francis: Philadelphia, 1985).

3) J.A. Rupley, E. Gratton and G. Careri, Trends in Biochem. Sciences $\underline{8}$, 18, (1983).

4) G. Careri, M. Geraci, A. Giansanti and J.A. Rupley, Proc. Natl. Acad. Sci. U.S.A. $\underline{82}$, 5342 (1985).

5) G. Careri, A. Giansanti and J.A. Rupley, Proc. Natl. Acad. Sci. U.S.A. $\underline{83}$, 6810 (1986).

6) G. Careri, A. Giansanti and J.A. Rupley, Phys. Rev. $\underline{A}$ $\underline{37}$, 2703 (1988).

7) G. Careri and G. Consolini - (to be published)

8) H.V. Grabert, U. Weiss and P. Hanggi, Phys. Rev. Lett. $\underline{52}$, 2193 (1984); Z. Phys. B, 56 171 (1984).

9) V.I. Goldanskii, Y.F. Krupyanskii and V.N. Flenrov, Physica Scripta $\underline{33}$ 527 (1986).

HOPPING OF CHARGE CARRIERS ON QUASI-ONE DIMENSIONAL CHAINS OF RANDOMLY ORIENTED PROTON SPINS: SELF-AVERAGING, CLUSTER AND FINITE SIZE EFFECTS IN PARAMAGNETIC RESONANCE

J. Köhler and P. Reineker
Abteilung Theoretische Physik, Universität Ulm
W. Forst
Abteilung Mathematik IV, Universität Ulm
Albert-Einstein-Allee 11, 7900 Ulm, F.R. Germany

M. Schreiber
Institut für Physik, Universität Dortmund
4600 Dortmund, F.R. Germany

ABSTRACT

The ESR line shape and the free induction decay of electrons moving by hopping transport with hopping rate γ on quasi-one-dimensional lattices of randomly oriented frozen proton spins giving rise to random magnetic fields of average strength A at the sites of the electrons is investigated theoretically. We show by exact numerical calculations that for a given chain of length N the free induction decay shows a crossover from an $\exp(-(\Delta\omega t)^{3/2})$ to an exponential decay with increasing time for a hopping rate $\gamma/A \ll N$. For $\gamma/A \gg N$ we find an exponential decay in the whole time domain. For intermediate hopping rates cluster effects are important.

1. INTRODUCTION

In recent years radical cation salts of the type $(Aren)_2^+ X^-$ e.g. $(fluoranthene)_2^+ SbF_6^-$ were investigated intensively. The determination of their structure [1,2] showed that the cations are arranged in stacks with a slight dimerization and relatively small intermolecular distances. The anions are sitting between the stacks. This structure indicates that charge carrier motion should take place predominantly in one dimension.

Many experimental investigations [3-6] on these materials were carried out using electron (ESR) and nuclear magnetic resonance (NMR). Measurements of the spin echo decay on $(FA)_2^+[(SbF_6)_{1-x}(PF_6)_x]^-$ ($x \approx 0.5$)[5] showed an $\exp(-(\Delta\omega t)^{3/2})$ behavior for short times and an $\exp(-\widetilde{\Delta\omega} t)$ law for longer times. On the other hand, the spin echo in $(FA)_2^+(PF_6)^-$ [6] decayed always exponentially. Different theoretical investigations also predicted different time behavior. In[7] the calculation of the free induction decay (FID) was predicted to follow an $\exp(-(\Delta\omega t)^{3/2})$ law whereas according to [8] the transverse spin polarization should decay exponentially. The main aim of our theoretical investigation therefore was to determine the free induction decay signal for electrons moving on a linear chain of randomly oriented proton spins.

2. FREE INDUCTION DECAY FOR HOPPING MOTION OF CHARGE CARRIERS IN ONE DIMENSION

2.1 Linear response theory

In linear response theory [9] the ESR line shape $I(\omega)$ is determined by the imaginary part of the magnetic susceptibility $\chi''(\omega)$. Using the fluctuation-dissipation theorem and assuming that the Hamiltonian of the system does not change spin orientation, $\chi''(\omega)$ is given by

$$\chi''(\omega) \propto (1 - e^{-\beta\omega}) \, Re \int_0^\infty dt \, e^{-i\omega t} \langle S^- S^+(t) \rangle_0. \tag{1}$$

The free induction decay is described by the correlation function $F(t) = \langle S^- S^+(t) \rangle_0$. In the frame work of the memory-formalism [10] $F(t)$ is calculated from

$$\dot{F}(t) = i\Omega F(t) - \int_0^t dt' \, M(t-t') F(t'). \tag{2}$$

The frequency Ω and the memory-function $M(t) = M'(t) + iM''(t)$ are given by the following expectation values ((...,...): Hilbert-Schmidt-scalar product, L, L': Liouville operators of the Hamiltonians (5,6), Q: projection operator on irrelevant subspace)

$$\Omega = (S^+, LS^+) \cdot (S^+, S^+)^{-1} \tag{3}$$

$$M(t) = (iQL'S^+, \vec{T} \, e^{i \int_0^t d\tau \, QL(\tau)} iQL'S^+) \cdot (S^+, S^+)^{-1} \tag{4}$$

2.2 Model

For the evaluation of the expressions of the previous section we consider a model in which the charge carrier moves via a hopping process on a one-dimensional lattice of randomly oriented nuclear spins. The Hamiltonian is given by

$$H = \sum_{n,n'} h_{n,n'}(t) a_n^\dagger a_{n'} + \omega_0 S^z + H' \tag{5}$$

$$H' = A\sum_n \vec{S}\vec{I}_n a_n^\dagger a_n = A\sum_n (S^z I_n^z + S^+ I_n^- + S^- I_n^+) a_n^\dagger a_n \to \sum_n \omega_n a_n^\dagger a_n S^z \tag{6}$$

The first term of (5) describes the incoherent electron motion by a stochastic Hamiltonian [11]. We assume the stochastic process to be a δ-correlated Gaussian process and take into account only nearest neighbor interaction. The correlation function then is

$$\langle h_{n,n\pm 1}(t) h_{n,n\pm 1}(t') \rangle = \langle h_{n,n\pm 1}(t) h_{n\pm 1,n}(t') \rangle = 2\gamma \delta(t-t'). \tag{7}$$

2γ is the hopping rate between nearest neighbors. The second term in (5) is the Zeeman energy of the electron and the last term describes the hyperfine-structure-interaction, which is written in more detail in (6). In the numerical evaluation we have replaced the nuclear spins by frozen, random, local magnetic fields with dichotomic or Gaussian distribution (mean value 0 and variance 1).

2.3 Analytical results

A more detailed analytic discussion of the model is given in [12,13]. Upon Laplace transformation of the Mori-equation (2) it is obvious that the ESR lineshape is Lorentzian if the shift in the line position $M''(\omega)$ and the linewidth $M'(\omega)$ are independent of ω. The discussion in the time domain shows that the ESR lineshape is inhomogeneously broadened for slow electron motion. In the case of rapid electron motion $F(t)$ decays exponentially and the line shape is Lorentzian if the time integral over the memory-function exists. For one-dimensional hopping motion on a chain of infinite length $F(t)$ decays according to $\exp(-(\Delta\omega t)^{3/2})$.

2.4 Numerical results

For the numerical evaluation we assume that the interaction Hamiltonian H' is given by the last term in (6). The correlation function $F(t)$ can then be expressed by coefficients $c_n(t)$ in the following way:

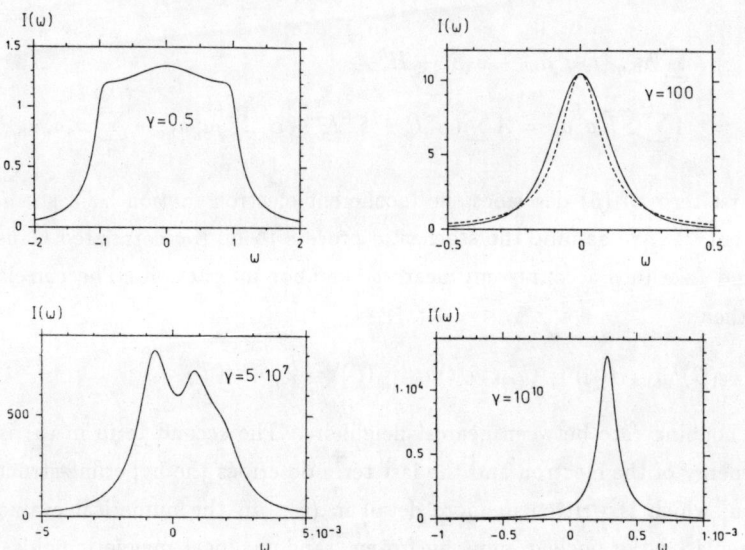

Fig 1. ESR line shapes for a one-dimensional lattice of $N = 10^7$ sites and $\gamma = 0.5$ (a), $\gamma = 100$ (b), $\gamma = 5 \cdot 10^7$ (c), $\gamma = 10^{10}$ (d).

$$F(t) = \sum_n \sum_{s=\uparrow\downarrow} \langle ns|S^- \langle \overrightarrow{T} \exp\{i \int_0^t d\tau \, L(\tau)\} S^+ \rangle_{RW} \, \rho_0 |ns\rangle \qquad (8)$$

$$= \frac{1}{N} \sum_n \langle n \downarrow | \langle \overleftarrow{T} \exp\{-i \int_0^t d\tau \, L(\tau)\} S^- \rangle_{RW} | n \uparrow \rangle$$

$$= \frac{1}{N} \sum_n c_n(t), \qquad (9)$$

The coefficients $c_n(t)$ are determined from the following system of differential equations ($\omega_0 = 0$)

$$\dot{c}_n(t) = (i\omega_n - 4\gamma) c_n + 2\gamma (c_{n+1} + c_{n-1}) \tag{10}$$

$$\dot{\vec{c}} = \overleftrightarrow{A}\, \vec{c} \tag{11}$$

where (11) defines the vector $\vec{c}$ and the matrix $\overleftrightarrow{A}$. The system of equations (11) was solved after a Laplace transformation with the help of continued fractions.

Figs. 1a-d show ESR line shapes $I(\omega) = Re \int_0^\infty dt \exp(-i\omega t) F(t)$ for a one dimensional lattice of $N = 10^7$ sites for various values of the hopping rate 2γ. For very small values of 2γ, i.e. $2\gamma/A \ll 1$, a situation not shown in the figures, one obtains two narrow lines at the positions of the hyperfine-interaction. For $2\gamma/A \approx 0.5$ (Fig. 1a) we observe a broad ESR line. With increasing γ (Fig. 1b) the ESR line gradually becomes narrower. For comparison the dashed curve represents a Lorentzian line with the same maximum.

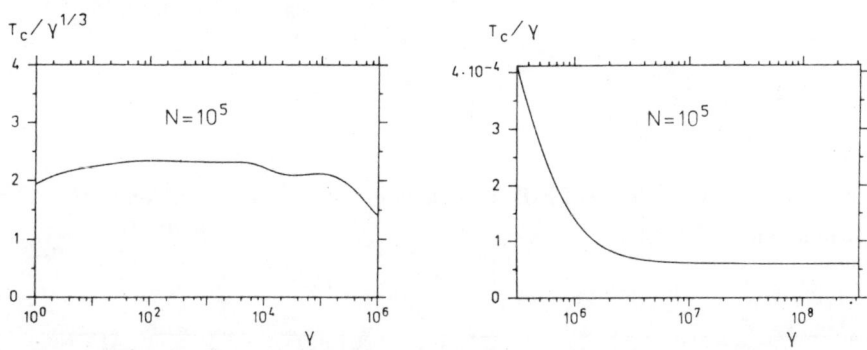

Fig. 2. Decay time of FID as a function of γ for a chain of 10^5 sites normalized to $\gamma^{1/3}$ (a) and to γ (b).

It will be shown that in this range of γ values the ESR linewidth decreases proportional $\gamma^{-1/3}$ and the correlation function decays mainly according to $\exp(-(\Delta\omega t)^{3/2})$ as seen in Figs. 3. In Fig. 1c γ/A is comparable to the number of lattice sites. The structures in the ESR line have their origins in spin clusters [12),13)]. For such values of γ/A the ESR line depends on the specific realization of spin arrangements. For still larger values of γ, i.e. $\gamma/A > N$, we finally obtain a Lorentzian line (Fig. 1d) whose width depends on γ/A and on the specific realization of the spin configuration.

2.5 Analysis of the numerical data

For the analysis of the line shape Figs. 2a and 2b show the free induction decay time $\tau_c = \int_0^\infty dt\, F(t)$, in Fig. 2a normalized to $\gamma^{1/3}$, in Fig. 2b normalized to γ. For a chain with 10^5 sites, τ_c was determined as the maximum of the numerically calculated line shape. The representations show that for small hopping rates $\tau_c \propto \gamma^{1/3}$, for large hopping rates $\tau_c \propto \gamma$. From the curve it is clear that the crossover occurs for $\gamma/A \approx N$.

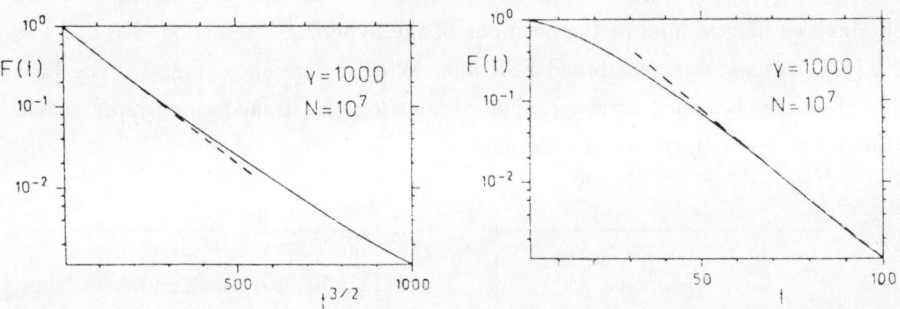

Fig. 3. Time dependence of FID for a chain of 10^7 sites as a function of $t^{3/2}$ (a) and a function of t (b).

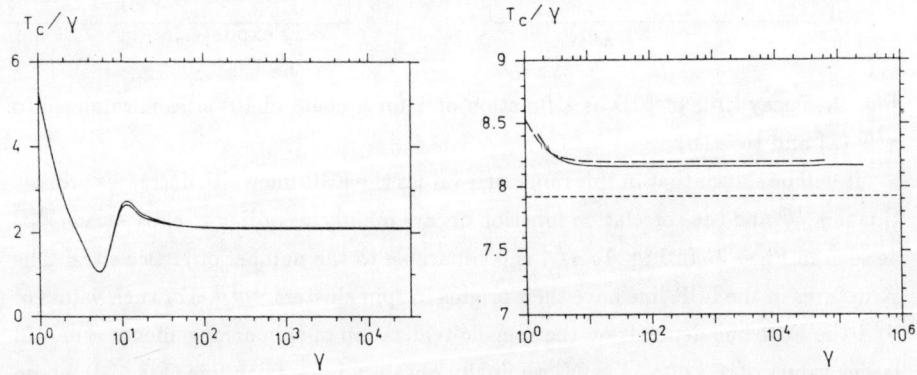

Fig. 4 Normalized decay time of FID τ_c/γ as a function of γ for (a) a 2-dimensional lattice of size 300×300 and two realizations drawn out of a Gaussian ensemble (b) Gaussian distribution (full line) and dichotomic distribution (long dashes) for a 3-dimensional lattice of size $28 \times 28 \times 28$; results according to Mori-formalism (dashed line).

Figs. 3a and 3b show the time dependence of the free induction decay, which was determined by numerically Fourier transforming the line shape in a semi-logarithmic plot as a function of $t^{3/2}$ and t, respectively. The hopping rate is in both cases $\gamma/A = 1000$. From the figures we see that for short times the FID signal decays according to $\exp(-(\Delta\omega t)^{3/2})$, whereas in the asymptotic region the decay follows an exponential law in agreement with [8]). Figs. 4a and 4b show the normalized decay time τ_c/γ of the free induction decay for a two- and three-dimensional lattice. The figures show that — in contrast to the behavior of τ_c for the linear chain — the usual motional narrowing, $\tau_c \propto \gamma$, is achieved for rather low hopping rates.

3. SUMMARY

In our model we consider an electron which moves on a one dimensional lattice of given length via a hopping process. At each site a local magnetic field was allowed to be oriented at random in positive or negative direction. The discussion of the ESR line shape and of the FID shows the linear chain behaves rather different from two- and three-dimensional lattices and that five ranges of the hopping rates can be made out: for small hopping rates we obtain single narrow ESR lines corresponding to an inhomogeneous line broadening. For $\gamma/A \approx 1$ the various lines have merged resulting in a broad ESR line. Increasing γ/A further the line narrowing starts; the FID shows an $\exp(-(\Delta\omega t)^{3/2})$ behavior with a purely exponential decay in the asymptotic region. For $\gamma/A \approx N$ we have structures in the ESR line on account of spin clusters. Finally, for $\gamma/A > N$, we arrive at a Lorentzian line.

ACKNOWLEDGEMENT

The support of the Stiftung Volkswagenwerk is gratefully acknowleged.

REFERENCES

1. V. Enkelmann, B.S. Morra, Ch. Kröhncke, G. Wegner, Chem. Phys. **66** (1982) 303

2. J.P. Pouget in: Low Dimensional Conductors and Superconductors, ed. D. Jérome, L.G. Caron, Plenum Press New York 1987, p. 17

3. M. Mehring in: Low Dimensional Conductors and Superconductors, ed. D. Jerome, L.G. Caron, Plenum Press New York 1987, p. 185

4. E. Müller, J.U. von Schütz, H.C. Wolf Mol. Cryst. Liq. Cryst. **93** (1983) 407

5. J. Sigg, Th. Prisner, K.P. Dinse, H. Brunner, D. Schweitzer, K.H. Hausser, Phys. Rev. **B 27** (1983) 5366

6. G. Denninger, W. Stöcklein, E. Dormann, M. Schwörer, Chem. Phys. Lett. **107** (1984) 222

7. M.J. Hennessy, C.D. McElwee, and P.M. Richards, Phys. Rev. **B 7** (1973) 930

8. R. Czech and K. W. Kehr, Phys. Rev. **B 34** (1986) 261

9. R. Kubo, M. Toda, N. Hashitsume, Statistical Physics II, Springer Series in Solid State Sciences **31**, Springer Verlag (1985)

10. D.Forster, Frontiers in Physics **47**: Hydrodynamic Fluctuations, Broken Symmetry, and Correlation Functions. The Benjamin/Cummings Publishing Company (1975)

11. P. Reineker in: Exciton Dynamics in Molecular Crystals and Aggregates, Springer Tracts in Modern Physics Vol. **94**, Springer, Berlin 1982, p. 111

12. J. Köhler, Ph.D. Thesis, University of Ulm 1989

13. J. Köhler, P. Reineker, M. Schreiber, to be published

SUBJECT INDEX

A

AC conductivity, 309, 472
Aharonov-Bohm oscillations, 146, 169, 171
Antiferromagnetic state, 61

B

Biological function, 523
Bipolaron, 349
    Bose condensation, 366
    superconductivity, 364

C

Cluster simulations, 112, 309
Coherence length, 198, 244
Compensation, 111, 121, 183, 287
Computer studies, 112, 125, 131, 330, 501
Conductance fluctuations, 260, 263, 274
Correlations, 22, 23, 27, 130, 136
Coulomb gap, 3, 26, 49, 85, 94, 97, 122, 141, 182, 210
Critical exponent, 111
Critical magnetic field, 66
Current oscillations, 283

D

Density of states, 4, 10, 122, 436, 492
    Coulomb gap, 3, 26, 50, 122
    Efros-Shklovskii DOS, 5, 12
    relaxation, 19
Differential conductance, 283, 327
Diffusion
    electronic, 501
    free volume, 464, 511
    hydrogen, 442, 450, 453
    time-dependent, 499, 508
Dipolar interaction, 125
Dispersive transport, 431, 450, 499
Drift mobility, 431, 435, 498
Dynamic disorder, 459
    hopping, 474

E

Effective medium theory, 96, 299, 303, 309, 318, 333, 495
Electric field, high, 182, 283

Electron-electron interaction, 3, 93, 111, 117, 122, 130, 190
Electron glass, 3, 5
Electron-phonon coupling, 130, 318
    Fröhlich interaction, 356
Energy relaxation, 408, 432, 491
Entropy, 133
Exchange interaction, 50, 492
Exciton, 408, 420, 492
Extended pair approximation, 77, 79

F

Ferroelectrics, 352, 358
    anisotropic, 358
Ferromagnetic state, 61, 72
Fractal dimension, 314
Free volume concept, 464, 511
Frequency dependent, 309, 472

G

Geminate recombination, 491, 507
Glass transition, 443, 451, 465, 511
Granular materials, 86, 93, 95, 207

H

Hall coefficient, 27, 34, 41, 43, 62, 209, 378, 385
High electric fields, 283, 317
Hydrogen diffusion, 442, 450, 453
Hydrogen glass model, 441
Hopping
    ac, 78
    dynamic disorder, 474
    frequency dependence, 309
    high field, 283, 317
    interference effects, 139, 155, 169, 200, 207
    length, 88, 101
    multi-electron, 26, 94, 129
    nonlinear, 299
    phonon-assisted, 26
    polymers, 461
    protons, 521
    proton spins, 527
    self-trapped hole, 393
    small polaron, 377, 385

I

Impurity conduction, 26
   ϵ_2-activation, 26
   ϵ_3-activation, 285
   saturation regime, 289
Inelastic diffusion length, 199
Interference phenomena, 139, 155, 169, 181, 198, 273

L

Localization
   Anderson, 40, 44, 318, 339, 416
   length, 37, 39, 62, 88
   high electric fields, 317, 337
   strong, 121, 152, 263
   weak, 148, 152, 171, 340

M

Magnetic gap, 51
Magnetic moment, 63, 70
Magnetic phase diagram, 74
Magnetic susceptibility, 57
Magneto resistance, 27, 38, 52, 61, 65, 71, 152, 158, 188, 257
   anisotropy, 67, 198
   electric field effect, 197
   negative, 54, 56, 65, 144, 159, 188, 194, 201, 257
   orbital, 512, 193, 204
   positive, 27, 38, 52, 257
Materials
   AgCl, 393
   Amorphous As_2Se_3, 446
                   As_2S_3, 446
                   Se, 446
                   $Ge_{1-x}Cr_x$, 53
                   Si:H, 425, 431, 442
                   Si:H:Au, 77
                   $Si_{1-x}C_x$:H, 422
   benzophenone glass, 505
   B_2O_3, 514
   borosilicate glass, 515
   cermets, 93, 95, 102
   CdSe, 86
   CdS_xSe_{1-x}, 408
   dipole glass, 447
   discontinuous metal film, 102
   GaAs, 181, 422
   granular, 86, 93, 102, 207
   In_2O_3, 161, 190, 193
   In_2O_3:Au, 201
   In_2O_{3-x} granular, 207
   La_2CuO_{4+y}, 61, 359
   Lysozyme enzyme, 522
   organic random solids, 491
   P_2O_5, 515
   perovskites, 359
   polyethylene oxide, 465
   polymer electrolytes, 459
   polymeric glass, 447
   polyphthalocyanine, 467
   polyvinylcarbazoie, 494
   polyimide films, irradiated, 55
   Si:B, 289
   Si:P, 25, 111, 289
   Si:Sb, 25, 289
   Si MOSFET, 243, 263
   spin glass, 447
   transmutation-doped Ge, 57
   UO_{2+x} and U_4O_{9-y}, 377
   vanadate glasses, 385
Mesoscopic effects, 139, 172, 234, 243, 263
   fluctuations, 140, 144, 172, 243, 259
   incoherent, 234
MOSFETs, 243, 263
   oscillations, 140, 144, 172, 259
Metal-insulator transition, 23, 25, 41
   critical concentration, 30
Microwave conductivity, polymers, 472

N

Negative differential conductance, 283, 327
Nonlinearity
   Coulomb gap model, 17
   nonohmic hopping conductance, 293, 317
   response, 15
   self-trapping, 353
Nonlinear hopping transport, 299
Nonohmic hopping, 283, 319

O

Optical dephasing, 414, 422
Optical excitation, 404

relaxation, 405
recombination, 404, 491, 507

P

Percolation, 41, 151, 244, 294, 299, 310, 318, 330, 466, 488, 521
  cluster, 313
  critical resistance, 310
  protonic, 522
  threshold, 39, 244, 469, 524
Photoemission, 5, 83, 94
Photoluminescence, 409, 504, 509
Picosecond spectroscopy, 411, 417
Polaron effect, 123, 319
  bipolaron, 349
  small polaron, 377, 385, 406
Polyimide films, irradiated, 56
Poole-Frenkel effect, 283, 507
Potential fluctuations, 96, 217, 264, 307, 407, 512

R

Random network, 311, 312
Random phase approximation, 35
Random potential, 96, 264, 340, 494, 508, 516
Random walk, 495, 516
Rate equation, 318, 323
  Markovian, 329
Recombination, 404, 491, 507
  geminate, 491, 507
Relaxation
  of energy, 408, 491, 500
  of phase, 403
  stretched exponential, 447, 511
  time-dependent, 449

S

Scaling, 81, 171, 175, 189, 310
Screening

dynamics, 3, 13
  length, 3, 8
  relaxation, 14
Self-consistent field method, 112, 126
Self-trapping, 355, 377, 385, 393
Simulations, 112, 113, 125, 131, 263, 274, 299, 302, 491, 495
  Monte Carlo, 491
Specific heat, 36, 111, 117, 133
Spin-orbit interaction, 43, 53, 162, 169, 200
Spin-spin interaction, 49, 61
Spin susceptibility, 112, 118
Superconductivity, 349, 364
  bipolaronic, 351, 364
  high temperature, 351, 364

T

Thermopower, 380, 387
  small polaron, 380, 387
Tunnel barrier
  fluctuations, 217
  punctures, 217
  transmittance, 219
Two-level tunneling modes, 454

V

Variable range hopping, 25, 49, 77, 101, 139, 151, 181, 207, 248
  Efros-Shklovskii, 85, 105, 210
  electric field dependence, 185, 300
  interference effects, 139, 169, 181
  Mott, 25, 41, 45, 49, 79, 85, 182, 207, 250
  one-dimensional, 243
Viscosity, 451, 512

W

Wannier-Stark states, 318
Weak localization, 30, 38

AUTHOR INDEX

Adamiya, Z. A. 283
Adkins, C. J. 93
Aladashvili, D. I. 283

Bässler, Heinz 491
Birgeneau, R. J. 61
Böttger, Harald 317

Careri, G. 521
Chen, C. Y. 61
Chicón, R. 121

Dai, Peihua 85
Druger, D. Stephen 459

Edrei, Itzhak 273
Emin, David 349
Entin-Wohlman, O. 151
Eto, Mikio 111

Fisher, R. 403
Först, W. 527
Freer, B. S. 61
Frost, J. E. F. 181

Gabbe, D. R. 61
Göbel, E. O. 403
Green, M. 263

Hansmann, L. 77
Hill, G. 181
Hunt, A. 309

Imry, Y. 151

Jenssen, H. P. 61
Jones, G. A. C. 181

Kakalios, James 441
Kamimura, Hiroshi 111

Kardar, M. 169
Kastner, M. A. 61
Kemp, M. 431
Köhler, J. 527

Lavdovskii, K. G. 283
Levin, E. I. 283
Levy, Miguel 85
Long, A. R. 77

Medina, E. 169
Mochena, M. 129
Monroe, Don 3

Nagels, P. 377, 385
Newbury, R. 181
Nissim, Meir 207
Noll, G. 403

Ochiai, Yuichi 25
Ortuño, M. 121
Ovadyahu, Z. 193

Peacock, D. C. 181
Pepper, M. 181
Picone, P. J. 61
Pollak, M. 129, 263, 299, 309
Popovic, Dragana 243
Preyer, N. W. 61

Raikh, M. E. 217
Ratner, Mark A. 459
Reineker, P. 527
Ritchie, D. A. 181
Rosenbaum, Ralph 207
Rowan, L. 393
Ruzin, I. M. 217

Sarachik, M. P. 85
Schreiber, M. 527

Shapir, Y. 169
Shklovskii, B. I. 139, 283
Shlimak, I. S. 49
Silver, M. 431
Sivan, U. 151
Slifkin, L. 393
Spivak, B. Z. 139

Talamantes, J. 299

Thio, Tineke 61
Thomas, P. 403
Tremblay, F. 181

Wang, X. R. 169
Wegener, Dieter 317
Weller, A. 403

Zhang, Youzhu 85